Normal and Pathological Anatomy of the Shoulder

Gregory I. Bain • Eiji Itoi
Giovanni Di Giacomo • Hiroyuki Sugaya
Editors

Normal and Pathological Anatomy of the Shoulder

International Society of Arthroscopy, Knee Surgery & Orthopaedic Sports Medicine • 1995 •

Springer

Editors
Gregory I. Bain
Department of Orthopedics and Trauma
Flinders University
Adelaide
South Australia
Australia

Eiji Itoi
Department of Orthopaedic Surgery
Tohoku University School of Medicine
Sendai
Miyagi
Japan

Giovanni Di Giacomo
Orthopaedic
Concordia Hospital for Special Surgery
Rome
Italy

Hiroyuki Sugaya
Funabashi Orthopaedic Sports
Med Cn Shoulder and Elbow Service
Chiba
Japan

Additional material to this book can be downloaded from http://extras.springer.com

ISBN 978-3-662-52560-9 ISBN 978-3-662-45719-1 (eBook)
DOI 10.1007/978-3-662-45719-1

Springer Heidelberg New York Dordrecht London

Softcover re-print of the Hardcover 1st edition 2015

Printed on acid-free paper

Springer-Verlag GmbH Berlin Heidelberg is part of Springer Science+Business Media (www.springer.com)

Foreword

This book is a product of the upper extremity committee of ISAKOS. While the anatomy of the shoulder has been studied well for many centuries and has not changed, the authors of this book highlight what has improved: our ability to evaluate and understand anatomy in health and disease in a dynamic and living body. In the past, researchers have been limited to cadaver dissections to uncover the secrets of functional anatomy. Those works have been important for our understanding, but they have limitations.

During the last 30 years, dramatic changes in the tools available for practitioners and researchers have become available. The arthroscope has been central to this effort but has had a great deal of help from imaging modalities including MRI, CT and ultrasound. These new tools have allowed greater understanding of how the anatomy is modified by disease processes. This book points us to a better understanding of normal anatomy and its variations as well as the changes that have occurred secondary to pathological processes. These insights will guide us to perform the necessary reconstructions to overcome functional losses. This enhanced understanding will also prevent us from having complications that will occur without the clear knowledge of how the pathological process has changed the anatomy.

We all are indebted to Drs Gregory Bain, Eiji Itoi and all the members of the upper extremity committee of ISAKOS for their work. The insights of this important manuscript will extend beyond the membership of ISAKOS to have an impact on the wider surgical community and the patients they manage. An additional benefit is that this book will form the basis for future research as we continue our quest to maintain the function of the musculoskeletal system in a minimally invasive and cost-effective way.

Gary G. Poehling, MD
Professor of Orthopedics
Wake Forest University Medical Center
Emeritus Editor in Chief Journal of Arthroscopy

Preface

Introduction

The principles of human gross anatomy have been developed for centuries and are the foundation of current medicine. Over the last two decades, there have been many advances in biomechanics, imaging and arthroscopy, which have enhanced our understanding of clinical, surgical and functional anatomy.

Why Anatomy and Pathology?

Pathology is the basic science of medicine, and anatomy is the basic science of surgery. Despite advances in both basic sciences, the concept of patho-anatomy is often overlooked. The way in which normal anatomy is affected by pathological processes such as trauma, disease and degeneration still requires further investigation.

The Monograph

The aim of this monograph is to bring together the newer concepts of shoulder anatomy and patho-anatomy. It commences with a discussion on comparative and developmental anatomy. For each clinically relevant anatomical area, there is an overview of gross anatomy; a discussion of the ultra-structure, imaging and arthroscopy and a review of how the anatomy is affected by pathological processes.

In creating this document, we have exchanged many concepts of applied, pathological and surgical anatomy of the shoulder. The relevant historical and latest literature has been analysed to develop new concepts, which are shared in this monograph. We trust that dissemination of this new understanding will advance the assessment and management of patients with disorders of the shoulder.

The Editors and Authors

The upper extremity committee of ISAKOS is enriched with many surgeons who have advanced the science of surgical anatomy over the last 20 years. The publication was developed and principally written by the members of the committee.

Acknowledgement

We sincerely thank the editors and authors for their time, effort and expertise in enabling this project to be completed. We acknowledge the significant contribution of the following individuals:

Editor Giovanni Di Giacomo, Italian Orthopaedic Surgeon and Anatomist, who also provided many wonderful images of cadaveric dissections from his book *Atlas of Functional Shoulder Anatomy* [1]

Henry V Crock AO, Australian Orthopaedic Surgeon and Anatomist, for providing the detailed vascular anatomical images that were reproduced from his book *An Atlas of Vascular Anatomy of the Skeleton and Spinal Cord* [2]

Mark Ross, Australian Orthopaedic Surgeon, for providing many excellent images from his cadaveric dissections

Pau Golano, Spanish Anatomist who tragically passed away at the time of preparation of this manuscript. His passing is a great loss to Orthopaedic Surgery. We were able to obtain a few of his images, which are beautifully demonstrated in the book.

Martin Langer, German Orthopaedic Surgeon, Artist and Anatomist, for his spectacular graphic illustrations

Ron Heptinstall, never quite retired registered nurse, photographer and graphic artist, for providing and bringing to life many of the graphics

Rebecca Lea and Enid Hillard from my private office and Amy Watts and Don Branwell for their assistance in copyright details, collating, referencing and editing

Gregory I. Bain
Editor
Deputy Chairman, Upper Extremity Committee, ISAKOS
Professor of Upper Limb Surgery and Research
Department of Orthopaedic Surgery
Flinders University
Adelaide, Australia

Eiji Itoi
Editor
Chairman, Upper Extremity Committee, ISAKOS
Professor, Department of Orthopaedic Surgery,
Tohoku University School of Medicine
Sendai, Japan

References

1. Di Giacomo G, Costantini A, Pouliart N, De Vita A, editors. Atlas of functional shoulder anatomy. Italia: Springer; 2008.
2. Henry V, Crock AO. An atlas of vascular anatomy of the skeleton and spinal cord. Published by Martin Dunitz; 1996. Henry V Crock AO maintains copyright of these images.

Contents

Part I

Introduction

1 Comparative Anatomy of the Shoulder

W. Jaap Willems

1.1 Introduction

Over a period of 370 million years, there has been an evolution from fish, through amphibians, reptiles and birds, to tetrapods. Fish have fins, which are rays with webbed or membranous processes. The evolution of fins in fishes to limbs in tetrapods was a very elegant progression, which has enabled the skeletal elements to be used for support, locomotion, followed by suspension and ultimately the ability to throw projectiles. The limbs in tetrapods are muscular appendages with well-defined joints.

In tetrapods, the limb *(chiridium)* is composed of three well-defined regions: the *autopodium* (wrist and fingers), *zeugopodium* (ulna and radius) and the *stylopodium* (humerus).

The pectoral girdle is the brace that supports the limbs [1]. Both pectoral girdles fused in the midline on the ventral surface of the body through a medium of the interclavicle.

In this chapter, the evolution of this pectoral girdle is described, with emphasis on the various tetrapods. Secondly, a comparison of the different animals models that can be used to study pathologies of the shoulder is described.

W.J. Willems, MD, PhD
Shoulder Unit, De Lairesse Kliniek,
Amsterdam, The Netherlands
e-mail: w.j.willems@xs4all.nl

1.2 Phylogeny of the Shoulder Girdle: Osseous Architecture

In the evolution of the pectoral girdle in fishes, two essential changes occurred. The pectoral girdle both in fishes (rhipdistians) and tetrapods (vertebrate with four limbs) consisted of dermal and endoskeletal elements.

1.2.1 Fish

In fishes, the pectoral girdle is attached to the skull and during evolution the skull becomes completely detached from the skull in vertebrates. After this detachment, the pectoral girdle consists of a ventral *dermal* part (*cleithrum* and *clavicle*) and *endoskeletal* (*scapula* and *procoracoid*) components.

While this latter originally was one element, in tetrapods it arises from two distinct embryonic centres of enchondral ossification, leading to two distinct bones, the scapula and procoracoid (Fig. 1.1) [1].

1.2.2 Amphibians

In amphibians, with the acquirement of terrestrial habits, the tripartite type of pectoral girdle made its first appearance: the coracoid became segmented into the anterior procoracoid and posterior coracoid, while the clavicle formed a connection to the procoracoid [2].

G.I. Bain et al. (eds.), *Normal and Pathological Anatomy of the Shoulder*,
DOI 10.1007/978-3-662-45719-1_1, © ISAKOS 2015

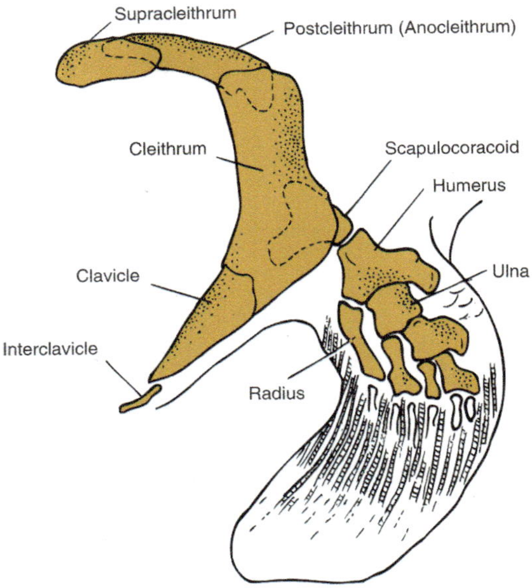

Fig. 1.1 Appendicular skeleton of the earliest tetrapods (rhipidistian, bony fish), where the earlier connection to the skull was lost leading to the formation of the pectoral girdle and an increase of the mobility of the head. The cleithrum and clavicula are located more ventrally, the scapulocoracoid more dorsally; the interclavicle connects both the clavicles (Reprinted with permission from: Kardong [1])

1.2.3 Reptiles

In reptiles, the pectoral girdle has migrated caudally from its intimate relation to the head, as seen in fish. The reptilian pectoral girdle comprised a scapula, a clavicle, which replaced the provoracoid, and a coracoid.

The evolution of the procoracoid varies by species. It becomes part of the anterior scapula around the glenoid fossa; in others, it is fused with the sternum, or is replaced by the clavicle. Although the scapula and coracoid process are anatomically united, genetic patterning of the coracoid and the scapula is under the control of different Hox genes, lending further support to the view that each is a separate phylogenetic derivative [2].

1.2.4 Birds

The avian pectoral girdle became specialized to enable flight. The clavicles became essential in suspending the limb away from the body and the coracoid became large and strong in response to the muscular demands of flight. Consequently, the scapula was small, curved and narrow to allow greater motion. The keeled attachment for the strong pectoral muscles used in flight. The sternum became keel shaped to provide attachment for the strong pectoral muscles used in flight.

1.2.5 Mammals

The coracoid in mammals tends to become greatly reduced, forming an insignificant process on the scapula. The only other vestige of the bone is the coracoid ligament, extending from the coracoid process to the bone, in which may be found isolated masses of cartilage. This arrangement frees the scapula from any bone attachment to the skeleton. In mammals without clavicles, the scapula has no bony attachments whatsoever. It becomes the sole support for limb and provides attachments for muscles necessary for a freely movable extremity. New functional demands on the girdle resulted in the development of a projection of bone on the dorsal surface of the scapula, the spine, which extends downwards and ends in the acromion [2].

The clavicle undergoes changes during evolution, when in tetrapods a change in limb posture arises. In a sprawled posture, the forces are medially directed toward the shoulder girdle, conferring on medial elements a major role in resisting these forces: In these animals the clavicles are interconnected, a so-called interclavicle. As the limbs are brought under the body, these forces are directed less toward the midline and more in vertical direction, leading to a less prominent role of the clavicles (Fig. 1.2) [1].

In mammals, which have acquired freedom of the forelimb to a marked degree, such as insectivores, primates and some marsupials and rodents, the clavicle is well developed. In others, including ungulates, carnivores and some rodents and marsupials, it is absent or rudimentary.

1.3 Phylogeny of the Shoulder Girdle: Musculature

Development of the shoulder and forelimb muscles in tetrapods comes from four sources: branchiomeric (jaw and pharyngeal muscles), axial, dorsal limb and ventral limb muscles (Fig. 1.3).

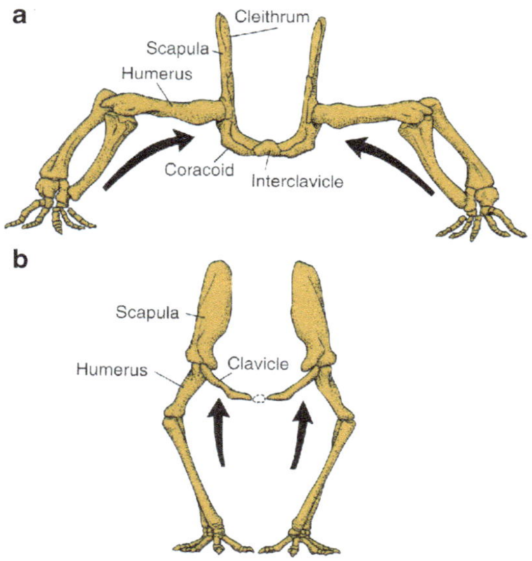

Fig. 1.2 Change in the role of the shoulder girdle with change of limb posture. (**a**) Sprawled posture with medially directed forces to the shoulder girdle. (**b**) As limbs are brought under the body, these forces are more directed more ventrally. Especially, the role of the clavicle is diminished with these changes (Reprinted with permission from Kardong [1])

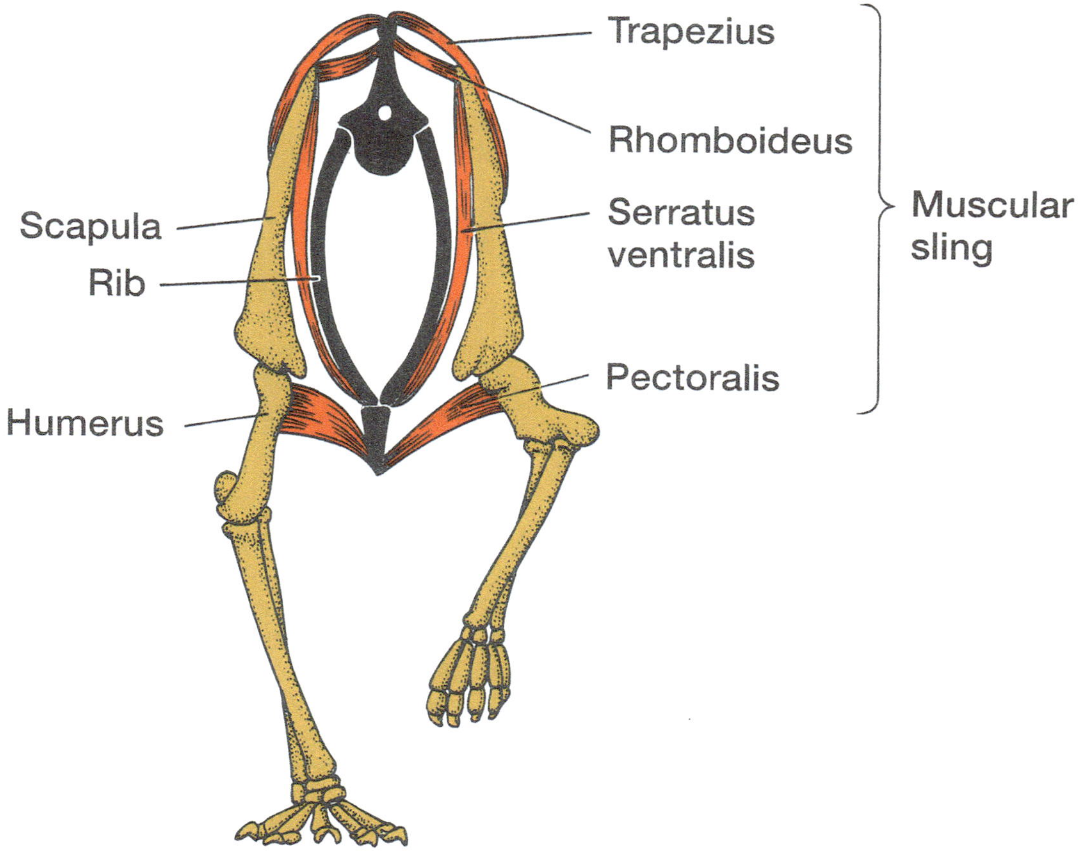

Fig. 1.3 Muscular slingv of mammals. Some of the muscles arise from the branchiomeric muscles (trapezius), some from the axial muscles (rhomboideus, serratus), and some from the ventral muscles (pectoralis) (Reprinted with permission from Kardong [1])

1.3.1 Branchiomeric Muscles

Branchiomeric muscles: The branchiomeric muscles contribute the *trapezius* and *mastoid* (including the cleidomastoid and sternomastoid).

1.3.2 Axial Muscles

Axial muscles: The axial musculature contributes the *levator scapulae, rhomboideus complex* and *serratus.* These three muscles together with the trapezius form the muscular sling that suspends the body between the two scapular blades. As the shoulder girdle became separated from the skull, the branchiomeric and axial muscles developed into serving as part of the muscular sling through which the forelimbs are attached to the body.

1.3.3 Dorsal Muscles

Dorsal muscles: The dorsal muscles insert on the humerus and function to oscillate the humerus during movement or fix it in position, when an animal stands; of these muscles only the *latissimus dorsi* originates outside the limb from the body wall. The other dorsal muscles that act on the humerus are the *teres minor, subscapularis* and *deltoideus,* which may form two distinct muscles. The *triceps* is also derivative of the dorsal muscles but it acts to extend the forearm.

1.3.4 Ventral Muscles

Ventral muscles: The *pectoralis* is an important ventral muscle. In early mammals, the pectoralis consisted of four separate muscles, in primates only two remained: *pectoralis major* and *pectoralis minor.*

The ventrally positioned *supracoracoideus* in mammals originates dorsally on the lateral face of the scapula. The bony scapular spine divides this muscle into the *supraspinatus* and *infraspinatus* muscles, which insert on the humerus as well. The *coracobrachialis* from the coracoid runs along the front of the humerus.

In mammals, the *biceps brachii* has two heads, representing the fusion of two muscles, which insert into the forearm and are responsible for flexion.

1.3.5 Rototor Cuff

Rototor Cuff: Sonnabend et al. studied the rotator cuff muscles in 22 different animals, including marsupials, carnivores, ungulates and other primates. He identified rotator cuff muscles in all animals and that in the majority of animals the tendons inserted independently onto the tuberosities [3].

He observed that in some animals the tendons blended together to form a true rotator cuff (common tendon) in some primates (e.g. baboon and hominoids such as chimpanzee and orangutan). There was one marsupial species, the tree kangaroo, which formed a common tendon. All other animals in his study, the rotator cuff tendons were not interconnected.

There was a strong correlation of the presence of a true rotator cuff, with interconnected tendons, and the ability to perform activities overhead or away from the sagittal plane [3].

1.4 Comparative Anatomy: Tetrapods to Humans

Inman et al. have studied extensively the evolutionary changes of the shoulder girdle from tetrapodal primates through arboreal (tree living) primates to bipedal species, including hominoids [4].

1.4.1 Scapula

During evolution, the scapula changed in form when functional demands changed, including both change of posture prehensile tasks (form follows function!).

In *reptiles,* the scapula was broad and massive; however, with increased efficiency in locomotion, there were a number of changes. These included the following:

1. Reduction in the size of the scapula
2. Orientation of the glenoid changes from lateral to posterior-inferior

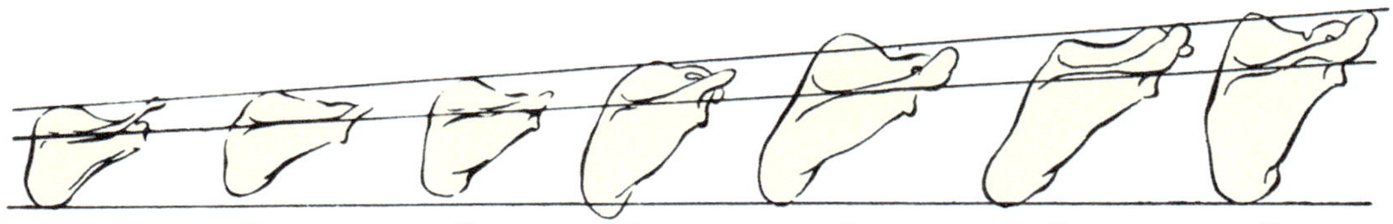

Fig. 1.4 Progressive decrease in scapular index in successive stages from the pronograde to the orthograde (Reprinted with permission from DePalma [10])

3. Development of the scapular spine
4. Disappearance of the procoracoid
5. Reduction of size of coracoid process
6. Rearrangement of muscles

The shape of the scapula is dependent on the posture: It is broad and massive in tetrapod animals that need a large serratus anterior muscle to support body weight. Alterations are brought about by change of posture from the pronograde (walking with the body horizontally) to the orthograde posture, as well as specialized requirements of a prehensile limb [2]. In orthograde posture, the position of the scapula changes from a lateral to a more dorsal position, also influenced by the flattening of the thorax. This change of position of the scapula also led to a lengthening of the clavicle.

The most impressive change in the scapula is in the length to width ratio. Pronograde forms have a long narrow scapula, while in homonoids the scapula becomes broader. Inman et al. introduced the scapular index, a ratio of the scapula width (base of the spine) to length (superior to inferior angle) [4]. The index decreases in successive stages toward the development of man (due mainly to an increasing length with minimal increase in width) (Fig. 1.4). The increase in scapula length occurs inferior to the spine, which also alters the relationship of the scapula axillary border and the working angle of the muscles on the humerus.

The deltoid has had an increasing role during the evolution of the primates. This is reflected in the change of the prominence of the outer end of the spine: In pronograde animals, the acromion is hardly developed; in orthogrades, it is a massive structure arching over the humeral head (Fig. 1.5).

1.4.2 Humerus

In *reptiles* with free motion of the forelimb, the upper extremity was brought under the body and the humerus became considerably smaller than the hindlimb. Two nodules appeared proximally, which evolved into mammalian tuberosities.

In *mammals* adapted for running (e.g. horse), the articular surface of both ends of the humerus function in the same sagittal plane: a line passing through the long axis of the humeral head is perpendicular to the line through the articular surface of the distal humerus.

In *primates* with an orthograde posture, this angle changes with an increase of internal rotation of the shaft which rotates the humeral head to match the altered scapula position described previously (Fig. 1.6).

Prehensile tasks, however, demand that the extremity as a whole functions anterior to the body, maintaining the elbow in the sagittal plane. To adapt to this situation, the humerus twists inwardly, while the articular surfaces at either end rotate in the opposite direction (Fig. 1.7) [2]. The angle of 90° in running mammals changed to 16–36° of retroversion.

During the evolution to enable prehensile tasks, another important change was the flattening of the glenoid and the lateralization of the tuberosities [4]. In pronograde forms, the biceps tendon runs over the centre of the head of the humerus and enters the groove in the same plane: In this position, the biceps is a strong flexor. In these animals, the tuberosities have the same size.

In orthogrades, the inward rotation of the shaft also rotates the groove more medially (30°) and there is also a decrease in size of the minor tuberosity. In this position, the long head of the biceps

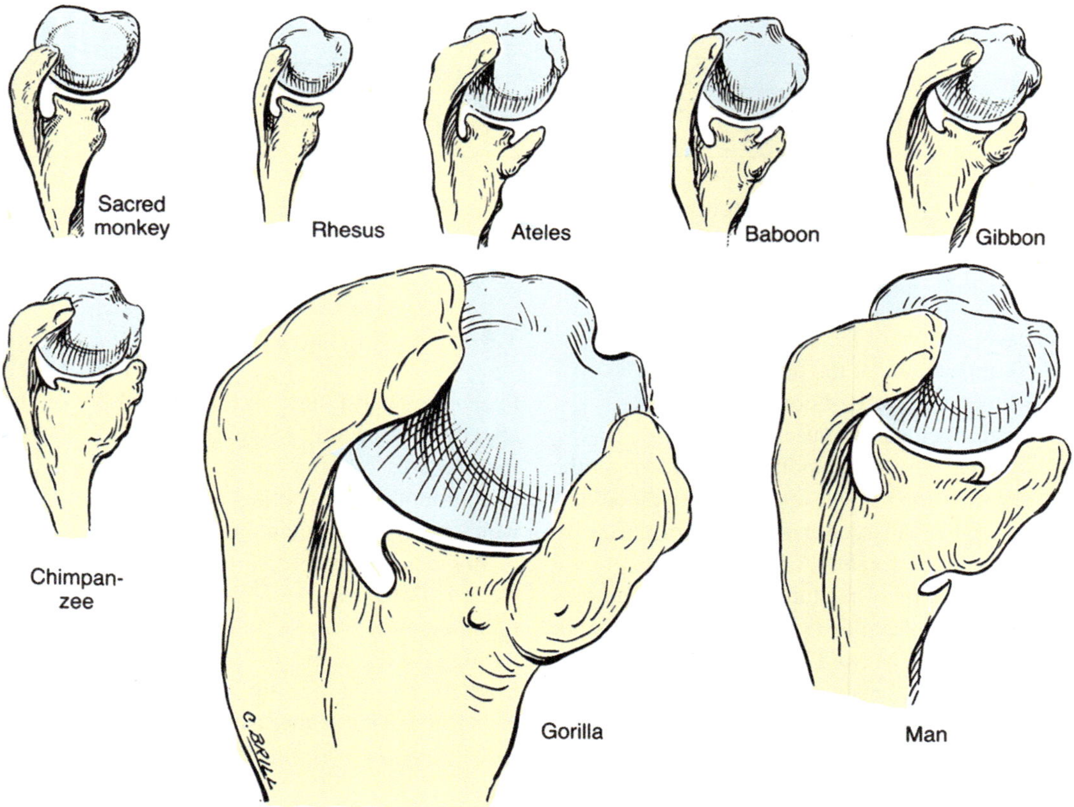

Fig. 1.5 Gradual increase in spine of the scapula and the acromion process during development from the pronograde to the orthograde. This change reflects the increasing importance of the deltoid muscle. Also note the increase in size of the coracoid process, the inequality of the two tuberosities of the head of the humerus and the inner displacement of the intertubercular sulcus in successive stages of development (Reprinted with permission from DePalma [10])

is less active as a shoulder flexor, other than when the arm is positioned in external rotation [4].

1.4.3 Muscles

The *deltoid* has a dominant role in higher primates due to the progressive shift of the insertion of the deltoid on the humerus (Fig. 1.8) and the deltoid shift of the humeral insertion.

1.4.3.1 Scapulohumeral

These muscles connect the scapula to the humerus and consist of the supraspinatus, infraspinatus, teres minor, subscapularis, deltoid and teres major. With evolution to a more mobile forelimb, the deltoid increases in size, while the supraspinatus decreases. The teres minor evolved from the deltoid to form a separate muscle passing from the inferior angle of the scapula. With the increasing size of the scapula below the spine, the mass of the infraspinatus progressively increased. The subscapularis is little affected during the evolution from the primitive to higher primates.

Due to the lengthening of the scapula below the spine, the lateral border of the scapula also increases in size. Therefore, the infraspinatus, teres minor and subscapularis originate from a more inferior position and, therefore, act as humeral head rotators as well as depressors.

Sonnabend et al. studied the rotator cuff muscles in 22 different animals, including marsupials, carnivores, ungulates and other primates.

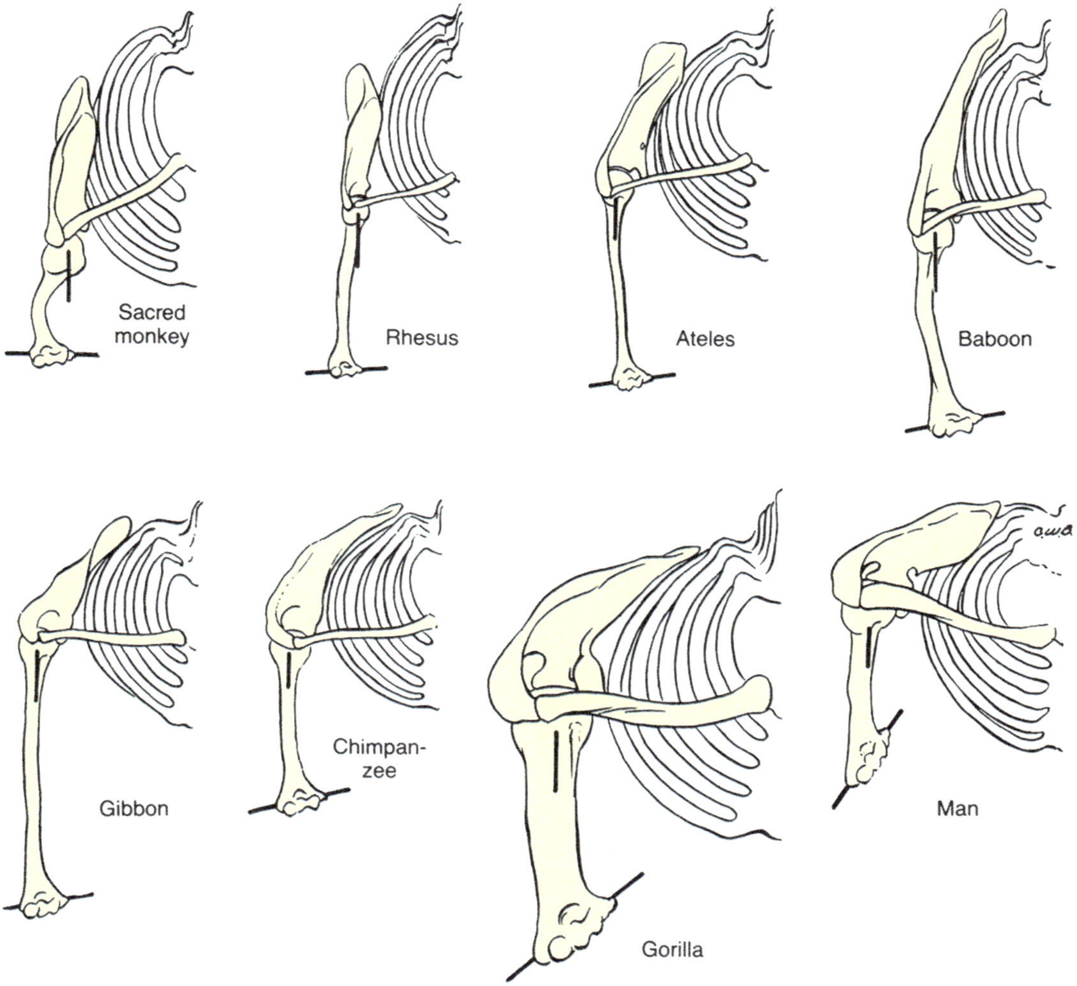

Fig. 1.6 Changes in successive stages from pronograde to orthograde locomotion: the thoracic cage flattens, the scapula migrates dorsally, the glenoid is directed more laterally, the clavicle lengthens (Reprinted with permission from DePalma [10])

He identified rotator cuff muscles in all animals and found that in the majority of animals the tendon inserted independently onto the tuberosities [3]. He observed that only those animals who perform activities overhead that the tendons blended together to form a rotator cuff (common tendon) (e.g. baboon and hominoids such as chimpanzee and orangutan).

1.4.3.2 Axiohumeral

These muscles connect the axial skeleton to the humerus and consist of the pectoralis major, pectoralis minor and latissimus dorsi. The pectoralis major muscle mass divided into a superficial and a deep layer. One part moved from the sternum to the clavicle (clavicular head). From the deep layer evolved the pectoralis minor.

The latissimus dorsi and teres major originate from a single muscle sheet, extending from the trunk, caudal to the scapula to the humerus. Later, the teres major split from the main muscle mass and now originates from the scapula tip. Both of these muscles were more developed in arboreal primates.

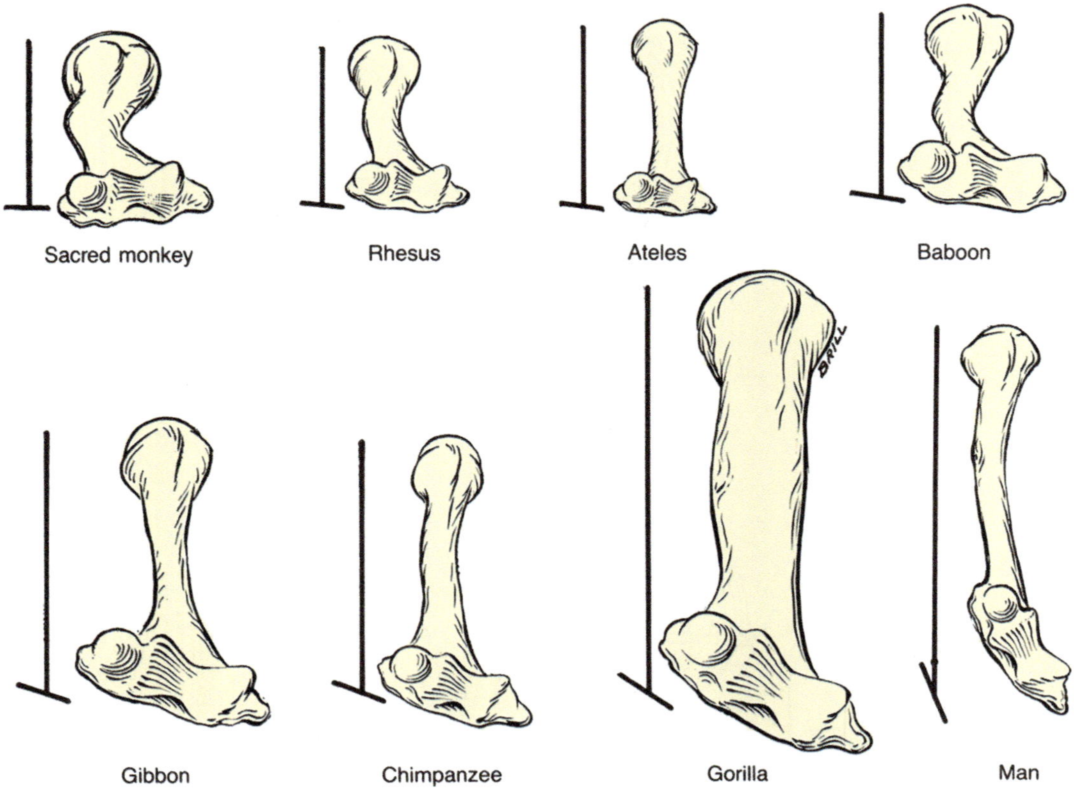

Fig. 1.7 Progressive change in torsion of the humeral shaft: the shaft rotates internally, the articular surface of the head rotates externally (from 90° of retroversion to about 16–36°), resulting in inward rotation of the bicipital groove (Reprinted with permission from DePalma [10])

1.4.3.3 Axioscapular

These muscles connect the axial skeleton to the scapula and consist of the serratus anterior, rhomboids, levator scapulae and trapezius. The first three muscles originate from the ribs or processes of the cervical vertebrae and insert into the vertebral border of the scapula. Their main role is to control the movements of the medial border of the scapula. The levator scapulae and serratus anterior originated from one muscle sheath; during evolution, the proximal and distal fibres increased in size, with reduction of the fibres in between, leading to the development of these two separate muscles. Now the scapula is elevated by levator scapulae and moved anteriorly by the serratus anterior. There has been little change in the trapezius in the evolution of primates.

1.4.3.4 Biceps and Triceps Muscles

Both these muscles evolved from ventral and dorsal brachial muscles, respectively. From the ventral brachial elements arose the biceps muscle by proximal migration along a fascial plane of brachial components to reach the scapula. In mammals other than primates, it is a single muscle. In animals like the horse, the powerful biceps and supraspinatus act as a single unit to lift the foreleg. However, in primates the biceps has two heads of origin (supraglenoid tubercle and coracoid

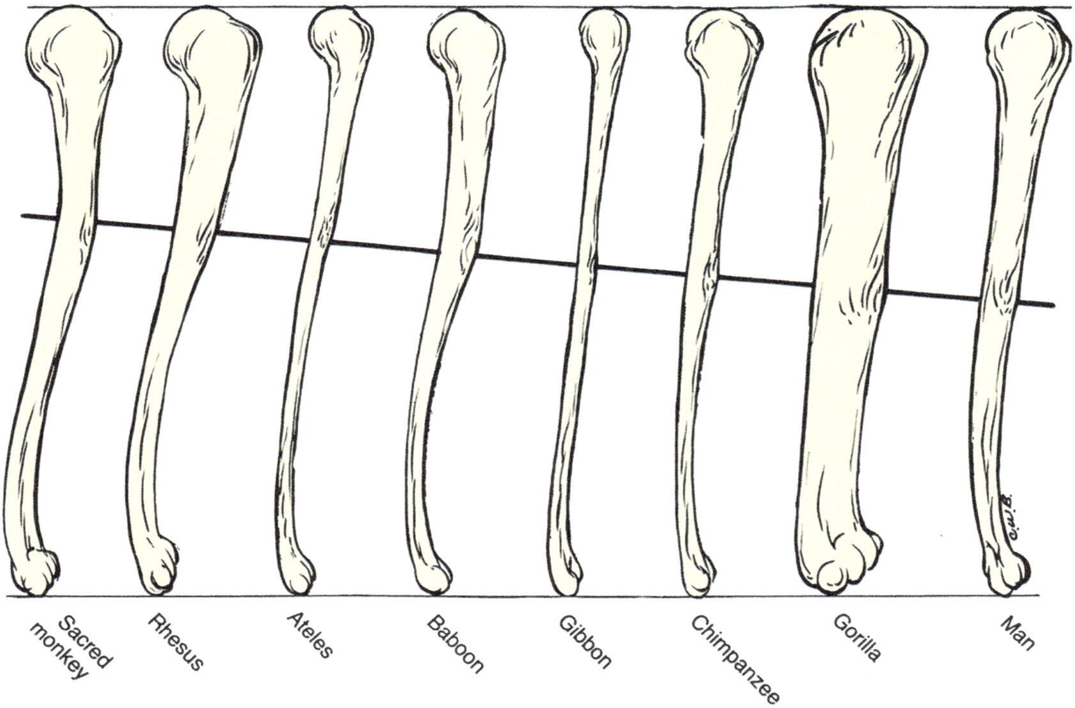

Fig. 1.8 Deltoid insertion migrates progressively to a lower level on the shaft of the humerus, indicating the significant role played by the deltoid in higher primates (Reprinted with permission from DePalma [10])

process). Medial displacement of the groove due to the torsion of the shaft makes the biceps less effective in elevating the human arm.

The triceps originated from a dorsal brachial muscle element. Like the biceps, the three heads migrated proximally, with the scapular head reaching the infraglenoid tubercle.

1.5 Animal Models

Animal models are very valuable in several fields of medical research and can be used to study anatomy, biomechanics and pathology both in vivo and in vitro.

Cadaveric studies provide an appropriate tool to study anatomy and biomechanics, muscle and tendon function, shoulder biomechanics and strength of various rotator cuff repair techniques. In vivo models provide the means to study tendon healing and degeneration, joint instability [5, 6, 8].

Animal models have been used to study shoulder contracture, arthroplasty, instability, and in recent years, focus has been on rotator cuff tears [5].

However, all of these pathological conditions lack a validated animal model.

In vivo and in vitro shoulder research has been performed on numerous animal models, involving mouse and rat, cats, rabbit, goat, sheep, dog, calf, cynomolgus monkey and baboons [6].

Soslowsky compared 33 animals and suggested that the rat is the most appropriate model to study the rotator cuff (Fig. 1.9) [9].

Plate compared the advantages and disadvantages of the several animal models used for shoulder research (Table 1.1).

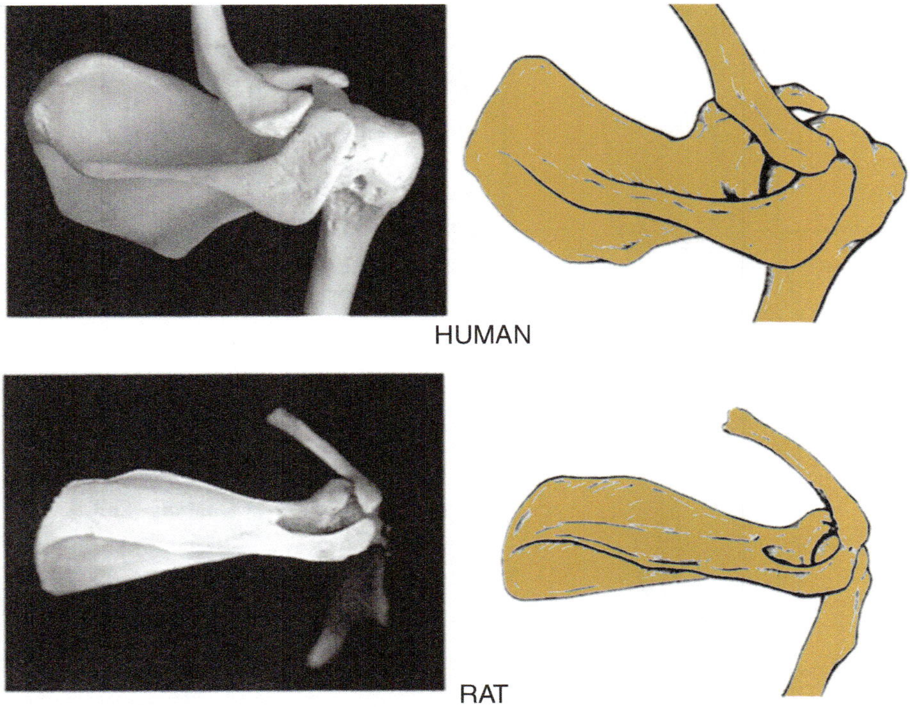

Photographs and schematics of the bony anatomy of the right human and rat shoulders from a postero-superior view demonstrating the similarity ot the acromion projecting anteriorly over the humeral head to the clavicle.

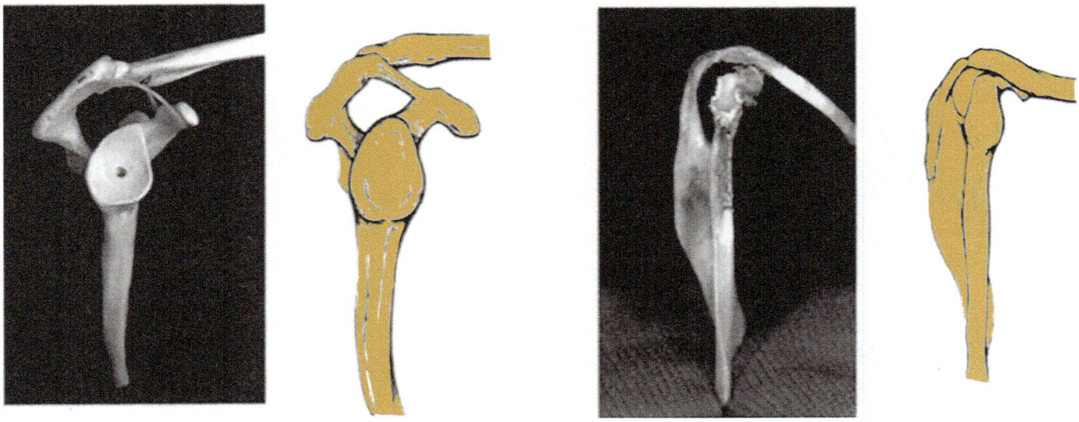

Photographs and schematics of the right human and rat shoulders from a lateral or "outlet" view with the humerus removed demonstrating the similar presence of an enclosed arch over the space for the supraspinatus tendon.

Fig. 1.9 Comparison of human and rat shoulder anatomy; posterosuperior and outlet view, demonstrating a closed arch over the supraspinatus in both shoulders (Reprinted with permission from Soslowsky et al. [9])

Table 1.1 Comparison of animal models utilized for shoulder research

	Rat/mouse	Rabbit	Dog	Sheep/goat	Nonhuman primates
Advantages	Comparable RC with SST below arch Available and cheap Lowest demand (care, facilities) Large sample size	Fibrofatty infiltration following injury Relatively inexpensive Low demand (care, facilities)	Assess T-BH Similar size, RC loads	RCR Assess T-BH Similar size	Similar anatomy, RC insertion and age-related degeneration RCR Multidirectional movement Assess T-BH
Limitations	Limited multidirectional movement Small-scale shoulder Significant fatty infiltration following RCT (but only in combination with SSN transection) Quadrapod No re-tears RC repair	Limited comparability (SubST, Quadrapod)	Limited multidirectional movement Different acromion and coracoid Quadrapod, Moderate demand (care, facilities) Expensive	Limited multidirectional movement Different IST Quadrapod, High demand (care, facilities) Expensive	Semi-terrestrial, weight-bearing forelimbs Highest demand (care, facilities) Highly expensive for longitudinal studies Ethics
Chronic RCT	Spontaneous healing with persistent partial tear, scar forms a "pseudo-tendon"	Chronic condition for muscular changes Spontaneous healing	Spontaneous healing with scar tissue	Spontaneous healing with scar tissue	Healing response to chronic injury not assessed
Outcome measures	Functional in vivo assessment Gait analysis Histo CT, MRI	CT, MRI, US Gait analysis Histo	CT, MRI, US Gait analysis Histo	CT, MRI, US Histo	CT, MRI, US Histo Assess upper limb activity, walking speed, functional and overhead activity
Areas of research	Mechanisms of age-related degeneration, e.g. intrinsic, extrinsic, impingement, overuse In vivo functional biomech. Molecular pathways Rehabilitation	Mechanisms of muscular changes Biomech. T-BH ± scaffolds	T-BH ± scaffolds Biomech. Mechanical strength RCR	In vivo biomech. of chronic RCT Mechanical strength RCR	Mechanisms of age-related degeneration In vivo T-BH Biomech. Molecular pathways Mechanical strength RCR

Modified with permission from Plate et al. [7]

RC rotator cuff, *SST* supraspinatous tendon, *T-BH* tendon to bone healing, *RCT* rotator cuff tear, *RCR* rotator cuff repair, *SSN* suprascaular nerve, *US* ultrasound, *Histo* histology analysis, *SubST* subscapularis tendon, *SST* supraspinatous tendon, *IST* infraspinatus tendon, *Biomech* biomechanical studies, *Scaffolds* scaffolds augmentation

Conclusion

The animal model is an important part of shoulder research; however, we need to use caution in translating the concepts to humans.

References

1. Kardong KV. The vertebrates. Comparative anatomy, function and evolution. 6th ed. Boston: McGraw-Hill International Edition; 2012.
2. DePalma AF. The classic. Origin and comparative anatomy of the pectoral limb. Surgery of the shoulder. Philadelphia: Lippincott Williams & Wilkins; 1950. p. 1–14. Clin Orthop Relat Res. 2008;466(3): 531–42.
3. Sonnabend DH, Young AA. Comparative anatomy of the rotator cuff. J Bone Joint Surg Br. 2009;91(12): 1632–7.
4. Inman VT, et al. Observations on the function of the shoulder joint. J Bone Joint Surg Am. 1944;26(1): 1–30.
5. Cavinatto L. Experimental models in shoulder research. A.G.E.S.A. G Milano (ed). London: Springer-Verlag. 2014.
6. Longo UG, et al. Animal models for translational research on shoulder pathologies: from bench to bedside. Sports Med Arthrosc. 2011;19(3): 184–93.
7. Plate JF, et al. Age-related degenerative functional, radiographic, and histological changes of the shoulder in nonhuman primates. J Shoulder Elbow Surg. 2013;22(8):1019–29.
8. Warden SJ. Animal models for the study of tendinopathy. Br J Sports Med. 2007;41(4):232–40.
9. Soslowsky LJ, et al. Development and use of an animal model for investigations on rotator cuff disease. J Shoulder Elbow Surg. 1996;5(5):383–92.
10. DePalma AF. Surgery of the shoulder. Philadelphia: Lippincott Williams & Wilkins; 1950. p. 1–14.

Developmental Anatomy of the Shoulder

2

Teresa Vázquez, Javier Calvo, Jose Sanudo, and Emilio Calvo

2.1 Introduction

From a molecular point of view, events of upper limb development include a secreted protein that binds to a specific receptor, which activates expression of specific target genes [1].

The notochord expresses sonic hedgehog, which is thought to regulate the initiation of limb bud formation. The limb bud is an outgrowth of somatic mesoderm (becomes muscles, nerves and vessels) and lateral plate mesoderm (becomes bone, cartilage and tendon) into the overlying ectoderm. This bulging ectoderm, called the apical ectodermal ridge, forms at the junction between the ventral and dorsal ectoderm.

T. Vázquez, PhD • J. Sanudo, MD, PhD
Department of Human Anatomy and Embriology, School of Medicine, Complutense University of Madrid, Madrid, Spain
e-mail: tvazquez@ucm.es; jrsanudo@ucm.es
URL: http://www.ucm.es

J. Calvo, MS
Institute of Health Research - Fundacion Jimenez Diaz Universidad Autonoma, Autónoma University, Madrid, Spain
e-mail: javier.calvot@estudiante.uam.es
URL: http://www.uam.es

E. Calvo, MD, PhD (✉)
Shoulder and Elbow Reconstructive Surgery Unit, Department of Orthopedic Surgery and Traumatology, Fundación Jiménez Díaz, Autónoma University, Madrid, Spain
e-mail: ecalvo@fjd.es
URL: http://www.fjd.es

The limb bud appears at 26 days after fertilization, when the embryo is only 4 mm long (crown to rump length). The limb buds appear as small elevations on the ventrolateral body wall. The upper limb buds appear before those from the lower limb and maintain the growth advantage throughout development. The limb bud is lined by the ectoderm, which subsequently becomes the nervous tissue, epidermis and its appendages. They also contain the mesodermal tissue, which forms cartilage, bone, connective tissue and muscle. Soon after the establishment of the bud, cartilage precursor cells form a chondrogenesis core, and other connective tissue cells (tendons and muscles) accumulate in the periphery.

The first step of musculoskeletal development is condensation of mesenchyme into a preskeletal blastema in the core of the bud. The next step is the overt differentiation of these cells into either cartilage forming chondrocytes (in endochondral skeletal elements) or bone forming osteoblasts (in membranous skeletal elements).

By 6 weeks, the first hyaline cartilage models foreshadowing the bones of the shoulder joint are formed by chondrocytes. When skeletal elements are formed, there are areas in between which do not undergo any change into cartilage or bone. These areas persist as joint interzones and are sites for development of future joints. Each interzone passes through a three-layered stage with two chondrogenous layers and an intermediate loose layer. The growth of cartilagenous elements compresses the central part of interzone and the cavity appears in

G.I. Bain et al. (eds.), *Normal and Pathological Anatomy of the Shoulder*,
DOI 10.1007/978-3-662-45719-1_2, © ISAKOS 2015

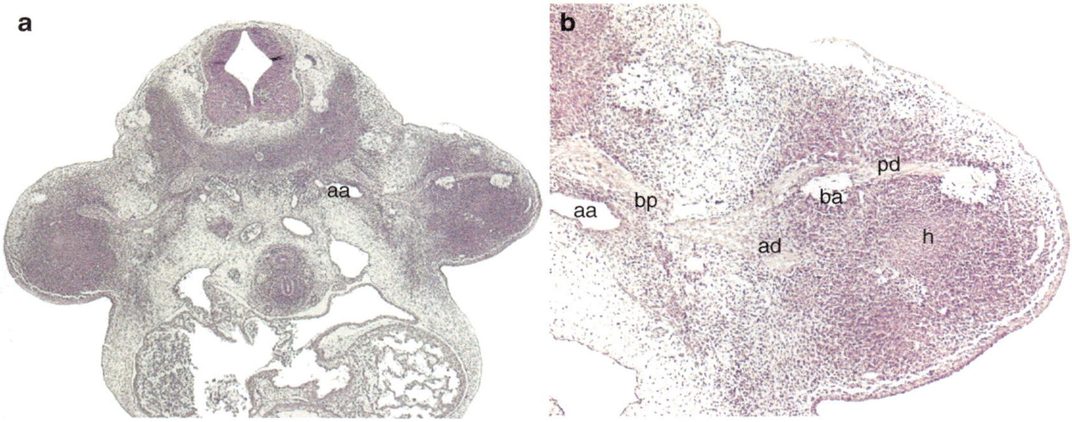

Fig. 2.1 Stage 15 (8.5 mm CRL) embryo transversally sectioned. Hematoxylin-Eosin stained. **a** (2×) and **b** (4×) area corresponding to the left upper limb magnified. *aa* axilary artery, *ad* anterior division of the brachial plexus, *ba* brachial artery, *bp* brachial plexus, *pd* posterior division of the brachial plexus, *h* humerus

its circumferential portion. The cavity expands gradually and extends towards the centre of joint. The cells in the area of cavitation undergo liquefaction. Macrophages present in the periphery of joint interzones produce lytic enzymes, which could be involved in the cavitation process [2].

In the upper limb, all bones have an endochondral development, with the exception of the clavicle, scapula body and the distal part of the distal phalanges, which are all developed with membranous . The process of in the upper limb commences proximal with the humerus (36 days after fertilization) and ending with the proximal parts of the distal phalanges (50 days after fertilization).

Development of joints occurs by repression of chondrogenesis at the sites of future joints. Three important proteins are highly expressed at these joint interzones: WNT4, WNT14 and growth and differentiation factor (which is also called cartilage-derived morphogenetic protein-1). The first joint interzone appears at 36 days at the shoulder, and the last interzones appear at 47 days in the hand.

2.2 Main Chronological Events

Some of the most important events described have been grouped following developmental stages for embryos described by O'Rahilly [3] and O'Rahilly and Müller [4] as well as the Crown Rump length (CRL) in millimetres and weeks of development (for fetuses). Postnatal development is commonly described in years.

2.2.1 Embrionary Development

Stage 13 (4–6 mm; 28 days), Stage 14 (5–7 mm; 32 days). The limb bud has continued its outgrowth. A larger axial trunk can be observed, originating from the dorsal aorta, but when it reaches the base of the limb bud it ramifies into capillaries throughout the whole limb bud. The nerves are still out the bud and no skeletal or muscular elements are visible [5].

Stage 15 (7–9 mm; 33 days), Stage 16 (8–11 mm; 37 days) (Fig. 2.1). The axial artery extends to include both the subclavian and axillary arteries. After crossing the neural plate, it ramifies into its capillary network. The nerves have begun to enter the limb bud, joining together to form the neural plate and then branching into anterior and posterior divisions.

The neural plate has divided into two divisions, anterior and posterior. The anterior one forms the musculocutaneous, ulnar and median nerves while the posterior one forms the radial nerve. It is clear that nerve ingrowth at the base of the bud begins in stage 16, but it is still not fully

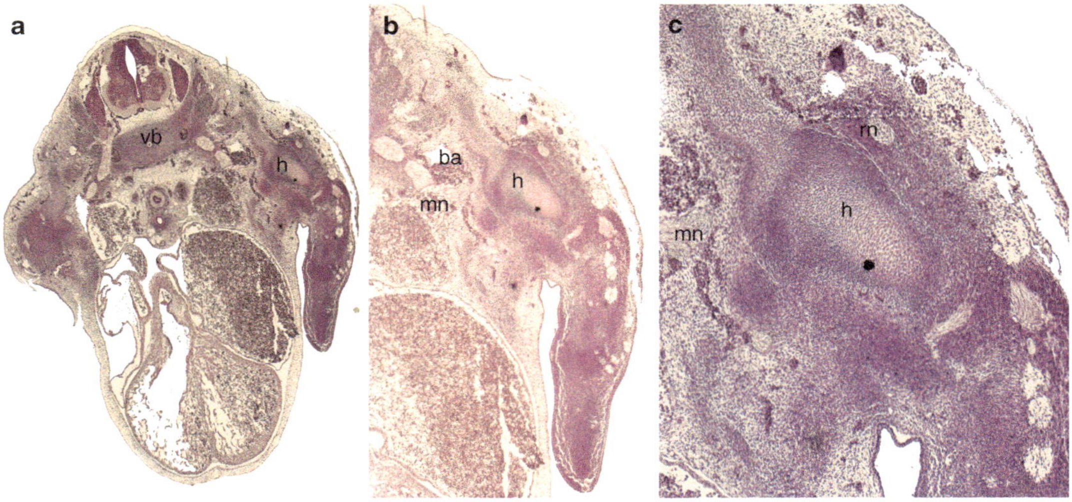

Fig. 2.2 Stage 17 (12 mm CRL) embryo transversally sectioned. Hematoxylin-Eosin stained. **a** (×2), **b** (×4) left upper limb magnified, **c** (×10) left shoulder magnified. *ba* brachial artery, *mn* median nerve, *h* humerus, *rn* radial nerve, *vb* vertebral body

understood how specific nerves are guided to supply specific muscles.

The first clavicle precursor appears at 11 mm, a condensed curved connective-tissue rod stretching from acromion inwards towards the first rib [6]. The mesenchymal tissue of the humerus has now begun to chondrify, and the ulna and radius are still represented by condensed mesenchymal tissue.

Stage 17 (11–14 mm; 41 days), Stage 18 (13–17 mm; 44 days) (Fig. 2.2). The nerves are easily recognizable as far as the hand. The mesenchyme of the humerus is in a chondrified state. Haines [1947] reported that the cartilages of the shoulder and elbow joints were adopting their characteristic shapes and their trilaminar interzones at 12 mm CRL [7].

The clavicle is still developing; an inner and an outer mass of precartilage is developed in the connective-tissue rod, the inner overriding above and in front the outer mass. The outer segment is at this time distinctly connected with the base of the coracoid process by the coracoclavicular ligament [2, 6].

Stage 19 (15–18 mm; 47 days) (Fig. 2.3). The skeletal elements have all chondrified, every upper limb muscle is identifiable and contains muscle fibres by the seventh week of development. As in all muscle development, the superficial muscles differentiate before deeper ones. Once again, proximal muscle masses differentiate before distal masses, paralleling the skeletal process. The nerves have achieved their definitive adult arrangement and can be easily recognized. The coracohumeral ligament has been described developing by the time of 6 1/2 weeks to 15 weeks.

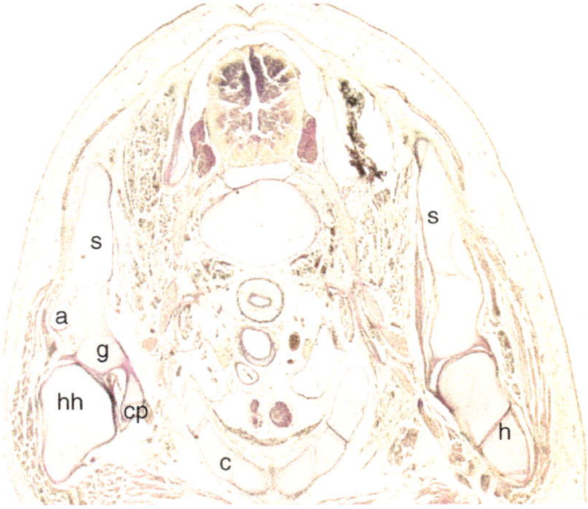

Fig. 2.3 Stage 19 (18 mm CRL) embryo transversally sectioned. Hematoxylin-Eosin stained. (×2). *a* acromion, *c* clavicle, *cp* coracoid process, *g* glenoid, *h* humerus, *hh* humeral head, *s* scapula

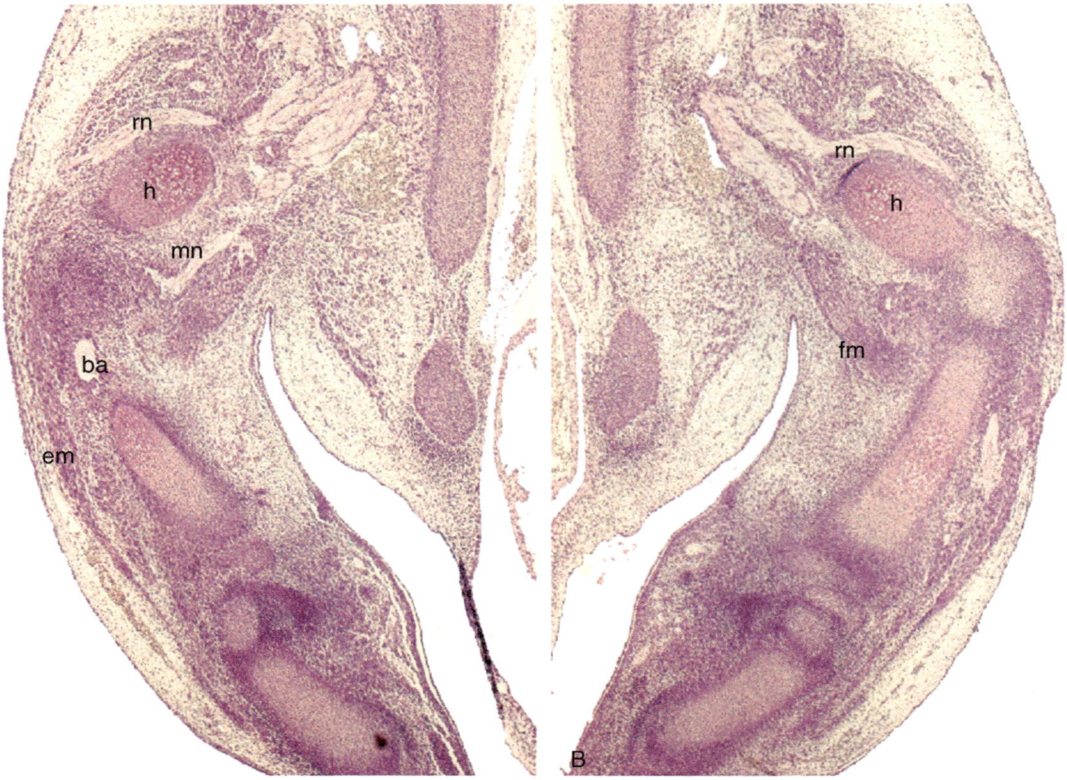

Fig. 2.4 Stage 20 (20 mm CRL) embryo transversally sectioned. Hematoxylin-Eosin stained. (×2). (**a**) Right upper limb. (**b**) Left upper limb. *ba* brachial artery, *em* extensor muscles, *fm* flexor muscles, *mn* median nerve, *h* humerus, *rn* radial nerve

Each precartilaginous mass for the future clavicle undergoes independent , the bone forming a central core to the precartilage; cartilage cells are seen in inner segment at this stage of development [6, 8, 9].

Stage 20 (18–22 mm; 50 days), stage 21 (22–24 mm; 52 days), stage 22 (23–28 mm; 54 days) and Stage 23 (27–31 mm; 56 days) (Figs. 2.4 and 2.5). From stage 20, the arterial pattern has already achieved its definitive morphology and during the 19 mm stage the two independent ossified clavicular centres fuse by a bony bridge resulting from the of the precartilaginous mass. Between 24 and 27 mm the clavicular development finishes, both chondral and perichondrial bone are formed. Cartilage is now present in the hollow at the outer end of the outer segment, and the anlage of the deltoid tubercle is present [6].

2.2.2 Fetal Development

Fetal CRL

30 mm (9 weeks), 53–58 mm (10–11 weeks) (Fig. 2.6). It has been described that the various components of the shoulder joint are discernable by 10 weeks. The head of humerus with its two tuberosities and intertubercular sulcus lodging the biceps tendon, acromion, coracoid process and spine of scapula are cartilagenous. Ossification occurs in the shaft which with due course of time extends higher up [9]. Periosteum of the humeral shaft has inner cellular and outer fibrous layers [2].

The humeral head is very primitive and appears as a small protrusion. The glenoid labrum is a very thin bright rim attached to the glenoid margin. The tendon of the long head of biceps muscle looks small rounded band like a cord, passed over the humeral head, attached to the superior part of

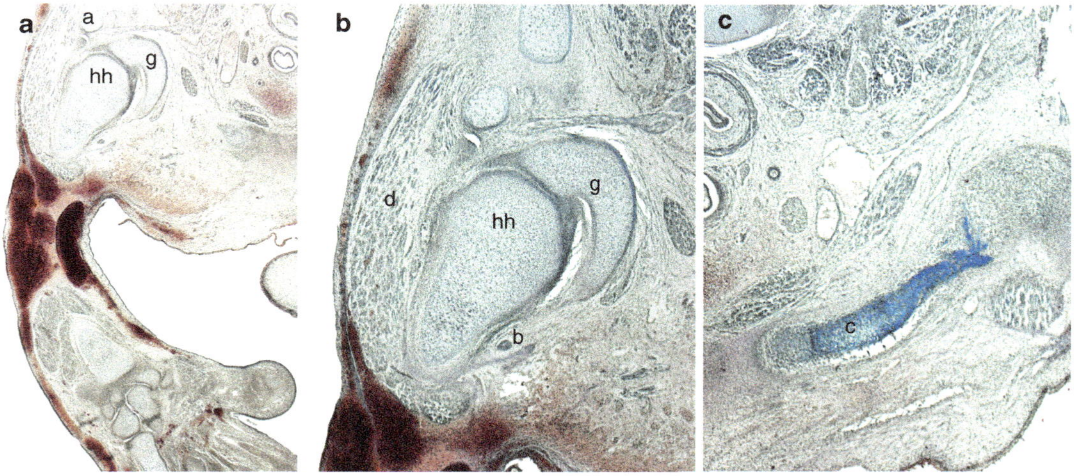

Fig. 2.5 Stage 20 (27 mm CRL) embryo transversally sectioned. Azan stained. **a** (×2). **b** and **c** (×4). *a* acromion, *b* biccipital tendon, *c* clavicle, *d* deltoid muscle, *g* glenoid, *hh* humeral head

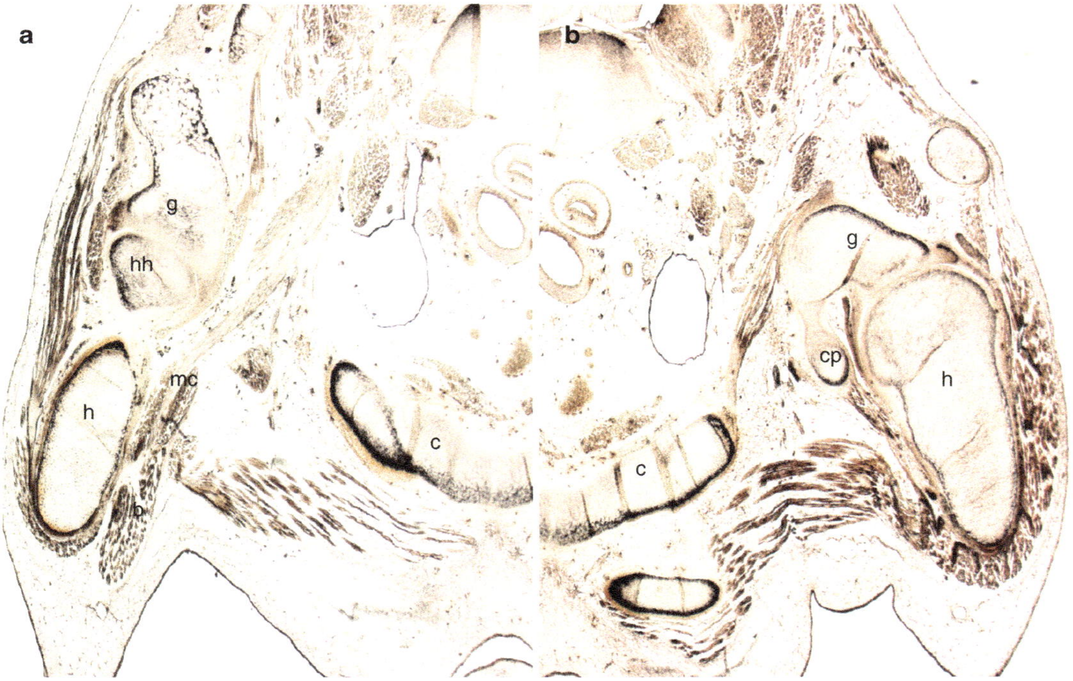

Fig. 2.6 Nine weeks development fetus (34 mm CRL) transversally sectioned . Bielchowsky stained. (**a**) Right upper limb (×2). (**b**) Left upper limb (×2). *c* clavicle, *cp* coracoid process, *g* glenoid, *h* humerus, *hh* humeral head, *mc* musculocutaneous nerve

the glenoid labrum and separated from the humeral head by a small cavity. The joint cavity is still narrow and small and the capsule very thin. No ligaments can be detected at this age [10].

The scapula shows a concave glenoid fossa and the neck can be differentiated. The coracoid process is larger in size than the acromion, which is still cartilagenous. The joint cavity can be

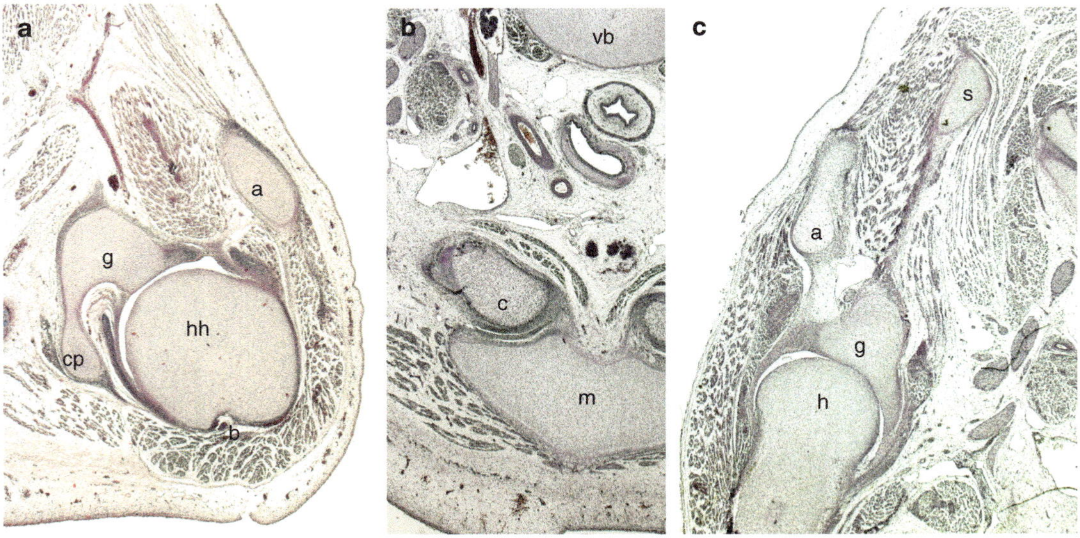

Fig. 2.7 Ten-eleven weeks development fetus (55 mm CRL) transversally sectioned. (**a**) Left shoulder (×2) Hematoxylin-Eosin stained. (**b**) Sternoclavicular joints level (×). Azan stained. (**c**) Right shoulder (×2). Azan stained. *a* acromion, *b* biccipital tendon, *c* clavicle, *cp* coracoid process, *g* glenoid, *h* humerus, *hh* humeral head, *s* scapule, *m* sternal manubrium, *vb* vertebral body

clearly visualized. The tissue lining the joint cavity is loose like synovial tissue and inferiorly it is reflected on the neck of humerus laterally and medially it is attached to glenoid labrum. The capsule is seen as continuation of perichondrium and is made of collagen fibres. It is more cellular than fibrous. Ossification of the humeral shaft progresses distal to the attachment of latissmus dorsi and teres major. The acromioclavicular joint can be recognized; it is lined by flattened cells. The acromial perichondrium extends to clavicle and serves the purpose of a capsular ligament. The lateral end of the clavicle is cartilagenous in nature [2].

The mesenchymal cells lining the articular surfaces and capsule of the shoulder joint are flattened and form a synovial membrane by 8–10 weeks while synovial villi develop by 11 weeks. The mesenchymal tissue surrounding the developing joint which is continuous with perichondrium forms a sleeve-like membrane which eventually transforms into a capsular ligament by 9 weeks. The capsule develops by 10 weeks and with increasing time the number of collagen fibres tends to increase. Coracohumeral and superior glenohumeral ligaments appear by the time of 10 weeks (Fig. 2.7).

The rotator cuff covering the humeral head appears initially as an insertion of the infraspinatus at 9 weeks. Similar to the biceps long tendon, the tendons of the supraspinatus, infraspinatus, and subscapularis are located together outside the joint cavity and separated from it by a thick membranous structure, possibly the joint capsule, covered by a primitive glenohumeral ligament. This primitive glenohumeral ligament appears to be established as a transient, but complete collateral ligament. At 12 weeks, however, it becomes modified so that the rotator cuff tendons become attached to the humeral head [11] (Fig. 2.8).

Fetal CRL

60 mm (12 weeks), 75–98 mm (12–14 weeks) (Fig. 2.9). The humeral head is greatly increased in size. Its shape becomes half a sphere, resembling that of the adult. The surgical neck can be identified. The glenoid fossa is progressively increased in size and depth, becomes pearshaped and has a depression on its anterior concavity like the adult form. The thickness of the glenoid labrum is progressively increased throughout its circumference except at its anterosuperior part, thus giving it the meniscus shape. It becomes

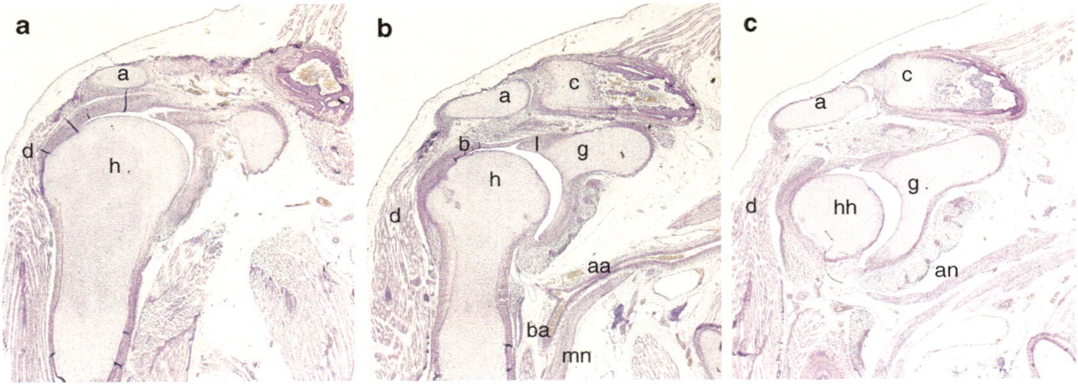

Fig. 2.8 (**a–c**) Ten-eleven weeks development fetus (57 mm CRL) Right shoulder oblique sectioned (×2). Hematoxylin-Eosin stained. *a* acromion, *b* biccipital tendon, *d* deltoid muscle, *g* glenoid, *hh* humeral head, *l* labrum, *aa* axilary artery, *an* axilary nerve, *ba* brachial artery, *c* clavicle, *h* humerus, *mn* median nerve

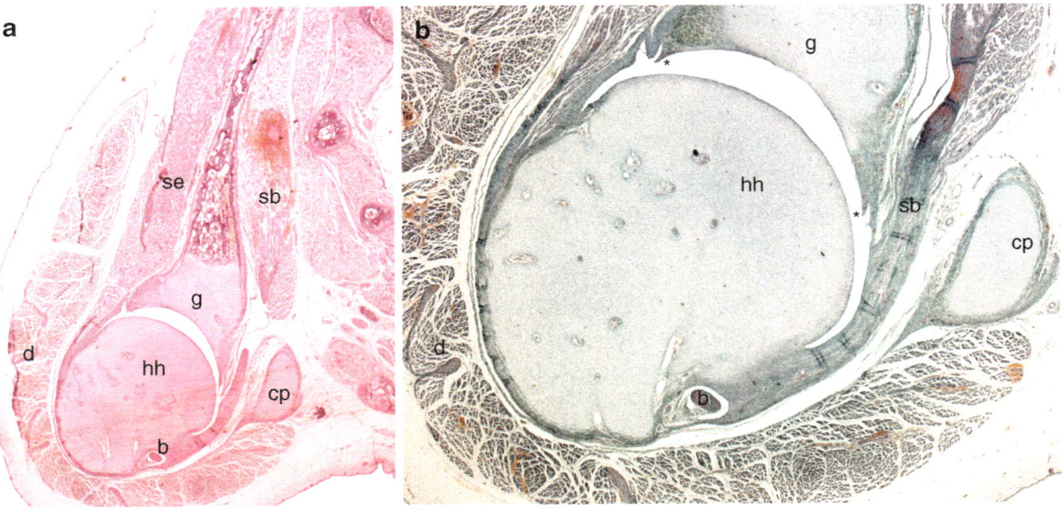

Fig. 2.9 Twelve-fourteen weeks development fetus (88 mm CRL) Right shoulder oblique sectioned (×1). (**a**) Hematoxylin-Eosin stained. (**b**) Azan stained. *b* biccipital tendon, *cp* coracoid process, *d* deltoid muscle, *g* glenoid, *hh* humeral head, *sb* subscapularis muscle, *se* spraspinatus muscle, (*) sinovial fold

fibrocellular by 12 weeks. The synovial tissue lines the whole capsule. The superior glenohumeral ligament is developed and appears as a thick band elevating the synovial membrane lining the anterior capsule and lay below and in parallel to the biceps tendon [10].

The inferior glenohumeral ligament develops by 14 weeks. The glenoid labrum, biceps tendon and glenohumeral ligaments form a complete ring that can be considered to be a prime factor in stability of shoulder joint.

Ossification in humerus is extended further in the shaft proximal to the attachment of latissmus dorsi and teres major muscle. The joint cavity is being surrounded by capsular ligament which is continuous with perichondrium of the humerus and scapula, and the capsular ligament is thickened superiorly below the attachment of the supraspinatus muscle between the lesser tuberosity and the glenoid labrum depicting the superior glenohumeral ligament. Inferiorly, between the lesser tuberosity and glenoid labrum, the middle glenohumeral ligament is seen as a thickening in the capsule. Frontal section of a 14 weeks fetus shows greater and lesser tubercles to be well distinguished. Ossification in the humeral shaft has extended up to a level slightly distal to inferior synovial reflection. An important observation is the presence of synovial villi in the joint cavity. The long head of the biceps tendon contains more collagen fibres and becomes densely collagenous in its centre. The space surrounding this tendon communicates with the joint cavity, hence proving that it is an extension of joint cavity [2].

The rotator cuff tendons are grossly evident by the time of 13–14 weeks and the glenoid labrum can be seen grossly as a distinct structure at 13 weeks [12].

The of the humeral head has been observed at 12 weeks, followed by ossification of the glenoid. A well-developed venous plexus is evident in the space between the supraspinatus tendon and the coracohumeral ligament. The intertubercular sulcus appears very deep at 12 and 15 weeks; however, no specific fibrous component is evident at the bottom of the sulcus in the humeral head [11].

Fetal CRL

120 mm (15–16 weeks). This stage shows an increase in of humeral shaft when the periosteum becomes thicker due to increase in number of collagen fibres. In the scapula, ossification has extended up to the neck and the joint cavity increases in size. The capsule has increased in thickness due to the increase in the number of collagen fibres, and it is being strengthened by the superior glenohumeral ligament. The ossification in the scapula has extended up to base of the acromion, but the acromion and the glenoid are still cartilagenous. The glenoid labrum is vascularised at its margin and has become more fibrous. The synovial membrane lining the joint cavity has numerous synovial villi and appears highly vascular [2]. The cord-shaped biceps tendon appears as an extension of the superior labrum. The middle and inferior glenohumeral ligaments are now developed. The middle ligaments appear below and parallel to the superior ligament and are separated from it by the aperture of the subscapular bursa.

The inferior ligament has a wide attachment to the anterior and inferior parts of the labrum.

Fetal CRL

132–142 mm (18 to 18-1/2 weeks). Ossification within the shaft has extended up to the level of the inferior synovial reflection and in scapula can be appreciated beyond the neck – almost up to the glenoid fossa. The joint cavity has increased in size with synovial villi within it and is now permeated by the biceps tendon demonstrating that this tendon is intracapsular and extrasynovial. The supraspinatus becomes vascularised. The glenoid labrum is also vascularised throughout its extent including the basal area. More fibrous tissue is present in the glenoid labrum and becomes fibrocellular rather than fibrocartilagenous [2].

Fetal CRL

143–168 mm (19–21 weeks). The extends in the shaft of humerus slightly beyond the level of inferior synovial reflection while in the scapula reaches the level of the scapular neck. The capsule has increased in thickness and the joint cavity in size. It is lined by a synovial membrane, and synovial villi can be seen inside. The shaft of humerus shows an increase in thickness of its bony collar [2].

There is a slightly fibrochondroid appearance of the labrum by 20 weeks. The transition between the fibrocartilaginous labrum and the cartilaginous glenoid fossa in the superior region becomes less distinct than the transition between the capsule/ligament and the cartilaginous glenoid fossa in the inferior region. The hyaline cartilage of the glenoid fossa is distinct from the labrum. There is no histologic evidence of any bare area in the hyaline cartilage of the glenoid fossa.

At this stage, the insertion of the subscapularis muscle has been described in two different ways: either closely united to the capsule or intra-articular near to the coracoid process at the same supero-inferior level as the lesser tuberosity. Moreover, the muscle insertion to the lesser tuberosity (the lowest part of the insertion in adults) is not identified in late-stage fetuses. Rearrangement between the muscle fibers and connective tissue seems to occur during postnatal development under mechanical demands as well as a result of the posterior shift of the scapular plate with flattening of the thorax [13].

Fetal CRL

172–240 mm (22–28 weeks). Synovial membrane lining the joint cavity has increased in thickness and becomes more vascular. Ossification of the humeral shaft extends beyond the inferior synovial reflection. The bony collar around the shaft of humerus has increased in thickness. Cartilage canals have reached articular surfaces of both humeral head and the glenoid fossa of the scapula.

Vascular invasion into the cartilaginous glenoid fossa has been observed earlier in the superior half of the glenoid fossa than in the inferior half. Vessels were seen in the superior half by 22 weeks, whereas similar vascular channels did not appear in the inferior half until 30 weeks [12]. Synovial villi develop collagenous cores at 26 weeks and show branching.

Fetal CRL

370 mm (40 weeks). All the intrarticular structures resemble the adult shape. The glenoid labrum looks markedly increased in thickness and forming a well-defined ring, so the concavity of the glenoid fossa has deepened. The thickness of the glenoid labrum is anteriorly still less than the thickness elsewhere. The biceps tendon becomes flat in shape [14].

The three glenohumeral ligaments are now clearly seen and exposed after cutting through the biceps tendon and most of the labrum. They appear as thick cords elevating the synovial surface of the anterior capsule. The superior and inferior ligaments are attached to adjacent parts of the glenoid labrum. They diverge as they pass towards the humerus and enclose a loose synovial tissue in between them. The middle ligament appears in parallel to the superior ligament and separated from it by the aperture of the subscapular bursa [10].

2.2.3 Postnatal Development

Clinical presentation of developmental abnormalities of the shoulder is rare. A genetic defect will alter the initial shape and development of the structure. An insult in the developing shoulder (in-utero or post-natal) will have a different effect, depending upon the age at which the insult occurs. Disorders of the nerves, such as cerebral palsy and Erbs palsy, will affect the muscle power, but also affects the growth of the osseous and soft tissue structures.

Since soft tissues change only in size after birth, the appearance and development of secondary centres are the main events in postnatal development of the shoulder.

2.2.3.1 Clavicle

The clavicle is the first bone to ossify, starting at the fifth week of gestation, with two separate centres (lateral and medial). The lateral and medial ossification centres ossify by endochondral ossification and are responsible for the longitudinal growth of the clavicle, while the increase in diameter is due to the intramembranous ossification of the diaphysis. The medial ossification centre is responsible for the majority of longitudinal growth of the clavicle. It commences ossification at 18 years, and fuses with the clavicle between 18 and 25 years.

2.2.3.2 Scapula

The scapula forms mainly by endochondral . There are eight ossification centres for the scapula (Fig. 2.10). The body and the spine of the scapula are the only areas ossified at birth. The coracoid is the first structure to ossify in the postnatal period. It shows two ossification centres which are located at the centre and the base of the process. They appear at the first and tenth years of life, respectively, and unite with the scapula at approximately 15 years.

The glenoid cavity is formed by two centres. The superior centre appears near the base of the coracoid and fuses at approximately 15 years.

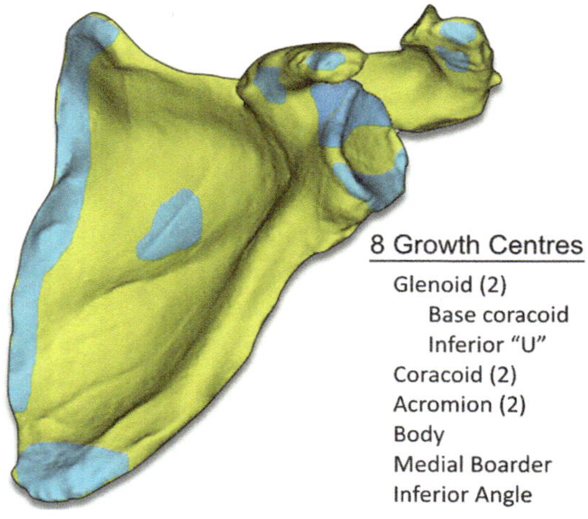

Fig. 2.10 Ossification centres of the scapula: Include superior and inferior glenoid, centre of the body, coracoid process, acromion, vertebral border and inferior angle (Used with permission from Landau [16], Elsevier)

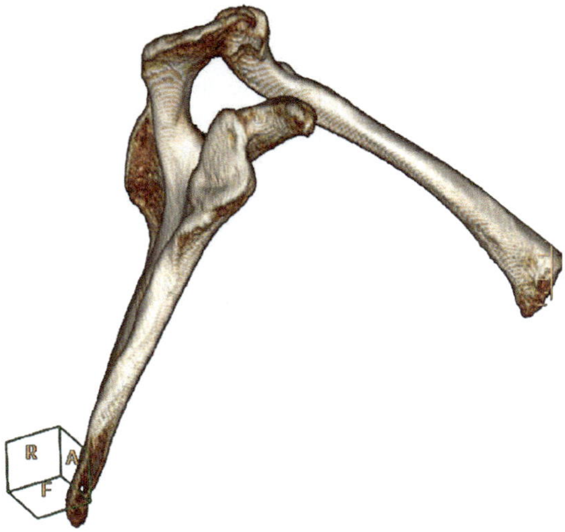

Fig. 2.11 3D CT scan of a dysplastic glenoid. Note that the coracoid and acromion are also dysplastic

The inferior ossification centre has a horseshoe shape and arises from the inferior portion and forms the lower three-fourths of the glenoid. The term "primary " is used to describe a rare developmental abnormality in which there is failure of ossification of the inferior glenoid (Figs. 2.11 and 2.12). It is often bilateral and associated dysplasia of the humeral head, coracoid or acromion. Glenoid dysplasia is usually asymptomatic and identified as an incidental finding on a chest X-ray [15]. However, it may also present as marked upper limb disability. The symptomatic presentation of glenoid dysplasia has two definite age-related peaks. The first peak is in adolescents and young adults that present with symptoms of instability during sport. The second is in the fifth or sixth decade when they present with degenerative changes of the gleno-humeral joint.

The acromion develops two (or three) centres at or during puberty and fuse at approximately 22 years. Failure to fuse these centres results in an os acromiale (see Chap. 4). Ossification centres at the vertebral border and the inferior apex of the scapula also appear at puberty and fuses at 22 years.

Proximal humerus. The proximal humerus is formed from three centres (humeral head, greater and lesser tuberosity). The ossification centre of the humeral head can be seen between the 4 and 6 months of fetal development, while the greater and lesser tuberosities do not form until the third and fifth years of life, respectively. The ossification centres of the tuberosities fuse at 5 years, and they fuse with the humeral head during the seventh year. The proximal humerus fuses with the shaft at 19 years, at which point longitudinal growth is complete.

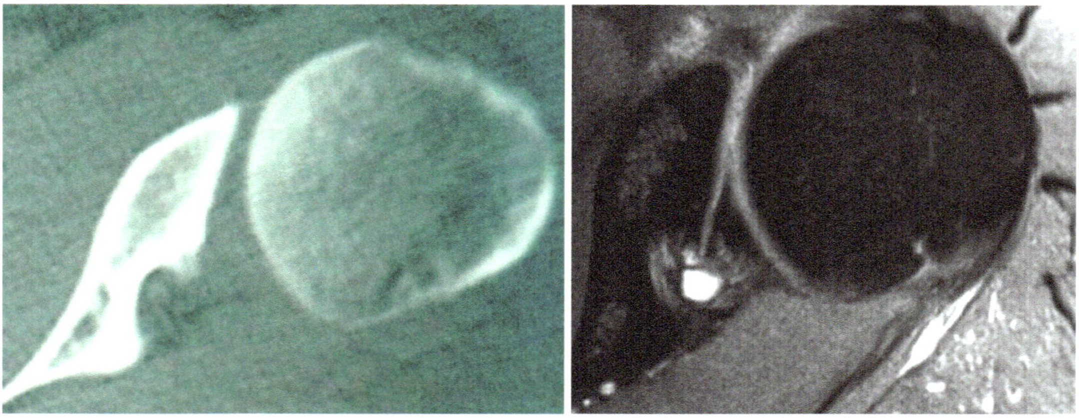

Fig. 2.12 MRI and computed tomography image of a scapula with . The CT demonstrates that the posterior-inferior glenoid is deficient, with posterior subluxation of the humeral head. The MRI demonstrates thick posterior cartilaginous tissue, which is unable to compensate for the deficient glenoid (Copyright Dr Gregory Bain)

References

1. Al-Qattan MM, Yang Y, Kozin SH. Embryology of the upper limb. J Hand Surg Am. 2009;34:1340–50.
2. Nazir SR, Bazir SZ. Histological development of human foetal shoulder joint. Int J Res Med Sci. 2014;2:293–9.
3. O'Rahilly R. Development stages in human embryos. Part A: embroys of the first three weeks. Washington: Carnegie Institution of Washington; 1973.
4. O'Rahilly R, Muller F. Developmental stages in human embryos: revised and new measurements. Cells Tissues Organs. 2010;192:73–84.
5. Rodriguez-Niedenfuhr MEA. Development of the arterial pattern in the upper limb of staged human embroys: normal development and anatomic variations. J Anat. 1999;1999(199):407–17.
6. Fawcett T. Development and ossification of the human clavicle. J Anat Physiol. 1913;47(Pt 2):225–34.
7. Haines RW. The development of joints. J Anat. 1947; 81(Pt 1):33–55.
8. Gardner E, Gray DJ. Prenatal development of the human shoulder and acromioclavicular joints. Am J Anat. 1953;92(2):219–76.
9. Gray DJ, Gardner E. The prenatal development of the human humerus. Am J Anat. 1969;124(4):431–45.
10. Aboul-Mahasen LM, Sadek SA. Developmental morphological and histological studies on structures of the human fetal shoulder joint. Cells Tissues Organs. 2002;170(1):1–20.
11. Abe S, et al. Early fetal development of the rotator interval region of the shoulder with special reference to topographical relationships among related tendons and ligaments. Surg Radiol Anat. 2011;33(7):609–15.
12. Fealy S, et al. The developmental anatomy of the neonatal glenohumeral joint. J Shoulder Elbow Surg. 2000;9(3):217–22.
13. Abe S, et al. Variation of the subscapularis tendon at the fetal glenohumeral joint. Okajimas Folia Anat Jpn. 2014;90(4):89–95.
14. Lapner PL, Lapner MA, Uhthoff HK. The anatomy of the superior labrum and biceps origin in the fetal shoulder. Clin Anat. 2010;23(7):821–8.
15. Smith SP, Bunker TD. Primary glenoid dysplasia: a review of 12 patients. J Bone Joint Surg Br. 2010;83: 868–72.
16. Landau JP. Genetic and biomechanical determinants of glenoid version. J Shoulder Elbow Surg. 2009; 18(4):661–7.

Part II

Osseous Structures

Proximal Humerus

3

Ronald L. Diercks

3.1 Osteology

The articular surface of the humeral head is ovoid in shape (although often described as spherical), with a diameter of 25–30 cm^2 which articulates directly with the glenoid (Fig. 3.1) [1]. The humeral head (caput humeri) is joined with the shaft by the anatomical neck (collum anatomicum). The physiological bare areas in the posterolateral aspect of the humeral head should not be considered as cartilage defects: As the proximal humeral head is slightly flattened postero-inferiorly, differentiation between the physiological posterolateral flattening of the humeral neck and the Hill Sachs lesion, a consequence of anterior shoulder dislocation, can be difficult. One way to distinguish the normal anatomy from pathological finding is that the Hill Sachs lesion is found at or above the level of the coracoid process. The anatomical neck is distally bordered by the greater and lesser tuberosity (Fig. 3.2). Several nutrient foramina are located in the anatomical neck, which supply the humerus head with blood [2]. The poor healing process of fractures located medial to the anatomical neck is a direct result of this anatomical feature.

The bicipital groove measures about 4 mm deep, through which the long head of the biceps tendon goes, and divides both tubercles, which form the insertion sites of the rotator cuff tendons: subscapular muscle on the minor tubercle, supraspinatus, infraspinatus and also teres minor on the major tubercle. Common findings on the humeral head are cysts, located dorsally at the insertion sites of the supraspinatus and infraspinatus tendons. In this location, humeral head cysts are not related to aging or rotator cuff tear [3]. The biceps sulcus is bridged by a band-like plate of connective tissue, the transverse humeral ligament (lig. transversum humeri), which is a continuation of the tendon of the subscapularis muscle. The only osseous fixation of this ligament lies at the medial margin of the greater tubercle of the humerus.

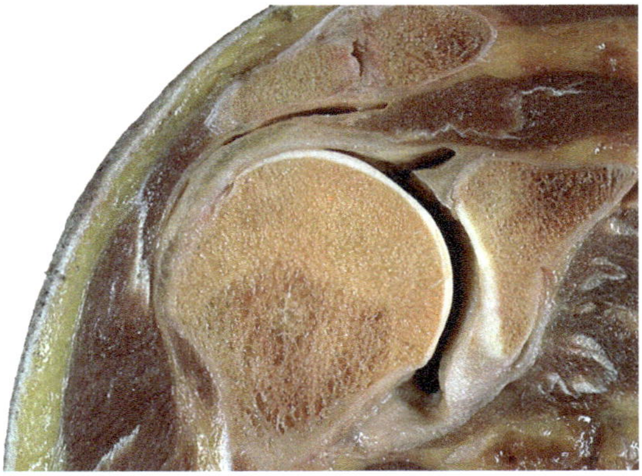

Fig. 3.1 Coronal cadaveric section of the right shoulder. Note the spherical head is contained by the glenoid, labrum and rotator cuff (Section prepared by the late Pau Golano)

R.L. Diercks
Department of Orthopedics, Sports Medicne Center, University Medical Centre Groningen, University of Groningen, Groningen, The Netherlands
e-mail: R.L.Diercks@umcg.nl

G.I. Bain et al. (eds.), *Normal and Pathological Anatomy of the Shoulder*, DOI 10.1007/978-3-662-45719-1_3, © ISAKOS 2015

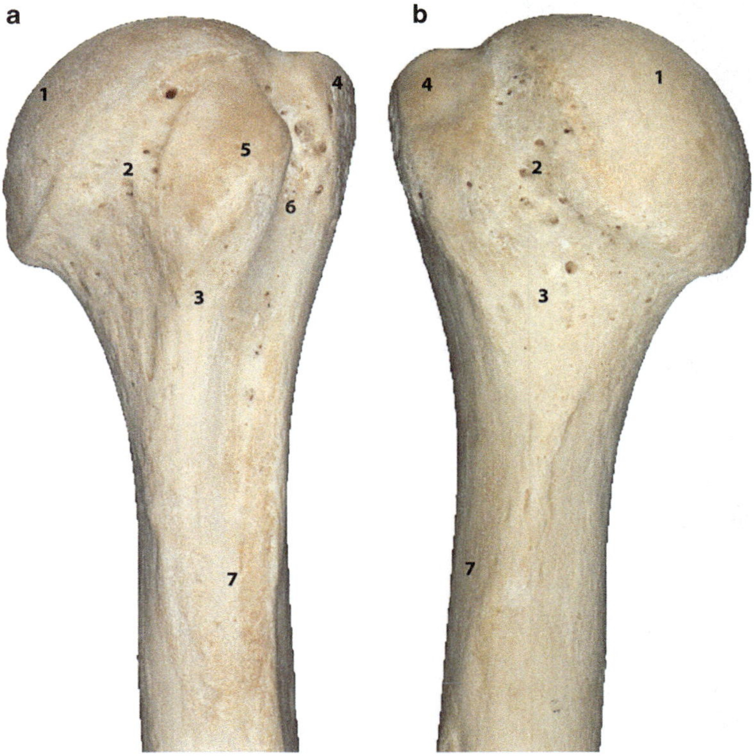

Fig. 3.2 Proximal left humerus. (**a**) Anterior aspect. (**b**) Posterior aspect. *1* head, *2* anatomical neck, *3* surgical neck, *4* greater tuberosity, *5* lesser tuberosity, intertuberous sulcus, *7* shaft, *8* radial groove (Modified with permission from Gray's anatomy, Fig. 49.9a, b, 39th Edition, Churchill Livingstone: Elsevier, 2004, ISBN 00443071683)

The surgical neck of the humerus is located distally from the major and minor tubercle. It is the region where most frequently fractures occur, due to the thin part of compact bone, which, in older age, is not anymore supported by the spongious metaphyseal bone: In older age, this is replaced by bone marrow, due to osteoporosis. The fine trabecular pattern can be visualised on the sectioned proximal humerus (Fig. 3.3).

There exists a rotatory difference between the direction of the head of the humerus and the elbow joint. There is a retroversion of 20° between the axis of the glenohumeral joint and the axis of the elbow joint.

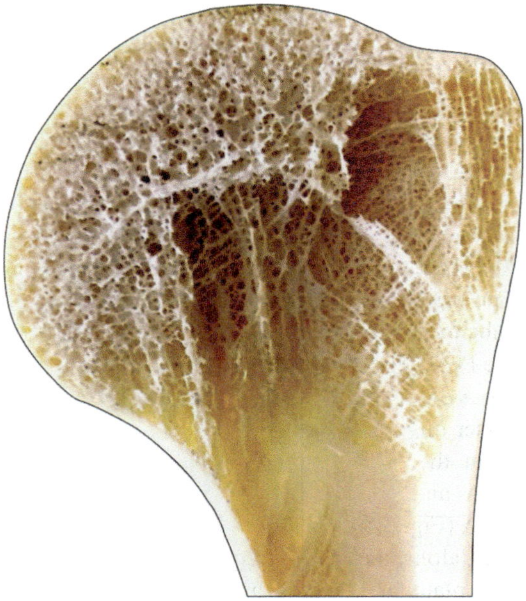

Fig. 3.3 Normal humeral head, demonstrating the trabecular pattern within the neck and head (Copyright Gregory Bain)

3.2 Vascularisation

The arterial blood supply of the proximal humeral epiphysis is known to derive mainly from the anterior humeral circumflex a. (ACA) (Fig. 3.4a). However, the posterior circumflex

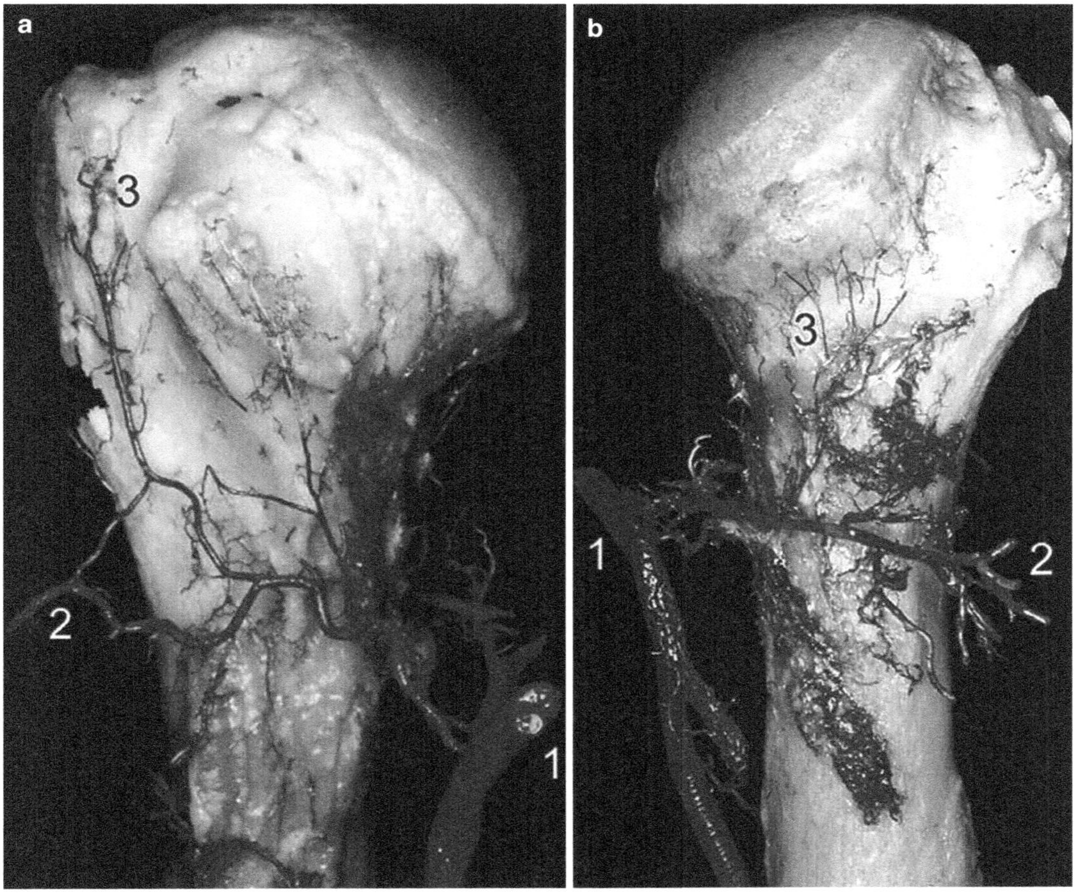

Fig. 3.4 Arterial supply of the right proximal humerus. (**a**) Anterior view. *1* axillary artery, *2* anterior humeral circumflex artery, *3* anterior humeral circumflex artery, ramus ascendens. (**b**) Posterior view. *1* axillary artery, *2* posterior humeral circumflex artery, *3* medial vascular group

artery (PCA) (Fig. 3.4b) is also considerably involved in the blood circulation of the bone [4]. The origin of the ACA and PCA is varied. The subchondral bone is predominantly vascularised the PCA. The apex of the head is vascularised by the ACA or the PCA equally, as is the head. The lesser tubercle derives its vascularisation mostly from the ACA, the greater tubercle from the PCA and the intertubercular groove from the ACA (Fig. 3.5a). The arcuate arteries are distributed along the metaphyseal side of the epiphyseal plate, and small branches cross the plate to reach the epiphyseal side and give numerous anastomoses to the branches of the ACA or the PCA (Fig. 3.5b). Therefore, irrespective of the choice of osteosynthesis treatment, fractures of the Neer II and 11-C AO types (fracture of the true neck) are those most vulnerable to the development of avascular necrosis. The roles of both the ACA and PCA are important and must be taken into account in evaluating the risk of necrosis after a fracture, by carefully considering the topography of the separation and the displacement of the different parts.

The tubercles receive multiple inflows from both circumflex arteries. Also the attached tendons and muscles protect these arteries so that even in cases of disruption of both tubercles there is always sufficient fragment perfusion.

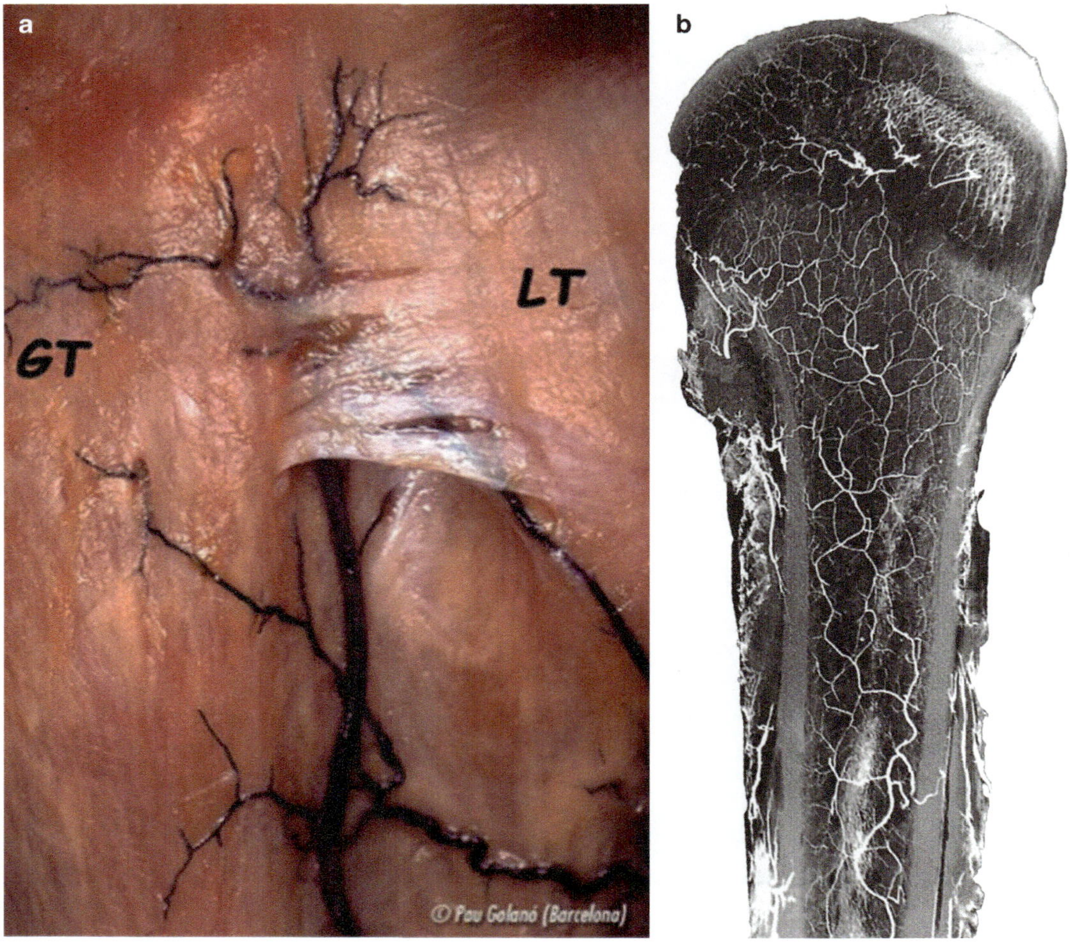

Fig. 3.5 Arterial supply of the proximal humerus. (**a**) Anterior view. Ascending branch of the anterior humeral circumflex artery, note it passes deep to the transverse ligament, supplied the biceps tendon and the humeral head (Section prepared by the late Pau Golano). (**b**) Arteriogram of a sectioned humerus of a child. Note the vessels in the metaphysis and medually canal. The humeral head has separate vessels (Copyright HV Crock AO)

3.3 Degenerative Arthritis

Degenerative arthritis of the proximal humerus takes place at two possible articulations: the glenohumeral and the acromio-humeral articulation. The last is typically a long-standing effect of a full rotator cuff tear and mostly named "cuff tear arthropathy". The degenerative changes in the glenohumeral joint can appear in different ways: as general thinning of the cartilage, with secondary signs of cyst formation in the head or the glenoid, as a clear osteophyte, mostly on the distal end of the humeral articular surface ("beard" osteophyte) or it affects the cranial end of the intertubercular sulcus and damages and encroaches on the bicipital tendon by osteophytes, which overlap or narrow the intertubercular sulcus [1] (Fig. 3.6).

3.4 Proximal Humeral Fracture

The large range of motion of the shoulder joint has some protective effect on fractures during the first three decennia of life. In youth and adolescence, epiphyseal fracture separation is seen.

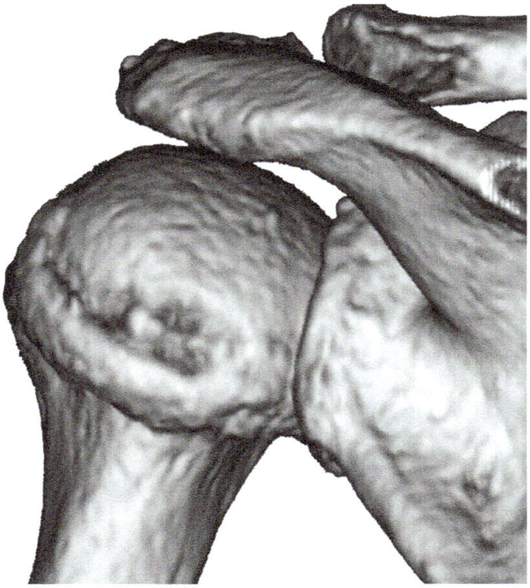

Fig. 3.6 Degenerative osteoarthritis with a beard osteophyte of the humeral head (Copyright Gregory Bain)

In adolescence and twenties, the most observed fracture of the proximal humerus is the impression fracture of the posterior side of the head, caused by indentation of the anterior glenoid rim after shoulder dislocation, the so-called Hill-Sachs lesion. Apart from high energy trauma, most fractures of the proximal humerus are age- and bone mineral density-related fractures, typically following a fall on the outstretched arm. Typically, the poor vascularisation of fractures located medial to the anatomical neck results in avascular necrosis or delayed healing. Fractures of the proximal humerus are very common in elderly people with an increasing incidence. Complex fracture patterns are observed. For radiological examination, x-rays in two planes are mandatory. Often an additive CT-scan can be useful for a better understanding of the whole fracture character.

Anatomic knowledge for fracture classification and treatment is necessary to perform the essential steps in operative reduction and internal fixation. Several classification systems are in use, all with a low inter-observer reliability. An anatomically-positioned humeral head opposite the glenoid, a stable fixation of the head with the humeral shaft and anatomical reduction of both tubercles to restore muscle balance are essential for stability and function.

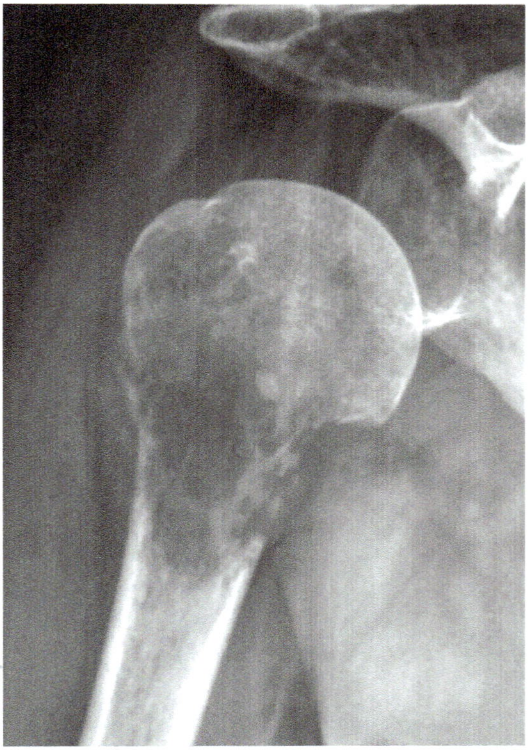

Fig. 3.7 Metastatic lesion of the humeral neck. Primary lesion was a renal cell carcinoma (Copyright Gregory Bain)

Pathological fractures most often present with pain or fracture of the surgical neck of the humerus (Fig. 3.7).

References

1. Prescher A. Anatomical basics, variations, and degenerative changes of the shoulder joint and shoulder girdle. Eur J Radiol. 2000;35(2):88–102.
2. Mutch J, et al. A new morphological classification for greater tuberosity fractures of the proximal humerus: validation and clinical implications. Bone Joint J. 2014;96-B(5):646–51.
3. Rudez J, Zanetti M. Normal anatomy, variants and pitfalls on shoulder MRI. Eur J Radiol. 2008;68(1):25–35.
4. Meyer C, et al. The arteries of the humeral head and their relevance in fracture treatment. Surg Radiol Anat. 2005;27(3):232–7.

4 Glenoid

Matthew T. Provencher, Rachel F. Frank, Daniel J. Gross, and Petar Golijanin

4.1 Glenoid Osteology

The scapula is a flat, triangular-shaped bone with three borders and three angles (Fig. 4.1). The three borders, superior, medial (or vertebral), and lateral, meet to form the angles of the scapula. The medial and lateral borders, as well as the superior and inferior angles, primarily serve as origin and insertion sites for 8 of the 17 muscles associated with the scapula.

In the resting position, the scapula is anteriorly rotated 30° in the axial plane [1] and lies at a 60° angle to the clavicle to accommodate the thoracic rib cage [2]. In the coronal plane, the scapula is rotated cephalad approximately 3–10°, and when viewed in the sagittal plane, it is antiverted 10–20° [3].

The scapula extends from the 2nd rib to the 7th, 8th, or 9th rib at the inferior angle (Fig. 4.2). The anterior surface of the scapula is referred to as the subscapular fossa, and its concave face accommodates the posterior thoracic rib cage to form the scapulothoracic articulation. The posterior surface of the scapula contains the scapular spine, which arises from the spinal trigone on the medial scapular border to divide the posterior surface into two separate fossae: the supraspinatus and infraspinatus fossa. These fossae communicate via an area at the lateral aspect of the scapular spine referred to as the spinoglenoid notch.

The lateral aspect of the scapular spine extends beyond the scapular borders to become

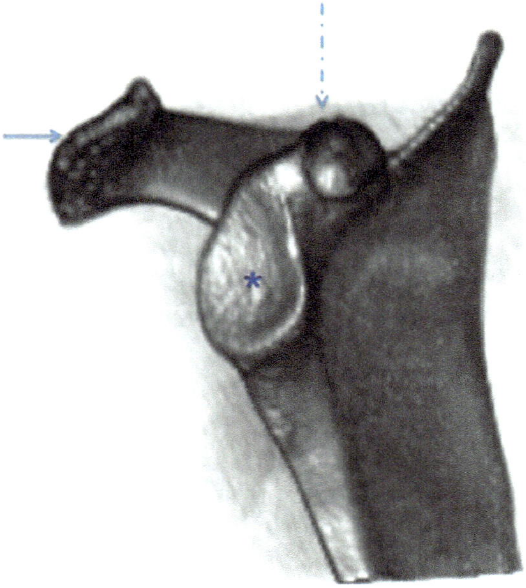

Fig. 4.1 En-face view of a three-dimensional reconstruction of the right scapula demonstrating the relationship of glenoid (*asterisk*) to the acromion (*solid arrow*) and coracoid process (*dotted arrow*)

M.T. Provencher, MD (✉) • D.J. Gross, MD
P. Golijanin, BS
Department of Sports Medicine and Surgery, Massachusetts General Hospital, Boston, MA, USA
e-mail: mattprovencher@gmail.com; danielgross23@gmail.com; golijaninp@gmail.com

R.F. Frank, MD
Department of Orthopaedics, Rush University Medical Center, Chicago, IL, USA
e-mail: rmfrank3@gmail.com

G.I. Bain et al. (eds.), *Normal and Pathological Anatomy of the Shoulder*,
DOI 10.1007/978-3-662-45719-1_4, © ISAKOS 2015

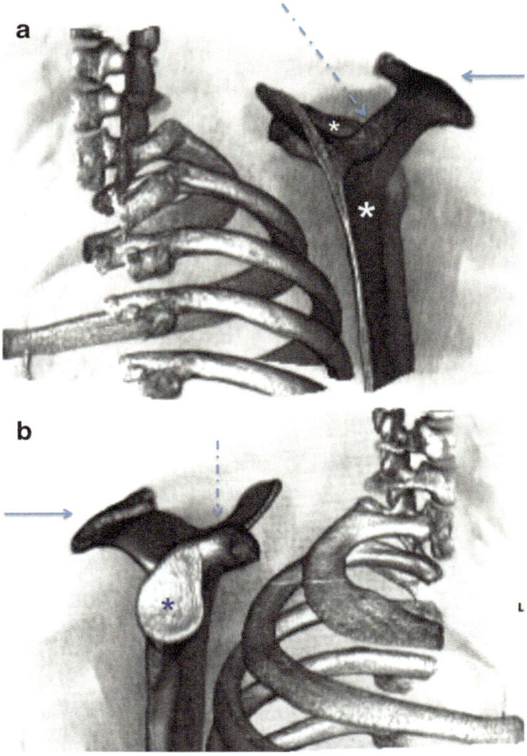

Fig. 4.2 (**a**) Medial view of a three-dimensional reconstruction of the right scapula demonstrating the relationship of the scapula to the thorax; note the infraspinatus fossa (*large asterisk*), supraglenoid fossa (*small asterisk*), scapular spine (*dotted arrow*), and acromion (*solid arrow*). (**b**) Lateral view of a three-dimensional reconstruction of the right scapula demonstrating the relationship of the scapula to the thorax; note the glenoid face (*asterisk*), acromion process (*solid arrow*), and coracoid process (*dotted arrow*)

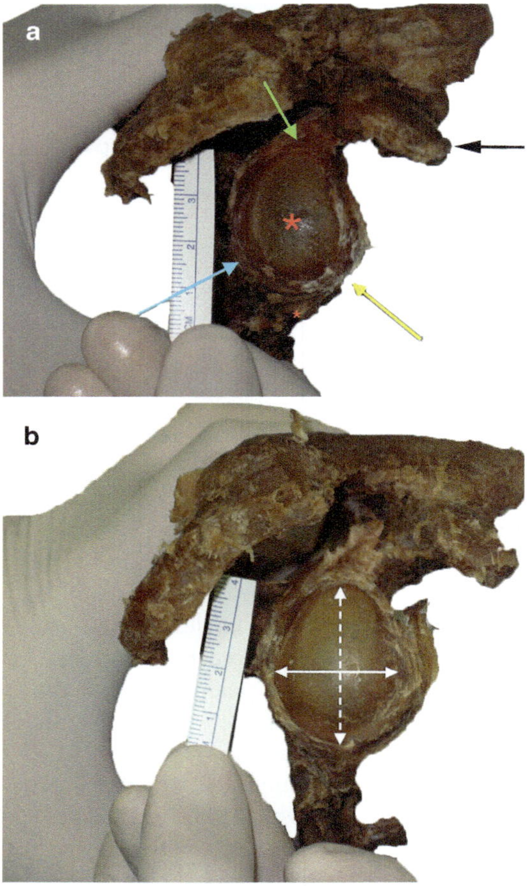

Fig. 4.3 (**a**) En-face view of a right cadaveric glenoid showing bare spot (*large* *), the projection of the coracoid (*white arrow*), superior labrum (*green arrow*), anterior inferior labrum (*yellow arrow*), posterior inferior labrum (*blue arrow*), and long head of triceps attachment (*small* *). (**b**) En-face view of a right cadaveric glenoid showing anterior to posterior distance (*solid arrow*) and superior to interior distance (*dotted arrow*)

the acromion. The acromial angle is the angle of reflection between the scapular spine and the acromion and has a mean angle of 78° (range 64–99°) [4].

The superior scapular border contains two unique morphologies: the suprascapular notch and the coracoid process. The suprascapular notch, which is traversed by the superior transverse ligament, allows for the passage of the suprascapular nerve within the groove and into the supraspinatus fossa [5, 6]. The suprascapular artery and vein pass above the transverse ligament.

The coracoid process is a beak-shaped structure that serves as an integral attachment for muscles and ligaments acting across multiple joints in the shoulder and upper extremity facilitating both movement and stability. The coracoid serves as the attachment of the conjoined tendon, which incorporates the coracobrachialis and the short head of the biceps brachii, as well as of the pectoralis minor.

The lateral angle of the scapula narrows to form the scapular neck, which then connects the glenoid to the scapula. The most lateral aspect of the glenoid is the glenoid cavity (Fig. 4.3), which is pear-shaped, concave fossa that narrows superiorly at the anterior glenoid notch, with the inferior 2/3 of the glenoid approximating a circle

[7–10]. The glenoid has two bony tubercles at the superior and inferior poles, the supraglenoid and infraglenoid tubercle, which serve as origin sites for the long head of the biceps brachii (LHB) and the long head of the triceps brachii, respectively.

4.2 Myology

A total of 17 individual muscles either originate or insert on the scapula, functioning across several joints in the shoulder and upper extremity, including the scapulothoracic joint, the glenohumeral joint, and the elbow.

The thoracoscapular muscles act to stabilize the scapula and provide scapulothoracic motion. These include the trapezius, the rhomboids, the levator scapulae, the serratus anterior, and the pectoralis minor. The rhomboids, levator scapulae, and trapezius muscles all originate from the cervical and thoracic spinous processes, but their insertion locations on the scapula result in distinct scapular movements.

The rhomboids insert on the superior angle and the posterior medial border of the scapula, acting to both retract (adduct) and downwardly rotate the scapula. The levator scapulae insert solely on the superior scapular angle, resulting in scapular elevation. The trapezius muscle inserts on the scapular spine and the distal third of the clavicle, opposing the rhomboids, and resulting in scapular elevation, upward/superior rotation, and abduction.

The serratus anterior muscle originates from ribs 1 through 8 or 9 and inserts on the anterior medial border and inferior angle of the scapula. The muscle acts to draw the scapula forward and to stabilize the medial scapular border to the thoracic ribcage. The serratus anterior also abducts the scapula via insertion on the inferior angle resulting in superior rotation of the glenoid cavity.

The scapulohumeral muscles primarily originate from the scapula and clavicle and insert on the humerus while acting across the glenohumeral joint to provide motion to the shoulder. The scapulohumeral muscles include the rotator cuff muscles, the teres major, the deltoids, and the coracobrachialis.

The rotator cuff consists of four muscles: the subscapularis, the supraspinatus, the infraspinatus, and the teres minor, which lie deep to the deltoid and confer both motion and stability to the glenohumeral joint. The subscapularis is the largest of the four muscles and originates within the anterior subscapular fossa, and inserts on the lesser tuberosity of the humerus where it acts to internally rotate the shoulder. The supraspinatus originates from the supraspinatus fossa on the posterior aspect of the scapula and inserts on the greater tuberosity of the humerus to abduct the shoulder in conjunction with the deltoid. The infraspinatus and teres minor work in concert to externally rotate the shoulder by inserting on the greater tuberosity. In addition to shoulder motion, the rotator cuff muscles (Table 4.1) also impart dynamic stability to the glenohumeral joint via three major mechanisms: (1) concavity compression of the humeral head into the glenoid socket, (2) coordinated muscle contraction, and (3) close association with and contribution to the glenohumeral capsule and ligaments [11].

Flexion of the shoulder is accomplished primarily via the coracobrachialis muscle, which is the central of three muscles attached to the coracoid process, along with the LHB (lateral coracoid attachment) and pectoralis minor (medial coracoid attachment).

Table 4.1 Musculature associated with the glenoid

Muscle	Origin	Insertion	Innervation	Function on scapula
Subscapularis	Subscapularis fossa	Humeral lesser tuberosity	Subscapularis nerve	Shoulder internal rotation
Supraspinatus	Supraspinatous fossa	Greater tuberosity of the humerus	Suprascapular nerve	Shoulder abduction, rotation
Infraspinatus	Infraspinatous fossa	Greater tuberosity of the humerus	Suprascapular nerve	External rotation of shoulder
Teres minor	Lateral border of scapula	Greater tuberosity of the humerus	Axillary nerve	External rotation of shoulder

4.3 Glenoid Labrum

The glenohumeral joint is often considered to be an incongruous joint due to the size discrepancy between the humeral head and the glenoid socket [2]. This size difference is slightly reduced by the glenoid labrum, a rim of fibrocartilage tissue that effectively enlarges the glenoid cavity and reduces the inherent instability of the shoulder.

The glenoid labrum enhances the stability of the glenohumeral joint through three primary mechanisms [11]. First, the labrum deepens the concavity of the glenoid up to 9 mm in the superior-inferior direction and also doubles the antero-posterior depth to 5 mm [12]. Second, the labrum increases glenohumeral stability by increasing the surface area through which the glenoid contacts the humeral head through an arc of motion. Finally, the labrum is the site of attachment for the various glenohumeral ligaments that confer static stability to the joint [13].

Overlying and attaching to the glenoid labrum and the scapular neck is the articular capsule, which is intimately associated with the glenohumeral ligaments. The glenohumeral capsuloligamentous complex consists of the articular capsule and three articular ligaments that serve as static restraints against excessive translation of the humeral head. The anterior band of the inferior glenohumeral ligament (AIGHL) attaches to the glenoid at the antero-inferior labrum and is the primary static restraint to anterior translation when the shoulder is in an abducted and externally rotated position [11].

4.4 Vascular

The vasculature of shoulder and upper extremity arise from the subclavian artery, itself a branch of the brachiocephalic trunk on the right and the aortic arch on the left. At the medial aspect, the subclavian artery begins as a retroclavicular vessel, and as it passes between the scalene muscles at the vertex of its arch, it emerges as the axillary artery [2].

The largest branch of the axillary artery is the subscapular artery, which emerges posteriorly to the pectoralis minor, and descends medially in front of the subscapularis muscle. The artery then divides into two terminal branches, the thoracodorsal artery and the circumflex scapular artery. The thoracodorsal artery travels behind the posterior axillary fold to its course along the lateral scapular border. The circumflex scapular artery continues through the axillary space and between the two teres muscles through the omotricipital triangle.

Medially, the thyrocervical trunk branches off of the subclavian artery, giving rise to the superficial and deep transverse cervical arteries. The deep branch of the transverse cervical artery crosses the brachial plexus moving posteriorly until it reaches the superior angle of the scapula where it gives rise to a descending branch to supply the posterior muscles of the scapula. Some variants of this artery branch directly off of the subclavian artery, in which case it is known as the dorsal scapular artery.

The suprascapular artery emerges from the thyrocervical trunk just below the transverse cervical artery and courses laterally in front of the anterior scalene muscle with the phrenic nerve. The artery continues behind the clavicle towards the superior scapula border where it passes posteriorly over the superior transverse scapular ligament into the supraspinatus fossa.

Eventually, the artery enters the infraspinatus fossa via the spinoglenoid. Together, along with the scapular circumflex artery and the descending branch of the transverse cervical artery, they form the scapular anastomoses.

The thoracoacromial artery emerges from the axillary artery at the level of the upper border of the pectoralis minor. The artery pieces the clavipectoral fascia and then divides into four branches that supply the muscles of the shoulder and proximal humerus. Of these four branches, the deltoid (or humeral) branch and acromial branch are the primary blood supply to the scapulohumeral muscles.

4.5 Neurologic

The muscles of the shoulder and upper extremity are innervated by the brachial plexus, which is formed from the branches of spinal roots C5-T1.

The brachial plexus is organized into roots, trunks, divisions, cords, and branches, with the branches emerging beyond the inferior border of the clavicle.

Specific to surgical scapular anatomy, the most important nerves are the suprascapular nerve and the axillary nerve. The suprascapular nerve arises from the superior trunk of the brachial plexus and passes through the suprascapular notch to enter into the supraspinatus fossa. As it passes under the superior transverse ligament, the nerve becomes most susceptible to injury via compression and sheering forces [1].

The axillary nerve runs anterior to the subscapularis, passing through the quadrangular space with the posterior circumflex humeral artery. The axillary nerve divides into three terminal branches, providing motor innervation to the deltoid and teres minor, and sensory innervation to the lateral shoulder via the superficial lateral cutaneous nerve. The anterior branch of the axillary nerve wraps around the surgical neck of the humerus approximately 3–5 mm inferior to the lateral acromion.

4.6 Arthrology

The shoulder is comprised of three primary joints (Fig. 4.4). Both the acromioclavicular and glenohumeral joints represent traditional diarthroidal joints. The scapulothoracic joint, while not a traditional linkage between bones, is integral to the shoulder's mobility. Scapular rotation allows for approximately 60° of the full 180° of shoulder abduction.

4.7 Imaging

Imaging studies for the evaluation of pathology about the scapula and glenohumeral joint include radiographs, computed tomography

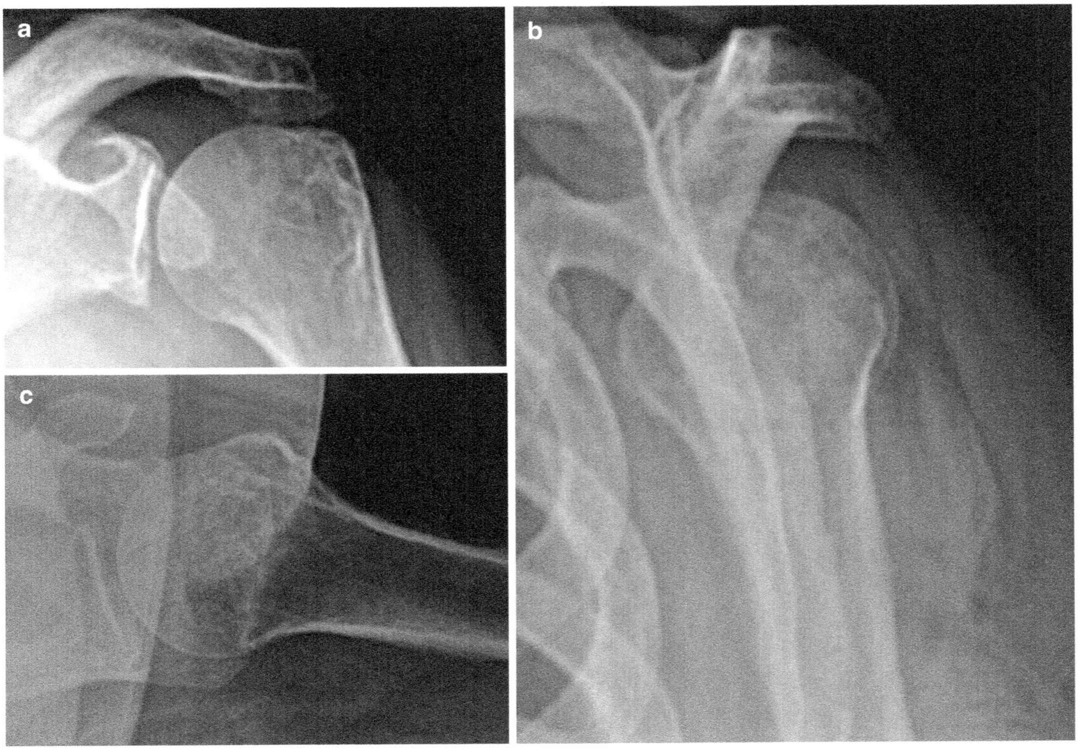

Fig. 4.4 (**a**) Anterior-posterior, (**b**) axillary, and (**c**) scapular-Y radiographs of a left shoulder demonstrating normal glenohumeral and acromioclavicular articulations

(CT), and magnetic resonance imaging/arthrography (MRI, MRA). For initial evaluation of the shoulder, standard radiographs are the preferred modality. Radiographs are useful in diagnosis of fractures and dislocations as these injuries may be apparent in a shoulder x-ray series. There are several views that are necessary for a shoulder instability series [17]. The serendipity view is a 40° cephalic tilt view of the SC joint and the medial 1/3 of the clavicle. It is indicated in SC joint separations and fractures of the clavicle. The Grashey view lines up the glenoid so that it is perpendicular to the plane of the x-ray allowing for evaluation of the glenohumeral joint space. The west point axillary view is a tangential view of the antero-inferior rim of the glenoid. This view is useful in identifying bony Bankart lesions or attritional bone loss of the antero-inferior glenoid. To evaluate Hill Sachs lesions after dislocation, the Stryker Notch view is used [14]. In this view, a cassette is placed under the affected shoulder and the palm of the hand of the affected extremity is placed on the forehead with the fingers pointing towards the back of the head. The x-ray tilts 10° toward the head, centered over the coracoid process. The axillary view is the best true lateral view of the shoulder. It allows for evaluation of anterior and posterior instability, glenoid fractures, and head compression fractures. In this view, the arm must be abducted and a cassette is placed on the superior aspect of the shoulder. The scapular Y view helps to confirm the diagnosis of a posterior shoulder dislocation. The anterior portion of the affected shoulder is placed against the x-ray plate while the other shoulder is rotated out approximately 40° and then the X-ray tube is placed posteriorly along the spine of the scapula. The Garth view, also known as the Apical Oblique view, is also used in patients with shoulder instability. It helps to evaluate the anterior and inferior glenoid rim for calcification and fractures subsequent to dislocation. In this technique, the X-ray beam is directed at a 45° angle through the glenohumeral joint toward the cassette, which is placed posterior and parallel to the scapular spine.

To allow for visualization of the glenohumeral alignment, while also detecting fractures, loose bodies, calcification and degenerative changes, the Didiee view is preferred. The patient lies prone with the arm abducted and slightly flexed at the elbow. The back of the patient's hand on the affected extremity should lie on the iliac crest. The film cassette is positioned under the shoulder and the X-ray is taken from a lateral standpoint directed towards the humeral at 45° [18, 19].

Computed tomography (CT) is useful in acute trauma settings. It can help to diagnose complicated fractures while also examining the presence of intra-articular bodies. CT is also indicated in evaluating more chronic lesions in addition to grading the degree of muscle atrophy or fatty infiltration in the setting of concomitant rotator cuff pathology [9]. Patients with a history of previous instability surgery and patients with a midrange of motion instability should undergo CT scan to evaluate for anterior or posterior glenoid insufficiency, engaging Hill-Sachs or reverse Hill Sachs lesions. Intra-articular contrast may be of benefit in future defining bony anatomy in chronic cases.

MRI is useful in evaluating the rotator cuff, labrum, glenohumeral ligaments, cartilage, and capsule. It provides superb detail of soft-tissue pathology, while also detailing the bone marrow. MRI with the administration of intra-articular gadolinium (MRA) results in improved sensitivity for detecting subtle pathology. For evaluation of the osseous integrity of the scapula and glenoid itself, radiographs and/or CT are the modalities of choice.

Diagnostic arthroscopy is also an excellent modality for assessing intra-articular and extra-articular pathology (Figs. 4.5, 4.6, and 4.7). This can be performed in either the beach chair or the lateral decubitus position and is typically performed in a consistent, routine fashion prior to performing any necessary repairs or reconstructions. It is critical to correlate any findings noted on arthroscopy with the patient's clinical symptoms in order to differentiate between those lesions that are symptomatic and those that are incidental in nature.

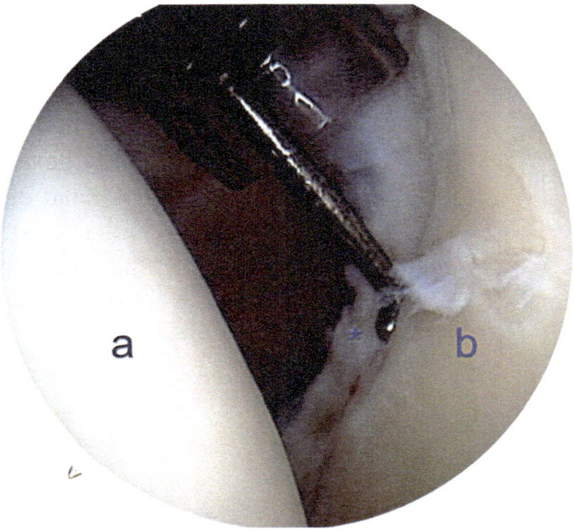

Fig. 4.5 Intraoperative arthroscopic photograph of the left shoulder in the beach chair position, viewing from the posterior portal (probe in the rotator interval portal), demonstrating the relationship between the humeral head (*a*), glenoid (*b*), and anterior labrum (*asterisk*)

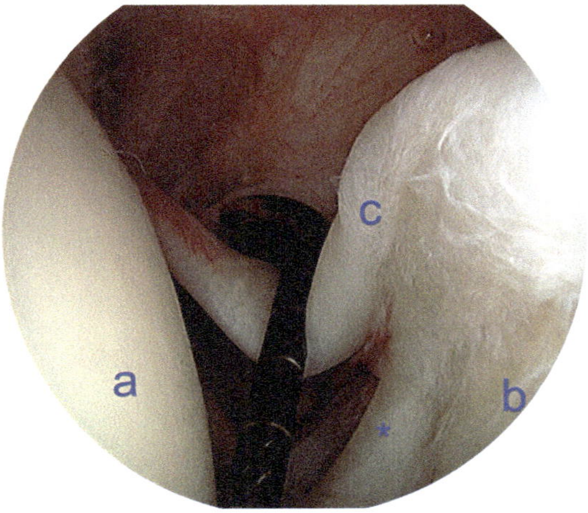

Fig. 4.7 Intraoperative arthroscopic photograph of the left shoulder in the beach chair position, viewing from the posterior portal (probe is superior to the long head of the biceps tendon retracting the tendon into the joint), demonstrating the relationship between the humeral head (*a*), glenoid (*b*), long head of the biceps tendon as it inserts onto the supraglenoid tubercle at the level of the superior glenoid (*C*), anterior labrum (*asterisk*)

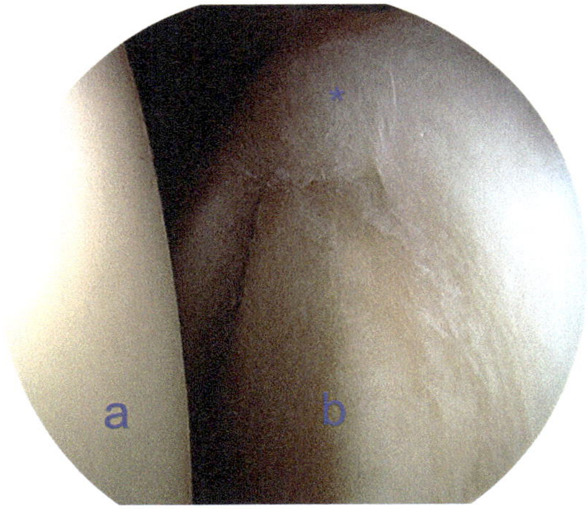

Fig. 4.6 Intraoperative arthroscopic photograph of the left shoulder in the beach chair position, viewing from the posterior portal, demonstrating the relationship between the humeral head (*a*), glenoid (*b*), and superior labrum (*asterisk*)

4.8 Pathoanatomy

Given the intimate relationship of the scapula with the both axial and appendicular skeleton, the variety of anatomical structures associated with the scapula are utilized during activities of daily living as well as activities of recreation. Injuries to some of these structures are common, either due to acute traumatic events (i.e., dislocation Figs. 4.8 and 4.9), or more commonly, as a result of repetitive microtrauma from overuse. In some cases, the anatomy of a given patient's scapula (i.e., glenoid retroversion) may make him/her more prone to injury (i.e., posterior glenohumeral instability, early onset glenohumeral arthritis) [15, 16, 20, 21].

Glenoid dysplasia is often bilateral, and diagnosis can usually be made via plain radiographs [22–24]. Clinical findings are pain with range of motion and limited abduction, and patients generally become symptomatic between the second and fifth decades of life [25]. Significant retroversion of the glenoid is a difficult clinical problem to manage (Fig. 4.10).

4.8.1 Glenohumeral Osteoarthritis

While age is often a contributing factor to osteoarthritis (OA), normal aging is not a pathologic

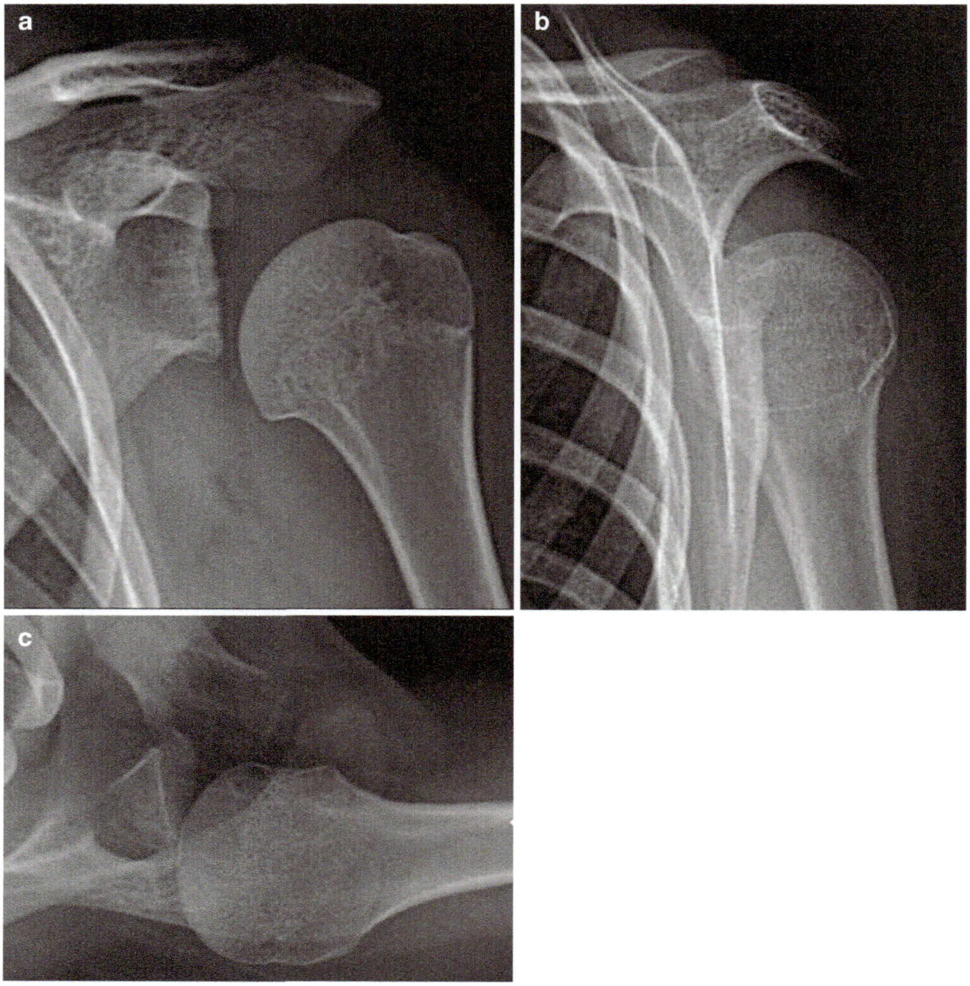

Fig. 4.8 (a) Anterior-posterior, (b) axillary, and (c) scapular-Y radiographs of a left shoulder demonstrating a posterior glenohumeral dislocation

process, and degenerative joint disease is distinct from OA. According to the National Health Interview Survey in the USA in 2011, more than 50 million people were diagnosed with some form of shoulder arthritis. Among different types of arthritis, the osteoarthritis (OA) is the most common one. OA is mostly predominant in the elderly and in the USA; it affects 32.8 % of the disabled population over the age of 60 [26]. OA can be classified as primary or secondary. When no predisposing factors that lead to OA are present, a patient is diagnosed with primary OA and it is caused by "wear and tear." However, when OA is caused as a result of chronic dislocations, trauma, surgery, recurrent instability, avascular necrosis, and massive RCT, it is referred to secondary OA [26–28]. Management of OA still remains controversial and decision making should depend on patient's age, level of activity, severity of symptoms, comorbidities, physical exam, and radiographic findings [29] (Fig. 4.11).

The large range of motion inherent to the shoulder is intimately associated with the significant stresses frequently experienced by the glenohumeral joint and surrounding soft tissue, and both the glenoid and labrum are susceptible to a wide spectrum of injuries. In the USA alone, shoulder instability has an incidence of 23.9 per 100,000 per year.

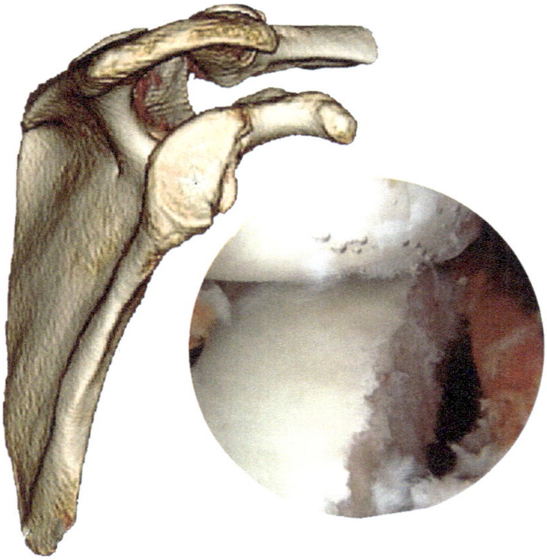

Fig. 4.9 3D reconstruction of a scapula with anterior bone loss and corresponding arthroscopic view (Image from, *Surgery of Shoulder Instability, Osteochondral Allograft Augmentation of the glenoid for Instability with Bone Deficiency*, 2013, p 78, Frank RM et al., Used With permission of Springer Science+Business Media)

4.8.2 Glenohumeral Rheumatoid Arthritis

Rheumatoid arthritis (RA) is an inflammatory arthritis in which the body's immune system attacks the joint, leading to hyperplastic synovial tissue, erosion of bone, and formation of a pannus [30]. The wear pattern in RA is also distinct from OA in that it often originates in the center of the glenoid, leading to medicalization of the gle-

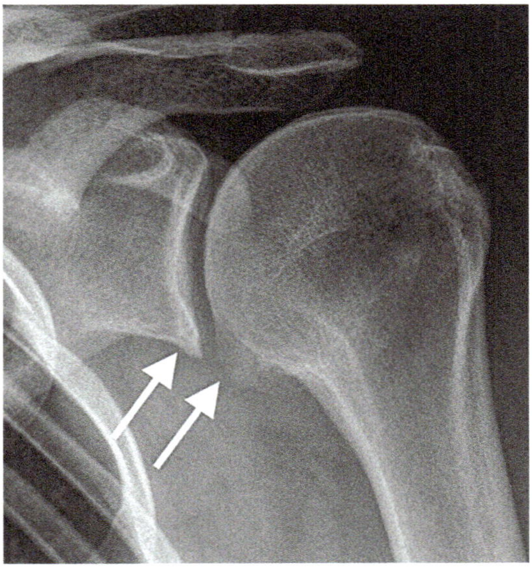

Fig. 4.11 X-ray imaging of a patient who was diagnosed with osteoarthritis (*white arrows*)

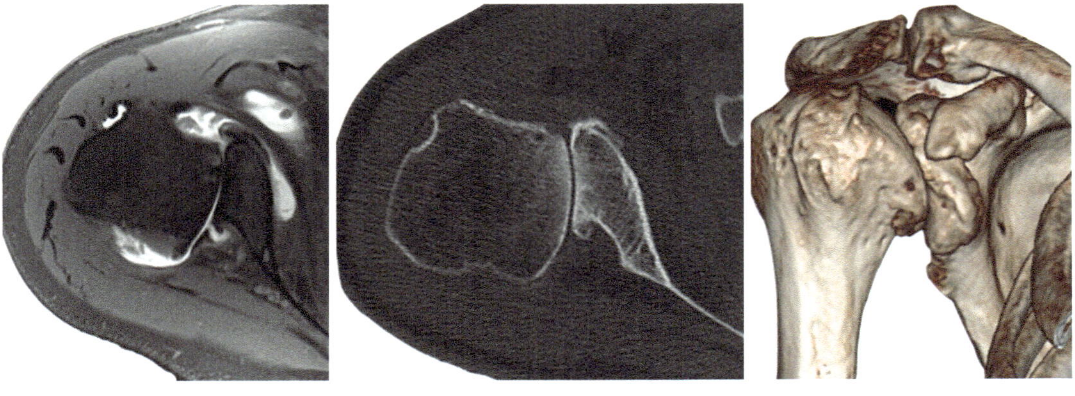

Fig. 4.10 MRI, 2D and 3D CT scan of a retroverted developmental dysplasia, with secondary OA (Copyright Dr Gregory Bain)

noid. Radiographic signs of RA include loss of both humeral head and glenoid bone density, leading to symmetric joint space destruction. RA not only affects the bone, but also the soft tissue, which in the shoulder often manifests as injury to the rotator cuff, which is present in up to 75 % of patients with RA [31, 32].

References

1. Dines D, Williams Jr GR, Laurencin C. Arthritis and arthroplasty: the Shoulder. Philadelphia: Elsevier Health Sciences; 2009.
2. Llusá M, Merí A, Ruano D. Surgical atlas of the musculoskeletal system. Rosemont: American Academy of Orthopaedic Surgeons; 2008. p. ix, 422 p
3. Ludewig PM, Cook TM, Nawoczenski DA. Three-dimensional scapular orientation and muscle activity at selected positions of humeral elevation. J Orthop Sports Phys Ther. 1996;24:57–65. doi:10.2519/jospt.1996.24.2.57.
4. Banas MP, Miller RJ, Totterman S. Relationship between the lateral acromion angle and rotator cuff disease. J Shoulder Elbow Surg. 1995;4:454–61. doi:10.1016/S1058-2746(05)80038-2.
5. Ghodadra N, Nho SJ, Verma NN, Reiff S, Piasecki DP, Provencher MT, Romeo AA. Arthroscopic decompression of the suprascapular nerve at the spinoglenoid notch and suprascapular notch through the subacromial space. Arthroscopy. 2009;25:439–45. doi:10.1016/j.arthro.2008.10.024.
6. Romeo AA, Ghodadra NS, Salata MJ, Provencher MT. Arthroscopic suprascapular nerve decompression: indications and surgical technique. J Shoulder Elbow Surg. 2010;19:118–23. doi:10.1016/j.jse.2010.01.006.
7. Huysmans PE, Haen PS, Kidd M, et al. The shape of the inferior part of the glenoid: a cadaveric study. J Shoulder Elbow Surg. 2006;15:759–63. doi:10.1016/j.jse.2005.09.001.
8. Lo IKY, Parten PM, Burkhart SS. The inverted pear glenoid: an indicator of significant glenoid bone loss. Arthroscopy. 2004;20:169–74. doi:10.1016/j.arthro.2003.11.036.
9. Sugaya H, Moriishi J, Dohi M, Kon Y, Tsuchiya A. Glenoid rim morphology in recurrent anterior glenohumeral instability. J Bone Joint Surg Am. 2003; 85-A:878–84.
10. Mologne TS, Provencher MT, Menzel KA, et al. Arthroscopic stabilization in patients with an inverted pear glenoid: results in patients with bone loss of the anterior glenoid. Am J Sports Med. 2007;35:1276–83. doi:10.1177/0363546507300262.
11. Provencher MT, Ghodadra N, Romeo AA. Arthroscopic management of anterior instability: pearls, pitfalls, and lessons learned. Orthop Clin North Am. 2010;41:325–37. doi:10.1016/j.ocl.2010.02.007.
12. Howell SM, Galinat BJ, Renzi AJ, Marone PJ. Normal and abnormal mechanics of the glenohumeral joint in the horizontal plane. J Bone Joint Surg Am. 1988;70:227–32.
13. Levine WN, Flatow EL. The pathophysiology of shoulder instability. Am J Sports Med. 2000;28:910–7.
14. Provencher MT, Frank RM, Leclere LE, Metzger PD, Ryu JJ, Bernhardson A, Romeo AA. The Hill-Sachs lesion: diagnosis, classification, and management. J Am Acad Orthop Surg. 2012;20:242–52. doi:10.5435/JAAOS-20-04-242.
15. Kuhne M, Boniquit N, Ghodadra N, Romeo AA, Provencher MT. The snapping scapula: diagnosis and treatment. Arthroscopy. 2009;25:1298–311. doi:10.1016/j.arthro.2008.12.022.
16. Frank RM, Ramirez J, Chalmers PN, McCormick FM, Romeo AA. Scapulothoracic anatomy and snapping scapula syndrome. Anat Res Int. 2013;2013:1–9. doi:10.1067/mse.2002.120807.
17. Piasecki DP, Verma NN, Romeo AA, Levine WN, Bach Jr BR, Provencher MT. Glenoid bone deficiency in recurrent anterior shoulder instability: diagnosis and management. J Am Acad Orthop Surg. 2009;17(8):482–93.
18. Saito H, Itoi E, Sugaya H, Minagawa H, Yamamoto N, Tuoheti Y. Location of the glenoid defect in shoulders with recurrent anterior dislocation. Am J Sports Med. 2005;33:889–93. doi:10.1177/0363546504271521.
19. Antoniou J, Duckworth DT, Harryman 2nd DT. Capsulolabral augmentation for the management of posterior instability of the shoulder. J Bone Joint Surg Am. 2000;82(9):1220–30.
20. Robinson CM, Aderinto J. Posterior shoulder dislocations and fracture-dislocations. J Bone Joint Surg Am. 2005;87:639–50.
21. Friedman LG, Griesser MJ, Miniaci AA, Jones MH. Recurrent instability after revision anterior shoulder stabilization surgery. Arthroscopy. 2014;30:372–81. doi:10.1016/j.arthro.2013.11.019.
22. Gardner E, Gray J. The prenatal development of the human shoulder joint. Surg Clin North Am. 1963; 43:1465–70.
23. Kozlowski K, Colavita N, Morris L, Little KE. Bilateral glenoid dysplasia (report of 8 cases). Australas Radiol. 1985;29:174–7.
24. Borenstein ZC, Mink J, Oppenheim W, Rimoin DL, Lachman RS. Case report 655: Congenital glenoid dysplasia (congenital hypoplasia of the glenoid neck and fossa of the scapula, with accompanied deformity of humeral head, coracoid process, and acromion). Skeletal Radiol. 1991;20:134–6.
25. Samilson RL. Congenital and developmental anomalies of the shoulder girdle. Orthop Clin North Am. 1980;11:219–31.

26. Chillemi C, Franceschini V. Shoulder osteoarthritis. Arthritis. 2013;2013:1–7. doi:10.2106/JBJS.J.00769.
27. Cole BJ, Yanke A, Provencher MT. Nonarthroplasty alternatives for the treatment of glenohumeral arthritis. J Shoulder Elbow Surg. 2007;16:231–40.
28. Boselli KJ, Ahmad CS, Levine WN. Treatment of glenohumeral arthrosis. Am J Sports Med. 2010;38:2558–72. doi:10.1177/0363546510369250.
29. van der Meijden OA, Gaskill TR, Millett PJ. Glenohumeral joint preservation: a review of management options for young, active patients with osteoarthritis. Adv Orthop. 2012;2012:1–9. doi:10.2106/JBJS.H.00318.
30. Cuomo F, Greller MJ, Zuckerman JD. Rheumatoid arthritis of the shoulder. J Am Acad Orthop Surg. 2003;11:12–24.
31. Gravallese EM. Bone destruction in arthritis. Ann Rheum Dis. 2002;61 Suppl 2:ii84–6.
32. Shiozawa S, Shiozawa K. A review of the histopathological evidence on the pathogenesis of cartilage destruction in rheumatoid arthritis. Scand J Rheumatol Suppl. 1988;74:65–72.

5 Coracoid Process

Benno Ejnisman, Bernardo B. Terra, and Alberto Costantini

5.1 Developmental Anatomy

The majority of the scapula is formed by intramembranous ossification. At birth, the body and the spine of the scapula have ossified, but not the coracoid process, glenoid, acromion, vertebral border, and inferior angle. The coracoid process has two and occasionally three centers of ossification. The first center appears during the first year of life in the center of the coracoid process. The second center arises at approximately 10 years of age and appears at the base of the coracoid process. The second ossific nucleus also contributes to formation of the superior portion of the glenoid cavity. These two centers unite with the scapula at approximately 15 years of age. A third inconsistent ossific center can appear at the tip of the coracoid process during puberty and occasionally fails to fuse with the coracoid. It is often confused with a fracture, just like the distal clavicular epiphysis (Fig. 5.1) [1–7].

B. Ejnisman (✉)
Department of Orthopedics,
Federal University of São Paulo, São Paulo, Brazil
e-mail: bennoale@uol.com.br

B.B. Terra
Department of Orthopedics,
Santa Casa Hospital, Vitória - ES, Brazil

A. Costantini
Department of Orthopedics,
Concordia Hospital for Special Surgery, Rome, Italy
e-mail: albertocostantini@iol.it

5.2 Description of Structure

Oriented superior, anterior, and lateral to the axis of the scapula, the coracoid process comes off the scapula at the upper base of the neck of the glenoid and passes anteriorly before hooking to a more lateral position as it projects from the superior body of the scapula.

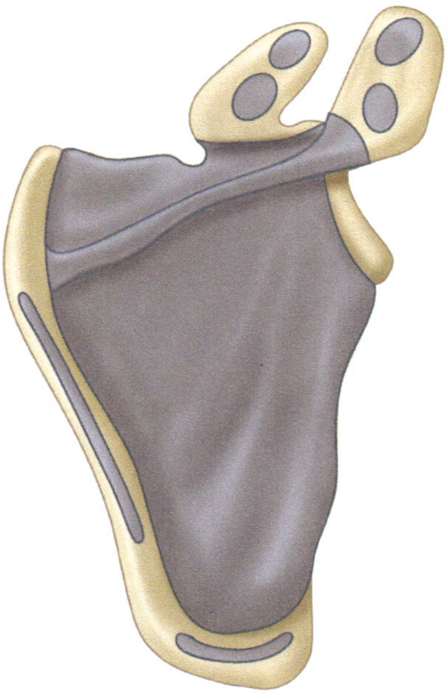

Fig. 5.1 Posterior view right scapula, demonstrating the primary and secondary ossifications centers

G.I. Bain et al. (eds.), *Normal and Pathological Anatomy of the Shoulder*,
DOI 10.1007/978-3-662-45719-1_5, © ISAKOS 2015

The coracoid process [8] showed in cadaveric study that the average length of the coracoid process is approximately 4.3 cm. The width and height at the tip is 2.1 and 1.5 cm, respectively (Table 5.1).

The distance separating the clavicle and the superior coracoid is 1.1–1.3 cm.

No statistically significant correlations were found between length or thickness of the coracoid process, prominence of the coracoid tip, coracoid slope, coraco-glenoid distance, or position of the coracoid tip with respect to the uppermost point of the glenoid. These anatomical characteristics were independent of the dimensions of the scapulae.

Three configurations of the coraco-glenoid space were identified.

Type I – Round bracket 45 %
Type II Square bracket 34 %
Type III Fish hook 21 %

Table 5.1 Coracoid process dimensions

Variable	Mean	SD	Range
Length (cm)	4.3	0.3	3.8–4.6
Tip width	2.1	0.2	1.8–2.4
Tip height	1.5	0.1	1.2–1.7

Modified from Terra et al. [8]
All measurements in centimeters

The lowest value of the coraco-glenoid distance were seen in Type I scapulae. Morphometric characteristics that might predispose to subcoracoid impingement were found in 4 % of Type I scapulae [9].

5.3 Muscle: Tendon Attachment

The coracoid process functions as origin of muscles and ligaments insertion (Fig. 5.2a, b).

The *ascending portion*, flattened from before backward, presents in front a smooth concave surface, across which pass the Subscapularis.

The *horizontal portion* is flattened from above downward; its upper surface is convex and irregular and gives attachment to the pectoralis minor; its under surface is smooth; its medial and lateral borders are rough; the former gives attachment to the pectoralis minor and the latter to the coracoacromial ligament; and the apex is embraced by the conjoined tendon of origin of the Coracobrachialis and short head of the Biceps brachii and gives attachment to the coracoclavicular fascia.

On the medial part of the root of the coracoid process is a rough impression for the attachment of the conoid ligament, and running from it obliquely forward and lateralward, on to the upper surface of the horizontal portion, is an

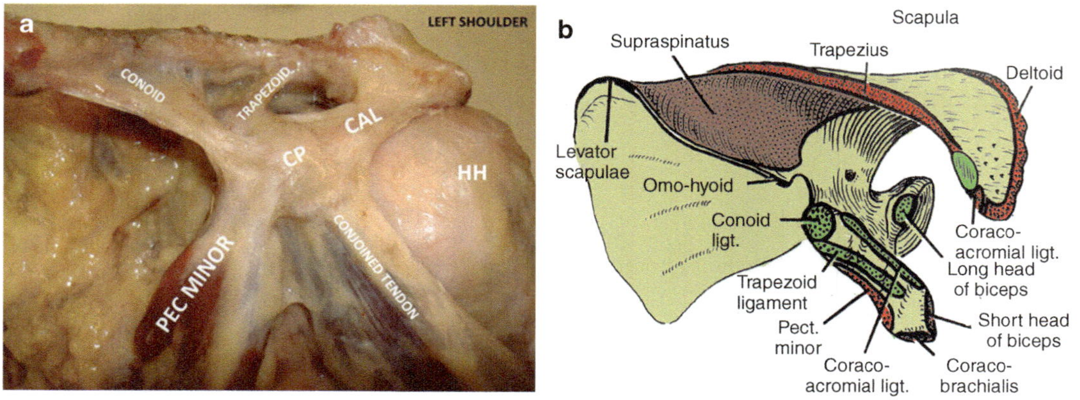

Fig. 5.2 (**a**) Attachments into the left coracoid process. *CAL* coracoacromial ligament, *CP* coracoid process, *CT* conjoint tendon, *CUL* coracohumeral ligament, *P Minor* pectoralis minor, *CC* ligs, trapezoid and conoid ligaments (Used with permission from Di Giacomo [39]). (**b**) Attachments into the left coracoid process (Used with Permission from Last [40])

elevated ridge for the attachment of the trapezoid ligament [10].

In summary:

Pectoralis minor muscle – from 3rd to 5th rib
Short head of biceps brachii muscle – to radial tuberosity
Coracobrachialis muscle – to medial humerus
Coracoclavicular ligament (conoid ligament and trapezoid ligament) – to the clavicle
Coracoacromial ligament – to the acromion
Coracohumeral ligament – to the humerus
Superior transverse scapular ligament – from the base of the coracoid to the medial portion of the suprascapular notch

The distance of these structures to the tip of the coracoid is shown in Table 5.2.

Although the coracoclavicular ligament complex functions as a single ligament, it is composed of two distinct ligaments.

The two (conoid and the trapezoid) attach the coracoid to the distal end of the clavicle and have an average length of about 1.3 cm [11]. The distance from the lateral edge of the clavicle to the center of the trapezoid and conoid tuberosities was 2.6±0.4 cm and 3.5±0.6 cm, respectively [12]. Several biomechanical studies have recently examined the function of the conoid and trapezoid ligaments in human cadaveric models [13–15].

The function to stabilize the clavicle to the scapula with the conoid ligament primarily preventing anterior and superior clavicular displacement. The trapezoid ligament is the primary constraint against compression of the distal clavicle into the acromion [13].

Table 5.2 Ligament footprint: distance from coracoid tip

Variable	Mean	SD	Range
Posterior CAL	2.8	0.33	2.0–4.0
Conjoint tendon	0.53	0.27	0.2–1.2
Anterior pect minor	1.2	0.10	1.0–1.4
Posterior pect minor	1.6	0.27	1.1–2.3
Conoid	3.7	0.35	3.0–4.2
Trapezoid	3.3	0.38	2.8–4.4
Coracohumeral ligament	1.7	0.32	1.2–2.4

Modified from Terra et al. [8]
All results are the distance from the coracoid process tip to the ligament footprint, measured in centimeters. Results rounded off to the nearest 0.1 cm
Descriptive measurements with 99 % of normality interval
SD standard deviation, *CAL* coracoacromial ligament

The trapezoid ligament (*ligamentum trapezoideum*), the anterior and lateral fasciculus, is broad, thin, and quadrilateral: It is placed obliquely between the coracoid process and the clavicle. It is attached, *below,* to the upper surface of the coracoid process; *above,* to the oblique ridge on the under surface of the clavicle. Its anterior border is free; its posterior border is joined with the conoid ligament, the two forming, by their junction, an angle projecting backward. The width of the clavicular origin of the trapezoid ligament was 1.2±0.1 cm.

The conoid ligament (*ligamentum conoideum*), the posterior and medial fasciculus, is a dense band of fibers, conical in form, with its base directed upward. It is attached by its apex to a rough impression at the base of the coracoid process, medial to the trapezoid ligament; above, by its expanded base, to the coracoid tuberosity on the under surface of the clavicle, and to a line proceeding medial ward from it for 1.25 cm. The conoid width at its clavicular origin was 2.5±0.5 cm. The broad conoid ligament was not reliably centered over the most prominent aspect of the conoid tuberosity.

These ligaments are in relation, in *front,* with the subclavius and deltoideus; *behind,* with the trapezius. The perform two major functions: (1) They guide synchronous scapulohumeral motion by attaching the clavicle to the scapula and (2) they strengthen the AC articulation.

5.3.1 Blood Supply

The vertical part of coracoid process is supplied by suprascapular artery and the horizontal part by branches of the axillary artery [16].

The blood supply is derived from vessels in muscles that have fleshy origin from the scapula.

Vessels cross these indirect insertions and communicate with bony vessels. The circulation of the scapula is metaphyseal; the periosteal vessels are larger than usual, and they communicate

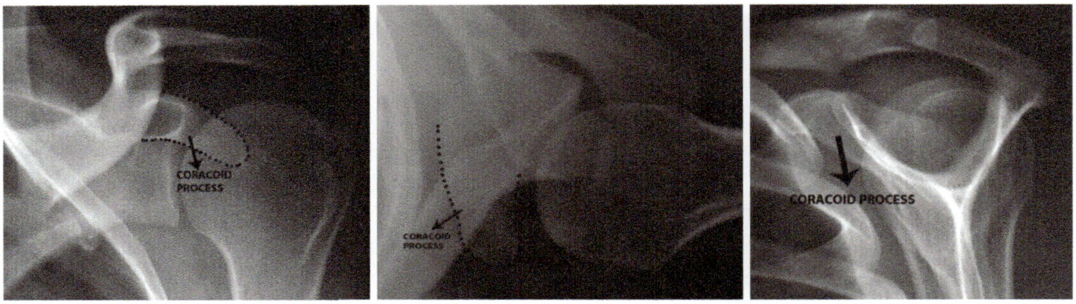

Fig. 5.3 Radiographs of normal left shoulder; *AP* axillary view and lateral views

freely with the medullary vessels rather than being limited to the outer third of the cortex. Such anatomy might explain why subperiosteal dissection is bloodier here than over a diaphyseal bone. The nutrient artery of the scapula enters into the lateral suprascapular fossa or the infrascapular fossa. The subscapular, suprascapular, circumflex scapular, and acromial arteries are contributing vessels.

This anatomical study has shown that the coracoid process had its own blood supply. During the Latarjet procedure, vascular sacrifices are mandatory to allow coracoid process transfer to the scapular neck. Such sacrifices could partially explain lysis or nonunion of the coracoid process after Latarjet procedure. Preservation of axillary artery branches supplying horizontal part of the coracoid process could be a possible solution to prevent nonunion and lysis of the bone transfer.

5.4 Variations

Several anomalies of the coracoid have been described. Pieper and colleagues [17] reported variations in the coracoacromial ligament in 124 shoulders. Their findings were two distinct ligaments in 60 % and one ligament in 26 % of shoulders. They were also able to identify a third band located more posterior and medial than the conoid in 15 % of shoulders. Very little variation was found in the dominant and nondominant shoulders of the same cadaver with regard to number of bands.

5.5 Imaging

The coracoid process is not easily visualized on a radiograph. Apart from the usual three-view trauma series, an AP tilt view (35° to 60°) and a Stryker notch view. A CT scan with three-dimensional reconstruction images will give more insight into the fracture pattern (Fig. 5.3).

5.6 Adjacent Structures

The coracoid process is readily palpable in the infraclavicular region just under the anterior head of the deltoid.

Lo et al. [18] showed that the portion of the coracoid tip which was closest to the neurovascular structures was the anteromedial portion of the coracoid tip. The distance from the anteromedial portion of the coracoid tip to each structure is as follows: axillary nerve 3.0 cm, musculocutaneous nerve 3.3 cm, lateral cord 2.9 cm, and axillary artery 3.7 cm. Similarly, the portion of the base of the coracoid that was closest to the neurovascular structures was its anteromedial portion. The shortest distance from the anteromedial aspect of the base of the coracoid to the axillary nerve, the musculocutaneous nerve, the lateral cord, and the axillary artery was 2.9, 3.7, 3.7, and 4.3 cm, respectively. Procedures about the coracoid are relatively safe procedures. The lateral cord of the brachial plexus is at greatest risk during dissection about the tip of the coracoid, and the axillary nerve is at greatest risk during dissection about the base of the coracoid. Pan et al. [19] demonstrated that the lateral cord moved closer to the coracoid process at 60° than at 30° of abduction under traction during simulated shoulder arthroscopy position using the lateral decubitus position.

The shape and size of the suprascapular notch (SSN) is one of the most important risk factors in suprascapular nerve entrapment. Five types of SSN were noted (See Chap. 33). In *type I*, maximal depth was greater than superior transverse diameter. *Type II* has equal MD, STD, and MTD. In *type III*, the superior transverse diameter was greater than the maximal depth. Scapulae with bony foramen were classified as *type IV* (4.7 %). In *type V*, a discrete notch was found. The frequency of type I and IV was lower in females than in males, but type III was more common in females than males [20].

Knowledge of the anatomical variations of the suprascapular notch should be helpful in both endoscopic and open procedures of the suprascapular region and also may increase the safety of operative decompression of the suprascapular nerve.

The suprascapular nerve always ran through the notch under the superior transverse scapular ligament. All shoulders had a single suprascapular artery, while multiple suprascapular veins appeared in 21.3 % [21].

The clavipectoral fascia, which is an offshoot of the axillary fascia, first envelops the pectoralis minor and then continues superiorly to surround the subclavius muscle and clavicle (Fig. 5.4). The pectoralis minor muscle runs in an inferior-to-medial direction to insert on the second through fifth ribs. The brachiocephalic veins exit the thorax behind the sternoclavicular joint and immediately divide into the internal jugular and subclavian veins.

It is the relationship to the pectoralis minor that divides the axillary artery into its three sections. The first part of the axillary artery is medial to the pectoralis minor muscle and has one branch, the thoracoacromial artery. Posterior to the pectoralis minor, the second part of the axillary artery has two branches: the thoracoacromial and the lateral thoracic arteries. Lateral to the pectoralis minor, the third section has three branches, the subscapular artery and the anterior and posterior circumflex.

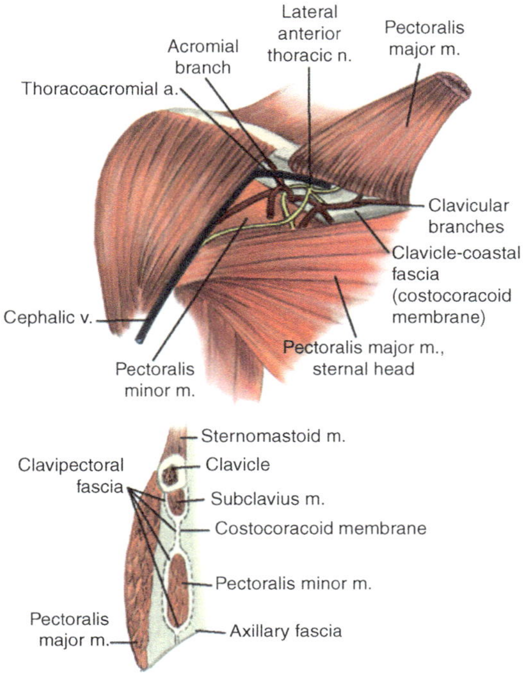

Fig. 5.4 Clavipectoral fascia, an offshoot of the axillary fascia, first envelops the pectoralis minor and then continues superiorly to surround the subclavius muscle and clavicle

5.7 Clinical Significance of the Anatomy

The coracoid process is palpable just below the lateral end of the clavicle (collarbone). It is otherwise known as the "Surgeon's Lighthouse" because it serves as a landmark to avoid neurovascular damage [22]. Major neurovascular structures enter the upper limb medial to the

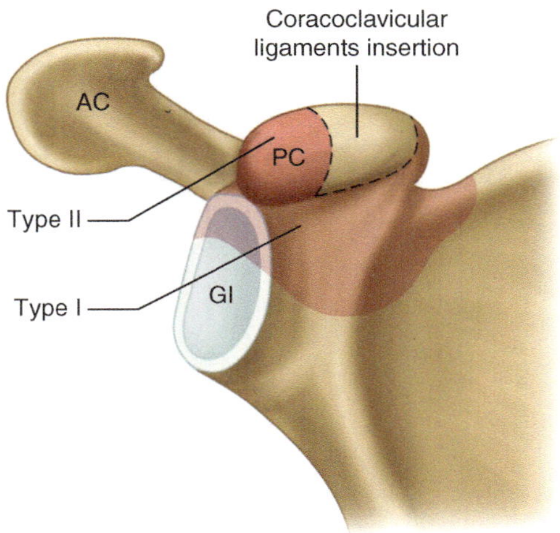

Fig. 5.5 Ogawa classification of coracoid fractures. Type I is situated proximal to the coracoclavicular ligament attachment and type II distal to these ligaments. *Gl* glenoid, *Ac* acromion, *CP* coracoid process (Rockwood and Matsen [7])

coracoid process, so that surgical approaches to the shoulder region should always take place laterally to the coracoid process.

5.8 Effect of Trauma

Ogawa [23], who simplified the classification scheme of Eyres, classified coracoid fractures into two different types. Type I is situated proximal to the coracoclavicular ligament attachment and type II distal to these ligaments (Fig. 5.5). Ogawa suggested that a type-I fracture may disturb the scapulothoracic connection. Type-I fractures were associated with a wide variety of shoulder injuries and consequent dissociation between the scapula and the clavicle. Treatment was usually by open reduction and fixation for type-I fractures and conservative methods for type II.

Fractures of the coracoid may be isolated, occurring from a direct blow to it or to the point of the shoulder. Coracoid fracture may also occur in association with or without acromioclavicular dislocation, with the remaining intact [24].

5.9 Coracoid Impingement

Coracoid impingement syndrome is a less common cause of shoulder pain. Symptoms are presumed to occur when the subscapularis tendon impinges between the coracoid and lesser tuberosity of the humerus and is thought to be an important factor in the development of degeneration and tears of the subscapularis tendon [25].

The mechanical impingement on the rotator cuff by the overlying acromial arch was postulated early in 1909 [26]. The same article also considered the possibility of rotator cuff impingement by its anteromedial portion, that is, the coracoid process.

Coracoid impingement was also described in 1937 [27] and in 1941 [28]. The coracoacromial interval includes the acromion, the coracoacromial ligament, and the tip of the coracoid process. It is the variation in the height and length of the coracoid process in most cases that is responsible for altering the size and shape of the space between the coracoacromial arch and the rotator cuff. A recent cadaveric study also suggests that the problem is functional with anterior instability leading to a narrowing of the coracohumeral distance [29]. There are no reports in the published literature of the incidence or prevalence of this condition.

Bone tumors of the coracoid process of the scapula are rare, and diagnosis and treatment often are delayed [30]. The coracoid process is the site with a markedly high proportion of chondrosarcomas while metastatic tumors are generally derived from breast or lung [31, 32]

5.10 Surgical Significance of the Pathoanatomy

Surgery about the coracoid is uncommonly performed but may be indicated in cases such as subcoracoid impingement. Surgical treatment of idiopathic subcoracoid impingement has generally aspect of the coracoid tip is resected to decompress the subcoracoid space. However, other authors have described an arthroscopic approach. In 2001, Karnaugh et al. [33] reported on a technique of arthroscopic coracoplasty through a subacromial approach. In 2003, Lo and Burkhart [34]

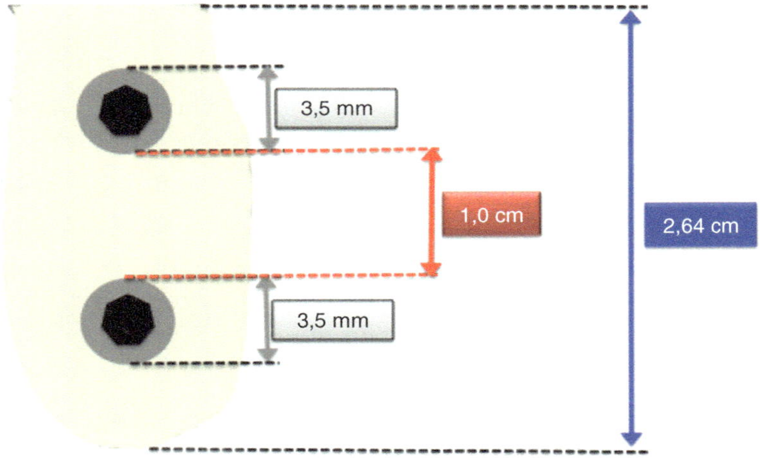

Fig. 5.6 The safety margin of the coracoid process (2.6 cm) with two 3.5-mm screws (From Terra et al. [8])

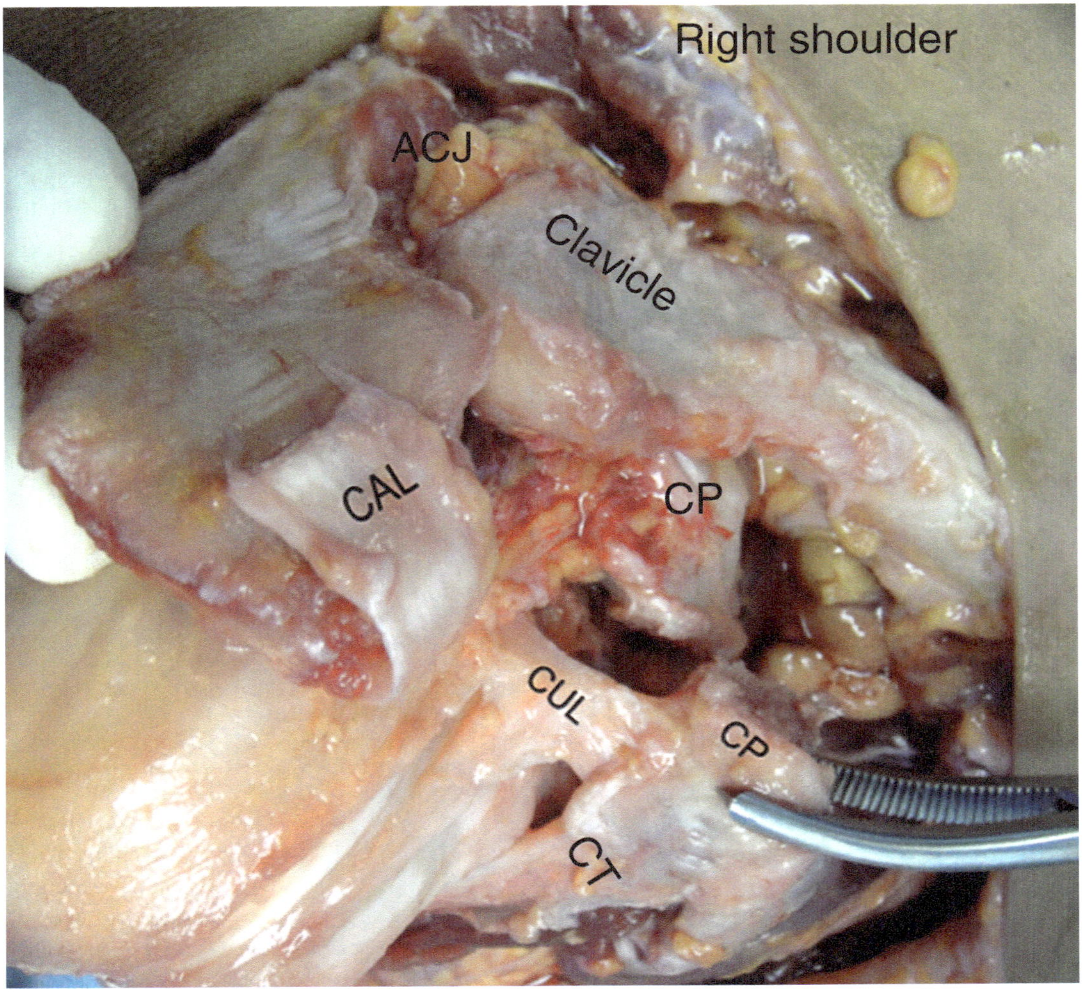

Fig. 5.7 Superior view of the right shoulder with the CAL released from the coracoid. A coracoid osteotomy has been performed to allow a Laterjet procedure to be performed. Note the remain intact. *CAL* coracoacromial ligament sectioned, *CP* coracoid process, *CUL* coracohumeral ligament, *CT* conjoint tendon, *ACJ* acromioclavicular joint (Terra et al. [8])

reported on arthroscopic coracoplasty through an intraarticular, transrotator interval approach. Although both open and arthroscopic techniques can effectively decompress the subcoracoid space, the relative safety of surgery about the coracoid is unknown. Furthermore, a number of rotator cuff mobilization techniques that involve releasing the rotator interval have recently been described [35]. These include the anterior interval slide (Tauro [36]), single and double interval slides, and the interval slide in continuity (Lo and Burkhart [37, 38]). These techniques generally involve a partial or complete excision of the coracohumeral ligament and may require exposure and dissection of the lateral aspect of the base of the coracoid.

In 2013, Ejnisman et al. [8] established that a coracoid process osteotomy 2.6 cm from the tip of the coracoid does not compromise the insertion of the and therefore does not cause any instability of the acromioclavicular joint. This is what has been defined as the safety margin that is applicable to procedures for the coracoid transfer process as in the Latarjet-Patte surgery, where we use two cannulated screws (4.0 mm). We recommended at least a 1.0 cm interval between the screws to avoid breaking the graft (Figs. 5.6 and 5.7).

References

1. Hsieh ET. A review of ancient Chinese anatomy. Anat Rec. 1921;20:97–127.
2. Huang-Ti N, Wen CS. The Yellow Emperor's classic of internal medicine. Los Angeles: University of California Press; 1966.
3. McGrew R. Encyclopedia of medical history. New York: McGraw-Hill; 1985.
4. Dung HC. Acupuncture points of the brachial plexus. Am J Chin Med. 1985;13:49–64.
5. Hoernle AFR. Studies in the medicine of ancient India. Part I. Osteology or the bones of the human body. Oxford: Clarendon Press; 1907.
6. Persaud TVN. From antiquity to the beginning of the modern era. Springfield: Charles C Thomas; 1984.
7. Rockwood Jr CR, Matsen FA. The shoulder. Philadelphia: Saunders Elsevier; 2009.
8. Terra BB, Ejnisman B, de Figueiredo EA, Cohen C, Monteiro GC, de Castro Pochini A, Andreoli CV, Cohen M. Anatomic study of the coracoid process: safety margin and practical implications. Arthroscopy. 2013;29(1):25–30. doi:10.1016/j.arthro.2012.06.022. Epub 2012 Nov 23.
9. Gumina S, Postacchini F, Orsina L, Cinotti G. The morphometry of the coracoid process—its aetiologic role in subcoracoid impingement syndrome. Int Orthop. 1999;23(4):198–201.
10. Gray H, Carter HV. Anatomy descriptive and surgical. London: John W. Parker and Son; 1858, Retrieved 16 Oct 2011.
11. Renfree KJ, Riley MK, Wheeler D, et al. Ligamentous anatomy of the distal clavicle. J Shoulder Elbow Surg. 2003;12(4):355–9.
12. Rios CG, Arciero RA, Mazzocca AD. Anatomy of the clavicle and coracoid process for reconstruction of the coracoclavicular ligaments. Am J Sports Med. 2007; 35:811–7.
13. Fukuda K, Craig EV, An KN, et al. Biomechanical study of the ligamentous system of the acromioclavicular joint. J Bone Joint Surg. 1986;68-A(3):434–40.
14. Harris RI, Wallace AL, Harper GD, et al. Structural properties of the intact and reconstructed coracoclavicular ligament complex. Am J Sports Med. 2000; 28(1):103–8.
15. Motamedi AR, Blevins FT, Willis MC, et al. Biomechanics of the coracoclavicular ligament complex and augmentations used in its repair and reconstruction. Am J Sports Med. 2000;28(3):380–4.
16. Hamel A, Hamel O, Ploteau S, Robert R, Rogez JM, Malinge M. The arterial supply of the coracoid process. Surg Radiol Anat. 2012;34(7):599–607.
17. Pieper HG, Radas CB, Krahl H, Blank M. Anatomic variation of the coracoacromial ligament: a macroscopic and microscopic cadaveric study. J Shoulder Elbow Surg. 1997;6(3):291–6.
18. Lo IK, Burkhart SS, Parten PM. Surgery about the coracoid: neurovascular structures at risk. Arthroscopy. 2004;20(6):591–5.
19. Pan WJ, Teo YS, Chang HC, Chong KC, Karim SA. The relationship of the lateral cord of the brachial plexus to the coracoid process during arthroscopic coracoid surgery: a dynamic cadaveric study. Am J Sports Med. 2008;36(10):1998–2001.
20. Polguj M, Sibiński M, Grzegorzewski A, Grzelak P, Majos A, Topol M. Variation in morphology of suprascapular notch as a factor of suprascapular nerve entrapment. Int Orthop. 2013;37(11):2185–92.
21. Yang HJ, Gil YC, Jin JD, Ahn SV, Lee HY. Topographical anatomy of the suprascapular nerve and vessels at the suprascapular notch. Clin Anat. 2012;25(3):359–65. doi:10.1002/ca.21248. Epub 2011 Aug 18.
22. Gallino M, Santamaria E, Tiziana D. Anthropometry of the scapula: clinical and surgical considerations. J Shoulder Elbow Surg. 1998;7(3):284–91.
23. Ogawa K, Yoshida A, Takahashi M, Ui M. Fractures of the coracoid process. J Bone Joint Surg Br. 1996; 78-B:17–9.
24. Kawasaki Y, Hirano T, Miyatake K, Fujii K, Takeda Y. Safety screw fixation technique in a case of coracoid base fracture with acromioclavicular dislocation and coracoid base cross-sectional size data from a computed axial tomography study. Arch Orthop Trauma Surg. 2014;134(7):913–8.

25. Okoro T, Reddy VRM, Pimpelnarkar A. Coracoid impingement syndrome: a literature review. Curr Rev Musculoskelet Med. 2009;2(1):51–5.
26. Goldthwait JE. An anatomic and mechanical study of the shoulder joint, explaining many of the cases of painful shoulder, many of the recurrent dislocations, and many of the cases of brachial neuritis. Am J Orthop Surg. 1909;6:579–606.
27. Meyer AW. Chronic functional lesions of shoulder. Arch Surg. 1937;35:646–74.
28. Bennett GE. Shoulder and elbow lesions of the professional baseball pitcher. JAMA. 1941;11:510–4.
29. Radas CB, Pieper HG. The coracoid impingement of the subscapularis tendon: a cadaver study. J Shoulder Elbow Surg. 2004;13(2):154–9.
30. Ogose A, Sim FH, O'Connor MI, Unni KK. Bone tumors of the coracoid process of the scapula. Clin Orthop Relat Res. 1999;358:205–14.
31. Benson EC, Drosdowech DS. Metastatic breast carcinoma of the coracoid process: two case reports. J Orthop Surg Res. 2010;5:22.
32. Deberne M, Ropert S, Billemont B, Daniel C, Chapron J, Goldwasser F. Inaugural bone metastases in non-small cell lung cancer: a specific prognostic entity? BMC Cancer. 2014;14:416.
33. Karnaugh RD, Sperling JW, Warren RF. Arthroscopic treatment of coracoid impingement. Arthroscopy. 2001;17(7):784–7.
34. Lo IK, Burkhart SS. Arthroscopic coracoplasty through the rotator interval. Arthroscopy. 2003;19: p667–71.
35. Gaskill TR, Braun S, Millett PJ. The rotator interval: pathology and management. Arthroscopy. 2011;27(4): 556–67.
36. Tauro JC. Arthroscopic repair of large rotator cuff tears using the interval slide technique. Arthroscopy. 2004;20(1):13–21.
37. Lo IK, Burkhart SS. Arthroscopic repair of massive, contracted, immobile rotator cuff tears using single and double interval slides: technique and preliminary results. Arthroscopy. 2004;20(1): 22–33.
38. Lo IK, Burkhart SS. The interval slide in continuity: a method of mobilizing the anterosuperior rotator cuff without disrupting the tear margins. Arthroscopy. 2004;20(4):435–41.
39. Di Giacomo G. Atlas of functional shoulder anatomy. Italia: Springer; 2008.
40. Last RL. Anatomy: regional and applied. Edinburgh: Churchill Livingstone; 1984.

6 Acromion and Coracoacromial Arch

Francisco Vergara and Nicolás García

6.1 Developmental Anatomy

The acromion is structured around the sixth week of embryonic development. The acromial base together with the body and the scapular spine begin their ossification at gestational week 8 [1, 2]. The rest of the acromion is cartilaginous throughout all the prenatal period and has a type II acromion shape as described by Bigliani (1986) [3], remaining constant and with no morphologic changes throughout this period [4].

Between 15 and 18 years of age, the secondary acromial ossifications centres appear. Four centres are described: pre-acromion (anterior centre), meso-acromion (mid centre), meta-acromion (posterior centre) and basi-acromion (located at the base of the scapular spine) [5] (Fig. 6.1). The ossification centres end up fusing between 22 and 25 years of age, usually from posterior to anterior [6].

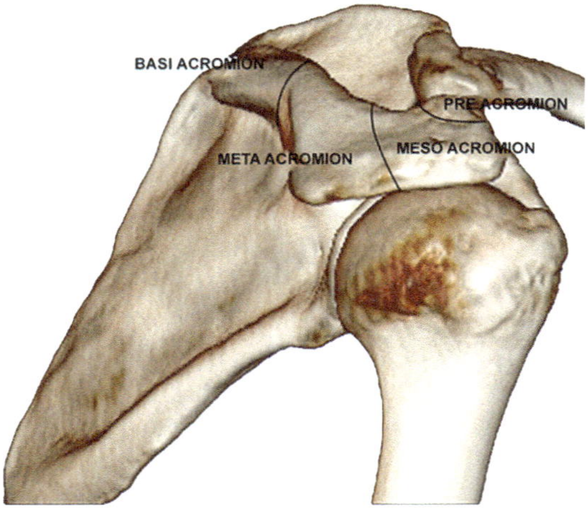

Fig. 6.1 3D CT scapulae reconstruction with a schematic representation of secondary acromial ossifications centres

6.2 Acromion Anatomy

6.2.1 Description

The acromion is a scapular structure projected anteriorly from the lateral border of the scapular spine, which represents continuity. It presents superior and inferior surfaces, and medial and lateral borders. The inferior lip of the crest of the spine is continued with the lateral acromion border, which is thick and irregular. On the other hand, the medial border corresponds to a prolongation of the scapular crest of the spine upper

F. Vergara (✉)
Department of Orthopaedics and Sports Medicine, Clínica Meds, Santiago, Chile
e-mail: franciscoverg@gmail.com

N. García
Department of Sports Medicine, Clínica Meds, Santiago, Chile
e-mail: nicolasgarciaa@gmail.com

G.I. Bain et al. (eds.), *Normal and Pathological Anatomy of the Shoulder*,
DOI 10.1007/978-3-662-45719-1_6, © ISAKOS 2015

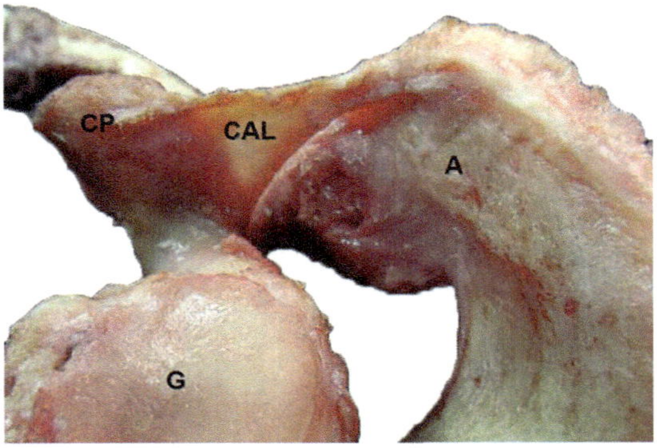

Fig. 6.2 Anatomic dissection of coracoacromial ligament (*CAL*), which is located between the lateral border of coracoid process (*CP*) and acromion (*A*). Glenoid fossa (*G*) is also shown (From Morphology Department, Universidad de los Andes, Santiago, Chile)

lip. It is short and is occupied anteriorly by an oval-shaped articular facet, directed medially and upwards, to join with the lateral end of the clavicle. Both acromial borders join anteriorly, forming a triangle called acromial angle [7, 8].

The superior acromial surface is subcutaneous and is covered only by skin and the superficial fascia. The lateral border is thick and irregular. The inferior surface is concave and smooth [7, 8].

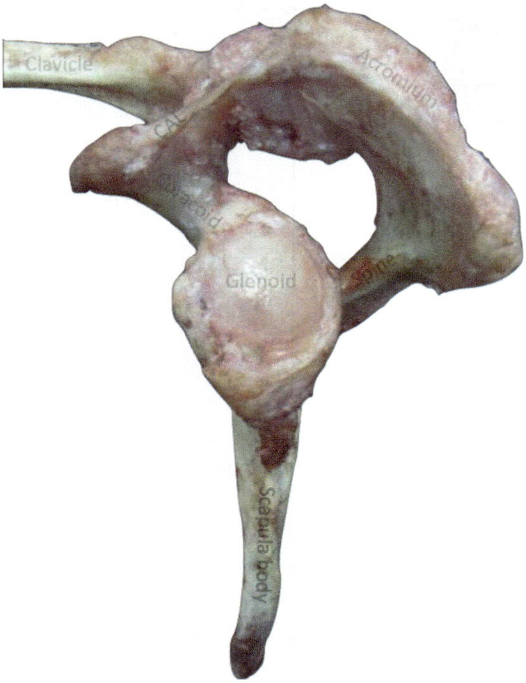

Fig. 6.3 Anatomic dissection of coracoacromial arch, formed by the coracoid process (*CP*), coracoacromial ligament (*CAL*) and acromion (*A*), establishing a curved structure over the glenoid process (*G*) (From Morphology Department, Universidad de los Andes, Santiago, Chile)

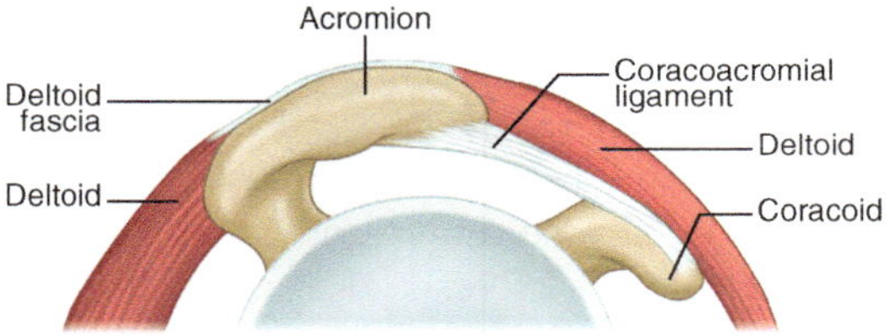

Fig. 6.4 Diagram of the acromion and coracoacromial ligament (Modified with permission, *Arthroscopic Rotator Cuff Surgery: A Practical Approach to Management*, Chapter 7, 2008, K. Yamaguchi, R. Tashjian Copyright Springer Science + Business Media)

6.2.2 Muscles and Ligaments Insertions

The mid portion of the deltoid muscle originates along the lateral border of the acromion, including the most anterior portion of the acromial process [8]. The deltoid insertion in the acromion has an average thickness of 5.4 mm, corresponding to approximately 74 % of the anterior acromion thickness [9]. These anatomic features must be considered when performing an acromioplasty, because a large resection could affect the deltoid origin.

On the other hand, the coracoacromial ligament is a triangular fibrous lamina, its apex is attached to the acromion and its base to the lateral border of the coracoid process (Fig. 6.2). The upper surface of the ligament is related to the surface of the deep deltoid muscle, while the inferior surface is oriented towards the glenohumeral joint and the periarticular muscles, from which it is separated by a synovial pouch (subacromial or subdeltoid pouch) [8]. The ligament has two bands: lateral (stronger and thicker) and medial (with a variable insertion in the acromion) [10].

6.2.3 Function

The acromion, together with the coracoacromial ligament and the coracoid process, form the cora-

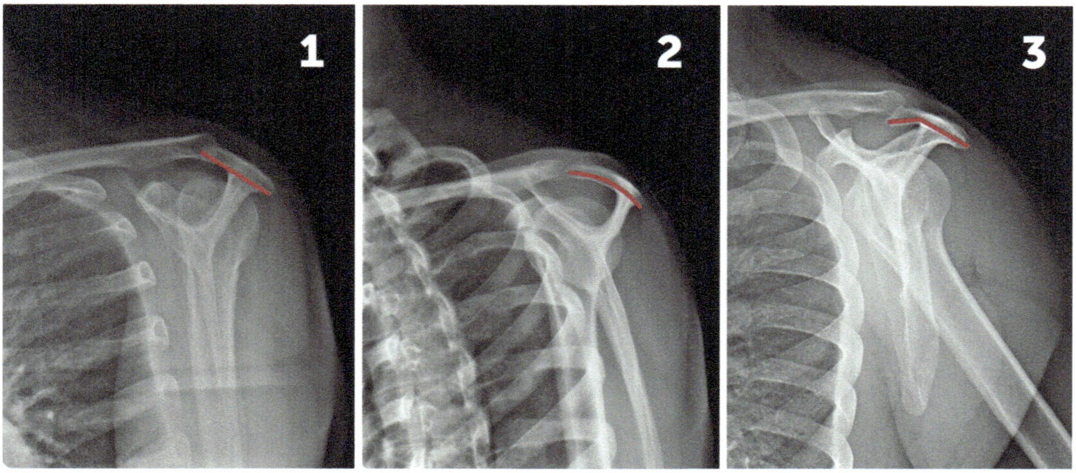

Fig. 6.5 Outlet views of shoulder radiographies with the three types of acromion described by Bigliani (1986): type 1 (flat acromion), type 2 (curved acromion) and type 3 (hooked acromion)

coacromial arch (Figs. 6.3 and 6.4), which is a curved structure meant to protect the glenohumeral joint. Specifically, the acromion and the coracoacromial ligament limit the upper translation of the glenohumeral joint [11].

6.3 Anatomic Variations and Changes with Age

Bigliani et al. (1986) described three types of acromion according to its inferior surface morphology and classified it as type I (flat acromion), type II (curved acromion) and type III (hooked acromion) [3] (Fig. 6.5). The prevalence described in the literature for each type depends upon the analysed population, but in general the evidence shows a higher frequency of type II acromion (near 50–55 %), followed by type III (25–30 %) and type I (15–20 %) [12]. In a healthy population, the incidence of type I acromion decreases with age, as type III increases [13]; therefore, it is considered that aging leads to secondary changes in the coracoacromial arch [14]. It has also been observed that with age the acromion presents degenerative changes characterized by an anterior spur which appears. In subjects below 50 years of age, the prevalence is considered to be 7 %, while above this age the prevalence rises to 30 % [15].

6.4 Acromial Morphology and Rotator Cuff Injury

Intrinsic and extrinsic factors intervene in rotator cuff injury. The evidence between acromial morphology and its association with rotator cuff injuries remains controversial. Neer (1983) established that 95 % of patients with a rotator cuff tear present a space conflict at the coracoacromial arch. Equally, Bigliani et al. (1986) showed that the incidence of type III acromion was 70 % in patients with rotator cuff injury, compared to 38 % in the general population [3, 16]. However, studies performed have not been able to show such association [17, 18], assuming that also age could be a confusion factor in this topic, especially taking into account the higher prevalence of type III acromion as years go by [13], even more when age is considered an independent risk factor for rotator cuff injury [19]. Therefore, acromial morphology would represent a degenerative process with aging rather than an innate anatomical feature [18]. Gill et al. (2002) found no significant association between acromial morphology and rotator cuff injury in patients over 50 years of age [18].

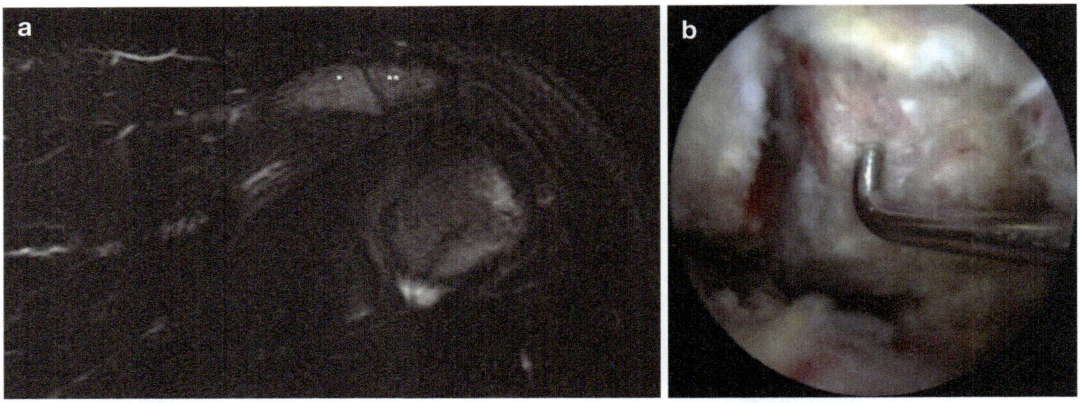

Fig. 6.6 (**a**) Shoulder MR, coronal plane of acromion with os acromiale between meso-acromion (**) and meta-acromion (*). (**b**) Arthroscopic view of os acromiale

6.5 Pathology: Os Acromiale

Os acromiale (OA) is an incomplete fusion of the secondary acromial ossification centres, thus resulting in an epiphyseal fragment which is separated from the rest of the bone structure [20]. The most usual location is between the meso-acromion and meta-acromion, named "common os acromiale" (approximately 76 % of the cases) (Fig. 6.6), while the separation between the pre-acromion and meso-acromion centres is less frequent ("terminal osacromiale", 15 % of the cases). The location between meta-acromion and basi-acromion is exceptional (1.8 %) [21, 22]. The degree of fusion varies from a fibrous union (complete or partial) up to a diartrodial joint [23].

The aetiology is not clear at all, currently two theories have been mentioned [24]. The first one says that OA would result from mechanical forces applied over a developing acromion, while the second one hypothesizes a genetic base. Other authors [25] hypothesize that OA would be a result of both.

The prevalence of OA in cadaver studies is considered 6.4 %, with racial variations (5.2 % in white race studies, 14.8 % in black race and 4.1 % for American Natives), while the prevalence in radiologic studies is somewhat less (4.2 %) [22] the unilateral presentation being more frequent.

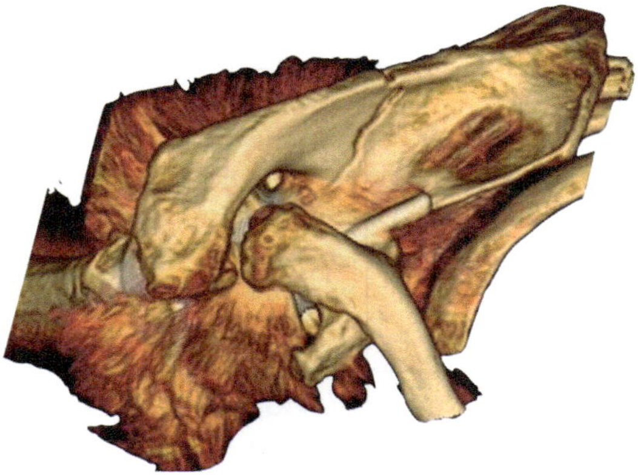

Fig. 6.7 Fracture of the scapula spine in a patient with a reverse total shoulder replacement. This type of fracture is becoming more common, especially in patients with a reverse total shoulder replacement (Copyright Dr Greg Bain)

Clinically, OA can be a cause of shoulder pain, being responsible for 4 % of this disease in primary attention [26]. When the conservative treatment fails, it can be treated with autologous graft associated with internal fixation, or else to opened or arthroscopic removal of the unstable fragment if the size is small [27]. Regarding its role in rotator cuff impingement, the evidence is not clear as to establish an association, although most of the studies show similar OA prevalences in subjects with and without rotator cuff injury

[28]. The recommendation for rotator cuff injuries associated with OA is to correct both injuries during surgery.

Fractures of the acromion are uncommon; however, they are increasing since the advent of the reverse total shoulder replacement. They are thought to be due to a stress riser from persistent traction force of the deltoid (Fig. 6.7).

References

1. Lewis W. The development of the arm in man. Am J Anat. 1901;1:145–83.
2. Gardner E, Gray DJ. Prenatal development of the human shoulder and acromioclavicular joints. Am J Anat. 1953;92(2):219–76.
3. Bigliani L, Morrison D, April EW. The morphology of the acromion and its relationship to rotator cuff tears. Orthop Trans. 1986;10:228.
4. Fealy S, et al. The developmental anatomy of the neonatal glenohumeral joint. J Shoulder Elbow Surg. 2000;9(3):217–22.
5. Folliason A. Un cas d'os acromial. Rev Orthop. 1933;20:533–8.
6. McClure JG, Raney RB. Anomalies of the scapula. Clin Orthop Relat Res. 1975;110:22–31.
7. Standring S. The anatomical basis of clinical practice. In: Gray's anatomy. 39th ed. London: Elsevier; 2005.
8. Rouviere H, Delmas A. Anatomia Humana: Descriptiva, Tropografica y funcional, vol. 8–9. 11th ed. Barcelona: Masson; 2005. p. 45–6.
9. Green A, Griggs S, Labrador D. Anterior acromial anatomy: relevance to arthroscopic acromioplasty. Arthroscopy. 2004;20(10):1050–4.
10. Fealy S, et al. The coracoacromial ligament: morphology and study of acromial enthesopathy. J Shoulder Elbow Surg. 2005;14(5):542–8.
11. Moorman CT, et al. Role of coracoacromial ligament and related structures in glenohumeral stability: a cadaveric study. J Surg Orthop Adv. 2012;21(4): 210–7.
12. Collipal E. The acromion and its different forms. Int J Morphol. 2010;28(4):1189–92.
13. Wang JC, Shapiro MS. Changes in acromial morphology with age. J Shoulder Elbow Surg. 1997;6(1): 55–9.
14. Speer KP, et al. Acromial morphotype in the young asymptomatic athletic shoulder. J Shoulder Elbow Surg. 2001;10(5):434–7.
15. Nicholson GP, et al. The acromion: morphologic condition and age-related changes. A study of 420 scapulas. J Shoulder Elbow Surg. 1996;5(1):1–11.
16. Neer 2nd CS. Impingement lesions. Clin Orthop Relat Res. 1983;173:70–7.
17. Hamid N, et al. Relationship of radiographic acromial characteristics and rotator cuff disease: a prospective investigation of clinical, radiographic, and sonographic findings. J Shoulder Elbow Surg. 2012;21(10):1289–98.
18. Gill TJ, et al. The relative importance of acromial morphology and age with respect to rotator cuff pathology. J Shoulder Elbow Surg. 2002;11(4):327–30.
19. Banas MP, Miller RJ, Totterman S. Relationship between the lateral acromion angle and rotator cuff disease. J Shoulder Elbow Surg. 1995;4(6):454–61.
20. Edelson JG, Zuckerman J, Hershkovitz I. Os acromiale: anatomy and surgical implications. J Bone Joint Surg Br. 1993;75(4):551–5.
21. Hunt D. The frequency of os acromiale in the Robert J. Terry collection. Int J Osteoarchaeol. 2007;17:309–17.
22. Yammine K. The prevalence of Os acromiale: a systematic review and meta-analysis. Clin Anat. 2014; 27(4):610–21.
23. Bigliani LU, et al. The relationship of acromial architecture to rotator cuff disease. Clin Sports Med. 1991;10(4):823–38.
24. Stirland A. Patterns of trauma in a unique medieval parish cemetery. Int J Osteoarchaeol. 1996;6:92–100.
25. Case DT, Burnett SE, Nielsen T. Os acromiale: population differences and their etiological significance. Homo. 2006;57(1):1–18.
26. Cadogan A, et al. A prospective study of shoulder pain in primary care: prevalence of imaged pathology and response to guided diagnostic blocks. BMC Musculoskelet Disord. 2011;12:119.
27. Harris JD, Griesser MJ, Jones GL. Systematic review of the surgical treatment for symptomatic os acromiale. Int J Shoulder Surg. 2011;5(1):9–16.
28. Jerosch J, et al. Arthroscopic subacromial decompression–indications in os acromiale? Unfallchirurg. 1994;97(2):69–73.

7 Scapular Body

Tom Clement Ludvigsen

7.1 Osseous Anatomy

The scapula is a triangular-shaped thin sheet of bone with two faces; the dorsal and the costal or anterior face, three margins; the medial, the superior and the lateral margin, and three angles; the superior, the inferior and the lateral angle (Figs. 7.1, 7.2 and 7.3). It is thicker at its superior and inferior angles, at its processes and along the medial margin. This reflects the attachment of strong muscles that connect the scapula to the body and the upper extremity to the scapula. In many ways, the scapula can be compared to the patella as a site of muscle attachment favouring the biomechanics of the upper extremity. It has formed three processes that also are thicker, the spine with the acromion, the coracoid and the glenoid. At the base of the spine is the spinoglenoidal notch (Fig. 7.3), and medial to the base of the coracoid at the superior margin is the suprascapular notch (Fig. 7.1). Superior to the spine on the dorsal side is the supraspinatus fossa, inferior the infraspinatus fossa (Fig. 7.2).

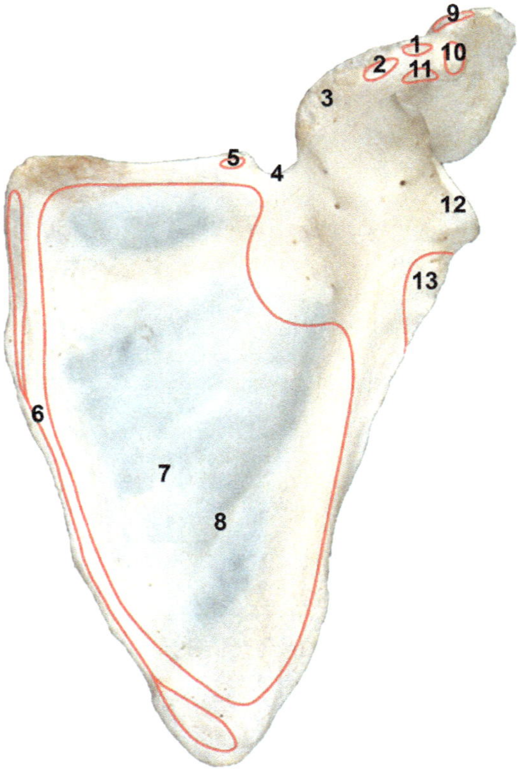

Fig. 7.1 Scapula – anterior (costal) aspect. *1* Trapezoid ligament attachment. *2* Conoid ligament attachment. *3* Acromion process. *4* Suprascapular notch. *5* Omohyoid (inferior belly). *6* Serratus anterior. *7* Subscapularis. *8* Ridge for intermuscular tendon of subscapularis. *9* Deltoid. *10* Biceps (short head) and coracobrachialis. *11* Pectoralis minor. *12* Glenoid fossa. *13* Triceps (long head) (Modified with permission from *Grays Anatomy*, 39th Edition, Strandring, S. Figure 49.4 Copyright Elsevier, 2005)

T.C. Ludvigsen
Orthopedic Department,
Oslo University Hospital, Oslo, Norway
e-mail: tomcl@getmail.no

G.I. Bain et al. (eds.), *Normal and Pathological Anatomy of the Shoulder*,
DOI 10.1007/978-3-662-45719-1_7, © ISAKOS 2015

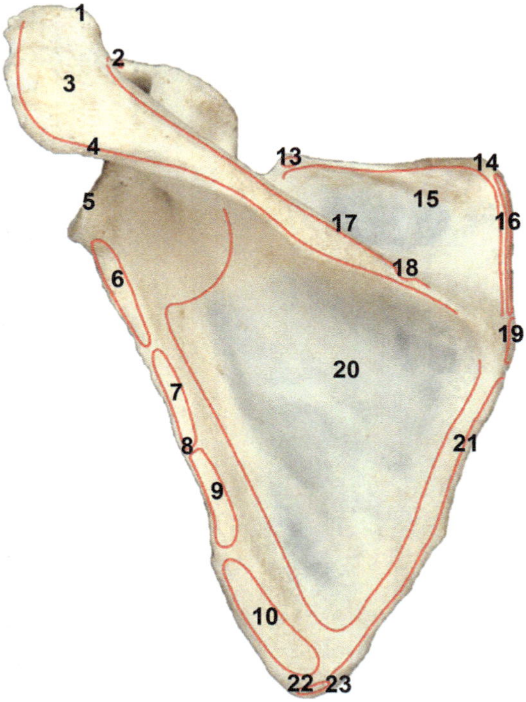

Fig. 7.2 Scapula – posterior (dorsal) aspect. *1* Clavicular facet. *2* Biceps (short head). *3* Acromion. *4* Deltoid. *5* Glenoid fossa. *6* Triceps brachii (long head). *7* Teres minor. *8* Groove for circumflex scapular artery. *9* Teres minor. *10* Teres major. *11* Conoid tubercle. *12* Coracoid process. *13* Omohyoid (inferior belly). *14* Superior angle. *15* Supraspinatus. *16* Levator scapulae. *17* Spine. *18* Trapezius. *19* Rhomboid minor. *20* Infraspinatus. *21* Rhomboid major. *22* Latissimus dorsi. *23* Inferior angle (Modified with permission from *Grays Anatomy*, 39th Edition, Strandring, S. Figure 49.3 Copyright Elsevier, 2005)

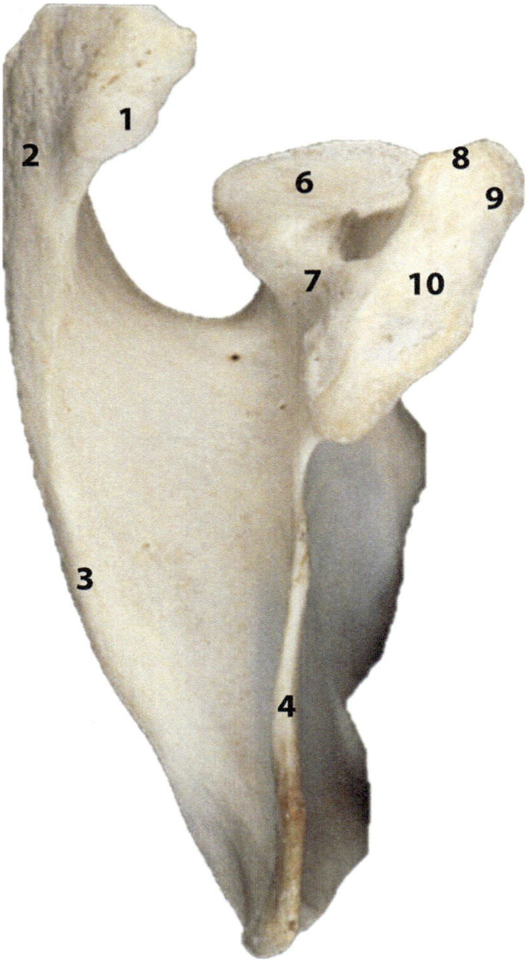

Fig. 7.3 Scapula – viewed from lateral view. Superior aspect of left scapula. *1* Facet for clavicle. *2* Acromial process. *3* Spine. *4* Superior border. *5* Head. *6* Glenoid fossa. *7* Neck. *8* Conoid tubercle (for conoid ligament). *9* Coracoid process. *10* Trapezoid ligament attachment (Modified with permission from *Grays Anatomy*, 39th Edition, Strandring, S. Figure 49.5 Copyright Elsevier, 2005)

7.2 Muscles and Tendons

Numerous muscles insert or originate at the scapula. Along the medial margin from superior to inferior insert the levator scapulae, the minor and major rhomboid. Most of the costal face (Fig. 7.1) gives the origin to the subscapularis muscle. On the costal face all along the medial margin and at the inferior angle insert the serratus anterior.

On the dorsal face (Fig. 7.2) along the lateral margin originate from superior to inferior the triceps, the teres minor and the teres major muscles. Sometimes also a few fibres of the latissimus dorsi insert at the tip of the inferior angle.

The long head of the triceps has its fibres partly attached to the distinct infraglenoid tubercle, a part of the inferior glenoid, and the inferior labrum, partly woven into the joint capsule and partly to the adjacent bone.

The dorsal face of the scapula is divided in two by the spine, below originates the infraspinatus muscle, and above the supraspinatus.

The trapezius muscle inserts with its upper part on the lateral two thirds of the clavicle, with its middle part on the acromion and scapular spine, and with its lower part at the base of the scapular spine. The deltoid muscle originates with its middle part on the acromion and its posterior part along the scapular spine.

At the superior margin, on the costal face just medial to the suprascapular notch is the origin to the omohyoid muscle, an important landmark for brachial plexus and cervical dissections. At the lateral angle of the scapula, anterior on the top of the glenoid originates the long head of the biceps.

7.3 Ligaments

Several ligaments attach to the scapula. Two of them, the coracoacromial ligament and the transverse scapular ligament, are special in the way that they attach to the same bone on both sides (Fig. 7.3). The transverse scapular ligament traverses the scapular notch and separates the underlying suprascapular nerve from the artery with the same name that runs cranial to the ligament. The shape of the notch shows great variability and that is also the case for this ligament that may be long or short, wide or thin, even split with the nerve in between in some cases (See Chap. 33). The ligament may ossify, and thus, become a foramen instead of a notch.

An inferior transverse scapular ligament has been reported in rare cases, running across the spinoglenoid notch.

The other scapular ligaments; the acromioclavicular ligament, the coracoclavicular ligaments, the coracohumeral ligaments and the glenohumeral ligaments will be described in different chapters.

7.4 Articulations

The scapula has two true articulations, the acromio-clavicular joint with the articular surface on the anteromedial aspect of the acromion, and the glenohumeral joint at the lateral angle [1, 2].

Considerable motion takes place between the scapula and the thoracic wall, the scapula gliding on the wall with only a bursa to assist in reducing the resistance.

7.5 Biomechanics and Function

The body of the scapula offers a mobile fixation point for the proximal upper extremity muscles. This means that scapulothoracic motion allows, among others, the deltoid muscle to remain in optimum position for effective contraction throughout the arch of arm elevation.

Motion of the upper extremity consists of a combined motion of the glenohumeral and the scapulothoracic joints. The relative contribution of each of these complex motion patterns in the total achieved varies with the position of the arm and also shows considerable variations between individuals and even by sex. Summarized, in the early phases of arm elevation a variably larger proportion occurs in the glenohumeral joint, whereas the last 60° occurs with about equal contribution of the two joints [3], It is generally accepted that the overall contribution of the glenohumeral to the scapulothoracic joint is a two-to-one relationship.

With arm motion the scapula undergoes a very complex rotation pattern in all planes. This is not only important for the total motion of the arm, but also for the stability of the glenohumeral joint. When one earlier on believed that shoulder joint stability mainly was achieved by the combined working of the passive stabilizing structures, ligamento-labral complex and joint capsule, and the dynamic stabilizers, muscles and tendons, convincing evidence now exists showing the important role of the dynamic positioning of the scapula in the complex biomechanics of the shoulder joint. This substantial knowledge has been popularized thanks to, among others, the work of Ben Kibler and co-workers. This explains why scapular winging from serratus anterior dysfunction can cause a subacromial pain syndrome and why throwing athletes can regain shoulder stability and performance by re-establishing the kinetic chain where scapular motion and control is crucial.

7.6 Vascularity

The main nutrient artery enters in the lateral supra- or infra-scapular fossa. The circumflex scapular, the subscapular and suprascapular arteries are all contributing, but a major part of the blood supply derives from smaller vessels in the muscles taking their origin from the bone (Fig. 7.4).

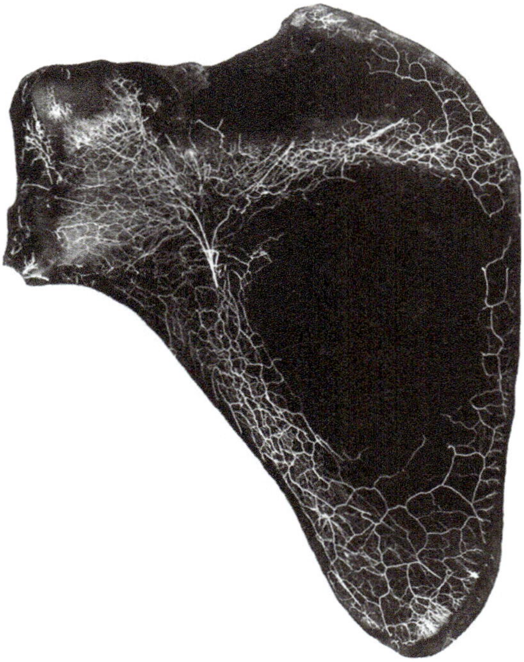

Fig. 7.4 Vascularity of the scapula in an 11-year-old boy. Coracoid and acromium have been excised. The nutrient artery enters the infraspinatous fossa in the lateral 1/3 and then divides into medial and lateral branches. The nutrient artery to the spine of the scapula and acromium enters the base of the spine near its inner 1/3. Periosteal vessels enter the inferior angle of the scapula, making it a site of vascularized bone grafts (Copyright HV Crock AO)

7.7 Nerve Supply

Included between muscular layers the scapular body mainly derives its nerve supply from the attached muscles.

7.8 Congenital Anomalies

7.8.1 Sprengel's Deformity

In the normal shoulder, the scapula descends from C5 at the 5th week in uterine. However, the superior medial corner of the scapula can be tethered to the cervical spine by undescended and rotated scapula (Figs. 7.5 and 7.6). The scapula remains high on the posterior chest wall, is rotated with its inferior angle pointing medially, the superior angle pointing forward to form a prominence in the supraclavicular region and the glenoid surface pointing downwards. It is usually hypoplastic and equilateral in shape.

In majority of cases, there is an omovertebral connection, and there are fibrous bands between the thoracic wall and the subscapularis muscle, which in combination with frequent muscular deficiencies and scapulospinous abutment severely impairs scapulothoracic motion.

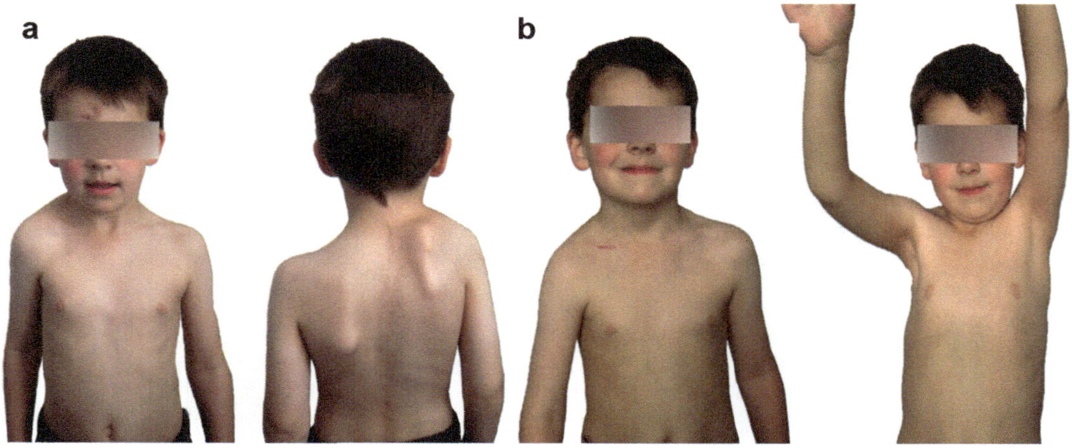

Fig. 7.5 (**a**) Sprengel's shoulder deformity. The failure of scapular descent produces a high-riding hyoplastic scapula, which restricts shoulder function. (**b**) Following surgery to resect the omo-vertebral bar, the patient is mobilizing the shoulder, and returning to function (Courtesy Dr Terri Bidwell, Auckland, New Zealand)

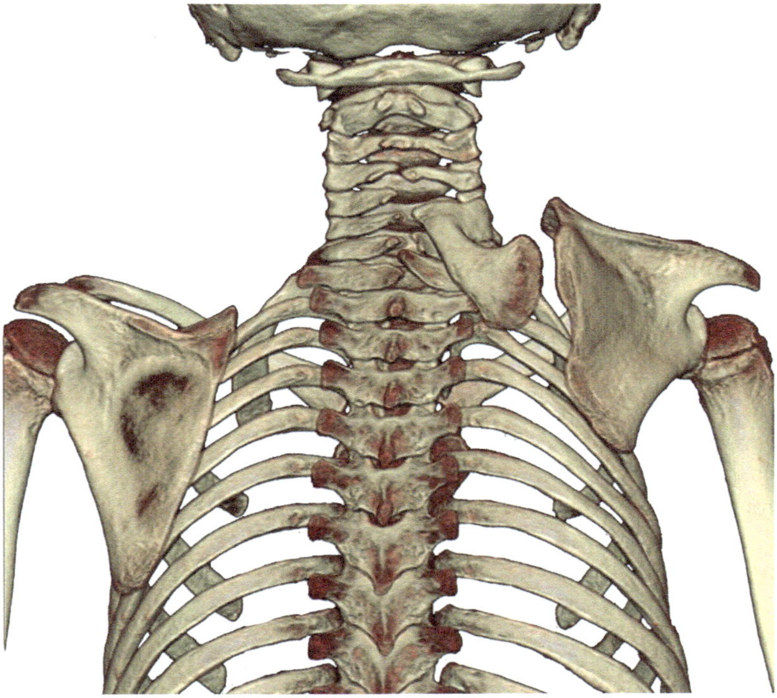

Fig. 7.6 Sprengel's shoulder deformity, with the ossification of the omo-vertebral bar that causes a failure of intrauterine descent of the scapula (Courtesy Dr Terri Bidwell, Auckland, New Zealand)

Cavendish (1972) gave name to a grading system 1–4 on the basis of appearance and recommended surgical treatment for grades 2, 3 and 4 [4].

7.8.2 Holt-Oram Syndrome

The term has been used to describe inherited associations of congenital heart disease and upper limb deformities. A prominent but hypoplastic and rotated scapula is the most common shoulder deformity seen, which produces protracted shoulders with decreased range of motion.

7.8.3 Ossified Transverse Scapular Ligament

A relatively common variation is the complete or partial ossification of the transverse scapular ligament, which transforms the notch to a foramen (See Chap. 33). In this case, the suprascapular nerve traverses the foramen and may become entrapped.

7.8.4 Ossification Disturbances

Clasp-like cranial margin and notched inferior angle are radiological variants, but thought to be of no clinical significance.

7.9 Acquired Anomalies

7.9.1 Fractures

Fractures of the scapula are relatively rare (1 % of all fractures). The scapular body is involved in about half of the cases (Fig. 7.7) [5]. Isolated body fractures usually have a good prognosis and are therefore treated conservatively. An international classification system has been developed [6].

7.9.2 Tumours

Malignant lesions as osteosarcoma are first of all located in the proximal humerus, but can rarely occur in the scapula and clavicle.

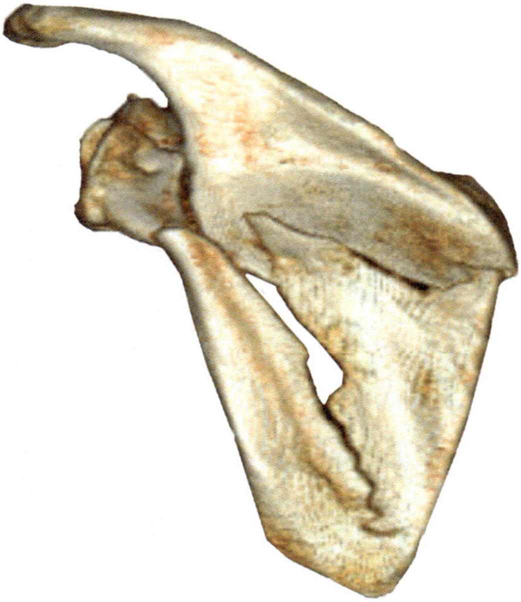

Fig. 7.7 Fracture of the scapular body

The scapula is the second most frequent location of chondrosarcomas [7].

Multiple myelomas may affect any part of the body, also the scapula, but the axial skeleton is the most common site of involvement.

Benign osseus lesions of the shoulder are uncommon, but both osteoid osteoma and osteoblastoma may occur, though they favour the proximal humerus or glenoid. In contrast, osteochondromas are frequently found in the shoulder region. Also these tumours, typically diagnosed in the skeletally immature individual, have a preference for the proximal humerus, but may arise from the scapular body. Located on the costal face, they can cause mechanical problems impinging on the chest wall with resulting functional impairment.

Fibrous dysplasia is a congenital dysplasia that may affect any bone, scapula included.

In the normal shoulder, the scapula glides smoothly and synchronously over the thoracic cage with shoulder motion. The periscapular muscles move the scapula with motion of the arm to maintain stability and loading of the glenohumeral joint. The aim is to maintain the stability of the glenohumeral joint in the different positions of the arm.

Abnormalities of scapula motion include winging of the scapula and snapping scapula. Winging of the scapula is where the scapula tilts in an abnormal way during shoulder motion and is usually due to imbalance of the periscapular muscles (see Chap. 29).

Snapping scapula is an uncommon condition, which is usually caused by a hooked superomedial angle or a Luschka's tubercle (bone or fibrocartilage prominence of the superomedial angle [8, 9]) (Figs. 7.8 and 7.9), or a true exostosis (Fig. 7.10) which impinges on the ribs with above-head activities. There is often accompanying bursitis due to the recurrent micro-trauma. The phases of inflam-

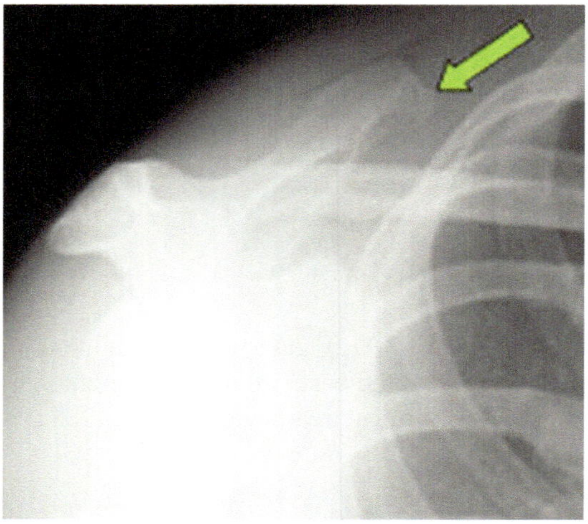

Fig. 7.8 Luschka's tubercle (*arrow*) on the superior medial aspect of the scapula, which impinges and produces a "snapping scapula"

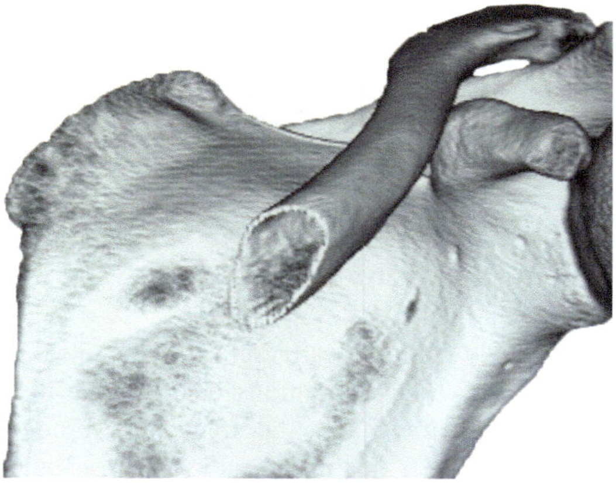

Fig. 7.9 Prominence of the superior medial aspect of the scapula, which impinges and produces a "snapping scapula" (Courtesy Dr Simon Bell, Melbourne, Australia)

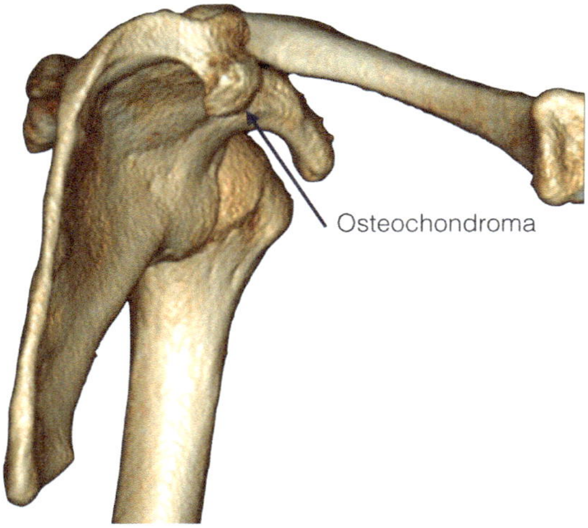

Fig. 7.10 Osteochondroma of the costal surface of the scapula

mation, reactive bursitis, and scarring can occur. These patients respond well to surgical excision of the exostosis/flattening of the underside of the superiomedial scapula, which can be performed as an endoscopic procedure [10].

Other pathological conditions that present as snapping of the scapula are extremely rare. The differential diagnosis includes osteochondroma of a rib or the anterior scapula occurs in adolescence and early adulthood. Scapular chondrosarcomas can occur in males 40–70 years.

Elastofibroma dorsi is a slow-growing, benign soft tissue tumour that presents in elderly women and athletes. It is usually located on the chest wall in the subscapular or infrascapular region, at the level of the rhomboid major and latissimus dorsi muscles.

References

1. Kuhn JE. The scapulothoracic articulation: anatomy, biomechanics, pathophysiology and management. In: Iannotti JP, Williams GR, editors. Disorders of the shoulder: diagnosis and management. Philadelphia: Lippincott Williams & Wilkins; 1999. p. 817–45.
2. Boinet W. Imperiale de Chir. Bull Soc. 1867;8(series 2):458.
3. Bergman A. Biomechanics and pathomechanics of the shoulder in Kibel et al: shoulder replacement. Berlin: Springer; 1987.
4. Cavendish ME. Congenital elevation of the scapula. J Bone Joint Surg Br. 1972;54(3):395–408.
5. Audige L, et al. The AO Foundation and Orthopaedic Trauma Association (AO/OTA) scapula fracture classification system: focus on body involvement. J Shoulder Elbow Surg. 2014;23(2):189–96.
6. Harvey E, et al. Development and validation of the new international classification for scapula fractures. J Orthop Trauma. 2012;26(6):364–9.
7. Rockwood Jr CA. The shoulder. 3rd ed. Philadelphia: Saunders; 2004.
8. Von Luschka H. Über ein Costo-scapular gelenk des Menchen. Vierteljahrsheft Prakt Heilkd. 1870;107:51–7.
9. Kuhne M, et al. The snapping scapula: diagnosis and treatment. Arthroscopy. 2009;25(11):1298–311.
10. Bell SN, van Riet RP. Safe zone for arthroscopic resection of the superomedial scapular border in the treatment of snapping scapula syndrome. J Shoulder Elbow Surg. 2008;17(4):647–9.

Clavice Anatomy

8

Joideep Phadnis and Gregory I. Bain

8.1 Development

The name 'clavicle' is derived from the Latin word *clavicula*, meaning 'small key' which in turn is derived from *clavis*, 'key' that refers to the 'S' shape of the bone. The clavicle is the first bone to ossify in the human skeleton at 5 weeks gestation. The diaphysial primary ossification centre ossifies by intramembranous ossification, which is unusual for a long bone. In contrast, the medial and lateral clavicular epiphyses (secondary ossification centres) ossify by endochondral ossification where the mesenchymal cells follow a stepwise cartilaginous ossification to ultimately become bone. The epiphyses are responsible for longitudinal clavicular growth whereas the diaphysis is responsible for increase in girth. The medial clavicular epiphysis is the more obvious of the two and begins to ossify at 18 years of age. It is the last epiphysis to fuse at up to 25 years of age. Sternoclavicular joint dislocations in the young adult population therefore tend to actually be physeal separations of the medial clavicular epiphysis. The lateral epiphysis appears as a thin calcified wafer and may be confused with a fracture [1].

8.2 Gross Anatomy

When viewed from superiorly, the clavicle is an S-shaped bone with two radii of curvature. The medial radius of curvature is greater in magnitude and convex anteriorly. This is thought to be to accommodate the neurovascular structures, which lie behind the medial clavicle before traversing beneath it. Indeed, in Sprengel's shoulder, a congenital deformity where the scapula does not descend, there is a less pronounced medial curvature, as there are no neurovascular structures to accommodate. The smaller lateral radius of curvature is concave anteriorly. In cross section the lateral clavicle is relatively flat whereas the middle and medial clavicle is more tubular. The mean cross-sectional diameter of the clavicle has been shown to be 23×22 mm at the sternal end, 12×12 mm in the mid-diaphysis and 21×11 mm in the wide, flat lateral end [2] (Fig. 8.1). Another study found significant

J. Phadnis, FRCS (Tr&Orth) (✉)
Department of Orthopaedics, Flinders University of South Australia, Bedford Park, SA, Australia
e-mail: joideep@doctors.org.uk

G.I. Bain, PhD, MBBS, FRACS, FA(Ortho)A
Department of Orthopaedics, Flinders University of South Australia, Bedford Park, SA, Australia

Department of Orthopaedic Surgery, Flinders Medical Centre, Adelaide, SA, Australia

Department of Orthopaedic Surgery, Flinders University, Adelaide, SA, Australia

University of Adelaide, Adelaide, SA, Australia

Department of Orthopaedics and Trauma, Royal Adelaide Hospital, Adelaide, SA, Australia

Department of Orthopaedics and Trauma, Modbury Public Hospital, Adelaide, SA, Australia
e-mail: greg@gregbain.com.au

G.I. Bain et al. (eds.), *Normal and Pathological Anatomy of the Shoulder*,
DOI 10.1007/978-3-662-45719-1_8, © ISAKOS 2015

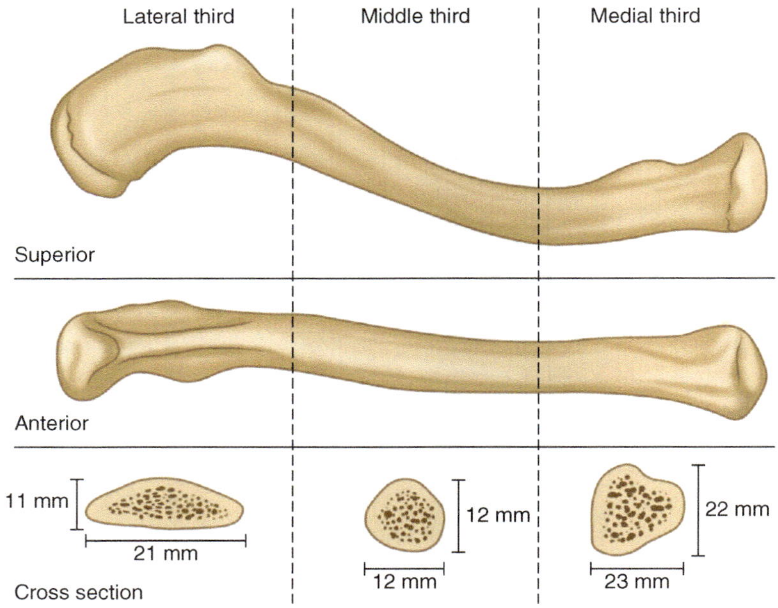

Fig. 8.1 Gross anatomy of the clavicle and mean cross-sectional dimensions (Modified Jeray [21]. April 2007 Copyright Lippincott Williams & Wilkins)

variation between races and between male and female subjects, with Caucasian female clavicles being the smallest studied [3]. The medial end of the bone has a 'Rhomboid fossa' on its inferior surface in 30 % of cases, which provides attachment for the costoclavicular ligaments [4]. The radiographic appearance of this fossa can mimic a lytic lesion of the medial clavicle. The middle third has an inferior subclavian groove where the subclavius muscle attaches. Laterally there are two further bony impressions for attachment of the coracoclavicular ligaments. The trapezoid line is along the mid-part of the inferior surface 25 mm from the AC joint and the conoid tubercle lies at the posterior apex of the lateral curvature 45 mm from the AC joint. De Palma and colleagues investigated the anatomic variation of 150 clavicles and found that no two clavicles had exactly the same characteristics [5]. He found a relationship between the clavicle length and the degree of the medial and lateral curvatures, which he was able to classify according to a clavicle curve index. Aside from the curve index these authors also found variation in the anterior torsion of the lateral clavicle and in the relationship between the curve index and the coronal plane slope of the AC joint. Three types of distal clavicle were defined based upon the degree of torsion and the resultant angulation of the AC joint. Type 1 clavicles had the most vertical AC joint and type 3 clavicles had the most oblique AC joint. Type 2 clavicles (intermediate orientation of AC joint) were the most common (48 %), followed by type 1 (41 %) and finally type 3 (11 %). Degenerative changes were most frequently found in the vertical articulations (type 1 clavicles). This was attributed to greater shear forces and decreased contact area at the AC joint.

Micro-computed tomography can be used to depict the internal trabecular structure of the clavicle (Fig. 8.2). This image demonstrates thick cortices and a paucity of trabecular bone within the diaphysis as compared to the medial and lateral ends where the cortices are thinner and the trabecular density is greater. This internal structure is similar to the other long bones in the skeleton.

Four muscles arise from, and two muscles insert into, the clavicle. The deltoid arises from the anterior surface of the lateral curvature and the clavicular head of pectoralis major arises from the anterior aspect of the medial curvature. There is a small hiatus between these two origins although this is difficult to discern during surgery.

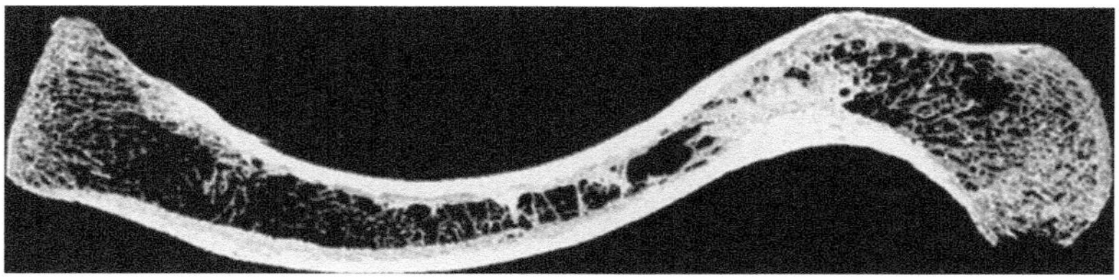

Fig. 8.2 Micro-CT scan of clavicle demonstrating trabecular and cortical bone patterns (Copyright Gregory Bain)

The clavicular head of sternocleidomastoid arises from the posterior aspect of the medial curve opposite pectoralis major. Just medial to the sternocleidomastoid is the origin of the clavicular part of sternohyoid. The trapezius is the major muscle inserting onto the clavicle. It does so opposite the deltoid on the posterior aspect of the lateral curvature. The subclavius has a fleshy insertion on the inferior aspect of the middle third at the subclavian groove. It is worth noting that subclavius is the only muscle that has its main attachment in the middle third of the clavicle. The relative paucity of soft tissue attachments in the middle third may have an influence on the incidence of fractures in this region and certainly influences the deformity after fracture.

8.3 Vascularity and Related Neurovascular Structures

Two cadaveric studies have demonstrated that the clavicular blood supply is predominantly periosteal with only sporadic, very short intra-osseous arteries in a few specimens [6, 7]. In contrast, Crock demonstrated that there are some nutrient vessels running within the clavicle as was evident on anatomic specimens with the periosteum removed [8] (Fig. 8.3). The main arterial supply came from branches of the thoracoacromial artery and suprascapular artery. The internal mammary artery also gave a contribution in one study. The periosteal blood supply was always on the anterior and superior surfaces of the clavicle and concentrated in the middle third of the bone. There was never any periosteal supply on the posterior or inferior surfaces. The lack of periosteal supply substantiates the increased risk of non-union in high-energy mid-shaft fractures where periosteal stripping and wide displacement of the bony ends occurs. It also highlights the importance of surgical technique when dissecting the anterior-superior aspect of the clavicle for plate fixation.

The subclavian artery and vein, and the divisions of the brachial plexus, lie posterior to the medial curvature of the clavicle and pass beneath the middle third of the clavicle where the divisions branch into cords and the subclavian vessels become the axillary vessels. Cadaveric data has demonstrated the subclavian vein to lie directly beneath the middle and medial third junction with the left subclavian vein marginally more medial than the right [2]. The subclavian artery lies lateral to the vein but is still medial to the midpoint of the clavicle. Both vessels are separated from the bone by the subclavius muscle although this muscle is fleshy and frequently torn in traumatic injuries. The median distance from the superior aspect of the clavicle to the subclavian artery in an intact clavicle was 26.1 mm (range 22–34 mm) [2].

Just proximal to the clavicle, the anterior and posterior divisions of the brachial plexus form the medial, lateral and posterior cords named after their relationship to the axilliary artery. Indeed, clavicular osteotomy can be performed to expose the plexus at this level. Of all the branches of the brachial plexus, the medial cord, which goes on to form the ulnar nerve, is closest to the middle and medial third junction and is hence the most commonly injured peripheral nerve in a clavicle fracture. The posterior and lateral cords lie posterior and lateral to the axillary artery, where they are relatively shielded.

Although the neurovascular structures are in close proximity to the medial half of the clavicle, they are protected from the bone by a continuous myofascial tissue layer extending from the omohyoid fascia in the neck down to the clavipectoral

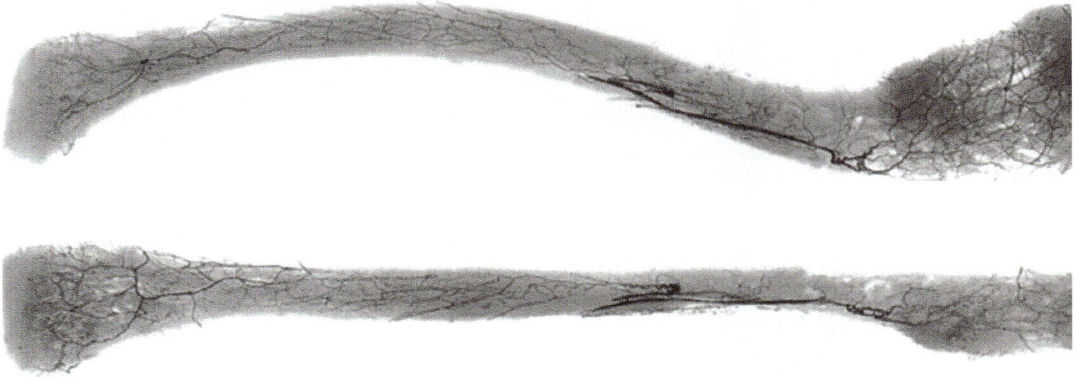

Fig. 8.3 Arterial supply of the clavicle. The nutrient artery is well seen in this specimen. Note the predominant periosteal blood supply, as has been identified by other authors [8] (Copyright HV Crock AO)

fascia which encases the pectoralis minor and subclavius muscles [9]. This layer is evident in comminuted fractures and retained in non-unions. Careful development of this plane is possible and recommended in all clavicle surgery in order to protect the adjacent neurovascular structures.

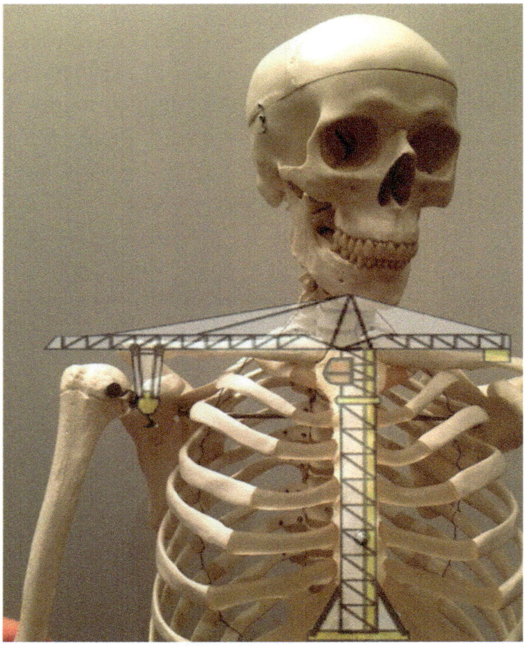

Fig. 8.4 The clavicle acts as a strut to suspend the limb away from the axial skeleton much like the arm of a suspension crane. See Chap. 39 for more detail

8.4 Function

The bony geometry of the shoulder girdle has developed and adapted according to the requirements of different species. A pronounced variation amongst mammals is the presence or absence of the clavicle. Mammals adapted for running and swimming such as horses or whales have lost their clavicle, which allows greater motion of the scapula. Mammals that fly have very long clavicles but a narrow small scapula. Humans and other brachiating mammals have strong well-developed clavicles and scapulae. This is an adaptation to perform tasks using the hands well away from the body and to climb, swing and grasp distant objects. These teleological differences help to define the function of the clavicle [4]. The primary role of the clavicle is to position the arm away from a trunk and provide a stable strut between the arm and the axial skeleton. Having a stable strut allows the muscles to generate adequate power to perform reaching and elevation tasks. In this way the clavicle behaves like the arm of a tower crane suspending the arm away from the body (Fig. 8.4). It is stabilized by the muscular and ligamentous attachments which are least yielding medially at the sternoclavicular ligaments. There is therefore relative rigidity of the clavicle medially compared to laterally, allowing the clavicle to rotate about its axis thereby working as a crankshaft to augment scapula motion. The scapula rotates a total of 60° during abduction of the arm [4]. Thirty degrees of this rotation comes purely from lateral clavicle rotation whereas the remaining 30° comes from angulation of the whole clavicle and scapula together at the sternoclavicular joint. The clavicle's role as a strut is not only important for activities performed away from the body but also when the arm is adducted and brought back into the body or across the trunk. The strut provides a fixed pivot for the arm to rotate and move around, without which this motion would be uncoordinated and inefficient. The clavicle's role as a strut for power generation and its role in range of motion have been questioned on the basis that some patients with severe shortening or even absence of the clavicle can have little functional deficit in arm elevation. However, other authors have reported weakness, fatigue, neurologic symptoms and inferior functional scores related to clavicular malunion with shortening [10]. The compensation seen in many patients with a clavicular abnormality may be related to the continuous myofascial sleeve that encircles the clavicle [9] although a stable bony framework to support these muscles must be mechanically advantageous and is likely to be more important in repetitive or heavy activities. Furthermore, the clavicle serves a protective role to the adjacent neurovascular structures. With shortening or malunion, these structures are more likely to suffer impingement and compromise as is often seen in non-unions and sometimes malunions [10] following fracture.

8.5 Clavicle Fractures

The clavicle is the most commonly fractured bone in the skeleton [4]. There are a variety of classification systems but most are based on the anatomic location of the fracture [4]. Eighty percent of fractures occur in the middle third of the bone. This is the region with the narrowest cross-sectional diameter and where compressive forces across the shoulder are concentrated [1]. In addition, the middle section has no muscular or ligamentous attachment, which makes it vulnerable to the bending moments exerted by the structures attached to the middle and distal thirds of the bone. Lateral third fractures occur in 15 % of cases. They are more common in elderly patients or those with co-morbidities, although they do occur in all age groups. Medial third fractures (5 %) are uncommon and are typically associated with high-energy trauma in middle-aged men [11]. Occult injuries to the thorax, spine and head should be searched for in these injuries.

The typical mechanism of injury leading to a fracture of the clavicle is a fall with direct impact to the outer aspect of the shoulder. In a middle third fracture, the lateral fragment is typically shortened, inferiorly translated and anteriorly rotated. The resultant uncoupling of the distal segment from the axial skeleton results in protraction of the scapula [1] (Fig. 8.5). The weight of the arm causes inferior displacement of the lateral fragment while the medial fragment is superiorly displaced by contracture of the sternocleidomastoid. The trapezius causes shortening due to its attachment on the distal fragment. These strong muscular forces and weight of the arm mean closed reduction and any form of sling immobilization is unsuccessful in reducing or maintain reduction of a middle third clavicle fracture.

8.5.1 Malunion

The vast majority of clavicle fractures unite with no intervention and little functional deficit although malunion is ubiquitous given that the fracture cannot be reduced and maintained by closed means. Recently there has been a growing appreciation that malunion of the clavicle may not be as benign as traditionally thought.

Malunion with significant shortening (>2 cm) also shortens the lever arm of the muscles that span the clavicle and results in fatigability, power loss and pain particularly during repetitive activity [12]. The altered scapula mechanics can lead to winging, periscapular muscle spasm and subacromial impingement. Shortening together with inferior displacement also reduces the size of the thoracic outlet and can result in persistent neurologic and vascular symptoms. For these select patients, clavicle osteotomy, length restoration and fixation has been performed with some success [10].

Despite these concerns, the fact remains that, in the vast majority of patients, malunion remains

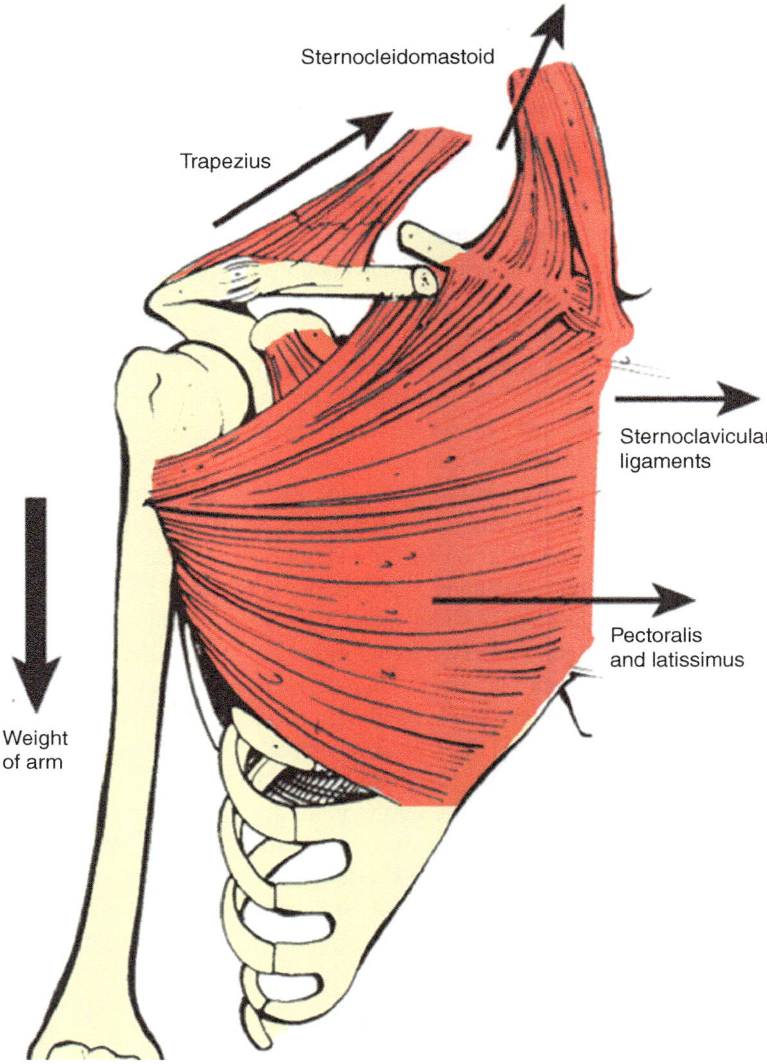

Fig. 8.5 Deforming forces on the middle third of a clavicle fracture (Modified with permission from Bucholz [1])

asymptomatic. This may be due to the deficits manifesting only in those performing repetitive heavy activities or that our understanding of the factors related to malunion are sub-optimal, with too much emphasis on shortening rather than on the three-dimensional mechanics of this entity.

8.5.2 Non-union

Predicting which patients with a clavicle fracture are likely to develop a symptomatic malunion is exceedingly difficult given that most remain asymptomatic. In contrast, most patients who develop a mid-shaft malunion tend to be symptomatic with pain, ache and weakness. The rate of non-union is now thought to be around 10–15 % of all displaced middle third clavicle fractures [13]. Several randomized control trials have shown improved functional outcomes with internal fixation of displaced mid-shaft clavicle fractures particularly compared to those patients who develop a non-union [14]. However between six and eight fractures need to be treated to prevent one non-union, thus it is more prudent to try and predict which patients may develop a non-union and select these for surgical fixation. Factors known to be associated with non-union include comminution, displacement of the main fracture ends by 2 cm in any direction, and smokers [13]. Why non-unions occur in the clavicle is multifactorial. The most important contributor is the lack of stability and consequent excess inter-fragmentary strain. This is because of difficulty in maintaining reduction of the fracture by closed means, as evidenced by the finding that clavicular non-unions tend to be hypertrophic in nature. Moreover, non-unions are known to heal after plating and application of compression to reduce strain. Other contributing factors towards non-union are soft tissue interposition between widely displaced fracture ends and fractures with an avascular component such as those sustained in high-energy mechanisms where there is significant soft tissue stripping and comminution. Moreover, the middle part of the clavicle has little muscular attachment meaning blood supply can be precarious in this area further predisposing to non-union. We would therefore recommend treating non-unions with compression plating and bone grafting to address all possible contributors to non-union. The key to a good result is restoration of clavicle length and alignment. This may require extensive mobilization of the bone ends in a non-union. Special care should be taken to protect the neurovascular structures. Dissection with care taken to maintain the myofascial layer around the clavicle will minimize risk.

Distal clavicle fractures have a higher incidence of non-union than mid-shaft fractures, with figures as high as 40 % reported [15]. Factors associated with non-union are increasing age and fracture displacement, but the majority of these (70 %) non-unions tend to be minimally symptomatic with no difference in functional score compared to if union had occurred [16]. In terms of pathoanatomy, distal clavicle fractures are very similar to AC joint disruptions with many of the stabilization techniques applicable to both injuries. The residual attachment of the conoid ligament (Chap. 17) in Rockwood grade 3 injuries is why we feel these injuries remain mostly asymptomatic. The same will apply to the Craig type 2b fracture of the distal clavicle where the fracture line runs between the conoid and trapezoid ligaments with minimal displacement of the medial clavicle shaft.

8.5.3 Paediatric Fractures

Clavicle fractures are the most common bony injury encountered during birthing. They manifest as a pseudoparalysis of the arm and must be differentiated from a brachial plexus palsy although the two can co-exist. To reduce the spasm from the sternocleidomastoid the infant may turn their head toward the fracture, which can be confused with acute torticollis. Additionally, an asymmetric Moro reflex may be present [17]. In older children clavicle fractures account for around 15 % of all paediatric fractures with the most common location being in the middle third, as in adults. Middle third fractures tend to be less displaced than in adults and, even in the presence of displacement, surgery is rarely indicated.

Distal third paediatric fractures are of interest as they represent periosteal sleeve injuries rather than pure bony injuries. They heal readily through the stripped but attached periosteal sleeve. The paediatric clavicle has medial and lateral epiphyses, which represent the secondary ossification centres. Fractures can often occur through these epiphyses and may be confused with sternoclavicular or acromioclavicular dislocations. These injuries rarely need treatment; although it is important to note the exception when there is a posterior displaced medial epiphyseal separation, which can threaten the vascular structures posterior to the sternoclavicular joint.

In the rare event where operative fixation of a paediatric clavicle fracture may be undertaken, such as for skin compromise or open fracture, reduction of the clavicle into its periosteal sleeve and repair of the periosteum is often all that is necessary [17].

8.6 Non-traumatic Disorders of the Clavicle

8.6.1 Infantile Cortical Hyperostosis (Caffey's Disease)

This condition of unknown aetiology presents within the first 6 months of life. The child has a painful lump over the clavicle, which shows a florid periosteal reaction on radiographs.

Although the clavicle is the most commonly affected bone, the mandible, scapula, humerus, tibia and femur may also be affected. Inflammatory markers are often elevated and more serious differential diagnoses such as non-accidental injury, infection and tumour must be excluded. The natural history is of resolution over the first 2 years of life with no residual dysfunction or pain.

8.6.2 Congenital Pseudoarthrosis of the Clavicle

This is a rare condition where there is failure of fusion of the medial and lateral ossification centres (Fig. 8.6). It occurs most commonly on the right side (90 %) and it is postulated this is because of pulsations from the right subclavian artery, which because of its origin from the brachiocephalic artery passes directly behind the sternoclavicular joint. This is thought to be contributory because in cases of dextrocardia where the great vessels and heart are reversed in orientation, pseudoarthrosis has been reported on the left side [18]. Rarely (10 %) it may be bilateral. The child presents with an unsightly lump over the clavicle, which is usually pain free. There is little functional deficit; however when surgical treatment is indicated, bone grafting and plating have high rates of union unlike pseudoarthrosis of the tibia for instance.

8.6.3 Chronic Recurrent Multifocal Osteomyelitis (CRMO)

This is an unusual condition that typically affects girls. Characteristically the clavicle is involved along with the metaphyseal regions of other joints. CRMO presents as a waxing and waning picture of recurrent musculoskeletal pain and fever. Treatment is with antibiotics although bone biopsy may be necessary to grow an organism. Usually no causative organism is detected; however *Propionibacterium acnes* has been implicated by some authors [19].

8.6.4 Cleidocranial Dysostosis

This is a rare autosomal dominant congenital condition. It primarily affects bones that undergo intramembranous ossification and the affected children have normal intelligence. The clavicles are characteristically either partially absent (90 %) or completely absent (10 %) resulting in hypermobility of the shoulders and the ability to touch the shoulders together in front of the chest (Fig. 8.7). Unstable remnants of the clavicles can

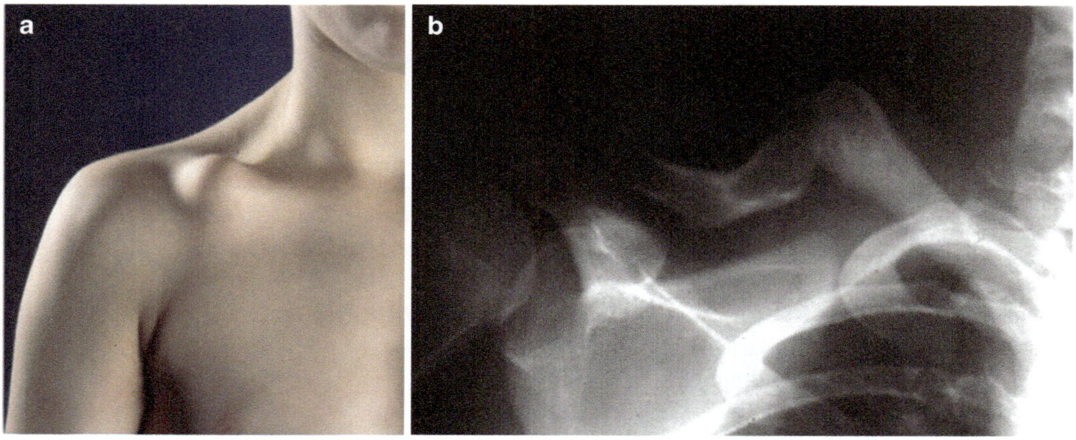

Fig. 8.6 (**a**) clinical photo and (**b**) x ray of pseudoarthrosis of the clavicle [18]

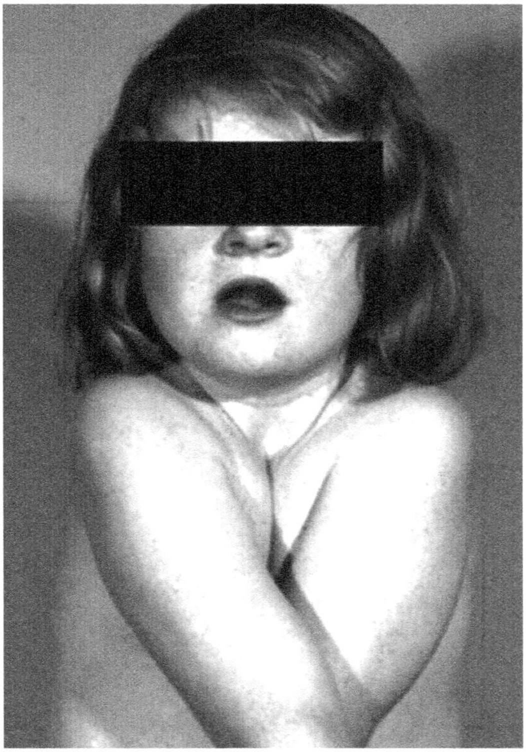

Fig. 8.7 Cleidocranial dysostosis – absence of the clavicle allows the child to bring the shoulders together in front of the chest (Beals et al. [18], April 2006, Copyright Lippincott, Williams & Wilkins)

cause neurovascular compression syndromes, which may need surgical decompression [4].

8.6.5 Freidrich's Disease

Otherwise known as osteonecrosis of the medial (most common) or lateral clavicle ends. Freidrich's disease presents in young adults spontaneously and is characterised by pain and inflammation. Radiographs show sclerosis at the site of osteonecrosis, and although the ESR may be elevated, other inflammatory markers tend to be normal. Non-operative management usually suffices until the symptoms settle although surgical debridement may be necessary in rare cases [18].

8.6.6 Distal Clavicle Osteolysis

This condition is thought to occur through repetitive micro-trauma of the AC joint and distal clavicle, although the aetiology is still debated. It typically occurs in weight lifters where extreme loading of the AC joint is thought to be the main cause. Histologic analysis has revealed high osteoblastic activity indicating an active remodeling process [20]. The diagnosis should be made after exclusion of other pathologies such as myeloma, infection, hyperparathyroidism and gout. Imaging shows erosive changes and lytic areas in the distal clavicle on plain radiographs. MRI scan demonstrates high signal on T2-weighted images with subchondral fissuring and adjacent oedema [20]. There is usually no instability associated. Treatment in is similar to that for AC joint arthritis where in the first instance activity modification, rest and analgesia should be used. Often, given the patient population this non-surgical management is not acceptable to them and AC joint resection may need to be performed.

References

1. Bucholz RW. Rockwood and Green's fractures in adults. 7th ed. Philadelphia: Wolters Kluwer Health/ Lippincott Williams & Wilkins; 2012.
2. Galley IJ, Watts AC, Bain GI. The anatomic relationship of the axillary artery and vein to the clavicle: a cadaveric study. J Shoulder Elbow Surg. 2009;18(5):e21–5. doi:10.1016/j.jse.2009.01.021.
3. Huang JI, Toogood P, Chen MR, Wilber JH, Cooperman DR. Clavicular anatomy and the applicability of precontoured plates. J Bone Joint Surg Am. 2007;89(10):2260–5. doi:10.2106/JBJS.G.00111.
4. Rockwood Jr CA, Matsen III FA, Wirth MA, Lippitt SB. The shoulder. Philadelphia: Elsevier Health Sciences; 2009.
5. De Palma AF. Degenerative changes in the sterno-clavicular and acromio- clavicular joints in various decades. Springfield: Ch.C.Thomas; 1957.
6. Havet E, Duparc F, Tobenas-Dujardin A-C, Muller J-M, Delas B, Fréger P. Vascular anatomical basis of clavicular non-union. Surg Radiol Anat. 2008;30(1):23–8. doi:10.1007/s00276-007-0278-1.

7. Knudsen FW, Andersen M, Krag C. The arterial supply of the clavicle. Surg Radiol Anat. 1989;11(3):211–4.
8. Crock HV. An atlas of vascular anatomy of the skeleton and spinal cord. London: Dunitz Martin Ltd; 1996.
9. Abbott LC, Lucas DB. The function of the clavicle; its surgical significance. Ann Surg. 1954;140(4):583–99.
10. McKee MD, Wild LM, Schemitsch EH. Midshaft malunions of the clavicle. J Bone Joint Surg Am. 2003;85-A(5):790–7.
11. Throckmorton T, Kuhn JE. Fractures of the medial end of the clavicle. J Shoulder Elbow Surg. 2007;16(1):49–54. doi:10.1016/j.jse.2006.05.010.
12. McKee MD, Pedersen EM, Jones C, et al. Deficits following nonoperative treatment of displaced midshaft clavicular fractures. J Bone Joint Surg Am. 2006;88(1):35–40. doi:10.2106/JBJS.D.02795.
13. Murray IR, Foster CJ, Eros A, Robinson CM. Risk factors for nonunion after nonoperative treatment of displaced midshaft fractures of the clavicle. J Bone Joint Surg. 2013;95(13):1153–8. doi:10.2106/JBJS.K.01275.
14. Lenza M, Buchbinder R, Johnston RV, Belloti JC, Faloppa F. In: Lenza M, editor. Surgical versus conservative interventions for treating fractures of the middle third of the clavicle. Chichester: Wiley; 1996. doi:10.1002/14651858.CD009363.pub2.
15. Banerjee R, Waterman B, Padalecki J, Robertson W. Management of distal clavicle fractures. J Am Acad Orthop Surg. 2011;19(7):392–401.
16. Robinson CM, Cairns DA. Primary nonoperative treatment of displaced lateral fractures of the clavicle. J Bone Joint Surg Am. 2004;86(4):778–82.
17. Rockwood CA, Wilkins KE, Beaty JH, Kasser JR. Rockwood and Wilkins' fractures in children. Philadelphia: Lippincott Williams & Wilkins; 2006.
18. Beals RK, Sauser DD. Nontraumatic disorders of the clavicle. J Am Acad Orthop Surg. 2006;14(4):205–14.
19. Girschick H. Chronic recurrent multifocal osteomyelitis in children. Orphanet Encyclopedia. 2002.
20. Schwarzkopf R, Ishak C, Elman M, Gelber J, Strauss DN, Jazrawi LM. Distal clavicular osteolysis: a review of the literature. Bull NYU Hosp Jt Dis. 2008;66(2):94–101.
21. Jeray K. Acute midshaft clavicle fractures. J Am Acad Orthop Surg. 2007;15:239–48.

Part III

Gleno-Humeral Joint

9 Glenoid Labrum

John Apostolakos, Justin S. Yang, Alexander R. Hoberman, Monica Shoji, Jeffrey H. Weinreb, Andreas Voss, Jessica DiVenere, and Augustus D. Mazzocca

9.1 Introduction

The glenohumeral joint allows more widespread range of motion than any other joint in the body, requiring a complex interplay between passive and active stabilizers in order to provide balanced, synchronous motion [35]. However, this extensive range of motion results in a propensity for decreased stability. While laxity can be a normal finding in the glenohumeral joint affected by a multitude of factors including age, gender, fitness level, and genetics, pathological laxity can ultimately lead to glenohumeral instability, resulting in excessive translation of the humeral head on the glenoid that clinically manifests in pain, weakness, or decreased performance [35]. Recurrent anterior shoulder instability as a result of force to the externally rotated and abducted arm is a common pathology to the glenohumeral joint with consequent treatment options and outcomes reported in the orthopedic literature dating back to the early twentieth century [23, 31, 40, 43].

Clinical instability can result from injury to the glenoid labrum, described in the literature as a triangular segment of fibrocartilaginous tissue that is circumferentially attached to the rim of the glenoid [26, 27, 35, 37]. It provides some passive stability to the humeral head by increasing the depth of the shallow glenoid fossa and serves as a primary attachment site for the glenohumeral ligaments, joint capsule, and long head of the biceps tendon [10, 43]. The labrum also functions to provide nutrition to the glenoid cavity and helps maintain joint lubrication [52]. Injuries to the glenoid labrum commonly occur due to acute trauma such as falling on a flexed abducted arm or overuse injuries such as repetitive overhead use frequently seen in athletes [16]. Most pathologic lesions occur in the anterior inferior portions of the labrum and the superior portions of the labrum [52]. In this chapter, we will discuss the normal clinical anatomy and physiology of the glenoid labrum and surround structures.

9.2 Gross Anatomy of the Labrum and Glenohumeral Ligaments

For the purposes of this chapter, we will discuss the anatomy pertaining to the stability of the glenohumeral joint. Specifically, we will discuss the physical properties of the glenoid labrum, insertion of the long head of the biceps, and the superior, middle, and inferior glenohumeral

J. Apostolakos, BS • J.S. Yang, MD • A.R. Hoberman, BA
M. Shoji, BA • J.H. Weinreb, BS • A. Voss, MD
J. DiVenere, BS • A.D. Mazzocca, MS, MD (✉)
Department of Orthopaedic Surgery, University of Connecticut Health Center, Farmington, CT, USA
e-mail: Mazzocca@uchc.edu

G.I. Bain et al. (eds.), *Normal and Pathological Anatomy of the Shoulder*,
DOI 10.1007/978-3-662-45719-1_9, © ISAKOS 2015

ligaments. Stability to the glenohumeral joint is a complex combination of osseous, ligamentous, and muscular structures that provide both passive and active stabilization which will be discussed in this section.

To begin the description of these structures, we must first mention the glenohumeral joint capsule and its accompanying structures (Fig. 9.1). The capsule is defined as forming superior, middle, and inferior glenohumeral ligaments with their physical descriptions and naming dating back to the early nineteenth century [13, 38, 45]. Understanding of these structures was further advanced in the early twentieth century when scholars began describing the ligament's contribution to the stability of the shoulder joint [11, 40]. In terms of physical description, the capsule is described as a thin capsule with nearly twice the surface area of the humeral head contributing to the great mobility of the joint [37]. The structure contributes to the passive stability of the shoulder and includes folds or thickenings which are termed the glenohumeral ligaments [8, 11, 13, 35, 37, 38, 45]. The capsule arises from the circumference of the glenoid and the bone surrounding it and inserts into the proximal anatomic neck of the humerus. The capsule is thickened in certain locations and takes on the appearance of ligaments which have been termed the coracohumeral ligament and anteriorly as the superior, middle, and inferior glenohumeral ligaments [13, 38, 45]. It is important to note that despite the description of these structures, they have been reported to be variable in terms of their existence and specific anatomic location [9]. In describing the physical location of these structures, we will define the most superior portion of the glenoid as the 12 o'clock position, the most inferior aspect of the glenoid as the 6 o'clock position, the most anterior aspect of the glenoid as the 3 o'clock position, and the more posterior aspect of the glenoid as the 9 o'clock position. Note the anterior and posterior reference points are in regards to a right shoulder.

The coracohumeral ligament originates proximally at the dorsolateral aspect of the coracoid as a fibrous brand extending superiorly over the head of the humerus before it attaches to the greater tuberosity [5, 37]. This ligament then blends in with the superior glenohumeral ligament (SGHL) inferiorly. The SGHL (arthroscopic view, Figs. 9.2 and 9.3) originates just anterior to the insertion of the long head of the biceps at the superior portion of the glenoid. The SGHL is described as originating between the 12 and 2 o'clock positions on the glenoid, just anterior to the insertion of the long head of the biceps, and runs inferiorly and laterally where it then inserts into the humerus at the fovea capitis which lies slightly superior to the lesser

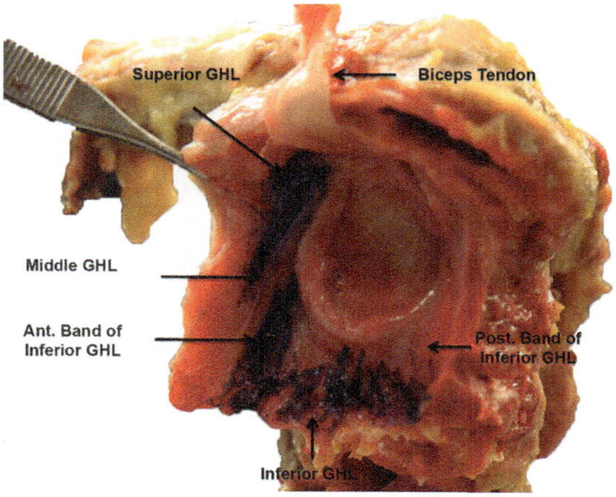

Fig. 9.1 Left cadaveric shoulder demonstrating the intra-articular ligaments

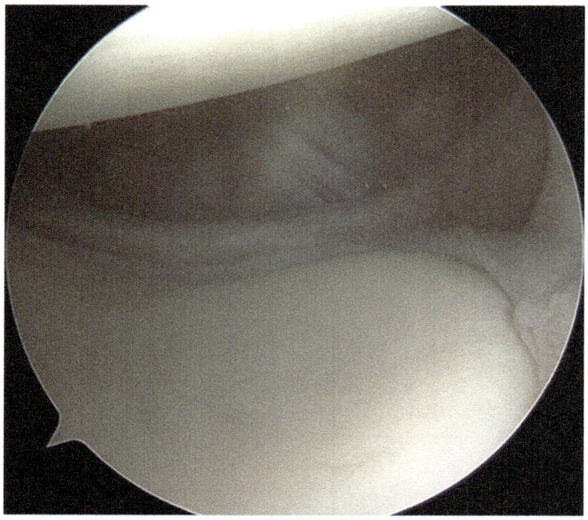

Fig. 9.2 Arthroscopic image of anterior-superior and anterior-inferior labrum with anterior capsule structures. Patient is in the lateral position, image taken from the posterior portal

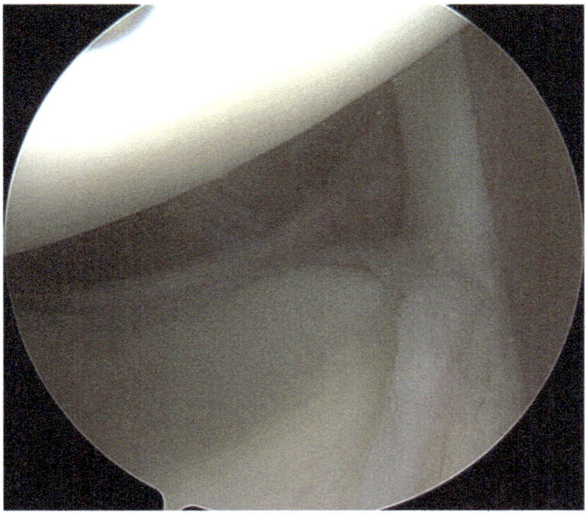

Fig. 9.3 Arthroscopic image of the superior biceps labral complex showing posterior and anterior superior labrum. Patient is in the lateral position, image taken from the posterior portal

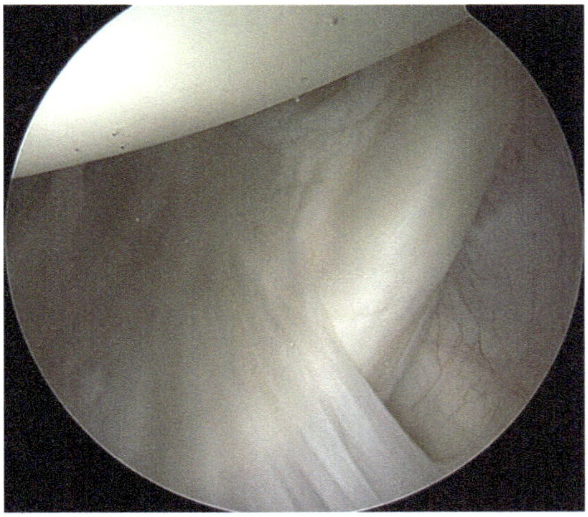

Fig. 9.4 Arthroscopic image of the middle glenohumeral ligament coming off at 90° angle to the subscapularis tendon. Patient is in the lateral position, image taken from the posterior portal

tuberosity [8, 45]. As we move counterclockwise around the glenoid, the middle glenohumeral ligament (MGHL, arthroscopic view Fig. 9.4) originates just inferior to the SGHL [5, 8]. It then inserts medial to the lesser tuberosity on the humerus. The MGHL often follows the superior marginal of the subscapularis tendon; however, it has also been described as the most variable ligament of the labral structures of the shoulder [5, 8]. Anatomically, the MGHL may appear to be flush to the SGHL, as a distinct structure that arises from the inferior aspect of the SGHL as described above, or may not even be clearly identified. The inferior segment of the capsule, the inferior glenohumeral ligament has been identified as the broad inferior glenohumeral ligament and is further broken down into an anterior band which originates from the 3 to 5 o'clock position, posterior band which originates from the 7 to 9 o'clock position which inserts onto the posterior articular margin of the humerus, and the axillary pouch in between [5, 8, 35].

The labrum of the glenoid circumferentially surrounds the outer portion of the glenoid and is the point of insertion of all the glenohumeral ligamentous structures described above. It acts to deepen the socket on average 9 mm in the superior-inferior dimension and 5 mm in the anteroposterior plane which acts to add stability to the joint [35]. The only caveat to these insertions is in regard to the IGHL which has two insertion sites. One insertion is into the anteroinferior aspect of the labrum with a second insertion into the anterior neck of the glenoid [8]. An additional structure that inserts into the superior aspect of the glenoid, just superior to the SGHL, is the long head of the biceps tendon. The biceps tendon originates distally at the radial tuberosity. It course up the humeral shaft in the bicipital groove located between the great and lesser tuberosities. The tendon finally inserts into the superior glenoid and is held in place by the coracohumeral ligament superiorly (arthroscopic view, Fig. 9.5) and the superior glenohumeral ligament inferiorly [46, 48]. This anatomic description has been termed the biceps pulley for the long head of the biceps.

The glenoid is a pear-shaped shallow socket with a wider inferior half [35, 37]. This structure is tilted anteriorly therefore making the joint more susceptible to anterior dislocation [37]. Specifically, the physical features of the glenoid bone are often variable but have been reported as 39 mm (±3.7 mm; range 30–48 mm) in the superior-inferior dimension, 29 mm (±3.1 mm; range 21–35 mm) in the anteroposterior dimension of the lower half, and 23 mm (±2.7 mm; range 18–30 mm) in the anteroposterior dimension of

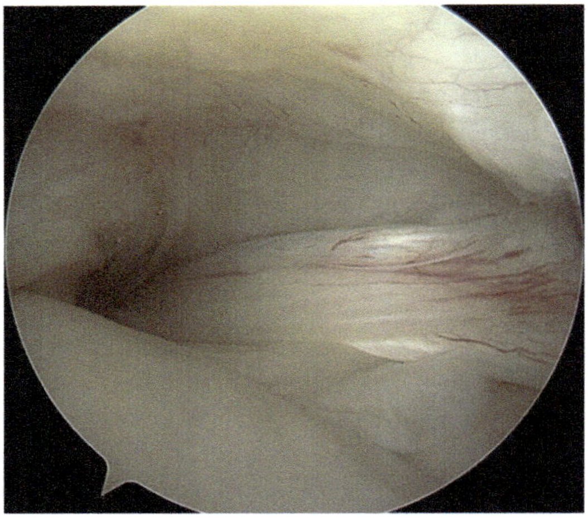

Fig. 9.5 Arthroscopic image of the biceps sling. Patient is in the lateral position, image taken with 70° scope in the posterior portal

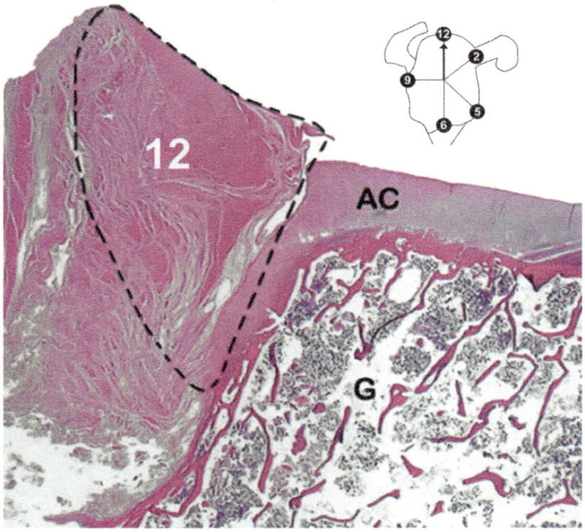

Fig. 9.6 Histological axial section at 12 o'clock. *AC* articular cartilage, *G* glenoid cancellous bone, *L* labrum. Note the labral attachment is off the face of the glenoid

the upper half with the fossa slightly inclined and retroverted [8, 35]. The variation of the size and angle of the glenoid is part of what makes pathology of this bone challenging to treat.

9.3 Histological Anatomy of the Labrum

Histology has provided a greater insight into the anatomic detail of the glenoid and labrum. At the 10–12 o'clock position, the labrum is positioned off the rim of the glenoid and has a concave surface to articulate with the humeral head [2]. There is also a synovial-lined recess between the labrum and the supraglenoid tubercle (approximately 6 mm off glenoid face), allowing the superior labrum to be relatively mobile (Fig. 9.6). The stability of the superior labrum and the glenohumeral joint is further enhanced by the biceps tendon (dynamic stability) and SGHL (static stability) [2, 12].

In contrast, the inferior labrum overlaps the glenoid face by 4 mm, with a convex rounded cross section which increases the depth by up to 50 % ([2, 18, 24] (Figs. 9.7 and 9.8). The inferior labrum is adherent to the articular cartilage and has a rigid bony foundation which prevents mobility of the labrum. The inferior labrum provides a bumper effect and is a fixed organ designed for compression. This anterior inferior labrum is critical to stability and is the common site of the Bankart labral tear seen in traumatic instability (Fig. 9.9).

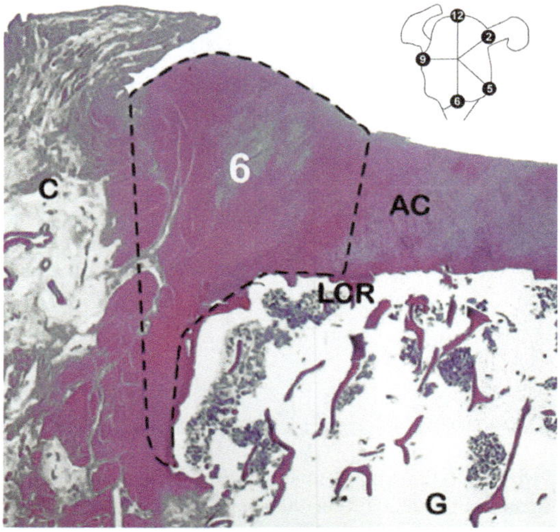

Fig. 9.7 Histological axial section at 6 o'clock. *AC* articular cartilage, *C* capsule, *G* glenoid cancellous bone, *L* labrum, *LCR* internal labral circumferential ridge. Note the bumper effect of the labrum, which is firmly attached on the face of the glenoid

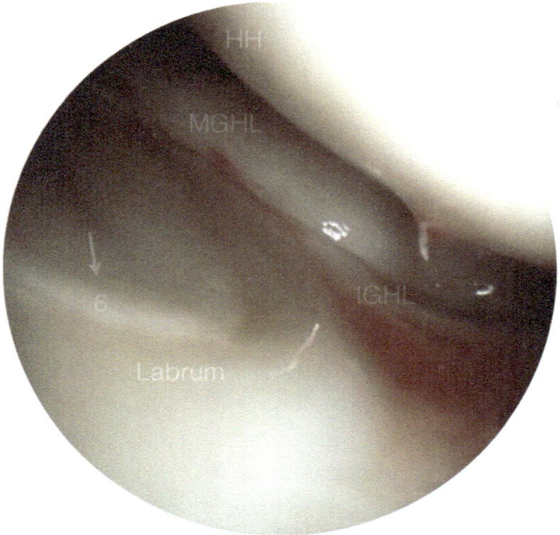

Fig. 9.8 Arthroscopic image of intact inferior labrum. *HH* humeral head, *MGHL* medial glenohumeral ligament, *IGHL* inferior glenohumeral ligament, and "6" correlating to the 6 o'clock position of the labrum

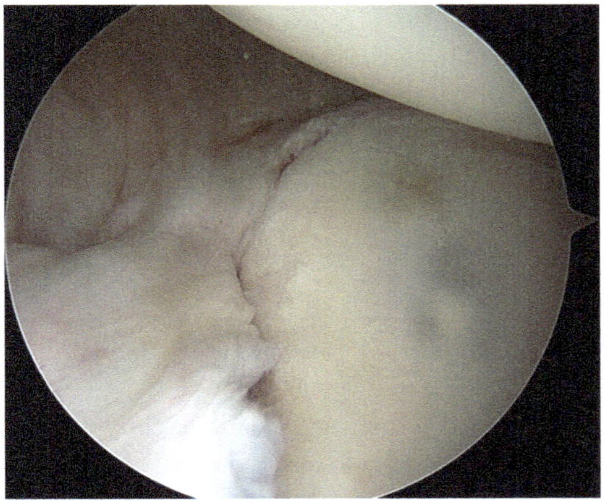

Fig. 9.9 Arthroscopic image of the torn anterior band of the inferior glenohumeral ligament. Patient is in the lateral position, image taken from the superior portal

There are three layers in the labrum: a thin mesh-like superficial layer, a stratified middle layer, and a circumferential deep layer [29]. The fibers that merge with the articular cartilage and capsule are radially orientated [2]. The radial collagen fibers at the superior labrum–articular interface are less densely packed allowing greater mobility of the labrum. In addition to the dense collagen fibers, there are chondrocytes and elastic fibers [24, 26, 27, 29]. With aging, there is a loss of chondrocytes [34].

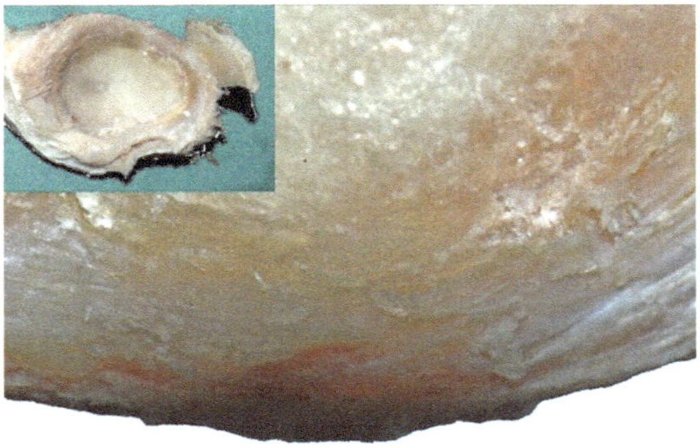

Fig. 9.10 The labrum has been stripped from the glenoid face to demonstrate circumferential striations from the Sharpey's fiber insertion

The labrum and bone interface is similar to a tendon insertion with a narrow zone of transition through uncalcified fibrocartilage, to calcified fibrocartilage incorporating Sharpey's fibers, to bone (Fig. 9.10). The labrum–capsule interface merges imperceptibly, which explains why the labral tears often occur between the labrum and bone. They are both composed of fibrous tissue with the labrum being much denser.

The biceps tendon fibers pass directly through the labrum to the bony insertion via the shortest possible route. The biceps tendon inserts both into the supraglenoid tubercle over a wide area (6 mm coronal section) and to the superior labrum. Approximately two thirds of the fibers insert into the supraglenoid tubercle and the remaining fibers attach to the superior labrum.

9.4 Functional Anatomy

The stability of the glenohumeral joint relies on a combination of muscles, bones, and other structures to maintain its integrity. Passive stability to the joint is provided by the labrum, capsule, and glenohumeral ligaments in normal shoulder function [11, 40]. The glenoid labrum is a fibrocartilaginous structure that encircles the edge of the glenoid fossa and acts to simultaneously protect the outer edges of the glenoid while also deepening the pocket to protect against dislocation [25–27, 32]. This labrum is continuous with the capsule, a thin structure that surrounds the joint by attaching to the glenoid medially and around the anatomic neck of the humerus laterally. The capsule allows for extensive range of motion. By itself, the thin, loose-fitting structure of the capsule does not offer much stability; however, the capsule is reinforced by the glenohumeral ligaments, which provide critical static restraints to dislocation [11, 40]. There are three glenohumeral ligaments that run within the capsule to provide support, all of which were described in the physical anatomy section. The functional significance of these structures will be described in detail within this section.

The superior glenohumeral ligament prevents inferior translation of the adducted shoulder and is an important restraint up to 50° abduction and during external rotation [5, 47]. The middle glenohumeral ligament limits lateral rotation from 45° to 75° abduction as well as supports the joint anteriorly [49]. The inferior glenohumeral ligament complex, which is composed of the anterior band, axillary pouch, and posterior band, is a vital stabilizer for the glenohumeral joint [30]. The anterior band is responsible for restraining anterior and inferior translation of the humerus [41, 45]. Support by the anterior band is most effective at increased (greater than 75°) abduction. This structure plays an important role in sports by preventing anterior humeral head migration when the arm is abducted to 90° and externally rotated. This is evident in activities such as freestyle stroke, tennis serve motion, and any overhead arm position [30]. Alternatively, the posterior band provides posterior stability primarily when the arm is in flexion and internal rotation [5].

Dynamic muscle function also acts to promote stability to the glenohumeral joint. Specifically, muscles of greatest importance include the four rotator cuff muscles, the deltoid, pectoral muscles, latissismus dorsi, and the teres minor [37]. The basis for dynamic stabilization is simply the idea that instability is attributed to end-range positions that place the shoulder its most vulnerable positions. An example of this is the apprehensive positioning of abduction and external rotation, which puts the shoulder at risk of anterior dislocation [22]. Therefore, understanding the muscles surrounding the shoulder is crucial to proper treatment. The concavity-compression mechanism is the theory that muscular activity of the shoulder provides stability by compressing the humeral head to the glenoid surface [17, 20, 21, 33]. This idea of dynamic compression is especially important during times of motion when the capsular tissue is lax making the muscles the primary form of stability during these activities [22]. Of significant importance are the rotator cuff muscles which have been reported in cadaveric studies to increase stability to the glenohumeral joint and consequently decrease humeral head translation [4, 19, 50].

Another structure associated with the glenohumeral joint is the long head of the biceps tendon (LHBT). The function of the LHBT remains a topic of debate. Some investigators believe that this is a vestigial structure, while some studies believe the tendon plays a small role in abduction of the shoulder [1]. In addition, studies have shown that the LHBT aids in medial rotation and flexion of the shoulder as well as stability of the glenohumeral joint in multiple directions [14, 15]. One way that this is thought to be achieved is by reducing the stress on the inferior glenohumeral ligament [36].

9.5 Imaging

The glenoid labrum can be an inherently difficult location to evaluate with magnetic resonance imaging (MRI). The best images are taken while the shoulder is in a neutral or externally rotated position as internal rotation of the shoulder may result in labral or anteroinferior capsule redundancy that can obscure labral tears [6]. Consecutive images in all planes should be reviewed to enable the most accurate reading and MR arthrography of the glenohumeral joint may also be employed [28]. Initially, the labrum was thought to be of low signal density in all MRI variants [52]. However, more recent evaluations have shown that at least one third of all normal labrum MRIs demonstrate high globular or linear signal intensity [28]. The magic angle effect, an artifact caused the structure of the articular cartilage collagen fibers, their orientation in the magnetic field, and the water–proteoglycan interaction that amplifies the prevailing orientation of the collagen fiber network, may also cause a signal with increased density [51]. An intermediate signal intensity may correlate with the transitional zone of fibrocartilage and should also not be mistaken as a labral tear [52].

Traditionally, three variants of the glenoid labrum have been described. The triangular type, which is the most common variant, is attached to the glenoid rim. The meniscoid type extends onto the glenoid articular surface and the bumper type is characterized by fibrous tissue outgrowth [7]. Several common anatomical variants can easily be misinterpreted as a labral tear. First, the space between the labrum and the anterior aspect of the middle glenohumeral ligament may be misinterpreted as a labral tear. Anatomical variations have been described in 13 % of anterior and anterosuperior labrum studies [44]. These anatomical variants may be imaged as linear high-density signals situated between the labrum and glenoid fossa [28]. A sublabral sulcus or recess is a common variant seen in as high as 73 % of cases [39]. The sublabral sulcus is described as a space between the superior labrum and glenoid fossa anterior to the biceps tendon insertion and may be continuous with the sublabral foramen. The sublabral sulcus may be determined by its location, contour, and orientation [28]. A thickened middle glenohumeral ligament may be associated with an absent anterosuperior labrum and is seen in 1.5 % of patients. This abnormality is known as the Buford complex [42]. The Buford complex may be misread as a sublabral foramen, detachment of the anterosuperior labrum, or a bucket-handle tear of the labrum on axial images [28].

Although not pathological, normal variants should be noted on MRI. These variants have been suggested to be associated with increased risk of labral pathology, may occur simultaneously with true labrum pathology, and unnecessary repair may result in worse outcomes [3, 52]. Normal anatomy of the glenoid labrum is demonstrated on T2 MRI scans in Figs. 9.11, 9.12, and 9.13.

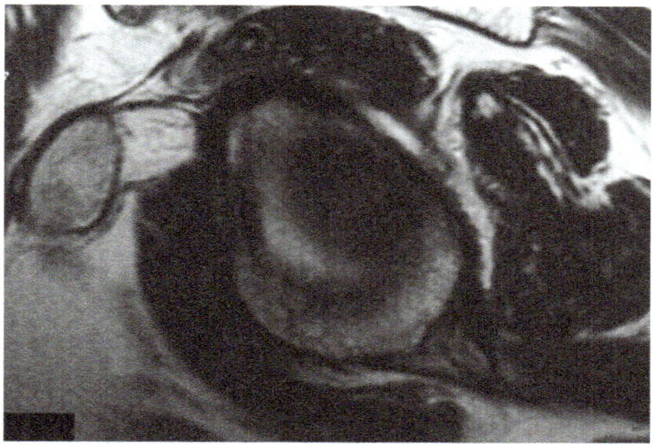

Fig. 9.11 T2 MRI scan sagittal oblique view of left shoulder with labrum intact

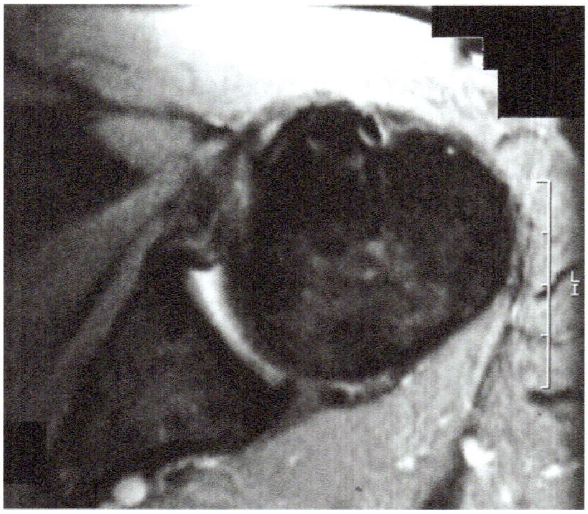

Fig. 9.12 T2 MRI scan axial view of left shoulder with labrum intact

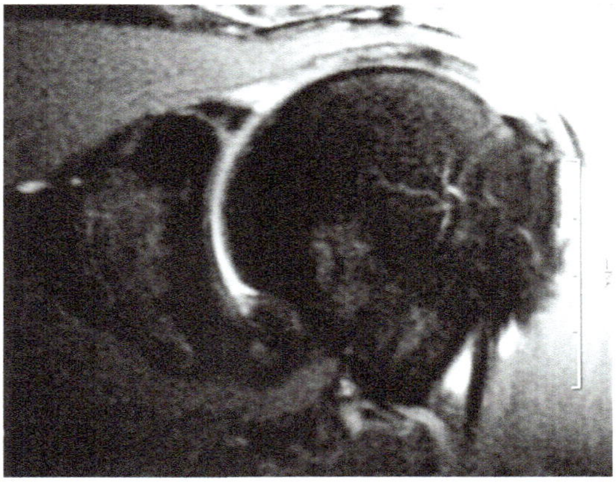

Fig. 9.13 T2 MRI scan coronal view of left shoulder with labrum intact

9.6 Surgical Significance

To fully mobilize the labrum during repair of a Bankart lesion, the surgeon should release the labral complex medially 7–8 mm from the glenoid rim up to the external capsular circumferential ridge [2]. To create an anatomic repair, the anteroinferior labrum should be repaired 4 mm onto the glenoid face to restore the normal bumper effect [2]. The superior labrum should be repaired off the glenoid face to restore the normal position of the labrum [2].

9.7 Summary

The stability of the glenohumeral joint is complex in nature and is determined by several structures including the labrum, shoulder capsule, and the superior, middle, and inferior glenohumeral ligaments as well as the dynamic muscular stabilizers. Due to the inherent flexibility, rates of dislocation in the shoulder joint are relatively common, especially in the younger athletic patient population. Additionally, high rates of variability pertaining to the physical anatomy of the glenohumeral joint's ligamentous structures, such as the medial glenohumeral ligament as well as the diameter and angle of the glenoid fossa, make reconstruction of these soft tissue and bony structures challenging.

References

1. Ahrens PM, Boileau P. The long head of biceps and associated tendinopathy. J Bone Joint Surg Br. 2007;89(8):1001–9.
2. Bain GI, Galley IJ, Singh C, Carter C, Eng K. Anatomic study of the superior glenoid labrum. Clin Anat. 2013;26:367–76.
3. Bents RT, Skeete KD. The correlation of the Buford complex and SLAP lesions. J Shoulder Elbow Surg. 2005;14(6):565–9.
4. Blasier RB, Guldberg RE, Rothman ED. Anterior shoulder stability: contributions of rotator cuff forces and the capsular ligaments in a cadaver model. J Shoulder Elbow Surg. 1992;1(3):140–50. PubMed PMID: 12072743.
5. Burkart AC, Debski RE. Anatomy and function of the glenohumeral ligaments in anterior shoulder instability. Clin Orthop Relat Res. 2002;(400):32–9.
6. Carroll KW, Helms CA. Magnetic resonance imaging of the shoulder: a review of potential sources of diagnostic errors. Skeletal Radiol. 2002;31(7):373–83.
7. Davidson PA, Rivenburgh DW. Mobile superior glenoid labrum: a normal variant or pathologic condition? Am J Sports Med. 2004;32(4):962–6.
8. DeLee J, Drez Jr D, Miller M. Orthopaedic sports medicine: principles and practice. Philadelphia: Saunders Elsevier; 2010.
9. DePalma A, Callery G, Bennett G. Variational anatomy and degenerative lesions of the shoulder joint. In: Instructional course lectures of the American Academy of Orthopaedic Surgeons. 1949. p. 255–80.
10. Dunham KS, Bencardino JT, Rokito AS. Anatomic variants and pitfalls of the labrum, glenoid cartilage, and glenohumeral ligaments. Magn Reson Imaging Clin N Am. 2012;20(2):213–28, x.
11. Dwight T, Hamann C, McMurrich J, Piersol G, White J. Human anatomy, including structure and development and practical considerations. Philadelphia: J.B. Lippincott Company; 1907.
12. Elser F, Braun S, Dewing CB, Giphart JE, Millett PJ. Anatomy, function, injuries, and treatment of the long head of the biceps brachii tendon. Arthroscopy. 2011;27:581–92.
13. Flood V. Discovery of a new ligament of the shoulder joint. Lancet. 1829–1830;13:672–3.
14. Gaskill TR, Braun S, Millett PJ. Multimedia article. The rotator interval: pathology and management. Arthroscopy. 2011;27(4):556–67.
15. Glousman R, Jobe F, Tibone J, Moynes D, Antonelli D, Perry J. Dynamic electromyographic analysis of the throwing shoulder with glenohumeral instability. J Bone Joint Surg Am. 1988;70(2):220–6.
16. Gottschalk MB, Karas SG, Ghattas TN, Burdette R. Subpectoral biceps tenodesis for the treatment of type II and IV superior labral anterior and posterior lesions. Am J Sports Med. 2014;42(9):2128–35.
17. Howell SM, Galinat BJ, Renzi AJ, Marone PJ. Normal and abnormal mechanics of the glenohumeral joint in the horizontal plane. J Bone Joint Surg Am. 1988;70(2):227–32.
18. Howell SM, Galinat BJ. The glenoid-labral socket. A constrained articular surface. Clinical orthopaedics and related research. 1989(243):122–25.
19. Itoi E, Newman SR, Kuechle DK, Morrey BF, An KN. Dynamic anterior stabilisers of the shoulder with the arm in abduction. J Bone Joint Surg Br. 1994;76(5):834–6.
20. Karduna AR, Williams GR, Williams JL, Iannotti JP. Kinematics of the glenohumeral joint: influences of muscle forces, ligamentous constraints, and articular geometry. J Orthop Res. 1996;14(6):986–93.
21. Kelkar R, Wang VM, Flatow EL, Newton PM, Ateshian GA, Bigliani LU, Pawluk RJ, Mow VC. Glenohumeral mechanics: a study of articular geometry, contact, and kinematics. J Shoulder Elbow Surg. 2001;10(1):73–84.

22. Labriola JE, Lee TQ, Debski RE, McMahon PJ. Stability and instability of the glenohumeral joint: the role of shoulder muscles. J Shoulder Elbow Surg. 2005;14(1):32S–8.
23. Mangels Jr M, Robinson CA. Modified Nicola operation for correction of recurring anterior dislocations of the shoulder. South Med J. 1947;40(9):784–6.
24. Mileski RA, Snyder SJ. Superior labral lesions in the shoulder: pathoanatomy and surgical management. J Am Acad Orthop Surg. 1998;6:121–31.
25. Moore K, Dalley A, Agur A. Clinically oriented anatomy. Philadelphia: Wolters Kluwer/Lippincott Williams & Wilkins; 2013.
26. Moseley H, Övergaard B. The anterior capsular mechanism in recurrent anterior dislocation of the shoulder. J Bone Joint Surg Br. 1962;44B(4):913–26.
27. Moseley H, Övergaard B. The anterior capsular mechanism in recurrent anterior dislocation of the shoulder. J Bone Joint Surg Br. 1962;44(4):913–27.
28. Motamedi D, Everist BM, Mahanty SR, Steinbach LS. Pitfalls in shoulder MRI: part 1–normal anatomy and anatomic variants. AJR Am J Roentgenol. 2014;203(3):501–7.
29. Nishida K, Hashizume H, Toda K, Inoue H. Histologic and scanning electron microscopic study of the glenoid labrum. J Shoulder Elbow Surg. 1996;5:132–8.
30. O'Brien SJ, Neves MC, Arnoczky SP, Rozbruck SR, Dicarlo EF, Warren RF, Schwartz R, Wickiewicz TL. The anatomy and histology of the inferior glenohumeral ligament complex of the shoulder. Am J Sports Med. 1990;18(5):449–56.
31. O'Connor SJ, Klaaren HE. Recurrent anterior superior dislocation of the glenohumeral joint. J Bone Joint Surg Am. 1956;38-A(4):889–90.
32. Perry J. Anatomy and biomechanics of the shoulder in throwing, swimming, gymnastics, and tennis. Clin Sports Med. 1983;2(2):247–70.
33. Poppen NK, Walker PS. Normal and abnormal motion of the shoulder. J Bone Joint Surg Am. 1976;58(2):195–201.
34. Prodromos CC, Ferry JA, Schiller AL, Zarins B. Histological studies of the glenoid labrum from fetal life to old age. J Bone Joint Surg Am. 1990;72:1344–8.
35. Provencher M, Romeo A. Shoulder instability: a comprehensive approach. Philadelphia: Saunders Elsevier; 2012.
36. Rodosky MW, Harner CD, Fu FH. The role of the long head of the biceps muscle and superior glenoid labrum in anterior stability of the shoulder. Am J Sports Med. 1994;22(1):121–30.
37. Rothman RH, Marvel Jr JP, Heppenstall RB. Anatomic considerations in glenohumeral joint. Orthop Clin North Am. 1975;6(2):341–52.
38. Schlemm F. Ueber die Verstarkungsbander am Schultergelenk. Arch Anat. 1853;22:45–8.
39. Smith DK, Chopp TM, Aufdemorte TB, Witkowski EG, Jones RC. Sublabral recess of the superior glenoid labrum: study of cadavers with conventional nonenhanced MR imaging, MR arthrography, anatomic dissection, and limited histologic examination. Radiology. 1996;201(1):251–6.
40. Snodgrass LE. Anterior dislocation of the shoulder-joint by abduction and internal rotation. Ann Surg. 1934;100(3):539–43.
41. Soslowsky LJ, Malicky DM, Blasier RB. Active and passive factors in inferior glenohumeral stabilization: a biomechanical model. J Shoulder Elbow Surg. 1997;6(4):371–9.
42. Stoller DW. MR arthrography of the glenohumeral joint. Radiol Clin North Am. 1997;35(1):97–116.
43. Townley C. The capsular mechanism in recurrent dislocation of the shoulder. J Bone Joint Surg. 1950;32(2):370–80.
44. Tuite MJ, Orwin JF. Anterosuperior labral variants of the shoulder: appearance on gradient-recalled-echo and fast spin-echo MR images. Radiology. 1996;199(2):537–40.
45. Turkel SJ, Panio MW, Marshall JL, Girgis FG. Stabilizing mechanisms preventing anterior dislocation of the glenohumeral joint. J Bone Joint Surg Am. 1981;63(8):1208–17.
46. Walch G, Nove-Josserand L, Levigne C, Renaud E. Tears of the supraspinatus tendon associated with "hidden" lesions of the rotator interval. J Shoulder Elbow Surg. 1994;3(6):353–60.
47. Warner JJ, Deng XH, Warren RF, Torzilli PA. Static capsuloligamentous restraints to superior-inferior translation of the glenohumeral joint. Am J Sports Med. 1992;20(6):675–85.
48. Werner A, Mueller T, Boehm D, Gohlke F. The stabilizing sling for the long head of the biceps tendon in the rotator cuff interval. A histoanatomic study. Am J Sports Med. 2000;28(1):28–31.
49. Wilk K, Reinold MM, Andrews J. The athlete's shoulder. Philadelphia: Churchill Livingstone/Elsevier; 2009.
50. Wuelker N, Korell M, Thren K. Dynamic glenohumeral joint stability. J Shoulder Elbow Surg. 1998;7(1):43–52.
51. Xia Y. Magic-angle effect in magnetic resonance imaging of articular cartilage: a review. Invest Radiol. 2000;35(10):602–21.
52. Zlatkin MB, Sanders TG. Magnetic resonance imaging of the glenoid labrum. Radiol Clin North Am. 2013;51(2):279–97.

10 Glenohumeral Capsule and Ligaments

Jiwu Chen and Joideep Phadnis

10.1 Introduction

Arthroscopy has provided us with a deeper understanding of the anatomy and significance of the glenohumeral joint capsule. Four structures have been identified as the main components of the glenohumeral ligament complex: the superior glenohumeral ligament (SGHL), the middle glenohumeral ligament (MGHL), the anterior and posterior inferior glenohumeral ligament (AIGHL, PIGHL), and the labrum (Fig. 10.1a–c) [1–4].

More recently, further anatomic research has led to an appreciation of the variations in these main ligaments and identification of further important ligaments. This chapter focuses on these new findings, and their clinical relevance.

J. Chen, MD, PhD (✉)
Huashan Hospital,
No 12 Wulumuqi Middle Road,
Shanghai 200040, China
e-mail: jeevechan@yahoo.com.cn

J. Phadnis, FRCS (Tr&Orth)
Department of Orthopaedics,
Flinders University of South Australia,
Bedford Park, SA, Australia
e-mail: joideep@doctors.org.uk

10.2 Glenohumeral Ligaments

10.2.1 Incidence of Glenohumeral Ligaments

Table 10.1 outlines how frequently the glenohumeral ligaments have been identified during cadaveric dissection [5, 6]. Of the three main ligaments, the SGHL was identified most frequently (94–98 %), the MGHL least frequently (68–84 %) and the IGHL 75–93 % of the time.

10.2.2 Variable Origins of the Glenohumeral Ligaments

The SGHL usually originates from the supraglenoid tubercle just anterior to the long head of biceps attachment. However, it has also been found to originate from the posterior aspect of the supraglenoid tubercle, the superior labrum, the long head of biceps, the MGHL or a combination of these structures [7]. The SGHL origin has been classified into three types, the most common of which was from the MGHL, long head of biceps tendon and superior labrum [8]. (Table 10.2) spans from the anterior superior glenoid to the lesser tuberosity.

Two common variations of the MGHL origin have been reported. The first is from the labrum and SGHL (43 %) and the second from the labrum alone, independent of the SGHL (57 %) [9–11] (Fig. 10.2).

G.I. Bain et al. (eds.), *Normal and Pathological Anatomy of the Shoulder*,
DOI 10.1007/978-3-662-45719-1_10, © ISAKOS 2015

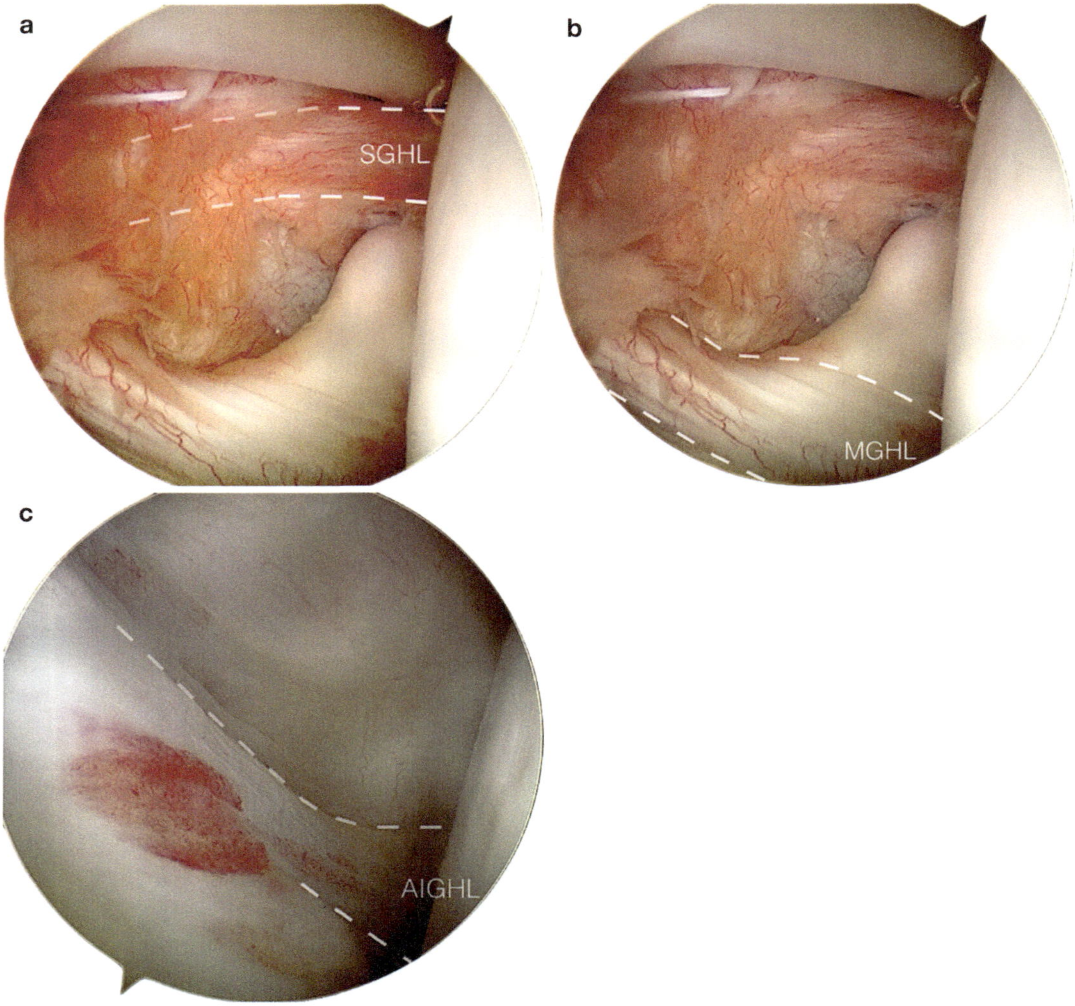

Fig. 10.1 Arthroscopic view of the shoulder ligaments: (**a**) SGHL within the rotator interval. (**b**) MGHL as it crosses the subscapularis tendon. (**c**) AIGHL coursing inferiorly and obliquely

Table 10.1 Incidence of glenohumeral ligaments [5, 6]

Glenohumeral ligament	Incidence (%)
SGHL	94–98
MGHL	63–85
IGHL	75–93

The AIGHL spans from the anterior glenoid and labrum to the lesser tuberosity and the surgical neck of the humerus [1, 2]. The AIGHL most commonly originates from the 3 o'clock position of the glenoid but can be as proximal as 2 o'clock and as distal as 5 o'clock (Table 10.3) [11–14].

Table 10.2 Variation in SGHL origin [8]

Types of SGHL origin	Location	Incidence (%)
A	MGHL, biceps tendon, superior labrum	76
B	Biceps tendon, superior labrum	21
C	Biceps tendon	1

The PIGHL usually arises from the posterior aspect of the glenoid between 7 and 9 o'clock. It is more difficult to appreciate than the AIGHL and is confluent with the thinner posterior capsular structures [14].

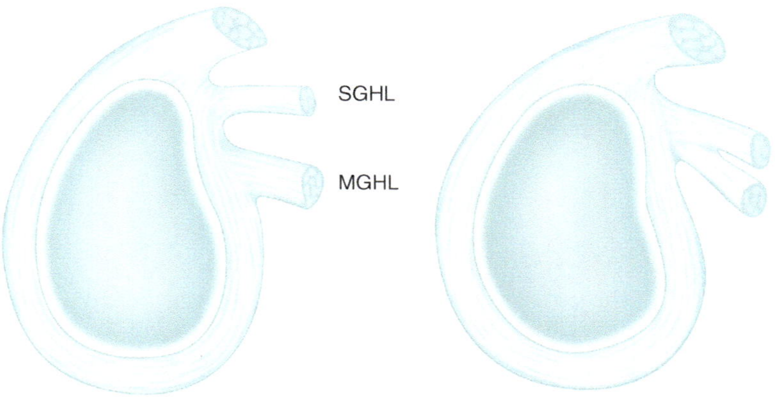

Fig. 10.2 Common variations in the origin of the MGHL

Table 10.3 Variation in AIGHL origin

AIGHL origin	O'Brien et al. [11] (%)	Ide et al. [9] (%)
2 o'clock	18	15
3 o'clock	73	64
4 o'clock	9	15
5 o'clock	0	7

10.2.3 Variations of the Middle Glenohumeral Ligament

The MGHL arises from the upper anterior aspect of the glenoid, runs obliquely and deep to the subscapularis tendon to insert onto the lesser tuberosity. A number of variants have been described including the sublabral foramen (12 %), cord-like MGHL (18 %) and the Buford complex (1–2 %) [1, 2, 9, 15, 16]. With a cord-like MGHL, the medial and lateral edges of the MGHL tend to have a curling shape. The Buford complex (Fig. 10.3a, b) is defined as an absent anterosuperior labrum with a cord-like MGHL, which originates from the superior labrum, and inserts into the humerus. The anterior labrum superior to the glenoid equator is variable in the degree of its attachment and the presence of a sublabral foramen is observed in around 12 % of arthroscopies [17]. This is a localised area where there is no labral attachment to the glenoid and should not be confused with a pathologic lesion such as a Bankart tear which has an irregular edge, associated synovitis, capsular tears and extends more inferiorly [15, 17].

10.2.4 Variations of the Inferior Glenohumeral Ligament

O'Brien et al. described three components of the IGHL: the anterior band (AIGHL), the posterior band (PIGHL), and the axillary pouch (AxIGHL), which is the hammock formed between the two bands.

Itoigawa et al. [12] reported that the mean length of the AIGHL attachment was 11.7 mm. The mean depth was 4.7 mm (1.6 mm on the articular cartilage and 3.0 mm on the glenoid neck) at the 2 o'clock position, 6.7 mm (2.4 and 4.3 mm, respectively) at the 3 o'clock position, 8.4 mm (3.0 and 5.4 mm, respectively) at the 4 o'clock position, and 6.8 mm (2.5 and 4.3 mm, respectively) at the 5 o'clock position.

Histology demonstrates that the AIGHL usually attaches to both the cartilage and bone (88 %), but can be attached to only the bone (12 %).

The humeral insertion of the inferior capsular fibres forms a "V" with a sharp or rounded apex (Fig. 10.3a, b) [12, 14, 18].

10.3 Rotator Interval and Its Contents

The rotator interval (RI) is a triangular space formed by the upper border of subscapularis inferiorly, the anterior edge of the supraspinatus superiorly and the coracoid process medially (Fig. 10.4) [19].

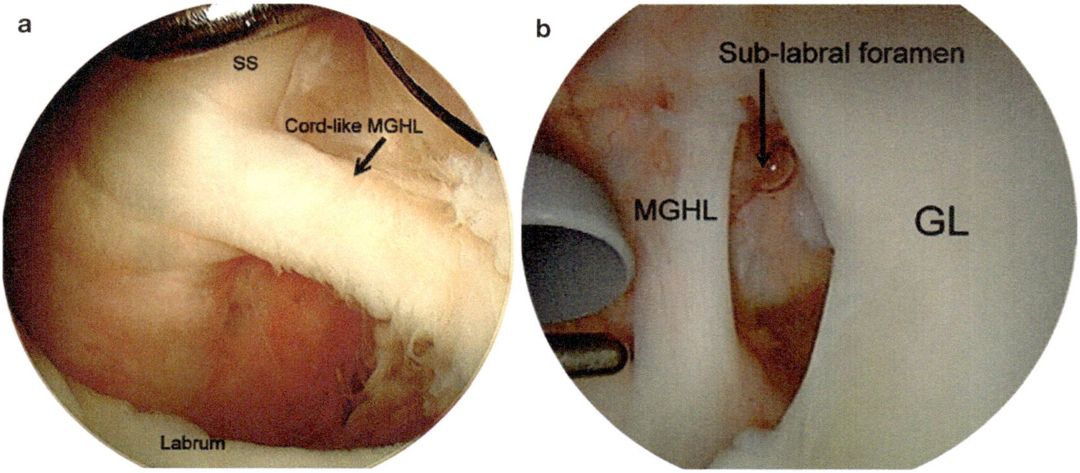

Fig. 10.3 (**a**) A cord like MGHL present in a Buford complex. (**b**) An anterior superior sublabral foramen. *SS* Subscapularis tendon, *GL* Glenoid (Copyright Yin, OCNA [17])

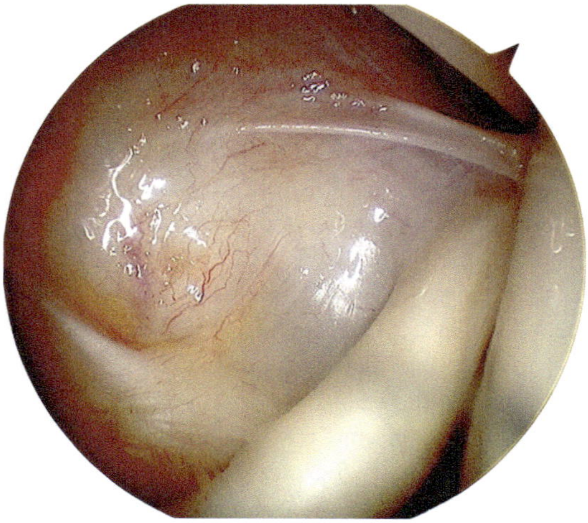

Fig. 10.4 Arthroscopic view of the rotator interval

The height of the RI is variable and ranges at the glenoid border from 2 to 8 mm and at its lateral margin from 13 to 25 mm [19].

From deep to superficial, the structures of the RCI include the SGHL, the interval capsule and the coracohumeral ligament (CHL). The CHL originates from the dorsolateral aspect on the base of the coracoid process, courses through the rotator interval and finally blends its fibres with those of the SGHL [20]. The rotator interval and its contents are thought to play a key role in the pathogenesis of adhesive capsulitis and multidirectional shoulder instability.

10.3.1 Superior Glenohumeral Ligament (SGHL)

The SGHL attaches to the fovea capitis of the humerus on the medial ridge of the intertubercular groove, curves anteriorly and superiorly toward the superior glenoid and partially attaches onto the anterior superior labrum. The remaining fibres of the SGHL insert onto the coracoid process [20].

The SGHL is composed of direct and oblique fibres. The direct fibres arise from the glenoid labrum and run parallel to the biceps tendon towards the lesser tuberosity. The direct fibres insert onto the lesser tuberosity and bridge over the bicipital groove forming the superior part of the transverse humeral ligament. The oblique fibres arise from the supraglenoid tubercle, run over the intra-articular part of the biceps tendon and attach into the humeral semicircular ligament [19–21] (Fig. 10.5).

10.3.2 Coracohumeral Ligament (CHL)

The anatomy of the CHL has been described with significant variation [20]. In general it is accepted that its origin is from the base and posterolateral aspect of the coracoid process although there is debate as to whether the origin is a single structure or has separate parts. As the CHL runs in the rotator interval, it is thought to run in two distinct parts although these parts are named differently

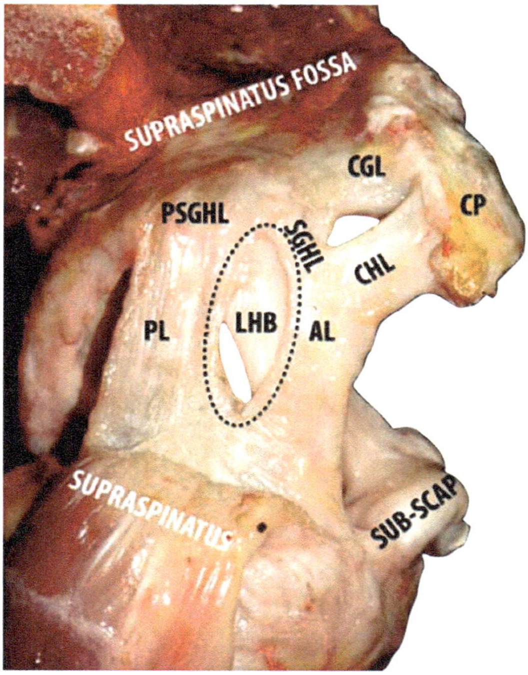

Fig. 10.5 Anatomic specimen showing the SGHL, relationship to the CHL and CGL. *PSGHL* Posterior superior glenohumeral ligament, *PL* Posterior ligament, *CP* coracoid process, *AL* Anterior ligament, *LHB* Long head of biceps tendon (Copyright Di Giacomo [20])

according to author. Regardless of nomenclature, it is apparent that some fibres of the CHL run towards the superior labrum, long head of biceps and capsule whereas the major fibres blend with the supraspinatus tendon, SGHL and rotator cable before inserting onto the greater tuberosity of the humerus where the CHL contributes to the biceps pulley system (Fig. 10.6) [22, 23].

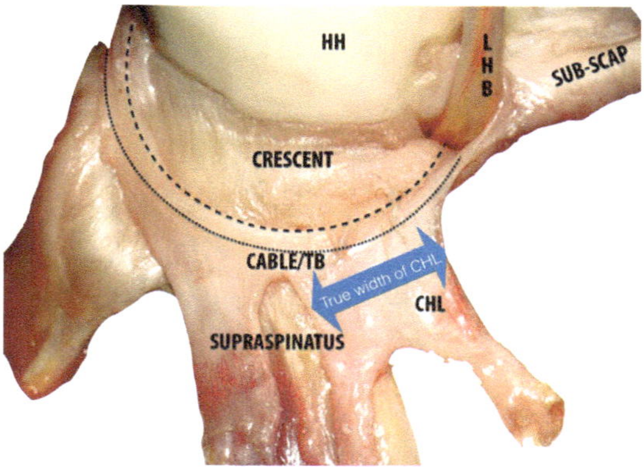

Fig. 10.6 Anatomic specimen showing the true width of the CHL and the rotator cable/crescent. *HH* Humeral head, *LHB* long head of biceps tendon, *CHL* coracohumeral ligament (Copyright Di Giacomo [20])

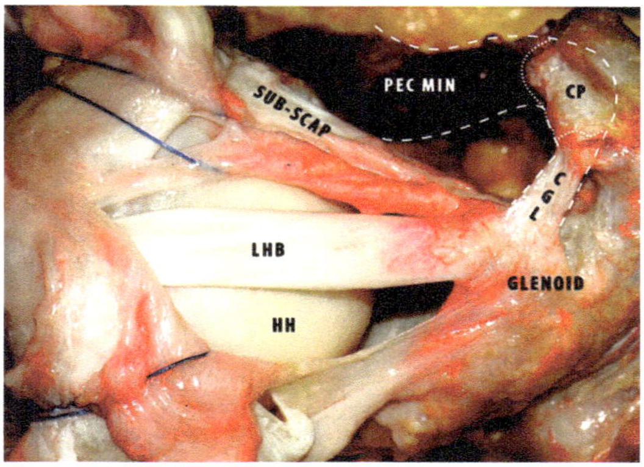

Fig. 10.7 Anatomic specimen demonstrating the CGL and its continuity with the pectoralis minor tendon. *CP* coracoid process, *CGL* coracoglenoid ligament, *LHB* Long head of biceps tendon, *HH* humeral head (Copyright Di Giacomo [20])

10.4 Other Ligaments

10.4.1 Coracoglenoid Ligament (CGL)

The CGL originates on the upper aspect of the coracoid process spanning to the posterior aspect of the superior glenoid tubercle (Fig. 10.7) [23]. The CGL is usually confluent with the medial margin of the coracohumeral ligament and may therefore assist in suspension of the humeral head. The CGL has also been shown to be in continuity with the pectoralis major fibres, which run up and over the coracoid process before merging with the CGL origin [20].

10.4.2 The Posterosuperior Glenohumeral Ligament

This structure originates on the posterior aspect of the glenoid neck medial and posterior to the biceps anchor [19]. It fans laterally and forms the posterior fibres of the rotator cable. This structure and the surrounding posterior superior capsule are variable in its composition. In the majority of cases, this tissue contains a gap, through which the intra-articular structures can be viewed [20]. This

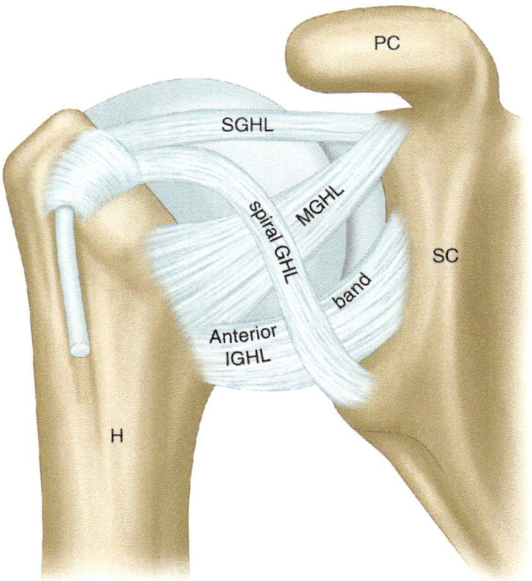

Fig. 10.8 Diagram of the spiral glenohumeral ligament. *PC* coracoid process, *SGHL* superior glenohumeral ligament, *MGHL* Middle glenohumeral ligament, *IGHL* inferiro glenohumeral ligament

reflects the relative thinness of the posterior capsular structures as compared to those anteriorly.

10.4.3 The Rotator Cable (Capsular Semicircular Band)

The rotator cable spans from the greater to the lesser tuberosity and is always present (Fig. 10.6) [21, 23]. Posteriorly, it arises from the greater tuberosity between the infraspinatus and teres minor muscle attachments, and has a contribution from the lateral spanning fibres of the posterosuperior glenohumeral ligament. The cable runs transversely to the longitudinal fibres of the supraspinatus and infraspinatus, until it inserts at the lateral edge of the bicipital groove where it merges with the CHL and transverse humeral ligament, covering the long head of the biceps [20]. In doing so, the rotator cable forms the medial boundary of the rotator cuff crescent, which is the terminal part of the rotator cuff and capsule as they insert onto the greater tuberosity. It is thought that the thick cable works like the cables in a suspension bridge to absorb and distribute load thereby reducing the stresses to the cuff insertion [24]. It may also have a role in limiting propagation of tears and allowing people with small tears within the crescent to maintain power and function.

10.4.4 Spiral Glenohumeral Ligament

The spiral glenohumeral ligament is a distinct and consistently present ligament that attaches to the infraglenoid tubercle and the long head of the triceps attachment. It courses obliquely upwards, crosses the underlying IGHL and establishing a tight connection with the MGHL, before inserting with the subscapularis tendon into the lesser tuberosity (Fig. 10.8) [13, 25–28]. In running transverse to the IGHL, it is likely to re-enforce the anterior inferior labrum and capsule.

References

1. Johnson LL. Arthroscopic surgical anatomy. In: Diagnostic and surgical arthroscopy of the shoulder. Philadelphia: Mosby; 1993. p. 189–230.
2. Detriasac DA, Johnson LL. Arthroscopic shoulder anatomy: pathological and surgical implications. Throughfare: Slack; 1986. p. 36–68.
3. Agur AMR. Grant's atlas of anatomy. 9th ed. Baltimore: Williams & Wilkins; 1991.
4. Tillmann B, Tondury G. Obere extremität. In: Leonhardt H, Tillmann B, Töndury G, Zilles K, editors. Rauber/Kopsch: Lehrbuch und atlas der anatomie des menschen, Bewegungsapparat, vol. 1. New York: Thieme; 1987.
5. DePalma AF. Regional, variational, and surgical anatomy. In: DePalma AF, editor. Surgery of the shoulder. 3rd ed. Philadelphia: JB Lippincott; 1983. p. 35–64.
6. Steinbeck J, Liljenqvist U, Jerosch J. The anatomy of the glenohumeral ligamentous complex and its contribution to anterior shoulder stability. J Shoulder Elbow Surg. 1998;7:122–6.
7. Ogul H, Karaca L, Emre C, et al. Anatomy, variants, and pathologies of the superior glenohumeral ligament: magnetic resonance imaging with three dimensional volumetric interpolated breath- hold examination sequence and conventional magnetic resonance arthrography. Korean J Radiol. 2014;15(4):508–22.
8. DePalma AF. Surgery of the shoulder. 3rd ed. Philadelphia: JB Lippincott; 1973.
9. Ide J, Maeda S, Takagi K. Normal variations of the glenohumeral ligament complex: an anatomic study for arthroscopic bankart repair. Arthroscopy. 2004;20:164–8.
10. O'Brien SJ, Arnoczky SP, Warren RF, et al. Developmental anatomy of the shoulder and anatomy

of the glenohumeral joint. In: Rockwood CA, Matsen FA, editors. The shoulder. Philadelphia: WB Saunders; 1990. p. 1–33.
11. O'Brien SJ, Neves MC, Arnoczky SP, et al. The anatomy and histology of the inferior glenohumeral ligament complex of the shoulder. Am J Sports Med. 1990;18:449–56.
12. Itoigawa Y, Itoi E, Sakoma Y, et al. Attachment of the anteroinferior glenohumeral ligament–labrum complex to the glenoid: an anatomic study. Arthroscopy. 2012;28:1628–33.
13. Delorme D. Die Hemmungsbänder des Schultergelenks und ihre Bedeutung für die Schulterluxationen. Arch Klin Chir. 1910;92:79–101.
14. Pouliart N, Gagey O. Reconciling arthroscopic and anatomic morphology of the humeral insertion of the inferior glenohumeral ligament. Arthroscopy. 2005;21:979–84.
15. Williams MM, Snyder SJ, Buford Jr D. The Buford complex: "Cord-like" middle glenohumeral ligament and absent anterosuperior labrum complex: a normal anatomic capsulolabral variant. Arthroscopy. 1994;10: 241–7.
16. Morgan CD. Anterior shoulder instability; arthroscopic anatomy and pathology. In: Esch JC, editor. Proceedings of the 15th Annual San Diego Meeting, Arthroscopic Surgery of the Shoulder. San Diego: San Diego Shoulder Arthroscopy; 1998. p. 119–23.
17. Yin B, Vella J, Levine WN. Arthroscopic alphabet soup:recognition of normal, normal variants and pathology. Orthop Clin N Am. 2010;41:297–308.
18. Ticker JB, Bigliani LU, Soslowsky LJ, et al. Inferior glenohumeral ligament: geometric and strain-rate dependent properties. J Shoulder Elbow Surg. 1996;5: 269–79.
19. Pouliart N, Somers K, Eid S, et al. Variations in the superior capsuloligamentous complex and description of a new ligament. J Shoulder Elbow Surg. 2007;16: 821–36.
20. Di Giacomo G, Poilart N, Costantini A, De Vita A. Atlas of shoulder anatomy. Milan: Springer-Verlag Italia 2008.
21. Kask K, Poldoja E, Lont T, et al. Anatomy of the superior glenohumeral ligament. J Shoulder Elbow Surg. 2010;19:908–16.
22. Yang H, Tang K, Chen W, et al. An anatomic and histologic study of the coracohumeral ligament. J Shoulder Elbow Surg. 2009;18:305–10.
23. Kolts I, Busch LC, Tomusk H, et al. Anatomy of the coracohumeral and coracoglenoidal ligaments. Ann Anat. 2000;182:563–6.
24. Burkhart SS, Esch JC, Jolson RS. The rotator crescent and rotator cable: an anatomic description of the shoulder's "suspension bride". Arthroscopy. 1993;9: 611–6.
25. Kolts I, Busch LC, Tomusk H, et al. Anatomical composition of the anterior shoulder joint capsule. A cadaver study on 12 glenohumeral joints. Ann Anat. 2001;183:53–9.
26. Merila M, Heliö H, Busch LC, et al. The spiral glenohumeral ligament: an open and arthroscopic anatomy study. Arthroscopy. 2008;11:1271–6.
27. Landsmeer JMF, Meyers KAE. The shoulder region exposed by anatomical dissection. Arch Chir Neerl. 1959;11:274–96.
28. Merila M, Leibecke T, Gehl HB, et al. The anterior glenohumeral joint capsule: macroscopic and MRI anatomy of the fasciculus obliquus or so-called ligamentum glenohumerale spirale. Eur Radiol. 2004;14: 1421–6.

Rotator Cuff Interval

11

Felix H. Savoie, Carina Cohen, and Katherine C. Faust

11.1 Introduction

The rotator interval is the triangular part of the glenohumeral joint between the anterior supraspinatus and superior subscapularis muscle borders, extending medially from the coracoid process and laterally the transverse humeral ligament which connects the greater and lesser tuberosities (Fig. 11.1). The term *rotator interval* originated from the landmark 1970 paper by Neer classifying proximal humerus fractures; he described it as a "ligamentous area between the tendons of the supraspiantus and subscapularis" [1, 2]. The superior and middle glenohumeral ligaments travel across the interval, along with the long head of the biceps tendon and the . The father of shoulder surgery, E. A. Codman, wrote that "The *coraco-humeral* and *gleno-humeral* ligaments should never have been described as entities," describing them as variable portions of the glenohumeral capsule [3]. However, with the advent of arthroscopic surgery and improved magnetic resonance imaging techniques, these portions of the rotator interval have proven consistent and useful anatomic landmarks.

F.H. Savoie (✉) • K.C. Faust
Department of Orthopaedics,
Tulane University, New Orleans, LA, USA

C. Cohen
Ortopedia e Traumatolgia, Cirurgia de Ombroe e Cotovelo, UNIFESP Hospital Israelita Albert Einstein, São Paulo, Brazil

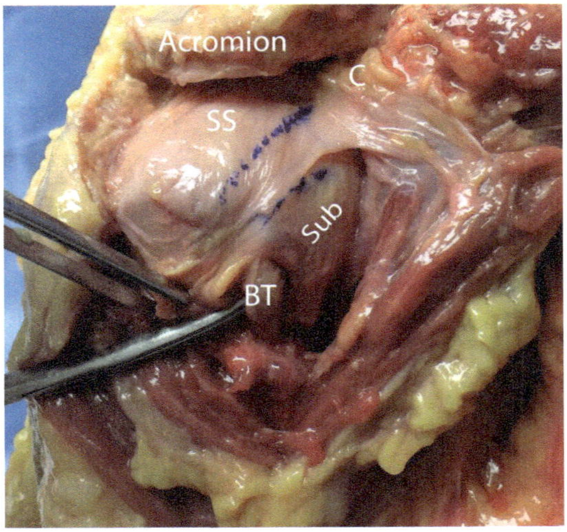

Fig. 11.1 Right shoulder from anterolateral with ink marks outlining the borders of the supraspinatus (*SS*) and subscapularis (*Sub*). The lateral and anterior portions of the deltoid have been detached exposing the anterior part of the humerus. The rotator interval extends from the coracoid (*c*) to lay over the long head of the biceps tendon (*BT*), which has been exposed in its groove

11.2 Rotator Interval Gross Anatomy

During dissection of the glenohumeral joint, the rotator interval is found between the supraspinatus and subscapularis muscles, whose fibers contribute to the interval [4]. The ligamentous tissue making it up consists of the superior glenohumeral ligament, arising from the supraglenoid

G.I. Bain et al. (eds.), *Normal and Pathological Anatomy of the Shoulder*,
DOI 10.1007/978-3-662-45719-1_11, © ISAKOS 2015

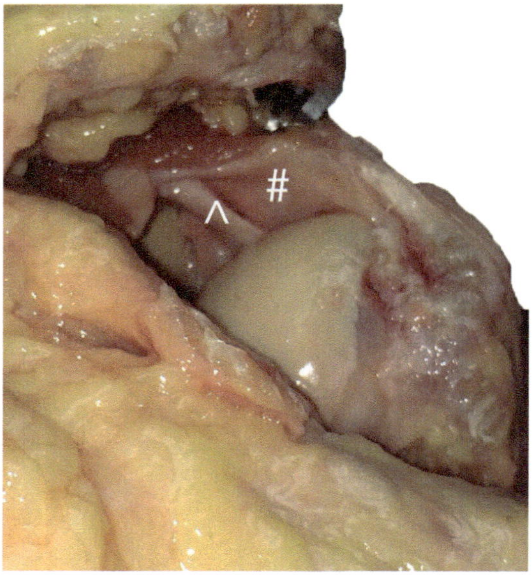

Fig. 11.2 Right shoulder, viewed from posterior to anterior after reflection of the infraspinatus and teres minor, to reveal the (#) and superior glenohumeral ligament (^), distinct entities that resist external rotation and inferior translation of the humeral head

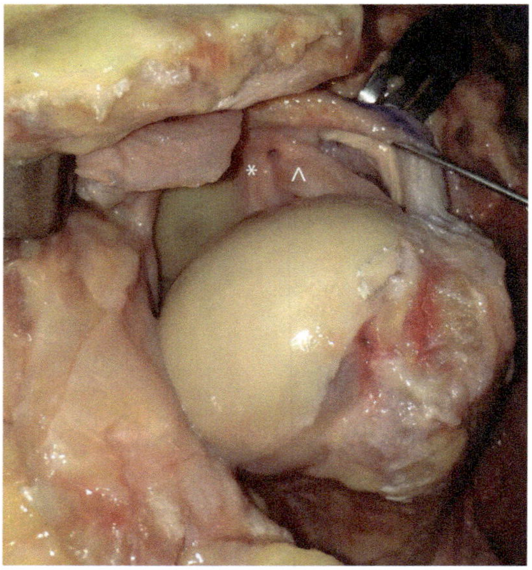

Fig. 11.3 Right shoulder looking from posterior to anterior after reflection of the infraspinatus and teres minor to allow visualization of the rotator interval. The superior glenohumeral ligament (^) can be seen extending toward the long head of the biceps tendon as it enters its groove. The middle glenohumeral ligament (*) can also be seen crossing the rotator interval from this view

tubercle and inserting on the greater tuberosity, and the , discussed in the next section, along with glenohumeral capsular tissue. These ligaments are distinct and blend with the bursal and deep fascia of the glenohumeral joint (Jost et al. [4]) (Fig. 11.2). The superior glenohumeral ligament runs deep to the coracohumeral ligament and can be better seen from inside the joint, while the coracohumeral ligament is better seen from outside (Figs. 11.3 and 11.4). The structure becomes loose with humeral head elevation and internal rotation, taut with humeral head depression and external rotation.

11.3 Coracohumeral Ligament Gross Anatomy

The has been described in Cunningham's and Morris' anatomy textbooks since their early editions as a ligament arising from the root and lateral border of the coracoid process, along with the conjoint tendon, and inserting on the greater tubercle [5, 6]. It consists of two bands, one

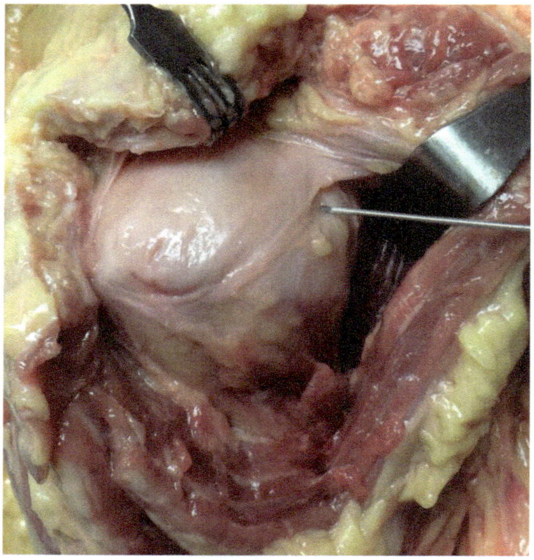

Fig. 11.4 Right shoulder with army-navy retractor subcoracoid and Senn rake around the acromion looking from anterior to posterior. The needle points out the broad as it travels from the coracoid to the greater and lesser tuberosities

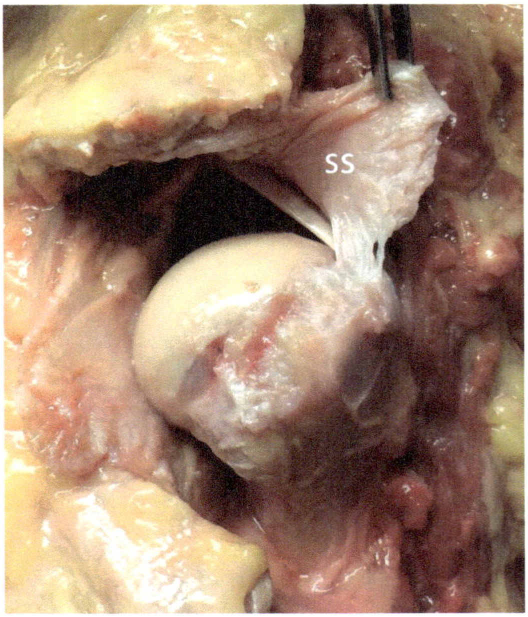

Fig. 11.5 Right shoulder viewed from lateral to medial with the supraspinatus (*ss*) reflected showing fibers of the investing the muscle's insertion and undersurface, along with the posterior sling of the biceps pulley

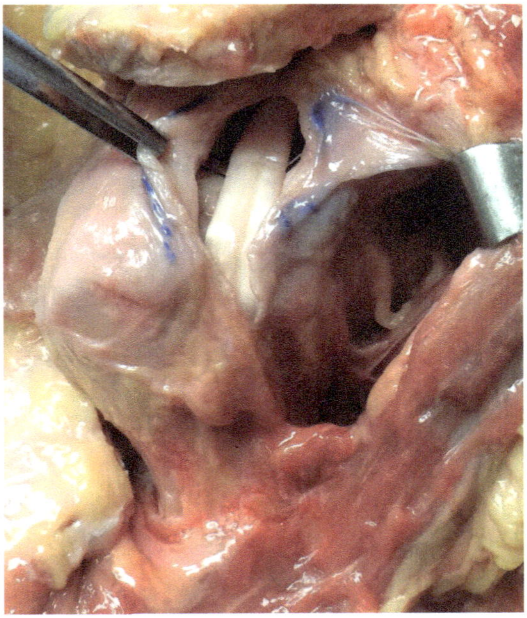

Fig. 11.6 Right shoulder viewed from the anterolateral aspect with the incised along the pulley it forms for the long head of the biceps tendon. The army navy retractor brings the conjoint tendon anteriorly. The forceps are retracting the posterior limb of the coracohumeral ligament, which is stout and fibrous in this region

which forms the anterior sling of the bicipital groove and invests in the subscapularis fascia and the other which forms the posterior sling and invests the supraspinatus fascia [7] (Figs. 11.5 and 11.6). Studies have shown that this anterior portion inserts on the lesser tuberosity [4]. It is this bicipital groove portion of the coracohumeral ligament that blends with the transverse humeral ligament to form the biceps pulley. The coracohumeral ligament also splits to incorporate both the superficial fascia of the muscles on their bursal side and the deep glenohumeral capsule [8, 9] (Fig. 11.7). The deep portion of the posterior sling thickens to form the that runs perpendicular to the supraspinatus and infraspinatus fibers. Morris' *Human Anatomy* describes the ligament from posterior as an "uninterrupted continuation of the capsule" and from anterior as "a fan-shaped prolongation" [5]. The ligament becomes taut with external rotation and lax with internal rotation of the humeral head (Fig. 11.8).

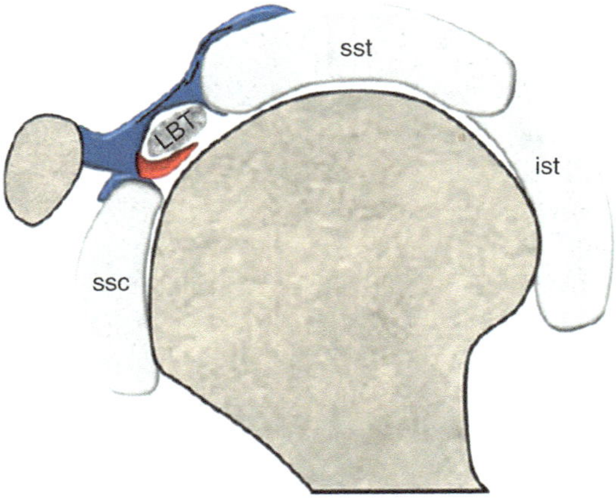

Fig. 11.7 Illustration of sagittal oblique view depicts distal portion of the (*blue*) and superior glenohumeral ligaments (*red*) and long head of the biceps brachii tendon (*LBT*). *ssc* subscapularis tendon, *sst* supraspinatus tendon, *ist* infraspinatus tendon (From Zappia et al. [9])

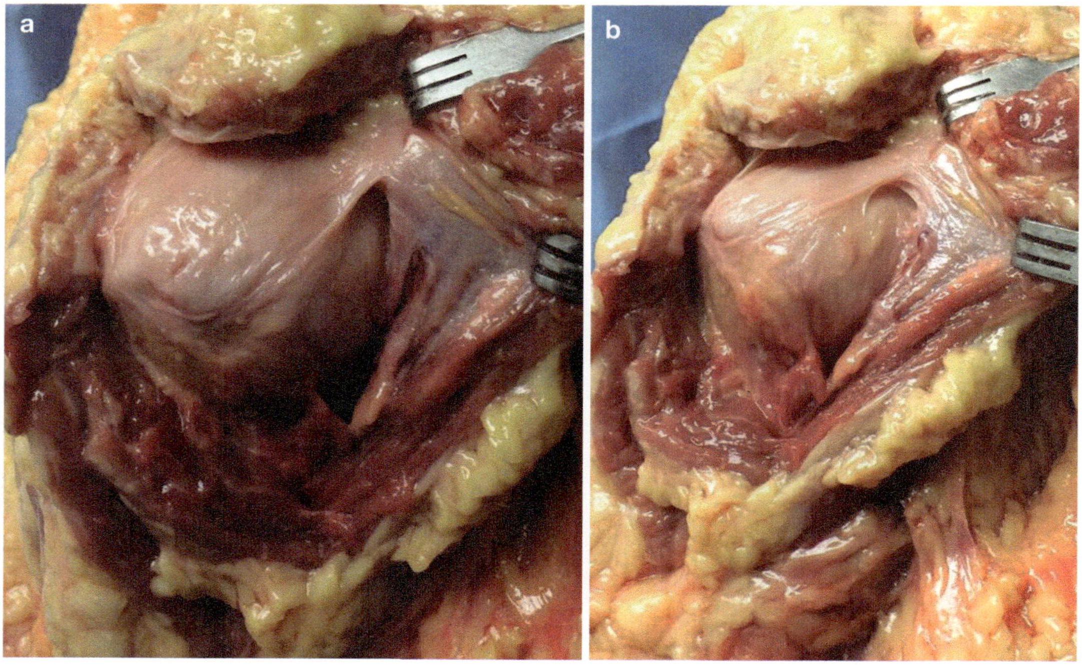

Fig. 11.8 Right shoulder viewed from anterolateral, with the arm at neutral in (**a**) and externally rotated in (**b**) showing the rotator interval becoming relatively taut in external rotation

11.4 Arthroscopic Appearance of the Rotator Interval and Coracohumeral Ligament

Arthroscopy of the glenohumeral joint reveals a majority of the rotator interval (Fig. 11.9a–d). The superior border of the subscapularis is readily visualized as the anterior border. Less well seen in most cases is the anterior edge of the supraspinatus. The superior glenohumeral ligament is a relatively large structure attached to the glenoid labrum just anterior to the biceps tendon (Fig. 11.10). The middle glenohumeral ligament can be seen traversing the rotator interval area but is not considered to be a part of the interval.

Visualizing the interval area subacromially demonstrates the borders of the tendons and the thick, well-formed coracohumeral ligament that lies between the coracoid and the supraspinatus tendon, with an extension laterally to the transverse humeral ligament where it crosses and stabilizes the long head of the biceps tendon.

11.5 Rotator Interval and Coracohumeral Ligament Pathologies

disorders can be grouped into three major groups: adhesive capsulitis, glenohumeral instability, and hidden lesions. As the rotator interval structures resist inferior translation and external rotation, contracture will further limit these motions, as is seen in adhesive capsulitis and tested by the inferior glide test (Fig. 11.11). Arthroscopic images of this pathology show synovitis and contracture (Figs. 11.12, 11.13, and 11.14). Surgical management of adhesive capsulitis invariably involves release of the rotator interval, especially the (Fig. 11.15). Conversely, laxity will lead to excessive motion and glenohumeral instability (Fig. 11.16); this pathology is evident in patients with multidirectional instability who have a persistent sulcus sign with external rotation of the arm. In these cases, arthroscopic examination will show a lax rotator interval (Fig. 11.17).

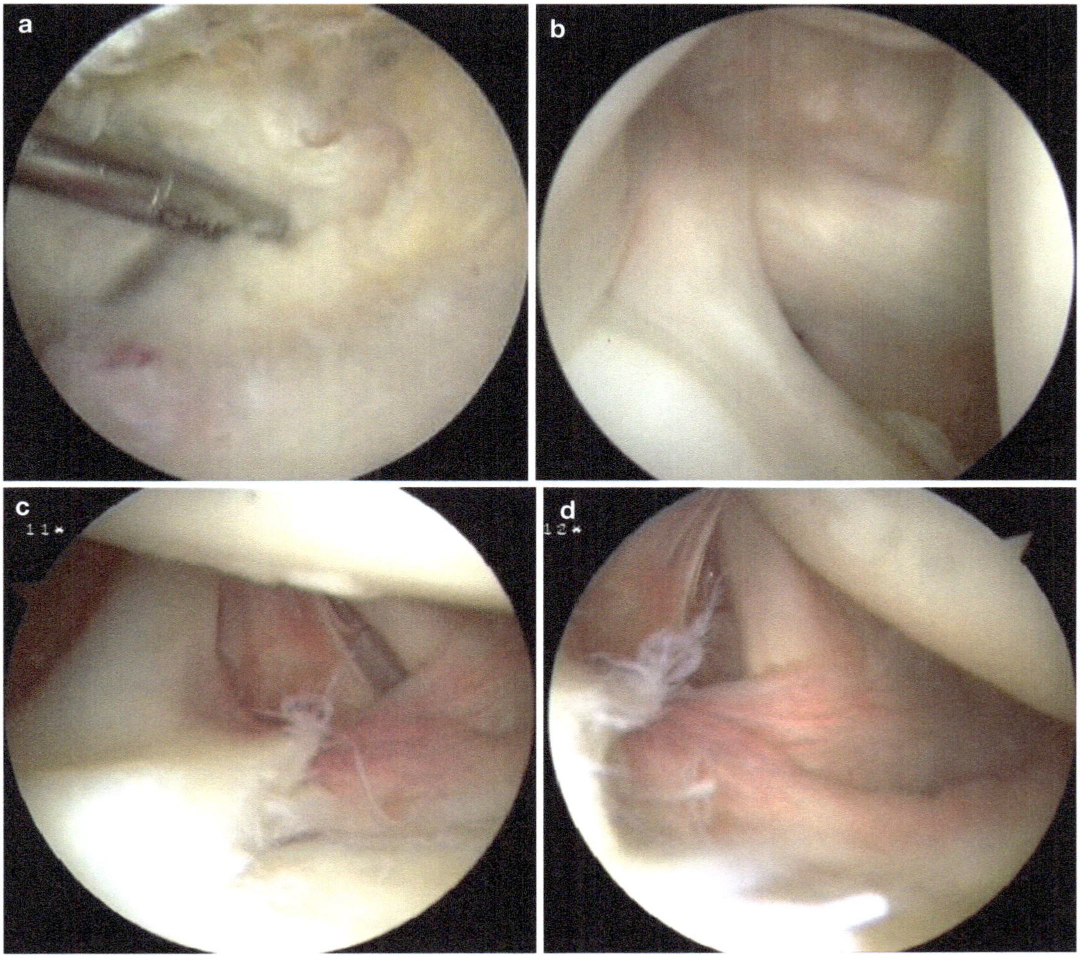

Fig. 11.9 Arthroscopic image of CHL (**a**) in a left shoulder in addition to the rotator interval (**b**), SGHL (**c**), MGHL (**d**) in a right shoulder

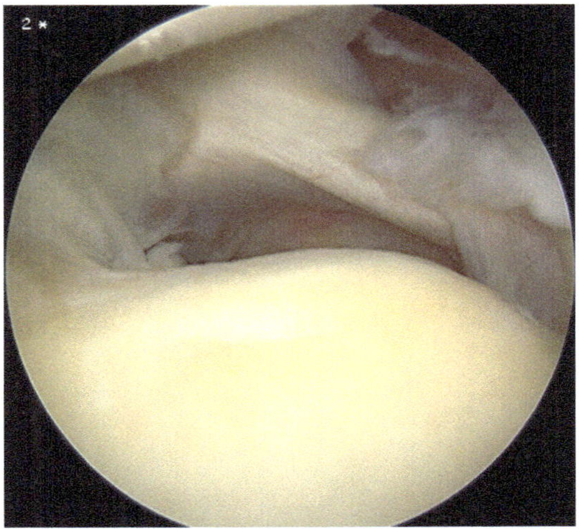

Fig. 11.10 Arthroscopic images of left shoulder showing cord-like MGHL with SGHL superior to glenoid and absent labrum

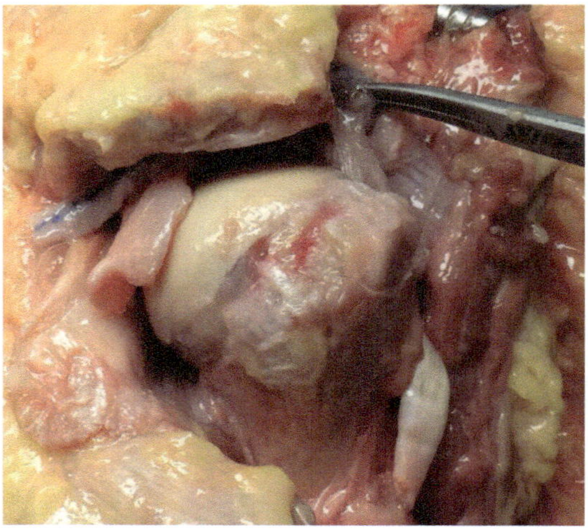

Fig. 11.11 Right shoulder viewed from laterally with hemostat "tightening" the rotator interval, specifically the humeral head elevates on the glenoid and, on exam, would resist external rotation more than normal

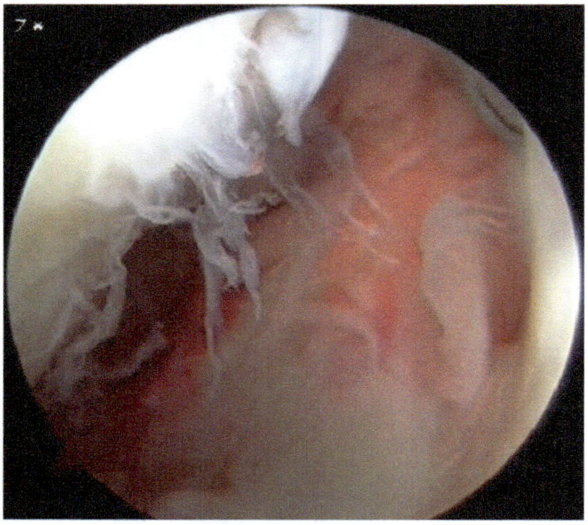

Fig. 11.12 Left shoulder arthroscopic visualizations of adhesive capsulitis with contracted SGHL. SGHL tightened by capsules

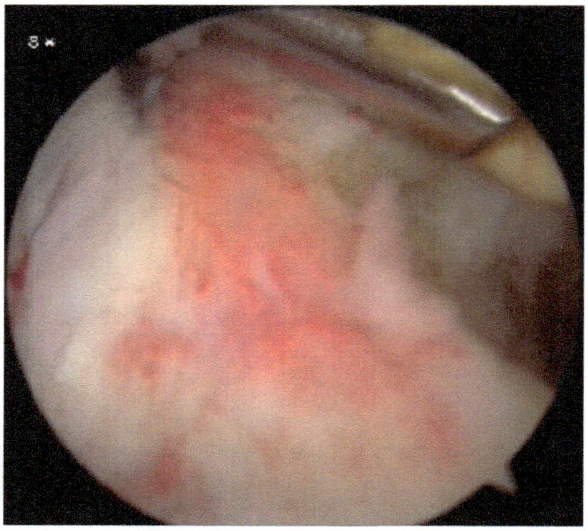

Fig. 11.13 Left shoulder SGHL synovitis depicted arthroscopically. Tight, inflamed SGHL

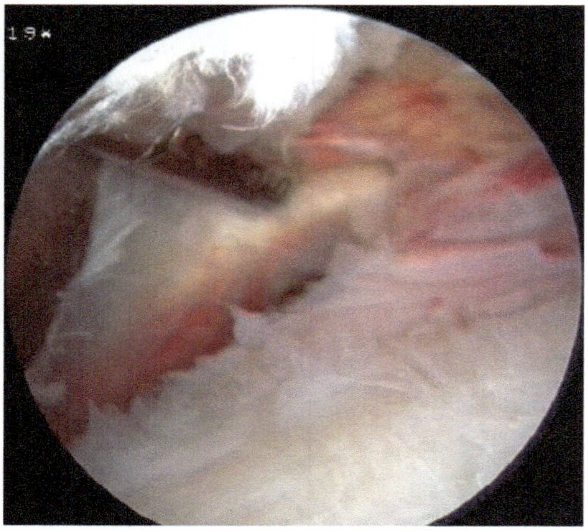

Fig. 11.14 Left shoulder inflamed CHL as seen arthroscopically in adhesive capsulitis

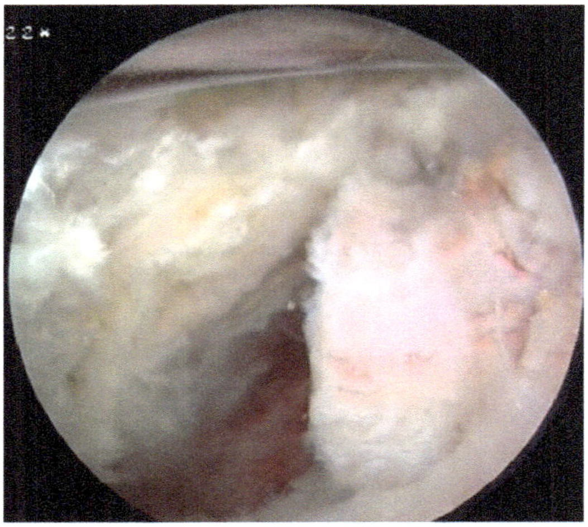

Fig. 11.15 Arthroscopic images after CHL release for adhesive capsulitis. Thick remnant of CHL detached from coracoid, still attached to interval

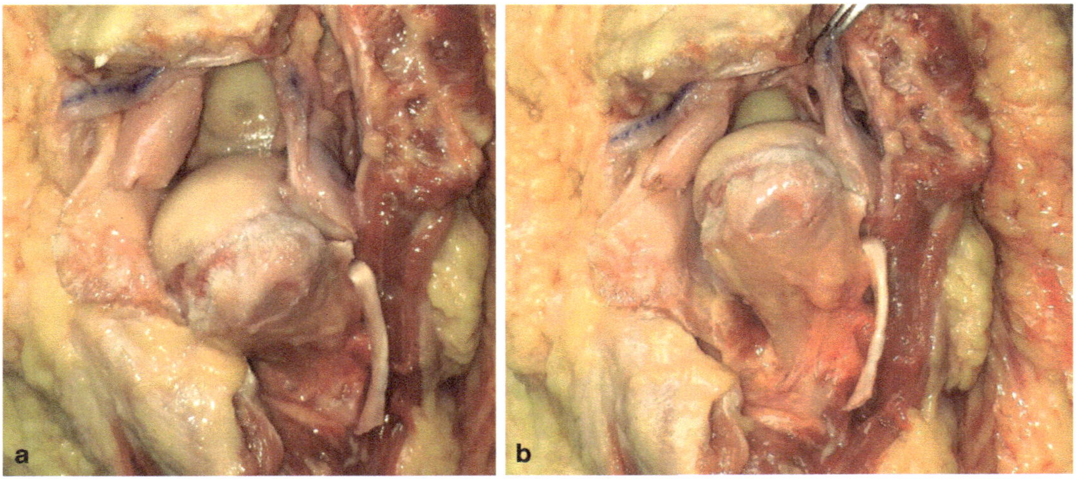

Fig. 11.16 (**a**) Right shoulder, viewed from laterally, after release of the entire rotator interval. The humeral head subluxates significantly on the glenoid. (**b**) Right shoulder, viewed from laterally. The hemostat reapproximates the rotator interval, demonstrating improved positioning of the humeral head on the glenoid in comparison to (a)

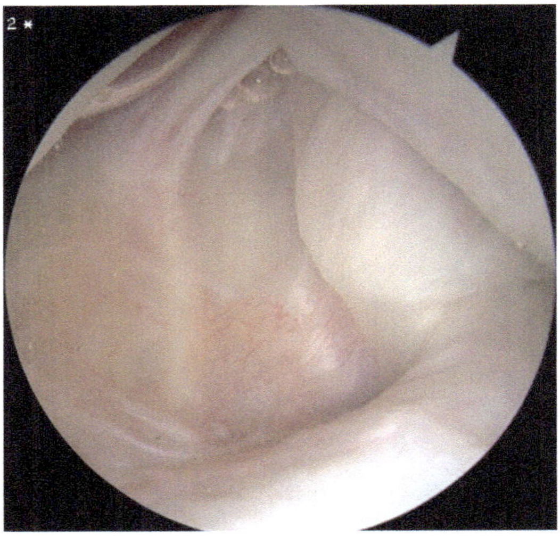

Fig. 11.17 Right shoulder arthroscopic image of rotator interval with laxity in a patient with multidirectional instability

With rupture or incompetence of the biceps pulley, termed a hidden lesion because it is often missed, long head of the biceps pain and subluxation can occur. Procedures for instability patients may include, in order to center the humeral head on the glenoid.

Conclusion

The coracohumeral and superior glenohumeral ligaments are consistent parts of the rotator interval, which acts as restraints to humeral head external rotation and inferior translation when the arm is at the side. The is the strongest of the rotator interval structures [4, 10]. Its anterior extent invests the subscapularis, its middle part forms the biceps pulley with the transverse humeral ligament, and its posterior portion blends with the supraspinatus and infraspinatus fascia to form the.

References

1. Neer 2nd CS. Displaced proximal humeral fractures. I. Classification and evaluation. J Bone Joint Surg Am. 1970;52(6):1077–89.
2. Hunt SA, Kwon YW, Zuckerman JD. The rotator interval: anatomy, pathology, and strategies for treatment. J Am Acad Orthop Surg. 2007;15(4): 218–27.
3. Codman E. The shoulder – rupture of the supraspinatus tendon and other lesions in or about the subacromial bursa. 2nd ed. Boston: Thomas Todd Company; 1934.
4. Jost B, Koch PP, Gerber C. Anatomy and functional aspects of the rotator interval. J Shoulder Elbow Surg. 2000;9(4):336–41.
5. Terry R. The shoulder-joint. In: Jackson CM, editor. Morris' human anatomy – a complete systematic treatise. 9th ed. Philadelphia: P. Blakiston's Son & Co Inc; 1933.
6. Blair D. The shoulder-joint. In: Robinson A, editor. Cunningham's test-book of anatomy. 6th ed. New York: Oxford; 1931.
7. Harryman 2nd DT, et al. The role of the rotator interval capsule in passive motion and stability of the shoulder. J Bone Joint Surg Am. 1992;74(1): 53–66.
8. Gyftopoulos S, et al. The rotator cable: magnetic resonance evaluation and clinical correlation. Magn Reson Imaging Clin N Am. 2012;20(2):173–85, ix.
9. Zappia M, et al. Long head of the biceps tendon and rotator interval. Musculoskelet Surg. 2013;97 Suppl 2:S99–108.
10. Boardman ND, et al. Tensile properties of the superior glenohumeral and coracohumeral ligaments. J Shoulder Elbow Surg. 1996;5(4):249–54.

Imaging of the Labrum

12

Eiji Itoi and Shin Hitachi

12.1 Imaging Modalities

12.1.1 Anteroinferior Labral Lesion/Labral Tears

The anteroinferior labral lesions, so called a Bankart lesion and its variations, caused by a traumatic anterior dislocation of the shoulder are best depicted by MR arthrography (MRA) because of the contrast between the contrast material (gadolinium [Gd]) and the labrum and other intra-articular structures [2, 18]. It is also true for skeletally immature and skeletally mature children [3]. Having said that, the importance of accurate history and clinical examinations need to be highlighted. Some warn that the need for MRA may not be as high as is currently believed [10].

The position of taking MRA is controversial. Some recommend the positioning of abduction and external rotation (ABER) such that the lesion is under tension and well visualized. They say the sensitivity and specificity of MRA taken in ABER is better than in conventional position of adduction [5, 6, 12]. However, others say there is no difference between them [20]. Among variations of Bankart lesion, Tian et al. reported that Perthes lesion was better depicted by MRA in ABER, but not the other types of lesions [21]. Wintzell et al. reported that indirect MRA taken in anterior apprehension test position of 90° of abduction and maximum tolerable external rotation was better than ABER [22]. The position during MR scanning is still controversial.

12.1.2 Superior Labral Anterior Posterior Tears, (SLAP) Lesion

Various imaging modalities have been used to identify glenoid labral lesions, including arthrography, CT arthrography, MRI, and MR arthrography. Among them, MRA is known to be the most accurate imaging modality, with high sensitivity and high specificity both in adults [7, 8, 19] and in children [3].

Bringing the arm from the neutral position to ABER position may translate the humeral head posteriorly relative to the glenoid. Patients with unstable SLAP lesions, which required repair, showed the posterior translation of the humeral head greater than 3 mm [4]. This may be used to evaluate the SLAP lesion and shoulder stability. According to Modi et al., additional sequences of MRA in ABER appear to improve diagnostic accuracy of MRA [14]. Applying downward traction to the arm during MR scanning is another method to improve diagnostic accuracy of this imaging modality [1].

E. Itoi, MD, PhD (✉)
Department of Orthopaedic Surgery,
Tohoku University School of Medicine, Sendai, Japan
e-mail: itoi-eiji@med.tohoku.ac.jp

S. Hitachi, MD, PhD
Department of Diagnostic Radiology,
Tohoku University School of Medicine, Sendai, Japan

G.I. Bain et al. (eds.), *Normal and Pathological Anatomy of the Shoulder*,
DOI 10.1007/978-3-662-45719-1_12, © ISAKOS 2015

Although US is a valuable diagnostic tool for rotator cuff lesions, it is less valuable for glenoid labral lesions [19]. This is probably because the labrum is located deeper than the rotator cuff tendons.

12.2 MRI

Although some say MRA is still superior to 3.0-T MRI [13], recent 3.0-T MRI has enabled us to see the articular cartilage and the labrum without injecting Gd into the glenohumeral joint (Fig. 12.1). The posterior labrum attaches to the glenoid through a thick cartilage, which may mimic a posterior labral tear from the glenoid. The anterior labrum also attaches to the glenoid rim through the articular cartilage. The anterior labrum attaches to both the cartilage and the bone in more than 80 % of cadaveric shoulders [9].

12.3 MR Arthrography

MR arthrography clearly depicts the contrast between Gd and the intra-articular structures such as the labrum, articular cartilage, and the joint capsule (Fig. 12.2). Between the anterior capsule and the anterior labrum is the middle glenohumeral ligament, which runs parallel to the anterior labrum.

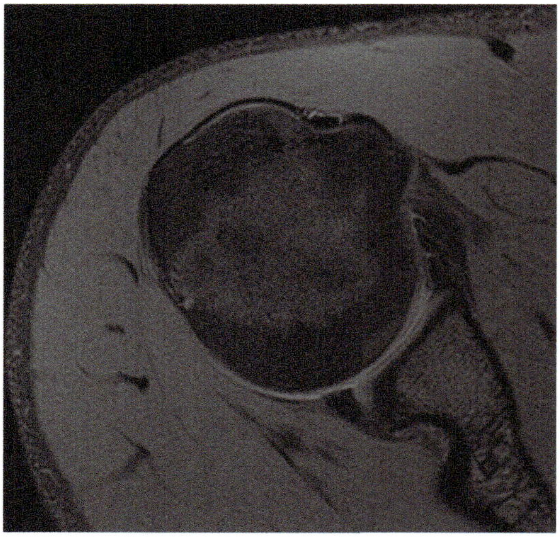

Fig. 12.1 Fast spine-echo proton density-weighted axial image. The articular cartilage and the labrum are clearly visualized without injecting Gd into the glenohumeral joint on this 3.0-T MRI. The posterior labrum attaches to the glenoid through a thick cartilage, which may mimic a posterior labral tear from the glenoid. The anterior labrum also attaches to the glenoid rim through the articular cartilage

For the purpose of assessing the labrum, the coronal oblique and axial images have limitations. When the slices are cut perpendicular to the labrum, its attachment to the glenoid rim is most clearly visible. In order to achieve this best image for any portion of the labrum, the radial-sequence MR imaging has been introduced [15]. The scout view of the glenoid shows the orientation of the slices (Fig. 12.3). Using these slices, any portion of the labral attachment to the glenoid is clearly visualized. The slice passing through 12–6 o'clock is equivalent to a conventional coronal oblique image (Fig. 12.4). On this image, the superior labral tear is observed. On the 2–8 o'clock slice, both the anteroinferior and posterosuperior labra are detached from the glenoid (Fig. 12.5). On the 4–10 o'clock slice, the anterosuperior labrum is detached from the glenoid (Perthes lesion) (Fig. 12.6). Other types of anterior labral lesion such as an ALPSA (anterior labroligamentous periosteal sleeve avulsion) lesion [16] (Fig. 12.7) and a GLAD (glenolabral articular disruption) lesion [17] (Fig. 12.8) are clearly depicted on these MR arthrograms. The posterior inferior part of the

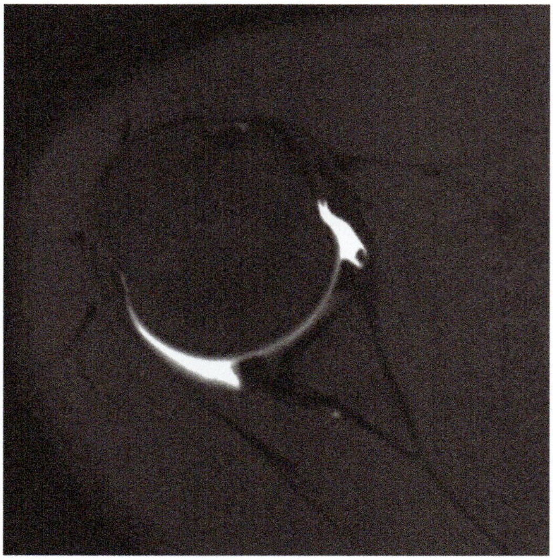

Fig. 12.2 Fast spine-echo T1-weighted axial image. The anterior and posterior labra, articular cartilage, the middle glenohumeral ligament, and the joint capsule were clearly visible with the help of the contrast material (Gd)

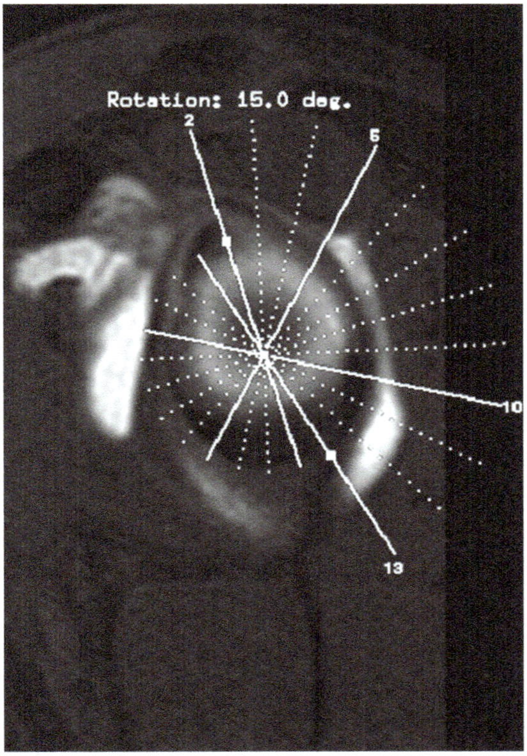

Fig. 12.3 Scout view showing the orientation of radial slices. The slices are drawn through the center of the glenoid with a 15-degree increment. The slice passing through 12–6 o'clock is equivalent to a conventional coronal oblique image, and the slice passing through 3–9 o'clock is equivalent to a conventional axial image

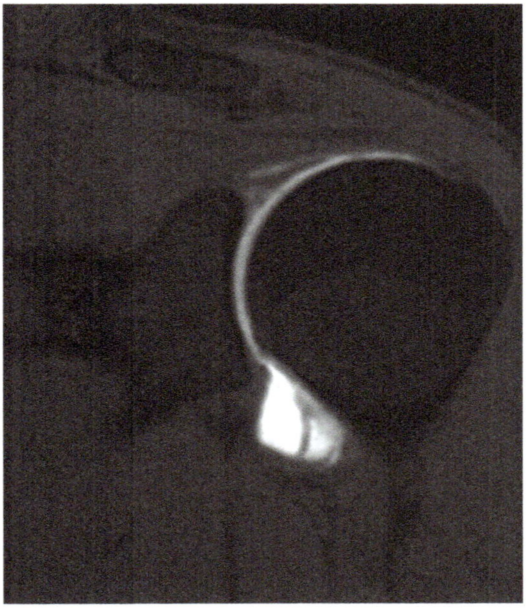

Fig. 12.4 Fast spine-echo T1-weighted 12–6 o'clock slice. There is a tear of the superior labrum at 12 o'clock position (left shoulder)

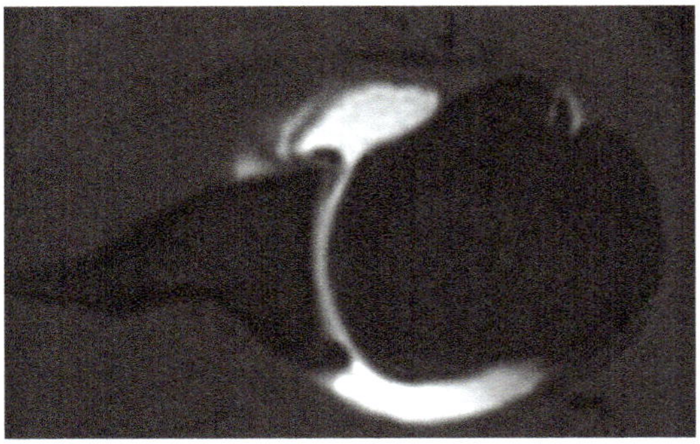

Fig. 12.5 Fast spine-echo T1-weighted 2–8 o'clock slice. The anteroinferior labrum is detached from the glenoid at 8 o'clock position, and the posterosuperior labrum is also detached at 2 o'clock (left shoulder)

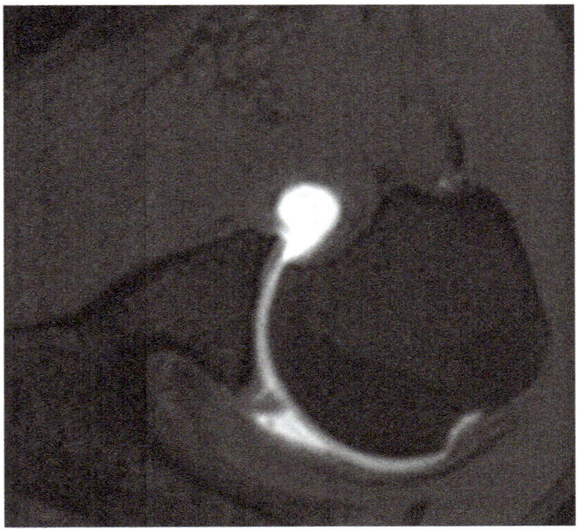

Fig. 12.6 Fast spine-echo T1-weighted 4–10 o'clock slice. There is a tear of the anterosuperior labrum at 10 o'clock position (left shoulder)

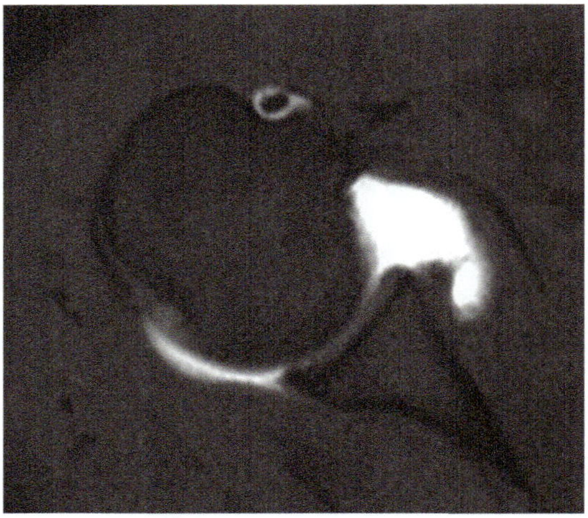

Fig. 12.7 ALPSA (anterior labroligamentous periosteal sleeve avulsion) lesion. The anterior labrum together with the anterior capsule including the inferior glenohumeral ligament is displaced medially. This is called an ALPSA lesion, which causes anterior instability of the shoulder

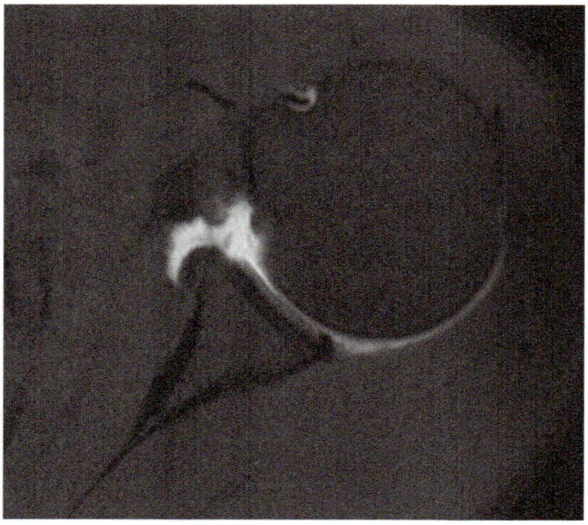

Fig. 12.8 GLAD (glenolabral articular disruption) lesion. The anterior labrum is detached and displaced medially with associated articular cartilage damage

labrum may have a concealed tear of the labrum in cases with multidirectional or posteroinferior shoulder instability (Fig. 12.9). This concealed tear is called a Kim lesion [11]. A new imaging technique, 3-D reconstruction of MR images, shows that the SLAP lesion extends all the way down to 8 o'clock anteriorly and to 2 o'clock posteriorly (type V SLAP lesion) (Fig. 12.10).

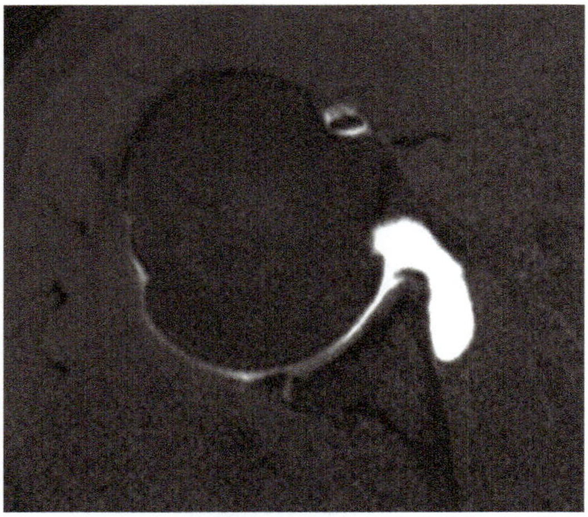

Fig. 12.9 Kim lesion. The posterior inferior part of the labrum may have a concealed tear of the labrum in cases with multidirectional or posteroinferior shoulder instability. This concealed tear is called a Kim lesion

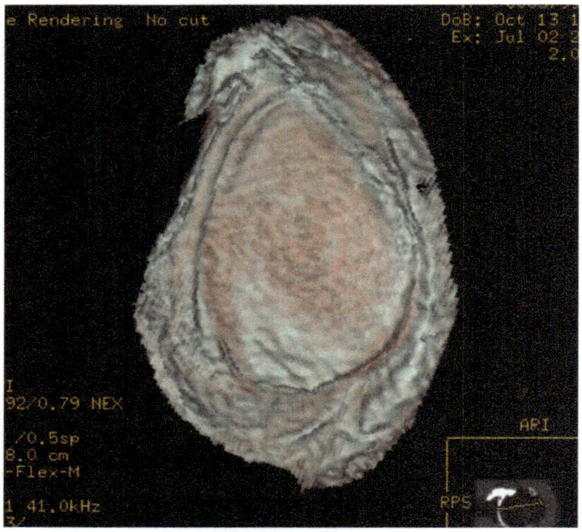

Fig. 12.10 3D reconstruction of an en face view of the glenoid and labrum. This is the left shoulder with 9 o'clock being anterior and 3 o'clock being posterior. The SLAP lesion extends all the way down to 8 o'clock anteriorly and to 2 o'clock posteriorly (type V SLAP lesion)

12.4 CT Arthrography

In order to make clear contrast among the soft tissues, CT arthrography is performed with use of contrast medium such as iodine contrast, air, or both (double contrast CT arthrography). This CT arthrogram with iodine contrast reveals that the

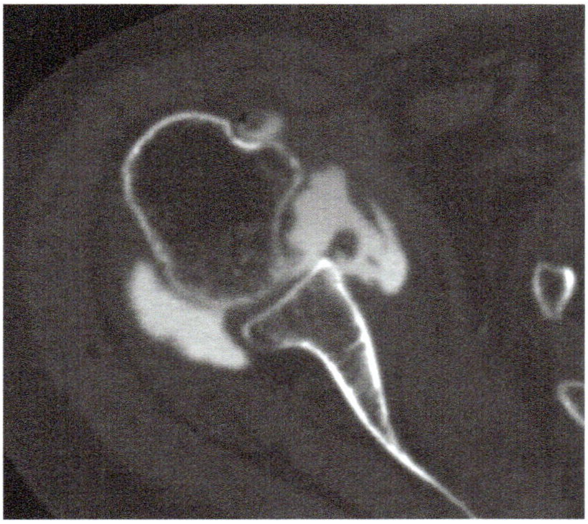

Fig. 12.11 CT arthrography. The anterior labrum is separated from the anterior capsule and from the glenoid rim (Bankart lesion). The anterior articular cartilage looks thinner than the posterior one due to recurrent anterior dislocation of the shoulder

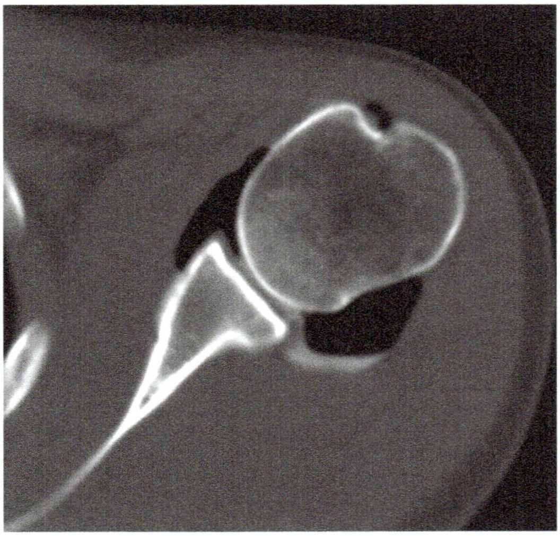

Fig. 12.12 Double-contrast CT arthrography. This is a case of recurrent anterior dislocation, with the anterior labrum and a part of the inferior glenohumeral ligament missing (Bankart lesion). The articular cartilage looks intact

anterior labrum separated from the anterior capsule and from the glenoid rim (Bankart lesion) (Fig. 12.11). The anterior articular cartilage looks thinner than the posterior one due to recurrent anterior dislocations of the shoulder. When both the iodine contrast and the air are used as contrast materials, it is called double-contrast CT arthrography (Fig. 12.12), which shows the surface of the intra-articular structures very clearly. This is also a case of recurrent anterior dislocation, with the anterior labrum and a part of the inferior glenohumeral ligament disappearing (Bankart lesion). The articular cartilage looks intact.

References

1. Becce F, Richarme D, Omoumi P, et al. Direct MR arthrography of the shoulder under axial traction: feasibility study to evaluate the superior labrum-biceps tendon complex and articular cartilage. J Magn Reson Imaging. 2013;37(5):1228–33.
2. Chandnani VP, Yeager TD, DeBerardino T, et al. Glenoid labral tears: prospective evaluation with MRI imaging, MR arthrography, and CT arthrography. AJR Am J Roentgenol. 1993;161(6):1229–35.
3. Chauvin NA, Jaimes C, Ho-Fung V, Wells L, Ganley T, Jaramillo D. Diagnostic performance of magnetic resonance arthrography of the shoulder in children. Pediatr Radiol. 2013;43(10):1309–15.
4. Chhadia AM, Goldberg BA, Hutchinson MR. Abnormal translation in SLAP lesions on magnetic resonance imaging abducted externally rotated view. Arthroscopy. 2010;26(1):19–25.
5. Choi JA, Suh SI, Kim BH, et al. Comparison between conventional MR arthrography and abduction and external rotation MR arthrography in revealing tears of the antero-inferior glenoid labrum. Korean J Radiol. 2001;2(4):216–21.
6. Cvitanic O, Tirman PF, Feller JF, Bost FW, Minter J, Carroll KW. Using abduction and external rotation of the shoulder to increase the sensitivity of MR arthrography in revealing tears of the anterior glenoid labrum. AJR Am J Roentgenol. 1997;169(3):837–44.
7. Genovese E, Spano E, Castagna A, et al. MR-arthrography in superior instability of the shoulder: correlation with arthroscopy. Radiol Med. 2013;118(6):1022–33.
8. Holzapfel K, Waldt S, Bruegel M, et al. Inter- and intraobserver variability of MR arthrography in the detection and classification of superior labral anterior posterior (SLAP) lesions: evaluation in 78 cases with arthroscopic correlation. Eur Radiol. 2010;20(3):666–73.
9. Itoigawa Y, Itoi E, Sakoma Y, Yamamoto N, Sano H, Kaneko K. Attachment of the anteroinferior glenohumeral ligament-labrum complex to the glenoid: an anatomic study. Arthroscopy. 2012;28(11):1628–33.
10. Jonas SC, Walton MJ, Sarangi PP. Is MRA an unnecessary expense in the management of a clinically unstable shoulder? A comparison of MRA and arthroscopic findings in 90 patients. Acta Orthop. 2012;83(3):267–70.

11. Kim SH, Ha KI, Yoo JC, Noh KC. Kim's lesion: an incomplete and concealed avulsion of the posteroinferior labrum in posterior or multidirectional posteroinferior instability of the shoulder. Arthroscopy. 2004;20(7):712–20.
12. Kwak SM, Brown RR, Trudell D, Resnick D. Glenohumeral joint: comparison of shoulder positions at MR arthrography. Radiology. 1998;208(2):375–80.
13. Magee T. 3-T MRI of the shoulder: is MR arthrography necessary? AJR Am J Roentgenol. 2009;192(1):86–92.
14. Modi CS, Karthikeyan S, Marks A, et al. Accuracy of abduction-external rotation MRA versus standard MRA in the diagnosis of intra-articular shoulder pathology. Orthopedics. 2013;36(3):e337–42.
15. Munk PL, Holt RG, Helms CA, Genant HK. Glenoid labrum: preliminary work with use of radial-sequence MR imaging. Radiology. 1989;173(3):751–3.
16. Neviaser TJ. The anterior labroligamentous periosteal sleeve avulsion lesion: a cause of anterior instability of the shoulder. Arthroscopy. 1993;9(1):17–21.
17. Neviaser TJ. The GLAD lesion: another cause of anterior shoulder pain. Arthroscopy. 1993;9(1):22–3.
18. Palmer WE, Brown JH, Rosenthal DI. Labral-ligamentous complex of the shoulder: evaluation with MR arthrography. Radiology. 1994;190(3):645–51.
19. Pavic R, Margetic P, Bensic M, Brnadic RL. Diagnostic value of US, MR and MR arthrography in shoulder instability. Injury. 2013;44 Suppl 3:S26–32.
20. Schreinemachers SA, van der Hulst VP, Jaap Willems W, Bipat S, van der Woude HJ. Is a single direct MR arthrography series in ABER position as accurate in detecting anteroinferior labroligamentous lesions as conventional MR arthography? Skeletal Radiol. 2009;38(7):675–83.
21. Tian CY, Cui GQ, Zheng ZZ, Ren AH. The added value of ABER position for the detection and classification of anteroinferior labroligamentous lesions in MR arthrography of the shoulder. Eur J Radiol. 2013;82(4):651–7.
22. Wintzell G, Larsson H, Larsson S. Indirect MR arthrography of anterior shoulder instability in the ABER and the apprehension test positions: a prospective comparative study of two different shoulder positions during MRI using intravenous gadodiamide contrast for enhancement of the joint fluid. Skeletal Radiol. 1998;27(9):488–94.

Pathoanatomy of Glenohumeral Instability

13

Seung-Ho Kim

13.1 Introduction

Glenohumeral instability is a common encounter in the shoulder problems. Although shoulder instability has been recognized as one the oldest disease in medicine, true anatomical knowledge on the pathologic entities has been proven lately. The glenohumeral joint provides the largest range of motion among all diarthrodial joints while the stability of the joint mostly relies on the soft tissue structures. Pathoanatomic changes in the anterior instability are more straightforward compared to the posterior or multidirectional instability. Wherever the direction of the instability is, failure generally occurs at the most restraint soft tissue structures, which varies depending on the arm position, humeral head version, vector of force applied on the shoulder, and age of the patients.

Shoulder arthroscopy allows us to understand various principal pathoanatomy responsible for the individual instability patterns and provides pathology-oriented approach during the surgical treatment. The Bankart lesion is the most common lesion in the traumatic anterior instability. However, other lesions are not uncommonly encountered, which may easily be underestimated especially in shoulder with combined lesions of Bankart and other lesions. Recent advances in the pathomechanism of the posterior and multidirectional instability owe the recognition of the pathoanatomic lesion of the instability which has long been underestimated. Thus, in this chapter, pathoanatomic lesions of the individual glenohumeral instability will be discussed.

13.2 Pathoanatomy of Anterior Glenohumeral Instability

The anterior glenohumeral instability mostly occurs with a traumatic episode. The trauma can be repetitive cumulative or more likely a single event. Therefore, the pathologic lesions are more distinctive than those in posterior or multidirectional instability. Also, recurrent instability develops multiple and complex pathology. In the traumatic anterior instability, failure can occur at any region of the anterior capsulolabral restraint. The anterior band of the inferior glenohumeral ligament is the primary static restraint in the abducted and externally rotated shoulder.

In young population, the most common location of the failure of the anterior glenohumeral restraint is at the anterior labrum–glenoid interface. The Bankart lesion is the avulsion of the anterior labroligamentous structures from the

Electronic supplementary material The online version of this chapter 10.1007/978-3-662-45719-1_13 contains supplementary material, which is available to authorized users.

S.-H. Kim, MD
Department of Orthopaedic Surgery, The Madi Hospital, 646 samsung-ro, Gangnam-gu, Seoul, Korea
e-mail: shk@madi.or.kr

G.I. Bain et al. (eds.), *Normal and Pathological Anatomy of the Shoulder*, DOI 10.1007/978-3-662-45719-1_13, © ISAKOS 2015

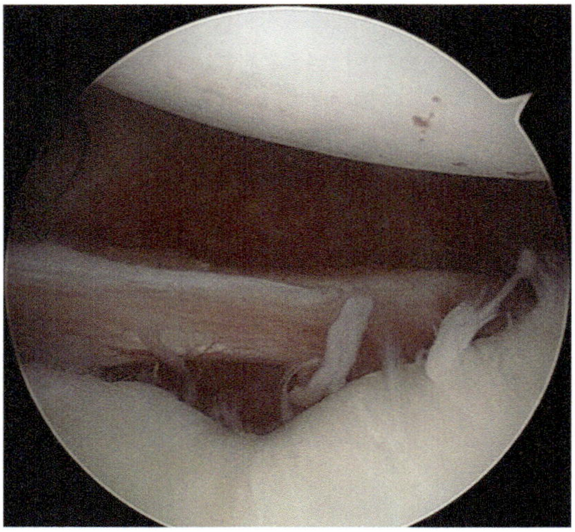

Fig. 13.1 Bankart lesion. The Bankart lesion is the avulsion of the anterior labroligamentous structures from the glenoid rim and the most common pathologic lesion in the traumatic anterior glenohumeral instability. The anterior labrum is disrupted from the glenoid along with disrupted periosteum of the scapular neck and it typically lies anterior to the glenoid

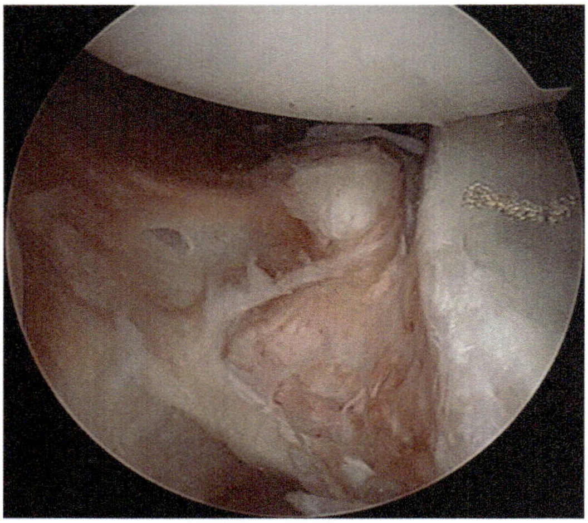

Fig. 13.2 ALPSA lesion. The torn labroligamentous structures are displaced and healed on the medial surface of the anterior glenoid neck

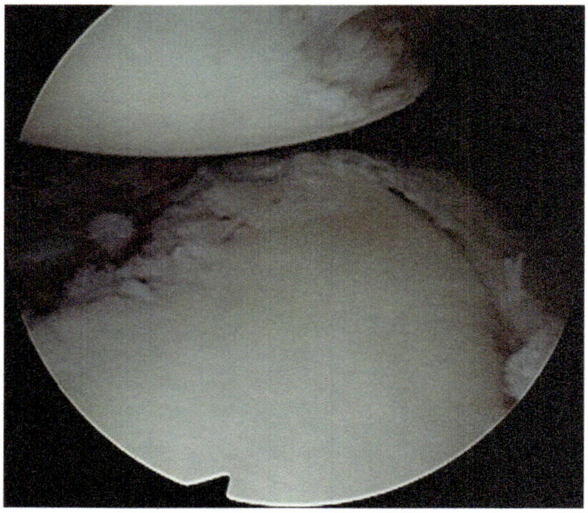

Fig. 13.3 Extended Bankart lesion. The anterior labral tear can be extended to the inferior and posterior area of the glenoid

glenoid rim and the most common pathologic lesion in the traumatic anterior glenohumeral instability. The anterior labrum is disrupted from the glenoid along with disrupted periosteum of the scapular neck and it typically lies anterior to the glenoid (Fig. 13.1). Approximately 90 % of all anterior instability has the Bankart lesion [1]. In the MRI and arthroscopic examination, an abnormal separation is present in the anterior margin of the glenoid.

ALPSA, Anterior Labroligamentous Periosteal Sleeve Avulsion, is first described by Neviaser in 1993 [2]. The torn labroligamentous structures are displaced and healed on the medial surface of the anterior glenoid neck (Fig. 13.2). The ALPSA lesion is often found in shoulders with recurrent dislocation [3]. Arthroscopically, it is difficult to identify from the standard posterior portals and often the lesion is not evident even from the anterior portal. When the ALPSA lesion contains a small bony fragment, the lesion is strongly healed to the glenoid neck and often difficult to mobilize during the arthroscopic procedure. ALPSA lesions would be repaired by elevating the labroligamentous complex of the scapular neck and bringing it back to the anatomic position on the glenoid face.

The anterior labral failure, either as the Bankart or ALPSA lesion, can be limited at the anterior glenoid area (Limited Bankart lesion) or it can be extended to the inferior and posterior area, which is coined as an extended Bankart lesion (Fig. 13.3).

GLAD, Glenolabral Articular Disruption, is also described by Neviaser in 1993 [4]. The GLAD

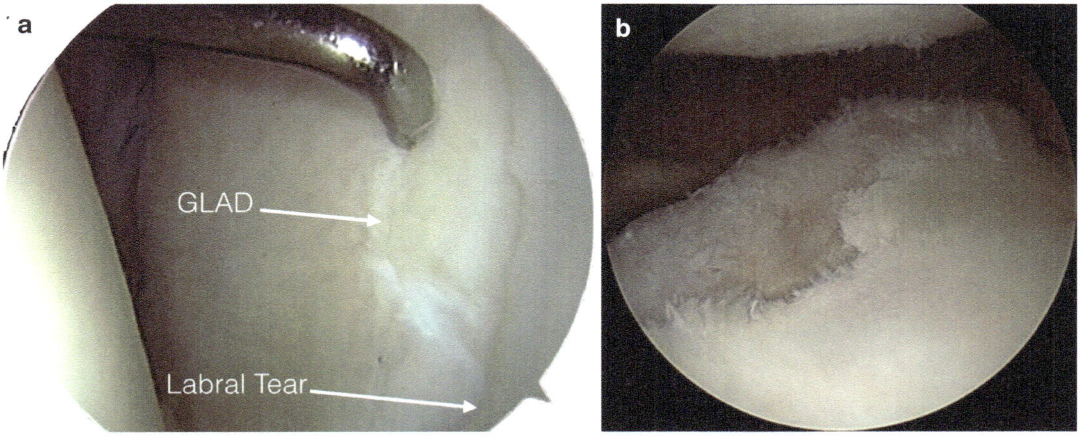

Fig. 13.4 GLAD lesion. The GLAD lesion is disruption of articular cartilage from the anterior inferior glenoid surface: (**a**) flap tear (courtesy of Dr Richard Page, Melbourne Australia); (**b**) cartilage loss

lesion is disruption of articular cartilage from the anterior inferior glenoid surface either as a flap tear or cartilage loss and is caused by a forced adduction injury to the shoulder from an abducted and external rotated position (Fig. 13.4a, b).

Bony Bankart lesion occurs with a significant injury, which causes fracture of the anterior inferior glenoid margin in varying sizes. The bony Bankart lesion is very common. Sugaya et al. evaluated the morphology of the glenoid rim and 90 % of the shoulder with traumatic anterior instability showed bony deficiency as a fragment type in 50 % and erosion type in 40 % [5]. The bony Bankart lesion is different from the glenoid erosion. The erosion of the anterior glenoid without bony fragment jeopardizes the stability of the glenohumeral joint even after Bankart repair. The glenoid defect greater than 21 % of the glenoid length or 25 % of the glenoid width requires reconstruction of the glenoid concavity [6].

HAGL, Humeral Avulsion of Glenohumeral Ligament, is detachment of the glenohumeral ligament from the humeral attachment and first described by Bach in 1988 [7]. Later in 1995, Wolf et al. coined the name "HAGL" and the incidence being 1–9 % of the shoulders with anterior glenohumeral instability [8]. The HAGL lesion has clinical significance in planning the type operative procedures. A small HAGL lesion can be managed by arthroscopic side-to-side repair while a large defect may require open surgical repair (Fig. 13.5).

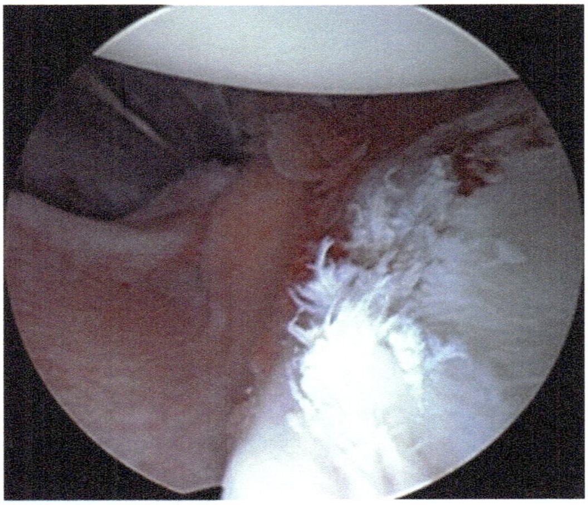

Fig. 13.5 HAGL lesion. HAGL, Humeral Avulsion of Glenohumeral Ligament, is detachment of the glenohumeral ligament from the humeral attachment

Posterior HAGL lesion is a typical finding in shoulders with traumatic posterior instability. We noted that the posterior HAGL lesion is also found in shoulders with traumatic anterior instability. The glenohumeral ligament is avulsed from the posterior humeral attachment. It closely lies in the area of the Hill–Sachs lesion. The posterior HAGL lesion is associated with a large Hill–Sachs lesion (Fig. 13.6).

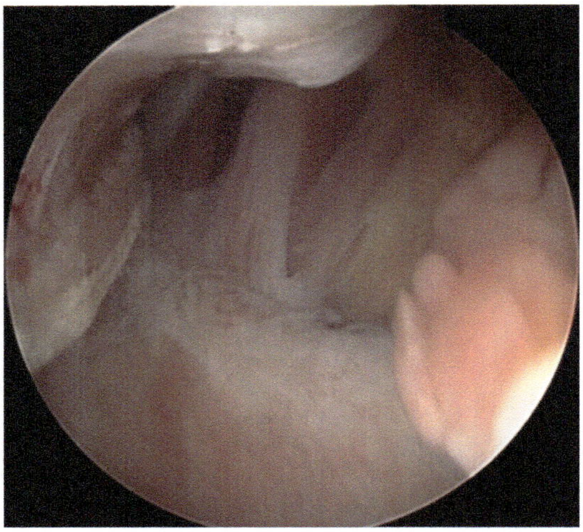

Fig. 13.6 Posterior HAGL. The glenohumeral ligament is avulsed from the posterior humeral attachment

Capsular lesion is not uncommon in shoulder with anterior glenohumeral instability. It is more common in elderly patients with recurrent dislocation in our experience. It can be an isolated entity. However, the capsular lesion is commonly associated with other pathology such as a Bankart or ALPSA lesion. The clinical significance of the capsular lesion is that the lesion is frequently masked by a thin layer of the fibrous scar tissue.

Bipolar lesion is a combination of any pathoanatomic lesion of the anterior instability.

A common form in ipsilateral bipolar lesion is a combination of Bankart lesion and HAGL or capsular lesion (Video 13.1). The contralateral bipolar lesion is the combined lesion of Bankart and posterior HAGL lesion (Video 13.2). We observed that the bipolar lesion is more common in patients over 30 years in age who have recurrent dislocation.

Hill–Sachs lesion is a bony defect in the posterolateral humeral head and is an impaction injury to the anterior glenoid rim. The incidence is more common in recurrent instability and greater than 20–25 % of the humeral head may develop clinical instability [9, 10]. During arthroscopic surgery, the arm is elevated to bring the shoulder in abduction and external rotation. In this position, if the Hill–Sachs lesion engages with the anterior rim of the glenoid, it is assessed as an engaging Hill–Sachs lesion [11, 12]. Glenoid tract is a zone of contact of the humeral head to the glenoid in the arm position of abduction and external rotation. When the Hill–Sachs lesion remains within the glenoid tract, the humeral head defect is in the risk of engagement and instability [13]. The extent of the Hill–Sachs lesion should be considered in association with the glenoid bone loss. With a small Hill–Sachs defect, a larger degree of glenoid bone loss will increase the risk of recurrent instability.

Rotator cuff tear in the recurrent anterior instability is usually partial-thickness tear of the articular surface of the tendon. However, in elderly patients with significant traumatic dislocation, full-thickness tears of entire supraspinatus and infraspinatus can develop anterior dislocation with or without Bankart lesion.

13.3 Pathoanatomy of Posterior and Multidirectional Instability

Posterior or multidirectional instability occurs usually by atraumatic or repetitive microtraumatic origin. There has been no universal agreement in the classification, terminology, and treatment options. The clinical presentation of the atraumatic instability is not as clear as traumatic anterior instability and many of patients with posterior instability are easily overlooked or treated under other diagnosis. Recent advances in the concept of the posterior instability have provided us reasonable insight into the pathology, pathogenesis, diagnostic examinations, and treatment options. The posterior instability very often presents as bidirectional posteroinferior instability, which has various degrees of inferior component of instability. Also the posterior instability is overlapping with the multidirectional instability in its diagnosis, clinical presentation, and management.

13.3.1 Pathogenesis

Several anatomic structures have been implicated including bony and soft tissue abnormalities. Bony abnormalities include increased humeral retroversion, glenoid retroversion, and glenoid hypoplasia. Although, several studies on the glenoid version have been focused on the bony glenoid measured, the stability of glenohumeral joint is an integral function of both bone and soft tissue stabilizer. Lazarus et al. showed a 65 % decrease in mechanical stability ratio and an 80 % reduction in the height of the glenoid associated with the creation of an anteroinferior chondrolabral defect [14]. Accordingly, the measurement of the glenoid version can be more ideal when the articular cartilage and labrum are considered as a whole. Soft tissue abnormality of the atraumatic instability has been an excessive capsular laxity. However, increased capsular ligamentous laxity alone cannot entirely explain the whole pathogenesis of the atraumatic instability, which often occurs in the mid-range of motion where normally the capsular ligaments become loose.

Kim et al. emphasized that loss of chondrolabral containment is a consistent finding in shoulders with atraumatic posteroinferior instability and is principally due to the loss of posterior labral height [15]. Kim et al. suggest that the loss of chondrolabral containment is a result of cumulative microtrauma to the posteroinferior glenoid labrum which initially has normal height and undergoes gradual change to retroversion by the rim-loading mechanism [15, 16]. With the loss of chondrolabral containment, the static restraint loses its function and the dynamic stabilizer of the shoulder becomes less effective in maintaining concavity compression of the glenohumeral joint. Bradley et al. similarly measured the posterior inferior chondrolabral version and bony glenoid version for each MR at the inferior one third of the glenoid rim [17]. In this study, there was increased bony and chondrolabral retroversion in the symptomatic group, which suggests that loss of anatomical containment predisposes to atraumatic instability (Video 13.3).

The concept of chondrolabral lesion in the atraumatic instability provides further insight to the cause of symptom development. Although there are two groups of people in which one group is asymptomatic and the other is symptomatic, it is interesting to know that the amount of increased translation either in posterior, inferior, or anterior direction is the same. Also asymptomatic people often become symptomatic over the time. Although, shoulder is loose in all three directions, concurrent production of symptoms is in one or multiple directions. There are evidences that the amount of translation is not fundamentally different between healthy subject who have asymptomatic laxity and those who need surgical intervention [18, 19]. Given these facts, there may be other pathology which is responsible for the shoulder symptom, rather than just an increased joint volume. The author found that majority of patients with asymptomatic jerk test in the posterior instability, which was represented by painless posterior clunk, were successful with the nonoperative treatment. However, patients with symptomatic jerk test, which was represented by sharp pain with posterior clunk, were not responding with the rehabilitation and invariably had posteroinferior labral lesion in the arthroscopic finding [20]. The author concluded that the jerk test was a hallmark for predicting the failure of nonoperative treatment in the posteroinferior instability. Shoulders with a painful jerk test have a posteroinferior labral lesion.

The labral lesions were classified into four types. Type I labral lesion is an incomplete detachment, in which the posteroinferior labrum is separated from the glenoid margin but not medially displaced. This type is more common in traumatic posterior instability than multidirectional instability. Type II lesion is a marginal crack, so-called Kim's lesion which is an incomplete and concealed avulsion of posteroinferior labrum (Video 13.4). Type III lesion is a chondrolabral erosion, and type IV lesion is a flap tear of the labrum (Fig. 13.7) [15, 21, 22].

The Kim's lesion refers to a superficial tearing between the posteroinferior labrum and the glenoid articular cartilage without a complete detachment of the labrum (marginal crack). The posteroinferior labrum lost its normal height and became a flat labrum, with subsequent retroversion of the chondrolabral glenoid. Probing the lesion demonstrates fluctuation of the posteroinferior labrum and reveals a loose attachment. These labral lesions are limited to the posteroinferior quadrant of the glenoid for shoulders with a pure posterior instability, typically present in 6–9 o'clock position for the right shoulder and 3–6 o'clock position for the left shoulder. However, the lesion is extended to entire inferior glenoid labrum

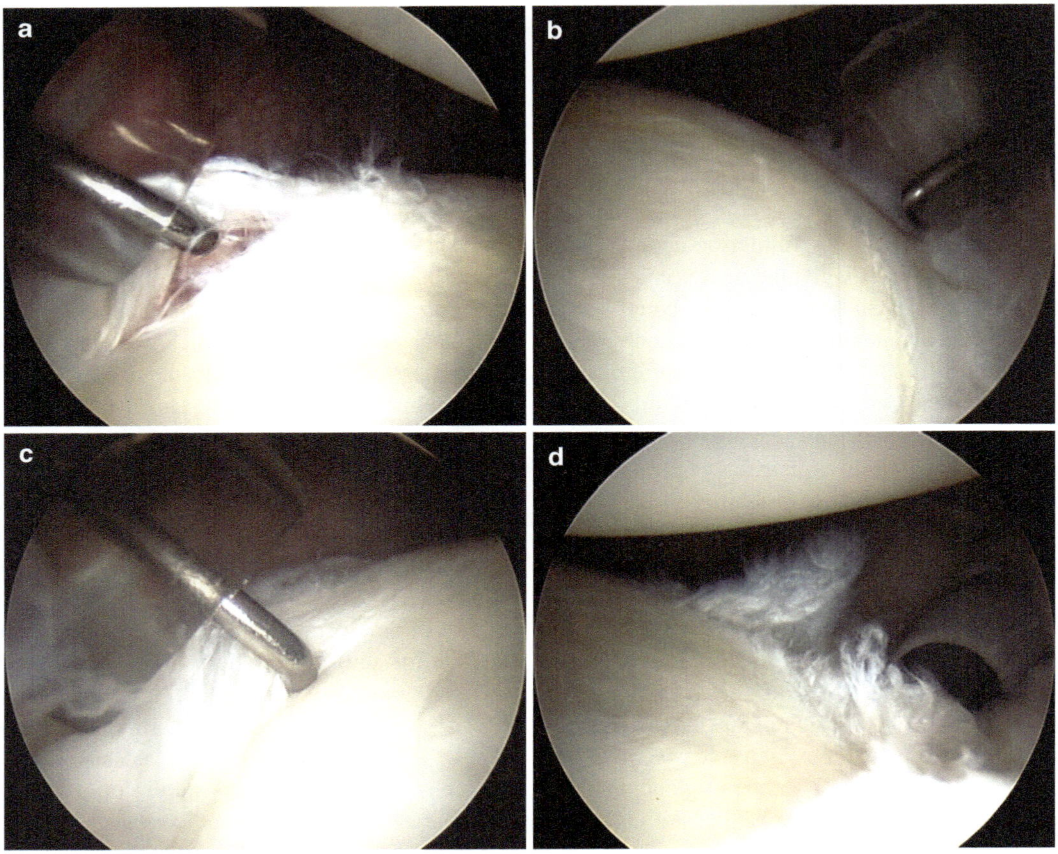

Fig. 13.7 Arthroscopic classification of the posterior and inferior labral lesion. (**a**) Type I: incomplete detachment. The posteroinferior labrum is detached from the glenoid but not displaced. (**b**) Type II: marginal crack or Kim's lesion. The labrum has marginal crack and retroversion. Deep portion is loose. (**c**) Type III: chondrolabral erosion. The labral surface has fraying and deep portion is loose. (**d**) Type IV: flap tear. The labrum has flap tear or multiple buck handle tear

from 4 or 5 to 9 o'clock in shoulders with posteroinferior multidirectional instability. When the superficial portion is incised with an arthroscopic knife, for 1 or 2 mm in depth, the lesion reveals detachment in the deep portion of the labrum from the medial surface of the glenoid. The Kim's lesion is quite similar to the intratendinous tear of the rotator cuff tendon which is often overlooked and unrecognized at the initial arthroscopic evaluation. Therefore, surgeon's insight to this hidden lesion is of paramount importance for the diagnosis of the pathology. The four types of labral lesions are a spectrum of severity of the instability. Perhaps, Kim's lesion may over time be converted into type I incomplete detachment when the marginal crack is extended to the deep portion tear.

It is believed that increased translation by the increased capsular laxity is initial lesion and underlying pathology of the posterior and posteroinferior multidirectional instability. This increased capsular laxity can be in-borne or developmental and asymptomatic or minimally symptomatic initially. In this stage, attempted translation does not produce symptoms. Also, jerk and Kim tests reveal posterior clunk without shoulder pain [20, 23]. However, repetitive subluxation over time overloads the posteroinferior glenoid labrum by the excessive rim loading of the humeral head. This excessive rim loading eventually develops posteroinferior labral lesion varying from simple retroversion to incomplete detachment. In this stage, the patient's symptom which is shoulder pain originates from the labral lesion when the humeral head glides over the pathologic labrum. The compressive force on the torn labrum in the jerk and Kim tests generates shoulder pain. The labral lesion is now the essential lesion which is responsible for the true shoulder symptom of the posterior and posteroinferior instability. Therefore, intact labrum does not produce shoulder pain no matter how lax the glenohumeral joint is. Increased translation alone produces asymptomatic posterior clunk until the repetitive rim loading eventually develops posteroinferior labral lesion (Table 13.1).

In the older patient, a tear of the rotator cuff or fracture of the greater tuberosity may occur with a shoulder dislocation (Fig. 13.8).

Table 13.1 Kim classification of posteroinferior labral lesion

Type	Arthroscopic findings	MR-arthrogram findings
I	Incomplete stripping	Type I: separation without displacement
II	Marginal crack	Type II: incomplete avulsion
III	Chondrolabral erosion	Type III: loss of contour
IV	Flap tear	Type III: loss of contour

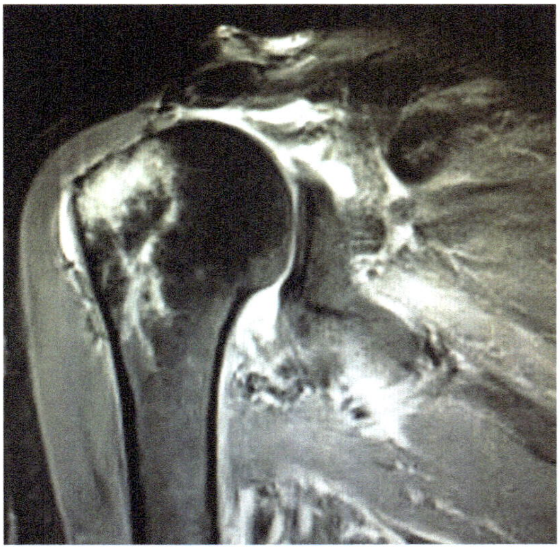

Fig. 13.8 MRI of shoulder of 70-year-old gentleman who dislocated his glenohumeral joint. Note the extensive rotator cuff tear, labral tear, and proximal humeral fracture with significant bone edema

References

1. Hintermann B, Gachter A. Arthroscopic findings after shoulder dislocation. Am J Sports Med. 1995;23(5):545–51.
2. Neviaser TJ. The anterior labroligamentous periosteal sleeve avulsion lesion: a cause of anterior instability of the shoulder. Arthroscopy. 1993;9:17–21.
3. Habermeyer P, Gleyze P, Rickert M. Evolution of lesions of the labrum-ligament complex in posttraumatic anterior shoulder instability: a prospective study. J Shoulder Elbow Surg. 1999;8(1):66–74.
4. Neviaser TJ. The GLAD lesion: another cause of anterior shoulder pain. Arthroscopy. 1993;9(1):22–3.
5. Sugaya H, Moriishi J, Dohi M, Kon Y, Tsuchiya A. Glenoid rim morphology in recurrent anterior glenohumeral instability. J Bone Joint Surg Am. 2003;85-A(5):878–84.
6. Itoi E, Lee SB, Berglund LJ, Berge LL, An KN. The effect of a glenoid defect on anteroinferior stability of the shoulder after Bankart repair: a cadaveric study. J Bone Joint Surg Am. 2000;82(1):35–46.
7. Bach BR, Warren RF, Fronek J. Disruption of the lateral capsule of the shoulder. A cause of recurrent dislocation. J Bone Joint Surg Br. 1988;70(2):274–6.
8. Wolf EM, Cheng JC, Dickson K. Humeral avulsion of glenohumeral ligaments as a cause of anterior shoulder instability. Arthroscopy. 1995;11:600–7.
9. Buhler M, Gerber C. Shoulder instability related to epileptic seizures. J Shoulder Elbow Surg. 2002;11:339–44.
10. Miniaci A, Berlet G. Recurrent anterior instability following failed surgical repair: allograft reconstruction of large humeral head defects. J Bone Joint Surg Br. 2001;83 Suppl 1:19–20.
11. Burkhart SS, De Beer JF. Traumatic glenohumeral bone defects and their relationship to failure of arthroscopic Bankart repairs: significance of the inverted-pear glenoid and the humeral engaging Hill-Sachs lesion. Arthroscopy. 2000;16:677–94.
12. Boileau P, O'Shea K, Vargas P, et al. Anatomical and functional results after arthroscopic Hill-Sachs remplissage. J Bone Joint Surg. 2012;94(7):618–26.
13. Yamamoto N, Itoi E, Abe H, Minagawa H, Seki N, Shimada Y, Okada K. Contact between the glenoid and the humeral head in abduction, external rotation, and horizontal extension: a new concept of glenoid track. J Shoulder Elbow Surg. 2007;16(5):649–56.
14. Lazarus MD, Sidles JA, Harryman 2nd DT, Matsen 3rd FA. Effect of a chondral-labral defect on glenoid concavity and glenohumeral stability. A cadaveric model. J Bone Joint Surg Am. 1996;78(1):94–102.
15. Kim SH, Noh KC, Park JS, Ryu BD, Oh I. Loss of chondrolabral containment of the glenohumeral joint in atraumatic posteroinferior multidirectional instability. J Bone Joint Surg Am. 2005;87(1):92–8.
16. Kim SH, Ha KI, Yoo JC, Noh KC. Kim's lesion: an incomplete and concealed avulsion of the posteroinferior labrum in posterior or multidirectional posteroinferior instability of the shoulder. Arthroscopy. 2004;20(7):712–20.

17. Bradley JP, Baker 3rd CL, Kline AJ, Armfield DR, Chhabra A. Arthroscopic capsulolabral reconstruction for posterior instability of the shoulder: a prospective study of 100 shoulders. Am J Sports Med. 2006;34(7):1061–71.
18. Lintner SA, Levy A, Kenter K, Speer KP. Glenohumeral translation in the asymptomatic athlete's shoulder and its relationship to other clinically measurable anthropometric variables. Am J Sports Med. 1996;24(6):716–20.
19. McFarland EG, Campbell G, McDowell J. Posterior shoulder laxity in asymptomatic athletes. Am J Sports Med. 1996;24(4):468–71.
20. Kim SH, Park JC, Park JS, Oh I. Painful jerk test: a predictor of success in nonoperative treatment of posteroinferior instability of the shoulder. Am J Sports Med. 2004;32(8):1849–55.
21. Kim SH, Ha KI, Park JH, Kim YM, Lee YS, Lee JY, Yoo JC. Arthroscopic posterior labral repair and capsular shift for traumatic unidirectional recurrent posterior subluxation of the shoulder. J Bone Joint Surg Am. 2003;85-A(8):1479–87.
22. Kim SH, Kim HK, Sun JI, Park JS, Oh I. Arthroscopic capsulolabroplasty for posteroinferior multidirectional instability of the shoulder. Am J Sports Med. 2004;32(3):594–607.
23. Kim SH, Park JS, Jeong WK, Shin SK. The Kim test: a novel test for posteroinferior labral lesion of the shoulder–a comparison to the jerk test. Am J Sports Med. 2005;33(8):1188–92.

14 Biceps Tendon

Vicente Gutierrez, Max Ekdahl, and Levi Morse

14.1 Introduction

The long head of the biceps (LHB) tendon has been recognized for years as a source of shoulder pain. It has been the subject of numerous studies to understand the pathology, function, anatomy, and its variations.

Instability of the tendon was probably the first recognized condition in 1829 [42] and 1841 [55]. Duplay implicated tendonitis of the LHB in his description of *periarthrite scapulo-humerale* in 1872 [24]. The first description of the biceps tendonitis was published by Pasteur in 1932 [48].

The tearing of the superior labrum near the insertion of the long head of the biceps tendon was described by Andrews et al. in 1985 [3], and later in 1990 Snyder et al. described a pattern of injury to the superior glenoid labrum beginning posteriorly and extending anteriorly and introduced the term SLAP lesion [54].

Recent advances have highlighted the clinical importance of instability of the tendon, and the association with pulley lesions and partial tears of the subscapularis tendon. There is also an association between SLAP lesions, rotator cuff pathology, and mechanical symptoms resulting from tendon hypertrophy [1].

A better understanding of the anatomy, pathology, and functions of the biceps tendon will help the clinicians to treat their patients in a more effective manner.

The interest in the biceps pathology and treatment has increased in recent times. The recognition and treatment of biceps lesions associated with rotator cuff disease has resulted in better patient outcomes. It is now appreciated that biceps tendon pathology is an important "pain generator" and that ignoring the biceps tendon may compromise the clinical outcome of operative or nonoperative treatments.

14.2 Anatomy

The LHB originates from the supraglenoid tubercle of the scapula and the superior glenoid labrum, with an intraarticular portion that passes over the humeral head before exiting the glenohumeral joint through the bicipital groove [25], at which point it becomes extraarticular.

The LHB tendon is approximately 9 cm long and 5–6 mm in diameter. Macroscopically, Vangsness et al., studying 100 shoulders, classified the attachment of the biceps tendon into four types: entirely posterior, posterior dominant, equal, and entirely anterior. The percentages of

V. Gutierrez, MD (✉) • M. Ekdahl, MD
Department of Orthopaedics and Traumatology, Clinica Las Condes, Santiago, Chile
e-mail: pacatoto@manquehue.net; maxekdahl@gmail.com

L. Morse
Department of Orthopaedic Surgery, Flinders University, Bedford Park, SA, Australia
e-mail: levimorse@yahoo.com

G.I. Bain et al. (eds.), *Normal and Pathological Anatomy of the Shoulder*, DOI 10.1007/978-3-662-45719-1_14, © ISAKOS 2015

Table 14.1 Insertion types of long head of tendon

	Vangsness (%)	Tuoheti (%)
Entirely posterior	22	28
Posterior dominant	33	55
Equal	37	17
Entirely anterior	8	0

"Rounded-off" data from Vangsness et al. [58] and Tuoheti et al. [57]

these four types were 22, 33, 37 and 8 %, respectively [58] Table 14.1.

Tuoheti et al. reported on 101 cadaveric shoulders macroscopically and then histologically found that the labral attachment of the long head of biceps tendon was posterior regardless of its macroscopic appearance, with 28 % entirely posterior, 55 % posterior dominant, and 17 % equal (Table 14.1). The variable macroscopic attachment pattern of the biceps tendon results from the differing attachment heights of the IGHL [57] Fig. 14.1.

The LHB slides passively on the humeral head during abduction and rotation. It slides up to 18 mm in and out of the glenohumeral joint during forward flexion and internal rotation Fig. 14.2.

Because of its location, the LHB is exposed to extraarticular constraints from possible subacromial impingement, and intraarticular restriction from the constant sliding of the tendon within the bicipital groove during elevation and rotation of the shoulder [1].

The LHB has an intraarticular and an extraarticular portion. The intraarticular portion is extrasynovial and is essentially static within the joint as the groove slides over the biceps during abduction and rotation [31]. This portion is flat and wide in shape with a length of 34.4 ± 4.2 mm.

The extraarticular portion is round and narrower with a length of 30.6 ± 5.7 mm [59]. The bicipital groove is an hourglass-shaped corridor between the greater and lesser tuberosities; this groove is narrowest and deepest at its midportion [47].

The reported approximate cross-sectional area of the tendon is

(a) Origin – 8 × 8 mm
(b) Entrance to groove 5 × 3 mm
(c) Musculotendinous junction 5 × 5 mm [18]

The contribution of the superior glenohumeral ligament (SGHL) and coracohumeral ligament (CHL) to the "biceps pulley" mechanism is critical to the stability of the tendon, with the failure of this mechanism leading to instability [1].

14.2.1 Restraints

The contours of tuberosities and a pulley mechanism formed by soft tissues maintain the biceps in its groove. There are two pulleys:

Medial (inferior or reflection pulley), formed by CHL, SGHL, and the superior border of the subscapularis

Lateral (superior pulley), formed by the anterior border of the supraspinatus and secondarily the rotator cuff cable Fig. 14.3

There exists a close relation among the LHB, the superiormost intramuscular tendon of the subscapularis muscle, and the SGHL [4] See Fig. 14.4.

14.2.2 Irrigation

The LHB blood supply is from the brachial artery. Three arteries supply blood to the bicipital tendon [17]. Blood supply to the proximal part of LHB is from the anterior circumflex artery, with the branches running along the bicipital groove in both cranial and caudal directions. There is a characteristic vascular pattern on the superficial surface of the tendon within the groove, while the deep "sliding" surface is avascular and composed of fibrocartilage [1]. Labral branches from the suprascapular artery supply the proximal tendon [21].

The distal portion of the tendon receives branches from the deep brachial artery [17] Fig. 14.5.

14.2.3 Innervation

Clinically, the LHB has been considered a pain generator in the shoulder. According to Alpantaki et al. [2], a rich innervation through a network of sensory sympathetic fibers may explain the bicipital pain. This pattern of innervation is asymmetrical, more concentrated at the biceps origin and less so at the musculotendinous junction. The musculocutaneous nerve is responsible for the motor innervation of the muscle.

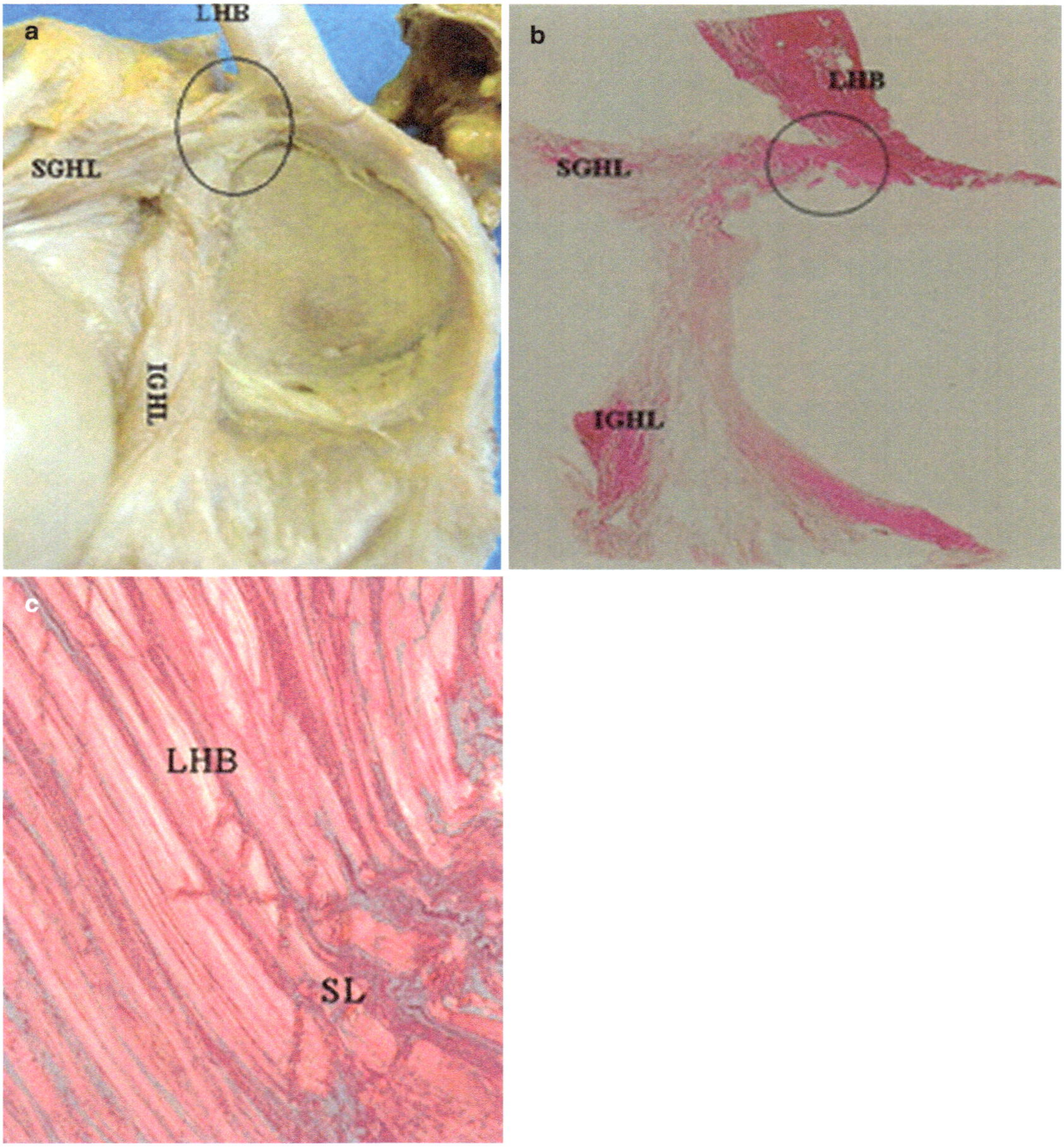

Fig. 14.1 (**a**) A posterior-dominant type. The LHB attaches to the SGT, mainly the posterior labrum. The IGHL attachment was high. The attachment of the MGHL was not clearly observed. (**b**) Histology (H & E) showed that the biceps attachment was entirely posterior. The IGHL linked to the SGHL on the superior labrum. (**c**) The polarized microscopic examination of the circled area in (b) showed that the fiber orientation of the biceps was totally posterior. *SL* superior labrum, *LHB* long head of biceps tendon, *SGHL* superior glenohumeral ligament, *IGHL* inferior glenohumeral ligament (Used with permission [57])

14.2.4 Anatomical Variations

Dierickx et al., in a review of 3,000 cases of shoulder arthroscopies, found 57 cases (1.91 %) with variations: the simple vinculum or pulley-like sling, the partial or complete mesotenon between the biceps and the capsule, the completely adherent LHB, the double tendon origin, the reversed-type split tendon, and the complete absence of the LHB. They suggested a classification of 12 variations of the intraarticular portion of the LHB [23].

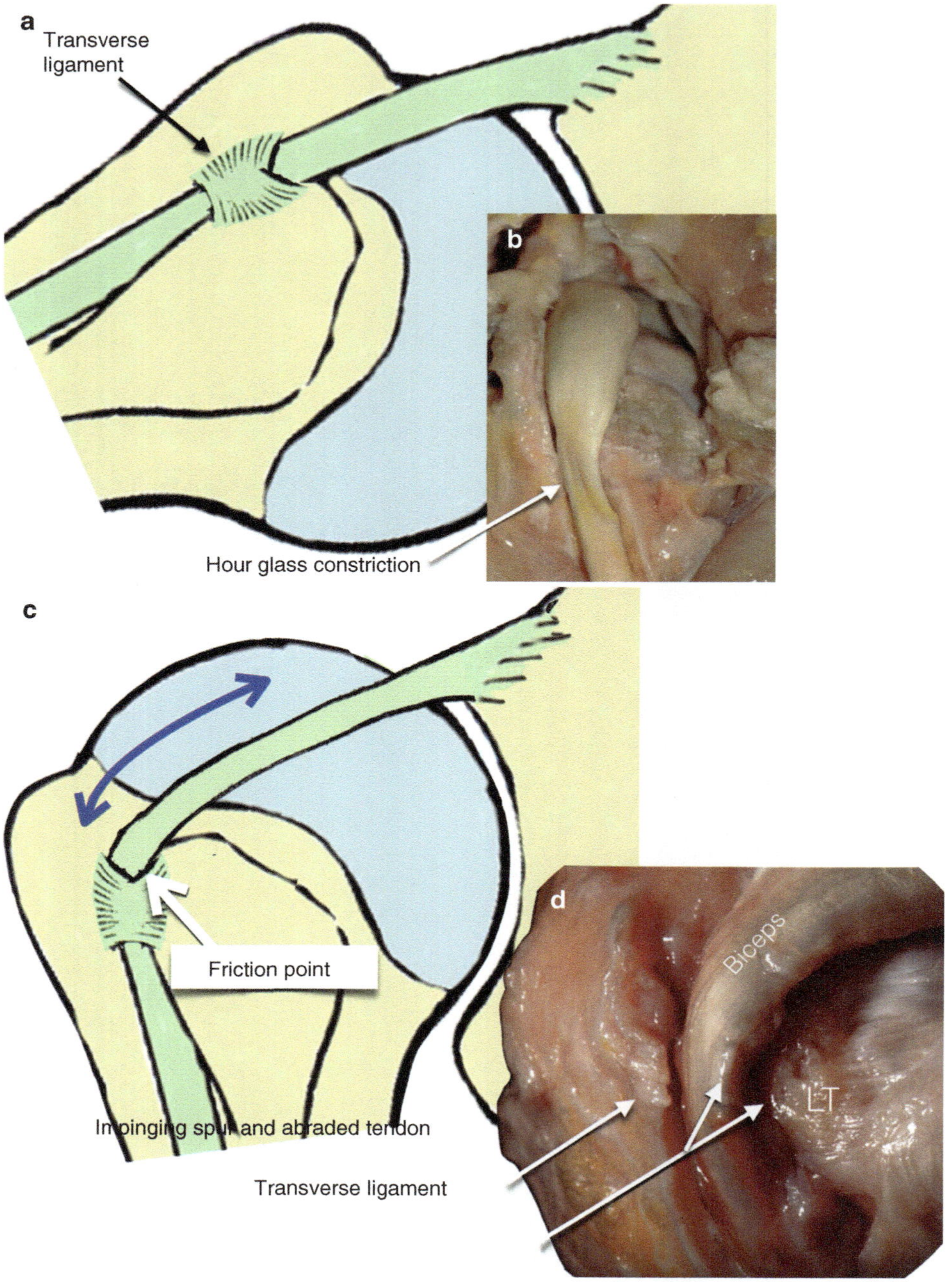

Fig. 14.2 LHB and humeral abduction. (**a**) Humeral abduction, the biceps tendon is stabilized by the transverse ligament. (**b**) The transverse ligament can impinge on the biceps tendon, producing an hour-glass constriction of the tendon. (**c**) Humeral adduction, LHB slides on the mobile humeral head up to 18 mm from forward flexion and internal rotation compared to neutral (Copyright Dr Gregory Bain), (**d**) The clinical photo demonstrates a spur and tendon abrasion (Copyright Di Giacomo [22])

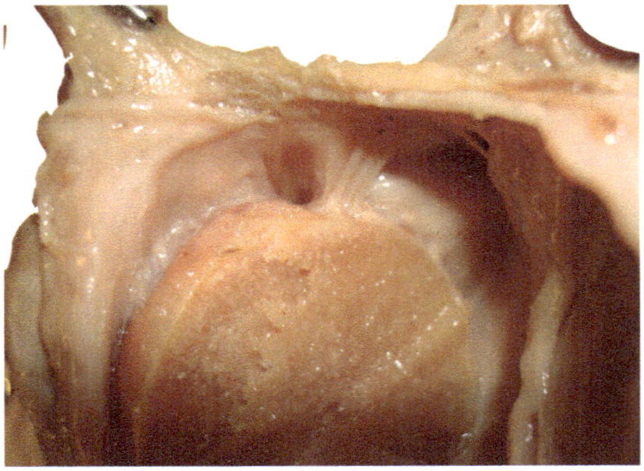

Fig. 14.3 Medial (inferior or reflection pulley) and lateral (superior pulley)

Four major families, each divided into 1–5 subgroups, were described: Fig. 14.6

1. Synovial mesentery (vinculum, pulley-like sling)
2. Adherent to rotator cuff
3. Split or bifid tendon
4. Absence of LHB

Many of these variations, such as the vinculum, the pulley-like sling, and the mesotenon-like adhesions, are mostly of a common kind. They refer to the human embryological evolution of the LHB tendon, which migrates intraarticularly after the fourth month of gestation. This process may apparently stop prematurely, which results in one of the mentioned conditions. These adhesions probably do not cause any pathology [23].

14.2.5 Comparative Anatomy

Comparative anatomy has demonstrated the evolutionary movement of the scapula to a more frontal plane with associated torsion of the humeral shaft, thus reducing the action of the LHB at the shoulder [15, 31, 35, 39].

Decreased retroversion of the proximal humerus has led to the groove no longer being centered in the plane of the humeral head, but lying at an angle of approximately 30°. As a consequence, the LHB is forced to bear on the lesser tuberosity at the medial wall of the groove, instead of at its middle. Such a position renders the tendon vulnerable to pathology, since the lesser tuberosity itself acts as a pulley [1].

14.3 Function

The function of the biceps tendon has been one of debate. Some suggest that it has no role and is only a vestigial structure, being the "appendix of the shoulder." Its role in glenohumeral kinematics is controversial. It is important for elbow function, as it is the prime forearm supinator and a significant elbow flexor.

For the shoulder, cadaveric studies have reported that it has a stabilizing effect on the glenohumeral joint in all directions. It is a humeral head depressor and is also an abductor in external rotation.

Despite the possible kinematic effect of the biceps tendon, in vivo studies are yet to establish its stabilizing effect on the shoulder. We know there is minimal functional impact of a biceps tenotomy or tenodesis. Furthermore, EMG studies have demonstrated little or no activation of LHB when the elbow is immobilized. During throwing, the biceps has been shown to function predominantly as an elbow flexor.

In summary, while important for elbow function, the biceps tendon has only a minor contribution to shoulder function. However, if a biceps lesion exists, it is often painful, and often has a significantly detrimental effect on shoulder function.

14.4 Imaging Study

Ultrasound has proven to be an accurate diagnostic method for LHB tears; however, it is less effective when studying other disorders, such as inflammatory changes or partial tears [52].

Magnetic resonance imaging (MRI) is useful for proximal biceps pathology, both for articular and extraarticular portions. Increased fluid in the synovial sheath is suggestive of tenosynovitis. In sagittal and coronal views, a hyperintense signal under the superior labrum is suggestive of injury of the labrobicipital insertion. The presence of paralabral cysts adjacent to the labrobicipital insertion is also suggestive of

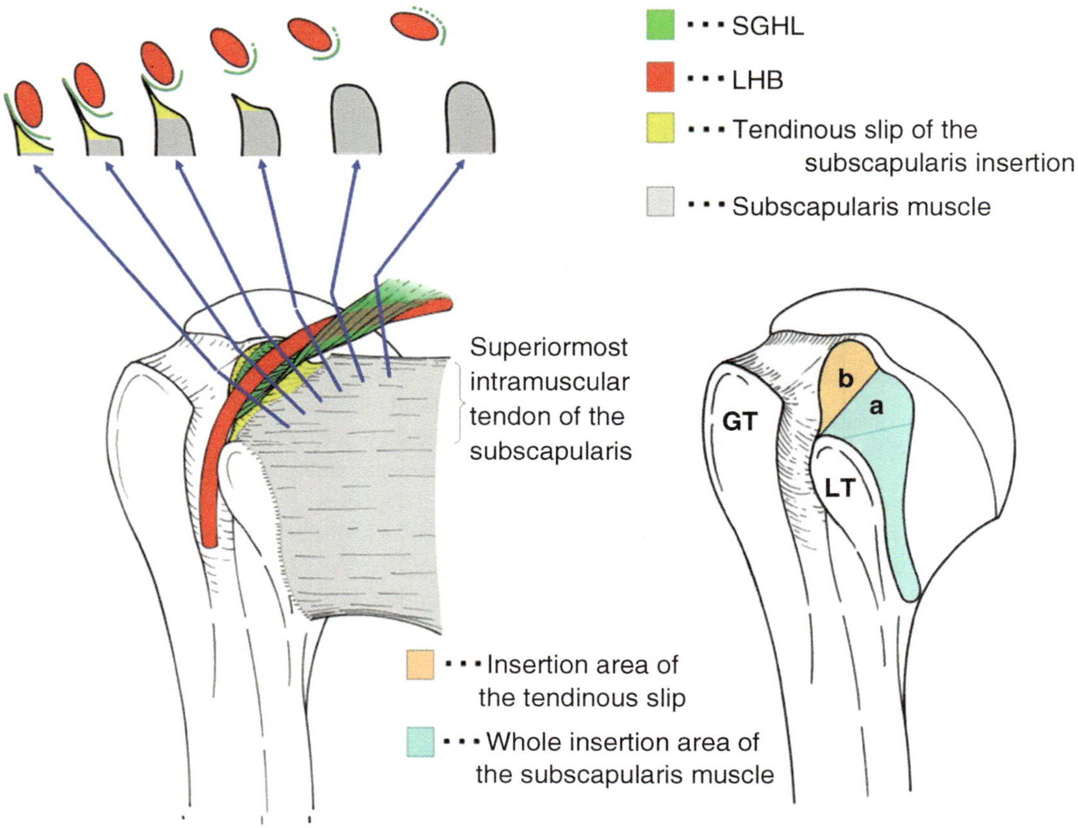

Fig. 14.4 Close relations of the long head of biceps tendon. Diagram of the relationships of the *LHB* (*red*), subscapularis (*gray*), and the *SGHL* (*green*). The intramuscular tendon of the subscapularis inserts into the superior footprint on the lesser tuberosity (*area a*). This insertion sends a thin tendinous slip to the fovea capitis of the humerus (*area b*). The SGHL passes from its glenoid attachment, passes laterally and wraps around the LHB in a spiral fashion, and attaches to the tendinous slip of the subscapularis insertion. The superior subscapularis tendon and the SGHL stabilize the LHB. The CHL and SGHL are continuous, and therefore also provide some stability for the LHB. *GT* Greater tuberosity, *LT* lesser tuberosity, *LHB* long head of the biceps tendon, *SGHL* superior glenohumeral ligament, *CHL* coracohumeral ligament (Used with permission [4])

injury. The use of gadolinium-enhanced MRI increases diagnostic accuracy; however, experienced radiologists must analyze it so as not to confuse normal anatomic variants such as the sublabral recess with pathological injuries [43]. The diagnostic accuracy of labrobicipital injuries is increased by abduction and external rotation of the arm, as it simulates the peel-back phenomenon, showing the medial labrobicipital complex caudal to the glenoid articular plane Fig. 14.7 [13]. The normal and pathologic anatomy of the biceps reflection pulleys may also be studied by MR arthrography. Oblique sagittal images and axial images are valuable for identifying the individual components of the pulley system [46].

14.5 Arthroscopic Biceps Anatomy

The LHB tendon is one of the most important reference points for orientation during the glenohumeral arthroscopy. The arthroscopic view starts from the posterior viewing portal, where its intraarticular portion and its origin at the supraglenoid tubercle by the labrobicipital union are easily visible. With the arm at the side and in neutral

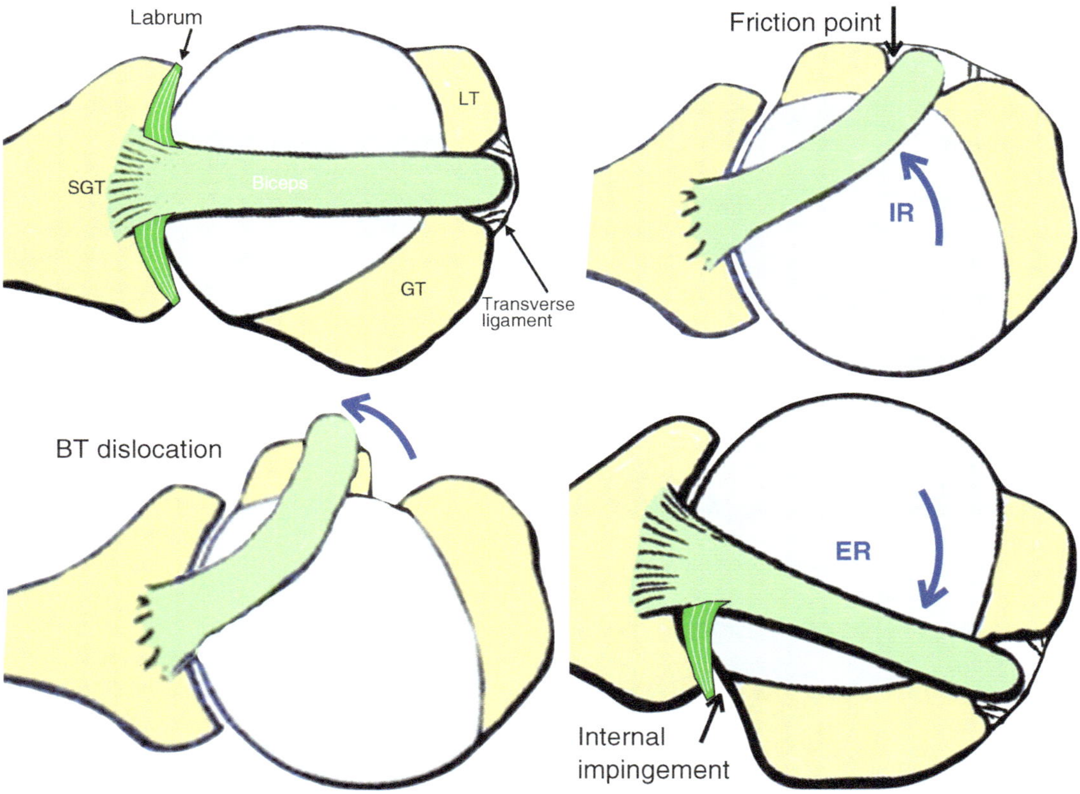

Fig. 14.5 LHBT and humeral rotation. (**a**) Medially, the LHBT is stabilized to the (*SGT*) supraglenoid tubercle, with extensions to the posterior +/− anterior labrum. Laterally, it is stabilized by the transverse ligament. Anterior and posterior stabilities are provided by the anteromedial (*AM*) and posterolateral (*PL*) pulleys. Each pulley has an osseous, ligamentous (static), and dynamic component. AM pulley- *LT*, subscapularis, SGHL. PL pulley- *GT*, supraspinatus. (**b**) Internal rotation of the humeral head will create a friction point, and the LHBT will abrade over the superior lesser tuberosity and superior subscapularis tendon. Note the AM pulley is dynamically supported by the subscapularis tendon. (**c**) Failure of the anterior pulley will produce a biceps tendon instability +/− subscapularis tendon tear. (**d**) With external rotation, especially with abduction ("cocking position"), there will be internal impingement, due to the posterior superior labrum abutting against the greater tuberosity and adjacent rotator cuff (Copyright Dr Gregory Bain)

rotation, a better visualization of the intraarticular portion is obtained. It is of major importance to assess the LHB extraarticular or intertubercular portion by applying downward traction with a probe, with an additional 3–5 cm view able to be obtained [26]. The extraarticular biceps portion is a common location for "lipstick synovitis," delamination, and partial tears. The medial and lateral pulleys complex can be seen with the scope from the posterior portal with the arm in 30° flexion and neutral rotation Figs. 14.8 and 14.9. Associated tears of the subscapularis tendon are common (Fig. 14.8b). Medial displacement of the LHB with internal rotation suggests an anteromedial pulley injury, whereas lateral displacement with external rotation of the arm suggests a posterolateral pulley injury. This is the so-called swinging test [9, 14].

14.6 Pathology

14.6.1 Tendinopathy: Tenosynovitis and Tendinosis

LHB synovial inflammation may be primary or secondary in origin. Primary tenosynovitis, characterized by the absence of other pathologic

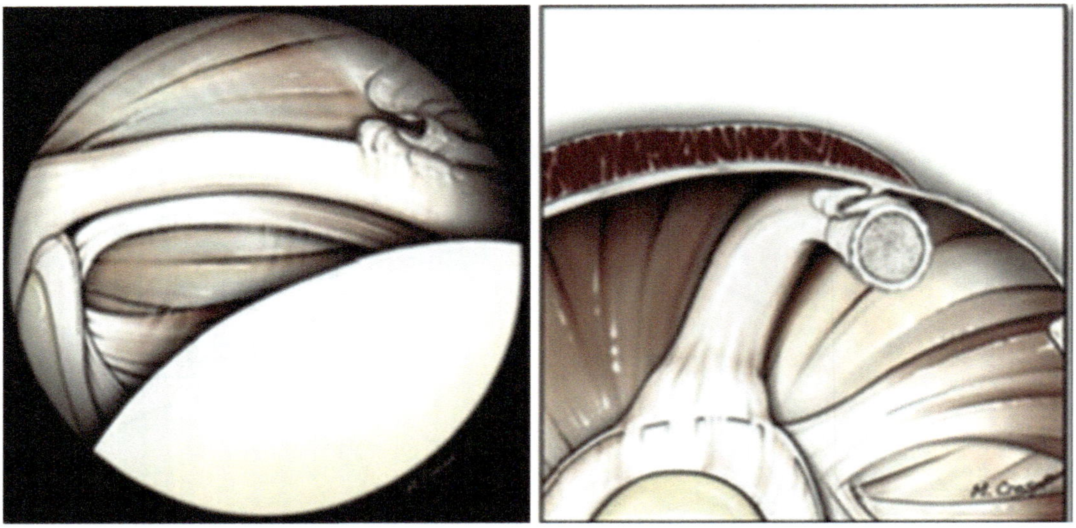

Fig. 14.6 Mesotenon (Vinculum) variation (Used with permission [23])

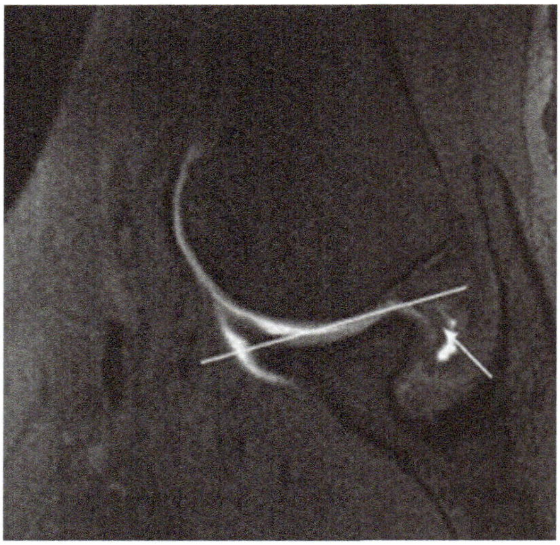

Fig. 14.7 MR arthrogram image in ABER, showing displacement of the posterosuperior labrum (*arrow*) medial/caudal to the glenoid articular plane (*line*) (Reproduced from Borrero et al. [13])

changes in the shoulder, is infrequent, corresponding to an estimated 5 % of the total incidence [5]. Secondary tenosynovitis is more frequent and is associated to disturbances of the adjacent osseous, ligamentous, and tendinous structures. Rotator cuff pathology is the most frequent association [6].

The LHB tendon degenerative process is usually associated with inflammatory changes of the synovial sheath, but not intratendinous; therefore, the most appropriate term is tendinosis. Histologically, intratendinous LHB degenerative changes include disorganization of the collagen fibers, tenocyte proliferation, and microtears [19].

In a series of 200 shoulder cases undergoing surgery for bicipital and rotator cuff pathology where LHB biopsy was obtained, Murthi demonstrated several changes such as chronic sheath inflammation, fibrosis, mucinous degeneration, vascular congestion, dystrophic calcification, and acute inflammation [45].

The clinical manifestation of biceps tendinopathy is characterized by anterior shoulder pain and the presence of symptoms with specific provocative tests. Pain directly on the bicipital groove is a frequent finding, but its sensitivity and specificity has not been studied. The Speed test is performed with the arm in forward elevation (flexion), the forearm in supination, and the elbow in extension. It is considered positive if pain in the bicipital groove appears when flexing the elbow against resistance. The specificity and sensitivity has been reported as 14 and 90 %, respectively [8].

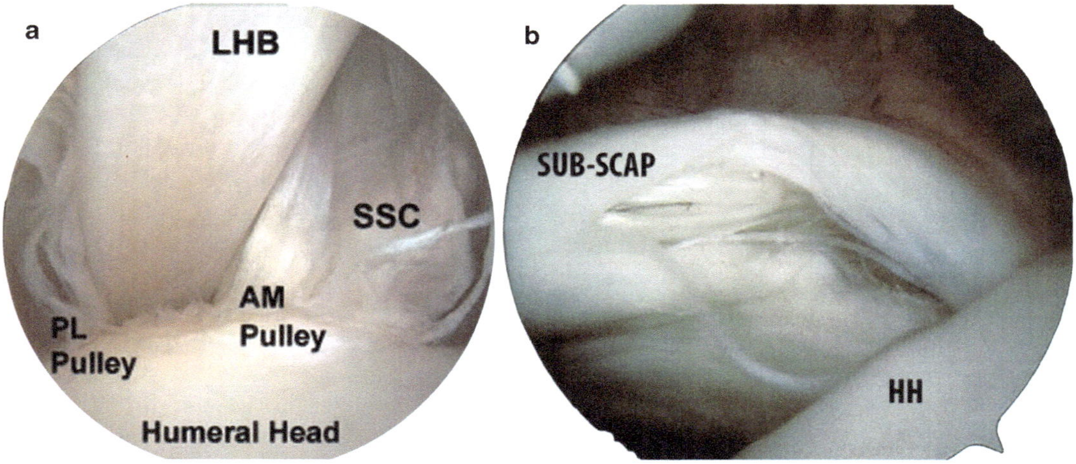

Fig. 14.8 (**a**) Left shoulder. Arthroscopic view of anteromedial (*AM*) pulley and the posterolateral pulley from posterior viewing portal. (**b**) Partial tear of the superior border of the right subscapularis tendon

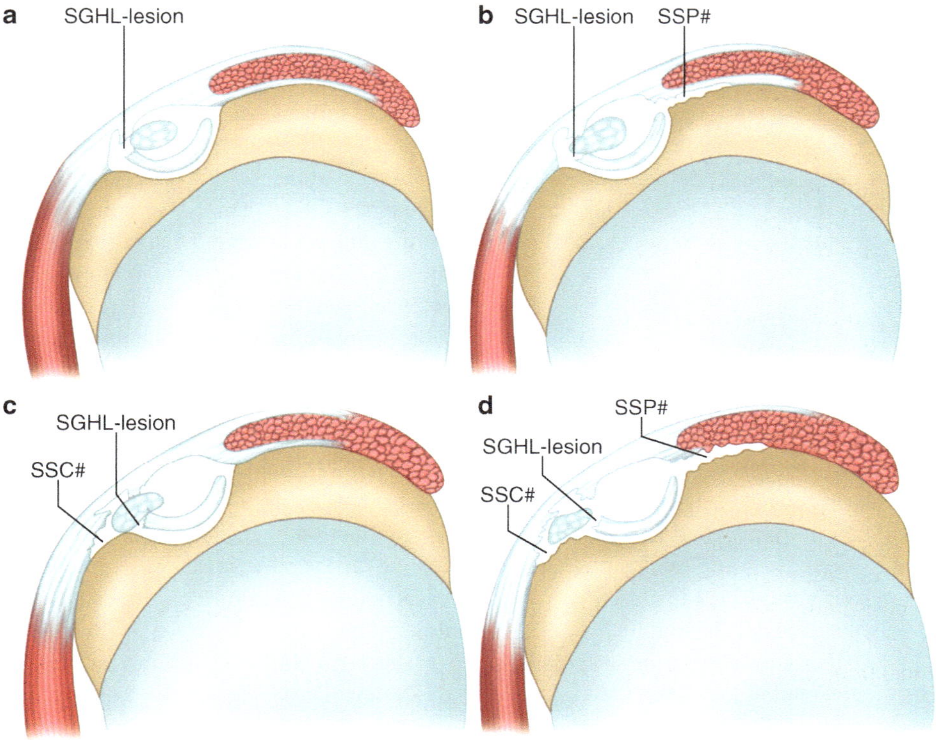

Fig. 14.9 Fig 14.9: Habermeyer LHB instability due to pulley lesions. (a) Group 1: isolated lesion of the SGHL (*arrow*). (b) Group 2: SGHL lesion and partial articular-side supraspinatus tendon tear (SSP#) (*arrows*). (c) Group 3: SGHL lesion and tear of the upper third of the subescapularis tendon (SSC#) (*arrows*). (d) Group 4: combined lesion of the SGHL, a partial articular tear of the supraspinatus (SSP#) and a tear of the upper third of the subescapularis tendon (SSC#) (*arrows*). (Reproduced from Habermeyer et al. [30])

14.6.2 Hourglass Biceps

Focal tendinosis associated with LHB tendon hypertrophy prevents appropriate excursion of the tendon within the bicipital groove during active and passive arm elevation Fig. 14.2a. This has been termed an hourglass biceps, originally described by Boileau et al. [10]. Arthroscopically, when performing passive arm elevation in the scapular plane with neutral rotation and elbow extension (intraoperative hourglass test), a buckling of the intraarticular portion of the biceps is observed.

14.6.3 Partial and Complete Tears

The most frequent LHB tear sites are its origin at the labrobicipital union and close to the musculotendinous junction. Complete or partial tears are present in tendons with chronic tendinopathy where histology shows the presence of mucoid degeneration, hypoxic degeneration, disorganized collagen fibers, tendolipomatosis, and calcific tendinopathy [37]. Spontaneous tears are associated with a Popeye deformity when the muscle belly is displaced distally. LHB ruptures occur more frequently in patients over 50 years of age.

14.6.4 Instability

As previously mentioned, LHB stability is maintained by a pulley complex formed by the SGHL, CHL, the supraspinatus, and subscapularis tendons [14]. According to Habermeyer, LHB instability due to pulley complex lesions can be classified in four groups [29] Fig. 14.9.

Group 1 – Isolated SGHL lesion

Group 2 – SGHL lesion and partial articular-side supraspinatus tendon tear (PASTA)

Group 3 – SGHL lesion and tear of the upper third of the subscapularis tendon

Group 4 – Combined lesion, SGHL, PASTA, upper subscapularis

LHB tendon instability can be medial or lateral, the former associated with injuries of the upper third of the subscapularis tendon and the latter with anterior supraspinatus pathology. In medial instability, the LHB tendon can dislocate over the subscapularis when it is intact or under the subscapularis when it is disrupted. The transverse humeral ligament does not significantly contribute to the LHB stability and may be intact in cases of tendon instability [34].

14.6.5 Labrobicipital Injuries

Andrews and colleagues were the first to describe the possible role of the labrobicipital injury as a cause of pain and dysfunction [3]. They postulated that eccentric biceps contraction in the follow-through phase in throwers could provoke a tensile overload at the superior labral insertion. Later, Burkhart et al. [16] postulated a different mechanism. The external hyperrotation observed in throwers generates a progressive contracture of the posterior glenohumeral capsule and an internal rotation deficit, which leads to a posterosuperior shift of the center of rotation of the humeral head and functional loosening of the anterior portion of the inferior glenohumeral ligament (IGHL). This torsional force occurring during the cocking and late cocking phases of throwing, with the arm in abduction and external rotation, generates a peel-back mechanism at the labrobicipital insertion. Once the superior labrobicipital complex is detached, more external rotation and posterosuperior shift may occur. An acute traumatic mechanism may also explain the labrobicipital union injury, the most frequent mechanism being a fall with the arm in flexion and abduction, which generates a compressive force over the superior labrobicipital union [54]. Clearly, more than one mechanism may explain this injury.

The term SLAP (Superior Labrum Anterior to Posterior) was coined by Snyder et al. [54], defining four different types of injury (I–IV). The Type II injuries were subclassified by Morgan, and the classifications were further extended by Maffet (V–VII) and Powell (VIII–X) [38, 44, 49] Figs. 14.10, 14.11, and 14.12.

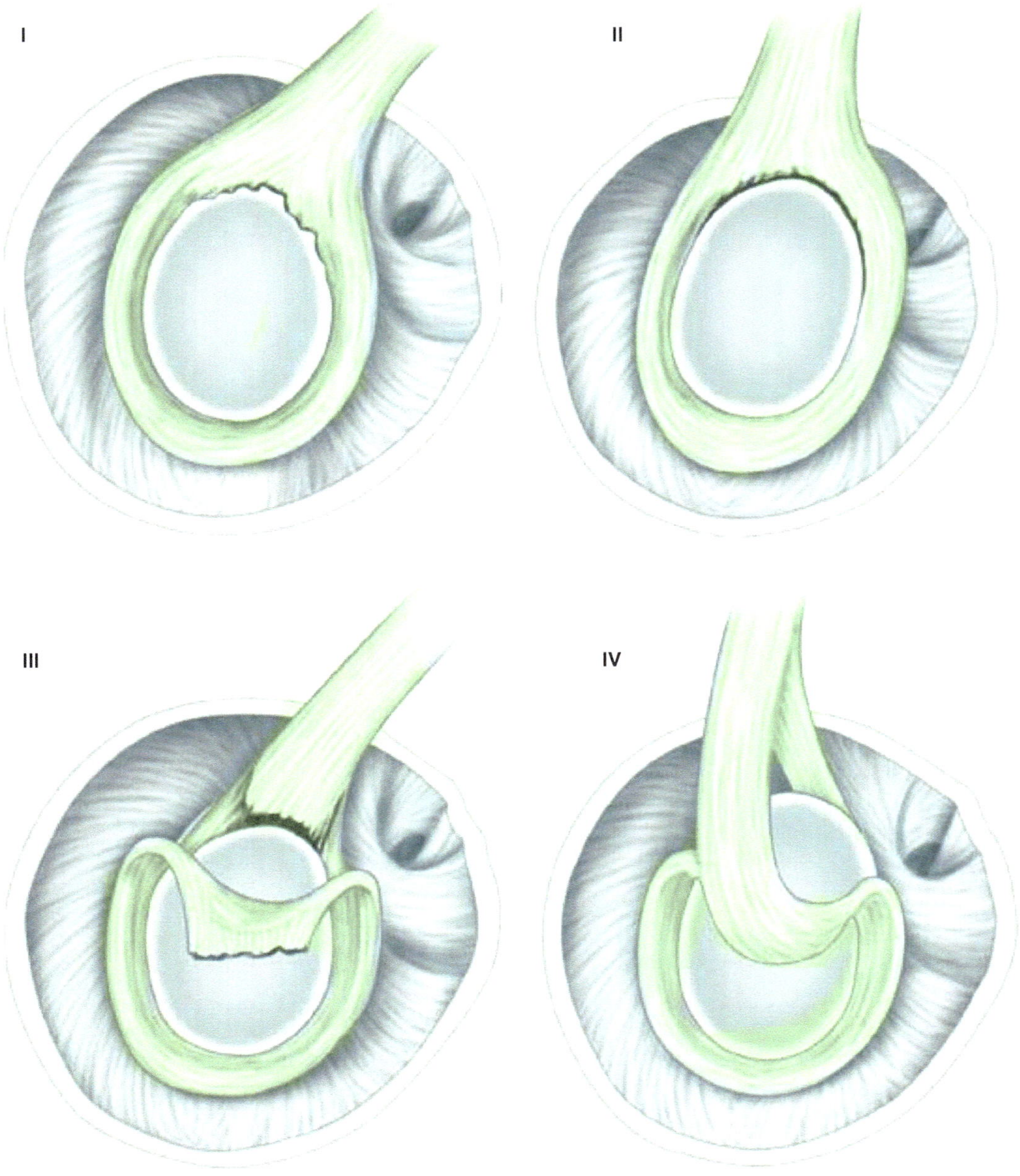

Fig. 14.10 Snyder classification of SLAP lesions. *I* Type-I SLAP lesion. *II* Type-II SLAP lesion. *III* Type-III SLAP lesion. *IV* Type-IV SLAP lesion (Reproduced from Habermeyer et al. [30])

Type I – Fraying and degeneration of the free border of the superior labrum.

Type II – Detachment of the superior bicipito-labral complex from the superior glenoid tubercle. Most frequent (41 %).

Type IIa – Predominantly anterior.

Type IIb – Predominantly posterior.

Type IIc – Combined anterior and posterior.

Type III – Bucket-handle tear of the superior labrum.

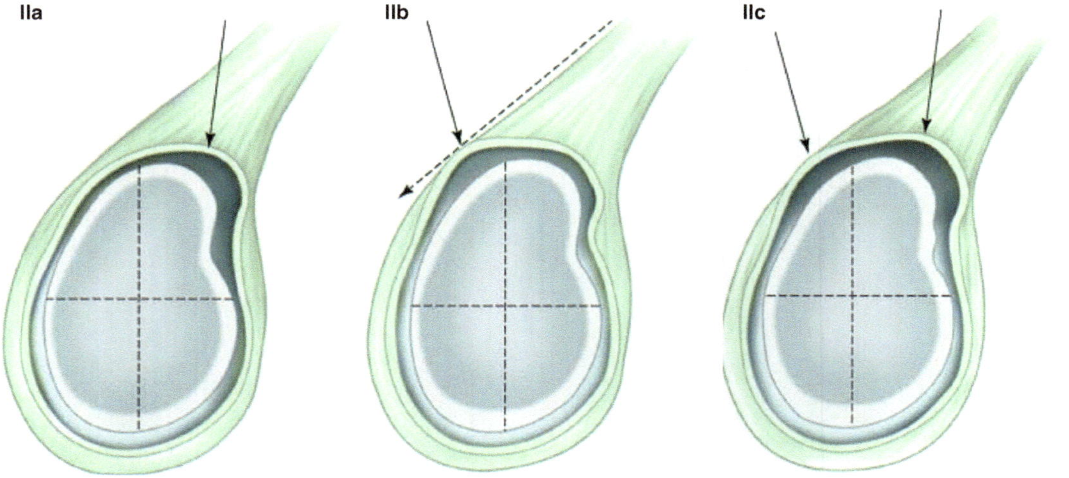

Fig. 14.11 Morgan classification of SLAP type-2 lesions. *IIa* Type-IIA SLAP lesion. *IIb* Type-IIB SLAP lesion. *IIc* Type-IIC SLAP lesion (Reproduced from Habermeyer et al. [30])

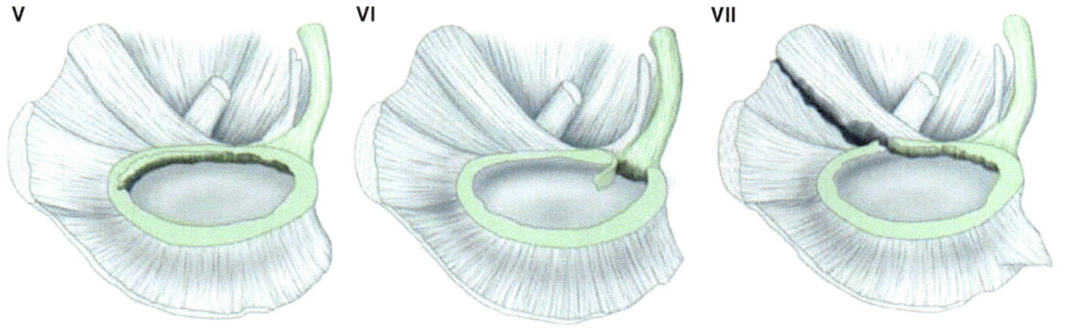

Fig. 14.12 Maffet modification of Snyder classification of SLAP lesions. *V* Type-V SLAP lesion. *VI* Type-VI SLAP lesion. *VII* Type-VII SLAP lesion (Reproduced from Habermeyer et al. [30])

Type IV – Bucket-handle tear that extends into the biceps tendon.

Type V – SLAP tear with a Bankart lesion that extends superiorly to the biceps attachment.

Type VI –SLAP tear with an unstable anterior or posterior labral flap.

Type VII – SLAP tear with an extension into MGHL.

Type VIII –SLAP tear with a posterior labral extension.

Type IX – SLAP tear with a circumferential labral tear.

Type X – SLAP tear with a posterior-inferior labral tear.

Diagnosis of a SLAP injury is challenging. It is rarely an isolated injury; in most of the cases, it is seen associated with rotator cuff injuries or glenohumeral instability; therefore, the clinical presentation can be extremely variable. Pain may be referred to the posterior, posterosuperior, or anterosuperior regions, or to the bicipital groove. For diagnosis, multiple clinical tests with variable results have been described in the literature, usually reporting acceptable sensitivity but low specificity values, such as the O'Brien test, Crank test, Speed test, compression rotation test, active compression test, anterior slide test, biceps load test I and II, and dynamic labral shear test [7, 13,

56]. In a diagnostic study, Cook et al. [20] compared the diagnostic accuracy of five clinical tests for SLAP: O'Brien test, biceps load II test, Speed test, dynamic labral shear test, and labral tension test. None of them proved useful when SLAP lesion is concomitant with other shoulder injuries; however, the biceps load II test showed a positive predictive value of 26 and negative predictive value of 93. Clinical history plus the combination of several provocative tests and the imaging study (see Sect. 14.4) are the best diagnostic tools.

Though arthroscopy is considered the gold standard for SLAP lesion diagnosis, even this tool is controversial. In independent studies, Gobezie et al. [27] and Wolf et al. [62] show a poor interobserver and intraobserver correlation with experienced shoulder arthroscopists for the arthroscopic diagnosis of SLAP injury, especially in differentiating type II lesions with normal insertions and between type III and IV lesions. Some arthroscopic findings suggestive of pathologic detachment of the labrobicipital complex include hemorrhage under the superior labrum, more than 5 mm displacement of the glenoid superior labrum, superior labrum displacement plus underlying hyaline cartilage abnormality, or the presence of granulation tissue under the superior labrum. Dynamic evaluation of the peel-back phenomenon with the arm in abduction and external rotation out of traction and the presence of a positive active compression test may be demonstrated. This test is performed with the arm in 90° of forward flexion, with the elbow extended; then, the arm is adducted 10° to 15° and internally rotated. In cases of an unstable SLAP lesion, the superior labrobicipital complex is displaced medially and inferiorly and is incarcerated in the joint [60].

14.7 Surgical Significance of the Pathoanatomy

Conservative management is the frontline in most of the LHB pathologies. Activity modifications, NSAIDs, physical therapy, and subacromial, intraarticular, or tendon sheath infiltrations are generally useful for symptom management.

Surgical treatment is generally reserved for chronic or hypertrophic tenosynovitis (hourglass biceps), SLAP lesions that do not respond to conservative therapy, biceps instability secondary to reflection pulley injuries, tenosynovitis, or partial rotator cuff tears.

In elderly and low-demand patients, tenotomizing the LHB near its insertion and allowing it to retract into the bicipital groove has proven to be effective in bicipital pain management [53]. Disadvantages with this procedure may include a Popeye cosmetic deformity, greater frequency of cramps, and reduced strength in forearm supination [61]. LHB tenodesis is the surgical treatment of choice to manage bicipital pathology in the active population and has had proven good results in terms of pain management with low frequency of cosmetic deformity (8 %) and strength preservation of forearm supination and elbow flexion [32]. How to best perform a biceps tenodesis is still controversial, both in its location as in the fixation method. There are more than 15 different tenodesis techniques described [50]. The tenodesis goal is to achieve a stable fixation and preserve the length/tension of the musculotendinous (MT) unit as anatomic as possible. Overtension may be associated to early failure and more postoperative pain, an insufficient tension to a potential strength decrease in flexion and supination, and a cosmetic deformity [36]. Knowing the anatomic location of the MT junction is important when surgical tenodesis is chosen. In a cadaveric study, Jarrett and colleagues [33] show that the MT junction is located on average 22 mm distal to the superior margin and 31 mm proximal to the inferior margin of the pectoris major Fig. 14.13. Also, LaFrance and colleagues [36] in a cadaveric study showed that the MT junction cannot be defined at one point, since it has an average length of 78.1 mm, and its proximal aspect is located 32.1 mm distally from the superior edge of the pectoralis major and its distal aspect extends distally 33 mm from the inferior edge of the pectoralis major. The frequency of inflammatory and degenerative changes in the intragroove portion of the LHB explains why the revision rates of the tenodesis techniques are lower when the humeral transverse ligament is released as compared to the techniques that preserve the ligament and maintain the tendon within the groove [50].

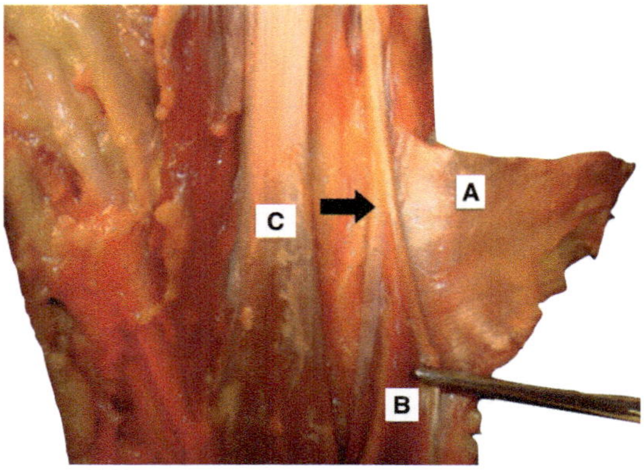

Fig. 14.13 Cadaveric specimen of a left shoulder with reflection of the pectoralis major tendon (*A*), showing the MTJ (*arrow*) of the LHB (*B*) and coracobrachialis (*C*) (Reproduced from Jarret et al. [33])

Our preferred tenodesis technique is the one described by Boileau et al. [11], which is performed arthroscopically, releasing the transverse ligament, exteriorizing the tendon, doubling it on itself to a prepared length of 25 mm, and fixating it with an interference screw to the bicipital groove 10–15 mm from its most proximal aspect. If we consider that the average length of the intraarticular portion is 35 mm, the fixation occurs 15 mm from the entrance of the groove and the tendon preparation implies 50 mm of tendon (25 mm when made double), the MT junction is maintained close to its anatomical location.

The surgical management of type II SLAP injuries that fail with conventional treatment is controversial. The options include anchoring fixation of the labrobicipital complex and LHB tenodesis. The arthroscopic fixation with anchors has demonstrated good results in selected populations; however, in patients older than 40 years and throwers, the results are less predictable [28, 51]. The decrease in external rotation is a described complication in the SLAP injury fixations, which can be associated to the low rates of return to sports in throwers. In cadavers, the anterior location of the fixation anchors has shown to cause restriction to the external rotation; so, when planning a repair, it is advisable to fixate the posterior component of the SLAP injury, especially in throwers [41].

As an alternative to anchor fixation, tenodesis has demonstrated good results. In a cohort study, Boileau and colleagues [12] showed an 87 % return to sports at a preinjury level in subjects undergoing tenodesis compared to 20 % for those undergoing a SLAP repair. Tenodesis is also a good option in those patients with failed SLAP repairs. In a case series study, McCormick et al. [40] demonstrated an 81 % rate of return to sports and good and excellent functional results in patients undergoing tenodesis after a failed SLAP repair.

References

1. Ahrens PM, Boileau P. The long head of biceps and associated tendinopathy. J Bone Joint Surg Br. 2007;89(8):1001–9.
2. Alpantaki K, et al. Sympathetic and sensory neural elements in the tendon of the long head of the biceps. J Bone Joint Surg Am. 2005;87(7):1580–3.
3. Andrews JR, Carson WG, McLeod WD. Glenoid labrum tears related to the long head of the biceps. Am J Sports Med. 1985;13(5):337–41.
4. Arai R, Mochizuki T, Yamaguchi K, Sugaya H, Kobayashi M, Nakamura T, Akita K. Functional anatomy of the superior glenohumeral and coracohumeral ligaments and the subescapularis tendon in view of stabilization of the long head of the biceps tendon. J Shoulder Elbow Surg. 2010;19:58–64.
5. Barber FA, Field LD, Ryu RK. Biceps tendon and superior labrum injuries: decision making. J Bone Joint Surg Am. 2007;89-A:1–14.
6. Beall DP, et al. Association of biceps tendon tears with rotator cuff abnormalities: degree of correlation with tears of the anterior and superior portions of the rotator cuff. AJR Am J Roentgenol. 2003;180(3):633–9.
7. Ben Kibler W, et al. Clinical utility of traditional and New tests in the diagnosis of biceps tendon injuries and superior labrum anterior and posterior lesions in the shoulder. Am J Sports Med. 2009;37(9):1840–7.
8. Bennett WF. Specificity of the Speed's test: arthroscopic technique for evaluating the biceps tendon at the level of the bicipital groove. Arthroscopy. 1998;14(8):789–96.
9. Boileau P. Long head of the biceps pathology, to tenodese or to repair. In: Instructional Course Lecture. American Academy of Orthopaedic Surgeons (AAOS) Congress; 2014.
10. Boileau P, Ahrens PM, Hatzidakis AM. Entrapment of the long head of the biceps tendon: the hourglass biceps, a cause of pain and locking of the shoulder. J Shoulder Elbow Surg. 2004;13(3):249–57.
11. Boileau P, et al. Arthroscopic biceps tenodesis: a new technique using bioabsorbable interference screw fixation. Arthroscopy. 2002;18(9):1002–12.
12. Boileau P, et al. Arthroscopic treatment of isolated type II SLAP lesions: biceps tenodesis as an alternative to reinsertion. Am J Sports Med. 2009;37(5):929–36.

13. Borrero CG, et al. Magnetic resonance appearance of posterosuperior labral peel back during humeral abduction and external rotation. Skeletal Radiol. 2010;39(1):19–26.
14. Braun S, et al. Lesions of the biceps pulley. Am J Sports Med. 2011;39(4):790–5.
15. Broca P. La torsion de l'humérus et le tropomêtre, instrument destiné à mesurer la torsion des os (rédigrée par L. Mamouvier). Revue d'Anthrop 2ême serie. 1881;4:193–1592.
16. Burkhart SS, Morgan CD. The peel-back mechanism: its role in producing and extending posterior type II SLAP lesions and its effect on SLAP repair rehabilitation. Arthroscopy. 1998;14(6):637–40.
17. Burkhead W, et al. The biceps tendon. In: Rockwood CA, Matsen FA, editors. The shoulder. Philadelphia: Saunders; 1998. p. 1009–63.
18. Burks R. Biceps tendon. In: Instructional Course Lecture 409, How about that proximal biceps tendon? New Orleans: American Academy of Orthopaedic Surgeons; 2013.
19. Claessens H, Snoeck H. Tendinitis of the long head of the biceps brachii. Acta Orthop Belg. 1972;58(1):124–8.
20. Cook C, et al. Diagnostic accuracy of five orthopedic clinical tests for diagnosis of superior labrum anterior posterior (SLAP) lesions. J Shoulder Elbow Surg. 2012;21(1):13–22.
21. Cooper DE, et al. Anatomy, histology, and vascularity of the glenoid labrum. An anatomical study. J Bone Joint Surg Am. 1992;74(1):46–52.
22. Di Giacomo G. Atlas of functional shoulder anatomy. Italia: Springer; 2008.
23. Dierickx C, et al. Variations of the intra-articular portion of the long head of the biceps tendon: a classification of embryologically explained variations. J Shoulder Elbow Surg. 2009;18(4):556–65.
24. Duplay S. La Périarthrite ScapuloHuméral. Gaz Hôp Paris, Arch Gen Med. 1872;69:571–3.
25. Elser F, et al. Anatomy, function, injuries, and treatment of the long head of the biceps brachii tendon. Arthroscopy. 2011;27(4):581–92.
26. Favorito PJ, Harding III WG, Heidt Jr RS. Complete arthroscopic examination of the long head of the biceps tendon. Arthroscopy. 2001;17(4):430–2.
27. Gobezie R, et al. Analysis of interobserver and intraobserver variability in the diagnosis and treatment of SLAP tears using the Snyder classification. Am J Sports Med. 2008;36(7):1373–9.
28. Gorantla K, Gill C, Wright RW. The outcome of type II SLAP repair: a systematic review. Arthroscopy. 2010;26(4):537–45.
29. Habermeyer P, et al. Anterosuperior impingement of the shoulder as a result of pulley lesions: a prospective arthroscopic study. J Shoulder Elbow Surg. 2004;13(1):5–12.
30. Habermeyer P, et al. Classifications and scores of the Shoulder. Berlin/Heidelberg: Springer; 2006. p. 35–40.
31. Hitchcock H, Bechtol C. Painful shoulder; observations on the role of the tendon of the long head of the biceps brachii in its causation. J Bone Joint Surg Am. 1948;30A(2):263–73.
32. Hsu A, et al. Biceps tenotomy versus tenodesis: a review of clinical outcomes and biomechanical results. J Shoulder Elbow Surg. 2011;20(2): 326–32.
33. Jarret CD, et al. Minimally invasive proximal biceps tenodesis: an anatomical study for optimal placement and safe surgical technique. J Shoulder Elbow Surg. 2011;20(3):477–80.
34. Khazzam M, George M, Churchill S, Kuhn J. Disorders of the long head of biceps tendon. J Shoulder Elbow Surg. 2012;21(1):136–45.
35. Krahl VE. Visible evidence of humeral torsion; the bicipital groove. Am J Phys Anthropol. 1948;6(2): 241.
36. LaFrance R, et al. Relevant anatomic landmarks and measurements for biceps tenodesis. Am J Sports Med. 2013;41(6):1395–9.
37. Longo UG, et al. Characteristics at haematoxylin and eosin staining of ruptures of the long head of the biceps tendon. Br J Sports Med. 2009;43(8):603–7.
38. Maffet MW, Gartsman GM, Moseley B. Superior labrum-biceps tendon complex lesions of the shoulder. Am J Sports Med. 1995;23(1):93–8.
39. Martin C. De la direction des axes du col et des condyles du femur et de l'humerus dans les mammiferes, les oiseaux et les reptiles. Compt Rend Acad Sci. 1857;44:1027–9.
40. McCormick F, et al. The efficacy of biceps tenodesis in the treatment of failed superior labral anterior posterior repairs. Am J Sports Med. 2014;42(4):820–5.
41. McCulloch P, et al. The effect on external rotation of an anchor placed anterior to the biceps in type 2 SLAP repairs in a cadaveric throwing model. Arthroscopy. 2013;29(1):18–24.
42. Monteggia G. Instituzione Chirurgiche. Milan: G Truffi; 1829. p. 170.
43. Morag Y, et al. MR arthrography of rotator interval, long head of the biceps brachii, and biceps pulley of the shoulder1. Radiology. 2005;235(1):21–30.
44. Morgan CD, et al. Type II SLAP lesions: three subtypes and their relationships to superior instability and rotator cuff tears. Arthroscopy. 1998;14(6): 553–65.
45. Murthi AM, Vosburgh CL, Neviaser TJ. The incidence of pathologic changes of the long head of the biceps tendon. J Shoulder Elbow Surg. 2000;9(5):382–5.
46. Nakata W, et al. Biceps pulley: normal anatomy and associated lesions at MR arthrography. Radiographics. 2011;31(3):791–810.
47. Nho SJ, et al. Long head of the biceps tendinopathy: diagnosis and management. J Am Acad Orthop Surg. 2010;18(11):645–56.
48. Pasteur F. Les algies de L'epaule et al physiothérapie, La téno-bursite bicipitale. J Radiol Electrol. 1932;16: 419–29.
49. Powell SE, Nord KD, Ryu RKN. The diagnosis, classification and treatment of SLAP lesions. Oper Tech Sports Med. 2004;12:99–110.

50. Sanders B, et al. Clinical success of biceps tenodesis with and without release of the transverse humeral ligament. J Shoulder Elbow Surg. 2012;21(1):66–71.
51. Sayde WM, et al. Return to play after Type II superior labral anterior-posterior lesion repairs in athletes: a systematic review. Clin Orthop Relat Res. 2012;470(6):1595–600.
52. Skendzel JG, et al. Long head of biceps brachii tendon evaluation: accuracy of preoperative ultrasound. Am J Roentgenol. 2011;197(4):942–8.
53. Slenker NR, et al. Biceps tenotomy versus tenodesis: clinical outcomes. Arthroscopy. 2012;28(4): 576–82.
54. Snyder SJ, et al. SLAP lesions of the shoulder. Arthroscopy. 1990;6(4):274–9.
55. Soden J. Two cases of dislocation of the tendon of the long head of the biceps humeri from its groove. Med Chir Trans. 1841;24:212–20.
56. Tennent TD, Beach WR, Meyers JF. A review of the special tests associated with shoulder examination. Part II: laxity, instability, and superior labral anterior and posterior (SLAP) lesions. Am J Sports Med. 2003;31(2):301–7.
57. Tuoheti Y, et al. Attachment types of the long head of the biceps tendon to the glenoid labrum and their relationships with the glenohumeral ligaments. Arthroscopy. 2005;21(10):1242–9.
58. Vangsness CT, et al. The origin of the long head of the biceps from the scapula and glenoid labrum. An anatomical study of 100 shoulders. J Bone Joint Surg. 1994;76(6):951–4.
59. Verma N. Long head biceps tendon: surgical techniques and complications. Instructional Course Lecture 271. Biceps Tendon: problems and surgical techniques. San Francisco: American Academy of Orthopaedic Surgeons; 2012.
60. Verma NN, Drakos M, O'Brien SJ. The arthroscopic active compression test. Arthroscopy. 2005;21(5): 634.e1–4.
61. Wittstein JR, et al. Isokinetic strength, endurance, and subjective outcomes after biceps tenotomy versus tenodesis: a postoperative study. Am J Sports Med. 2011;39(4):857–65.
62. Wolf BR, et al. Agreement in the classification and treatment of the superior labrum. Am J Sports Med. 2011;39(12):2588–94.

Part IV

Other Joints and Bursae

Subacromial Space

15

Stephanie C. Petterson, Allison M. Green,
and Kevin D. Plancher

15.1 Introduction

Subacromial impingement syndrome is a common cause of shoulder pain that afflicts both athletes and nonathletes alike, leading to decreased ability to participate in sports and daily activities. Impingement syndrome was first coined by Neer in 1972 to describe the trauma to the supraspinatus tendon encountered as it passes below the coracoacromial ligament and the anterior one-third of the acromion [1]. External impingement has been thought to be caused by the bony anatomy, specifically the shape of the acromion, as well as abnormalities in the surrounding soft tissues, such as the subacromial bursa and coracoacromial ligament, leading to a physical loss of subacromial space due to bony growth or inflammation. Secondary external impingement is often the result of altered scapulohumeral mechanics from glenohumeral instability and muscle imbalances [2]. Subacromial impingement syndrome can lead to a variety of sequelae, including rotator cuff tendinopathy, partial or full thickness rotator cuff tears, calcific tendinitis, and subacromial bursitis [3].

S.C. Petterson, MPT, PhD • A.M. Green, PhD
Department of Research, Orthopaedic Foundation,
Stamford, CT, USA

K.D. Plancher, MD (✉)
Department of Orthopaedic Surgery,
Albert Einstein College of Medicine,
New York, NY, USA

Plancher Orthopaedics and Sports Medicine,
1160 Park Avenue, New York, NY 10128, USA

Orthopaedic Foundation, Stamford, CT, USA
e-mail: kplancher@plancherortho.com

15.2 Description of Structure

Understanding the anatomy of the subacromial space, including the relationship between the bony anatomy and interposed subacromial bursa, is important in making the diagnosis of subacromial impingement syndrome and avoiding complications with surgical intervention. The subacromial space is defined by the coracoacromial ligament and acromioclavicular (AC) joint superiorly, the anterior edge and undersurface of the acromion, and the humeral head inferiorly. The rotator cuff tendons, subacromial bursa, long head of the biceps tendon, and coracoacromial ligament are located within this space. On average, the subacromial space, measured as the width of the space between the inferior surface of the acromion and the head of the humerus on anteroposterior radiographs, is 1–1.5 cm [3, 4].

15.3 Description of Important Parts

The bony structures of the subacromial space include the acromion, coracoid process, distal clavicle, acromioclavicular joint, and the greater

G.I. Bain et al. (eds.), *Normal and Pathological Anatomy of the Shoulder*,
DOI 10.1007/978-3-662-45719-1_15, © ISAKOS 2015

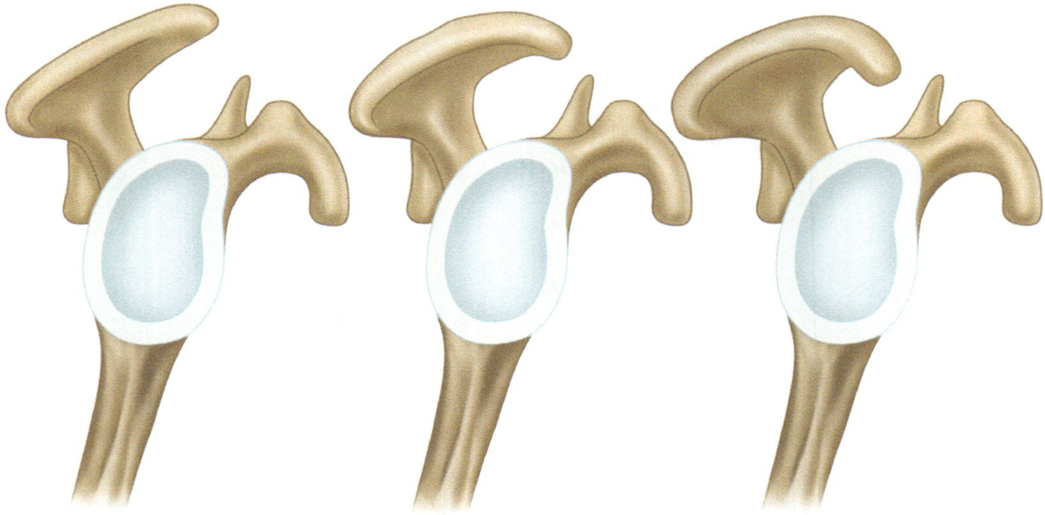

Fig. 15.1 Bigliani classification of acromial shape: type I (flat), type II (curved), and type III (hooked) (Figure adapted from Kevin D. Plancher, MD)

tuberosity of the humerus. The shape of the acromion dictates the size of the subacromial space and the room for the rotator cuff tendons. Bigliani et al. [5] classified three different acromion shapes. A type I acromion is flat. A type II acromion is curved. A type III acromion is hooked, which as a result decreases the subacromial space. A type III acromion has also been associated with a higher incidence of rotator cuff tears (Fig. 15.1)

The subacromial bursa lies between the undersurface of the acromion and the superior surface of the rotator cuff. The subacromial bursa does not have a sturdy ligamentous capsule, and the fascicle attachments are often quite thin. This anatomy can lead to fluid extravasation into the muscle and subcutaneous envelope of the shoulder. Surgical time should be limited and the fluid pump pressure and flow should be kept to a minimum when operating in this area.

Coracoid impingement, while less common, occurs between the anteromedial portion of the coracoid process and the lesser tuberosity of the humerus [6]. The coracoacromial ligament extends from the coracoid process to the anterior aspect of the acromion; thickening of the coracoacromial ligament can also reduce the size of this space.

15.4 Biomechanics

Functional range of motion of the shoulder can alter the dimensions of the subacromial space and contribute to clinical signs of impingement syndrome, specifically, shoulder abduction and rotation [7, 8]. As the shoulder moves from 30° to 120° of abduction, the distance between the humerus and the acromion significantly decreases by almost 50 % [8]. The minimum distance between the acromion and the humerus is also smallest with arm in external rotation at 90° [8]. Grachien and colleagues reported that when the arm is in 90° of abduction and 45° of internal rotation, the supraspinatus is closest to the anteroinferior border of the acromion [8]. Furthermore, while arm elevation leads to a decrease in subacromial space width, adduction muscle forces substantially increase the acromiohumeral distance and claviculohumeral distance compared to abduction muscle forces (138 % at 90° relative to abduction forces). The biomechanics support strengthening of the adductor muscles, including the latissimus dorsi, subscapularis, and teres major and minor, in both conservative and postoperative rehabilitation programs to avoid and lessen the symptoms of impingement syndrome [9].

Pressure in the subacromial bursa also changes with arm position as well as with changes in the demand of the activity. Sigholm et al. monitored pressure in the subacromial bursa using a micro-capillary infusion technique with the arm at rest by the side, with the arm abducted to 45°, and with the arm abducted to 45° holding a 1 kg weight [10]. Subacromial bursal pressure increased fourfold from 8 to 32 mmHg when changing the arm from 0° to 45° of abduction and increased sixfold when holding a 1 kg weight.

Scapular dyskinesis or dynamic scapular winging, seen on evaluation of the scapula during overhead range of motion, may contribute to clinical signs of impingement as a result of abnormal scapular muscle activity and subsequent abnormal scapular kinematics. Patients with impingement demonstrate decreased output force, muscle balance, electromyographical activity, and activation latency of the trapezius and serratus anterior muscles which stabilize the scapula and control scapular rotation. A study by Silva et al. [11] revealed that the subacromial space is smaller in patients with scapular dyskinesia than in control patients, and that the subacromial space undergoes greater reduction when the shoulder is moved from neutral abduction to 60° of elevation in patients with scapular dyskinesia than in control patients. Additionally, weakness of the rotator cuff can lead to abnormal glenohumeral and scapulothoracic kinematics and subsequent narrowing of the subacromial space [3].

15.5 Variations

There are several anatomic variants that can influence the development of subacromial impingement. The acromion has three cartilaginous growth centers that ossify during development. In up to 15 % of people, one or more of these growth centers do not ossify [12]. A persistent growth center in the acromion is referred to as the os acromiale. The presence of an os acromiale may increase the risk of subacromial impingement syndrome as failure of this growth plate to close allows for some motion of the acromion which may impinge on the rotator cuff tendons or bursa. Additionally, variations in the coracoacromial ligament, the direction of the slope or angle of the acromion, the inferior aspect of the acromioclavicular joint, and the shape of the acromion all contribute to narrowing of the subacromial space and outlet of the rotator cuff tendons [3].

15.6 Physical Examination: Special Tests

Several key maneuvers are essential to include in physical examination to aid in the diagnosis of subacromial impingement syndrome. The likelihood of a diagnosis of impingement is >95 % when a specific battery of tests are positive, including the Hawkins-Kennedy impingement

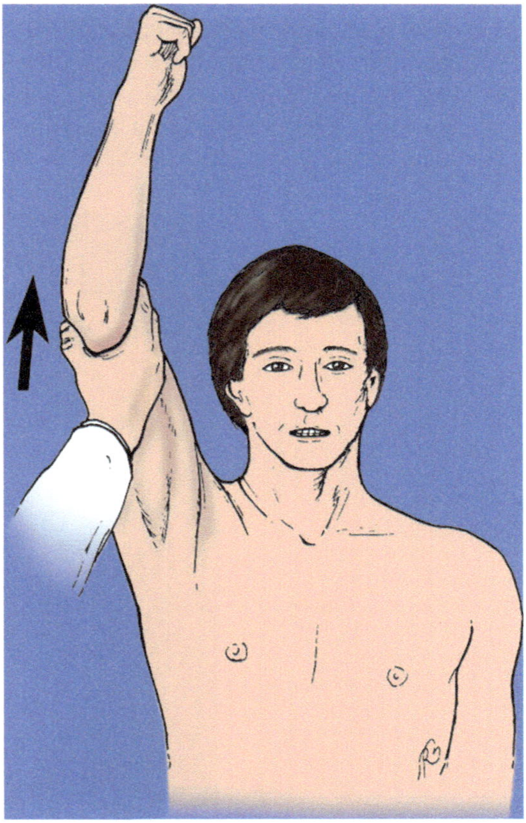

Fig. 15.2 Neer impingement sign. The examiner stabilizes the scapula and passively flexes the arm greater than 120° with the arm internally rotated. Provocation of pain at the anterolateral edge of the acromion is indicative of subacromial impingement (Copyright Kevin D. Plancher)

test, painful arc, and infraspinatus test [13]. When this battery of tests is negative, the likelihood of impingement is <24 %.

The Neer impingement sign causes provocation of pain at the anterolateral edge of the acromion when the examiner passively forward flexes the arm greater than 120° with the humerus internally rotated and the scapula stabilized (Fig. 15.2). The Neer sign has a sensitivity and specificity of 68.0 and 68.7 %, respectively [13]. Hawkins and Kennedy also described an alternative impingement test which elicits symptoms when the arm is placed in 90° forward elevation and then gently internally rotated (Fig. 15.3). The Hawkins-Kennedy sign has a sensitivity and specificity of 71.5 and 66.3 %, respectively [13]. These impingement tests place the greater tuberosity, rotator cuff, or biceps tendon against the undersurface of the acromion or coracoacromial ligament causing aggravation of an inflamed bursa.

A painful arc of motion between 60° and 120° of active forward elevation in the plane of the scapula is indicative of impingement. The patient often reports pain or painful catching in the shoulder. This test has a sensitivity of 73.5 % and specificity of 81.1 % [13]. Lastly, the infraspinatus muscle test performed with the arm at the side and the elbow flexed to 90° elicits pain when the patient resists against an internal rotation force.

Many disorders of the shoulder present similar to subacromial impingement syndrome. A diagnostic lidocaine anesthetic injection into the subacromial space can improve the accuracy of the diagnosis of subacromial impingement syndrome. We instill 10 mL of 1 % lidocaine using a 25-gauge, 1½-inch needle into the subacromial space through an anterior approach. Ultrasound can be used as an adjunct to guide the needle to ensure accuracy. Alternatively, the needle can be placed 1 cm inferior to the posterolateral corner of the acromion directed toward the coracoid. Provocative maneuvers should be performed following the injection to confirm the diagnosis. Alleviation of symptoms on impingement tests is highly indicative of subacromial impingement syndrome. The authors believe that a 1½-inch needle is essential if using a posterior approach to avoid a false negative result.

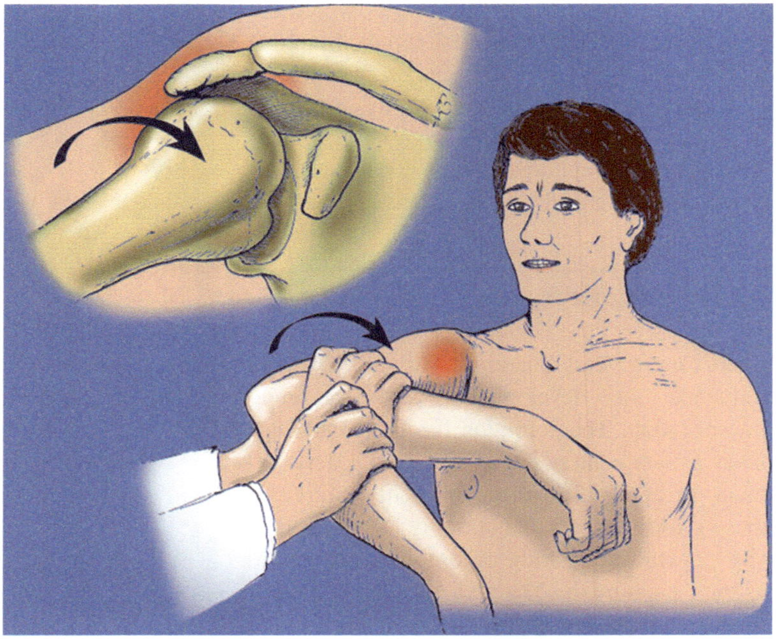

Fig. 15.3 Hawkins-Kennedy impingement sign. The patient's arm is positioned in 90° of forward elevation with the elbow flexed to 90°. The examiner then gently internally rotates the arm. Provocation of pain at the anterolateral edge of the acromion is indicative of subacromial impingement (Copyright Kevin D. Plancher)

15.7 Diagnostic Imaging

The specificity of special tests on physical examination is low; therefore, imaging of the shoulder should also be utilized in the diagnostic process in order to make an accurate and complete assessment of the underlying pathology [11]. Radiographs can aid in evaluating the concavity of the undersurface of the acromion, assessing for the presence of subacromial spurs, and for the presence of degenerative changes at the greater tuberosity, the acromioclavicular joint, or anterior acromion. A supraspinatus outlet view radiograph is best to evaluate acromial shape. The scapular outlet view, on the other hand, best evaluates the anteroinferior acromion (Fig. 15.4). This view is a true scapulolateral with the x-ray tube angled 5–10° caudally. An AP view of the shoulder with the x-ray tube angled 30° caudally can also be used to evaluate the anteroinferior acromion as well as for the presence of a calcified coracoacromial ligament (Fig. 15.5). This AP caudal tilt view has been shown to have the highest interobserver reliability [14].

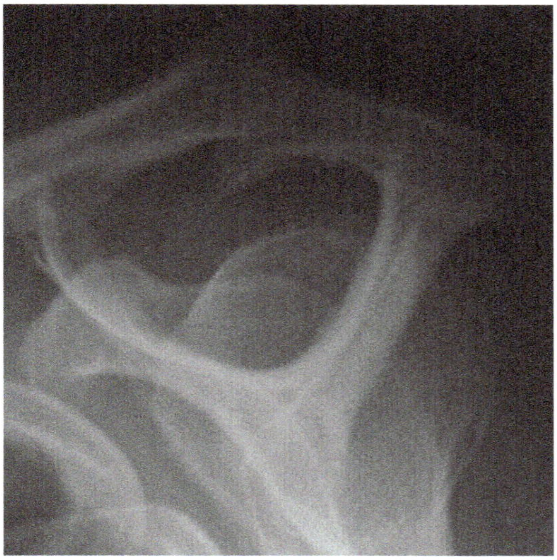

Fig. 15.4 Scapular outlet view demonstrating a type III, hooked, acromion. The scapular outlet view best evaluates acromial morphology (Copyright Kevin D. Plancher)

This radiographic series is extremely useful in surgical planning to determine the amount of undersurface of the acromion to be surgically resected to establish a flat acromion. A study by Kitay et al. demonstrated that the distance from the acromial cortex to the end of the acromial spur on x-ray significantly correlated with intraoperative spur length [14]. Acromial slope measured on the supraspinatus outlet view, which was shown to have less intraobserver reliability than the caudal tilt view, significantly correlated with intraoperative acromial thickness. Therefore, the authors believe these views should be included in routine radiographic evaluation and surgical planning when presented with suspected subacromial impingement or rotator cuff involvement prior to acromioplasty.

Magnetic resonance imaging (MRI) can also be useful to evaluate the bony pathology associated with rotator cuff pathology and assess the subacromial-subdeltoid bursa. Evaluation of subacromial spurs as well as for the presence of a type III or hooked acromion is best visualized on the coronal or sagittal oblique cuts (Fig. 15.6). Small spurs appear black (hypointensity) on T2-weighted images, whereas larger spurs appear as high signal on both T1-weighted and T2-weighted images because they contain marrow. Degenerative changes of the acromioclavicular joint can also be visualized on MRI, indicated by hypertrophy of the joint capsule as a medium signal intensity surrounding the acromioclavicular joint on pulse sequences with short repetition time (TR) and short echo time (TE). Changes in the subacromial-subdeltoid bursa and peribursal fat are signs of a rotator cuff tear as a complete tear allows extension of intra-articular fluid in the bursa. This is represented as high signal intensity or white within the bursa on T2-weighted images. The use of ultrasound, computed tomography (CT), and MRI have been shown to be reliable methods for measuring acromiohumeral distance [15]. Normal acromiohumeral distance is approximately 10.5–11 mm and is smaller in females compared to males [16, 17]. The distance is also dependent on arm position and has been

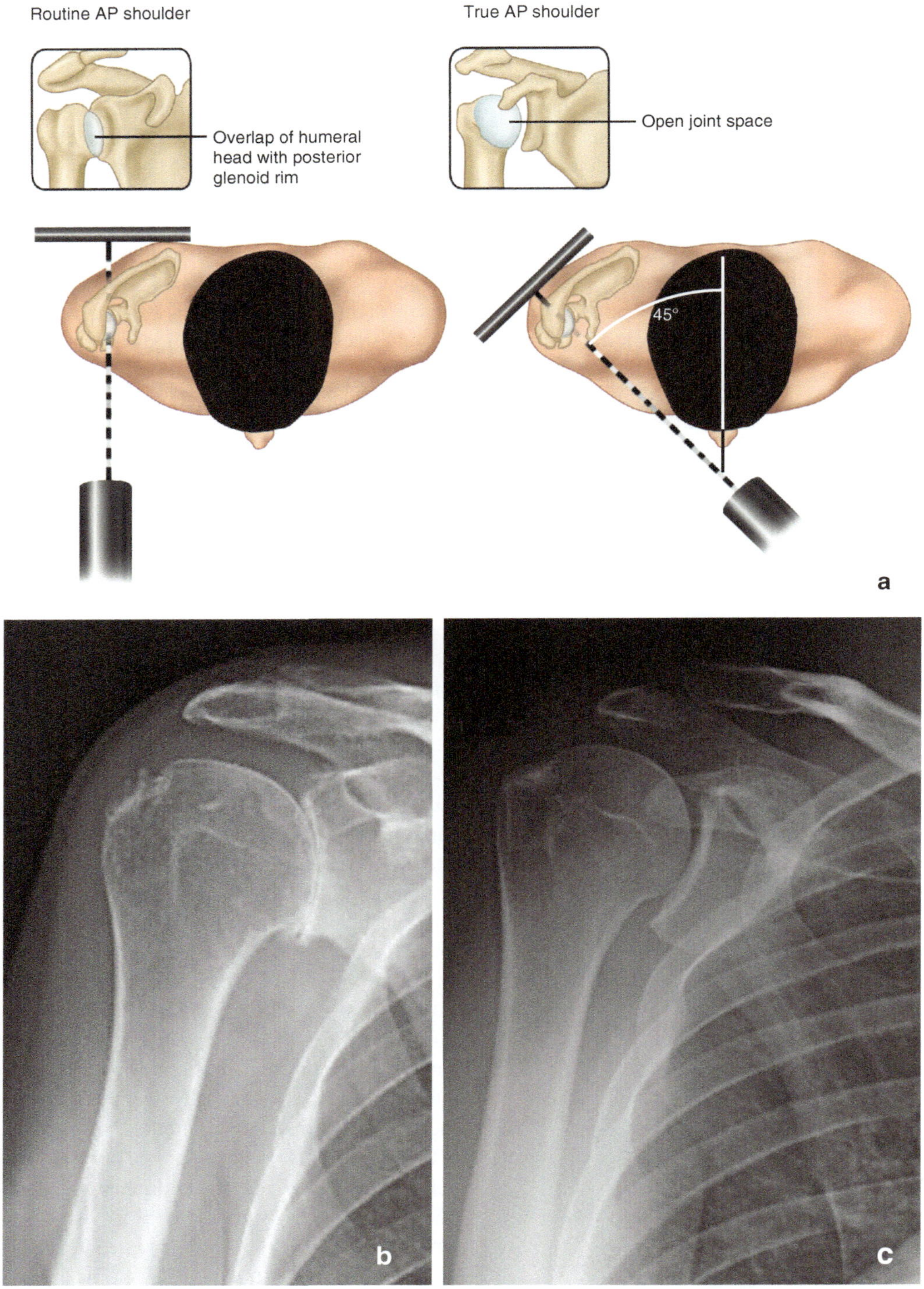

Fig. 15.5 (**a**) Artwork demonstrating the difference and correct way to obtain a true versus routine AP view of the shoulder (Figure adapted from Kevin D. Plancher, MD). (**b**) Routine AP shoulder. (**c**) True AP (Grashey) shoulder

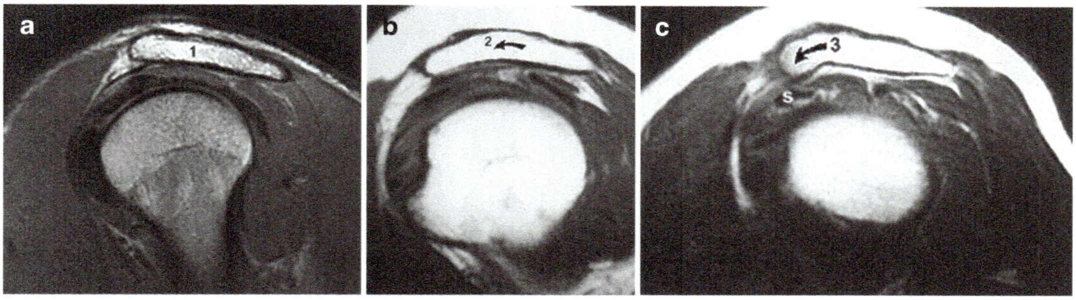

Fig. 15.6 MRI evaluation of acromial morphology. (**a**) Type I, flat. (**b**) Type II, curved. (**c**) Type III, hooked (Copyright Kevin D. Plancher)

shown to be smallest (8.1–9.9 mm) when the arm is flexed to 90° and in neutral rotation and is largest (11.2–12.2 mm) in positions of internal rotation [17]. Additionally, an acromiohumeral distance less than 7 mm has been correlated with a complete rotator cuff tear [18–20].

15.8 Sports Significance of the Anatomy

Overhead and throwing athletes often experience a loss of internal rotation range of motion in the dominant arm, ultimately affecting the biomechanics of the shoulder girdle. This loss of internal rotation has been correlated with loss of subacromial space, which may contribute to subacromial impingement syndrome in these overhead athletes known as internal impingement [21]. The cause may be related to either bony adaptations or inflammation in the shoulder or the result of instability, muscle imbalances, or abnormal scapulohumeral mechanics. This has been demonstrated in junior elite tennis players by Silva and colleagues [11]. They found that scapular dyskinesia was more prevalent among tennis players compared to a control group. Tennis players with dyskinesia had a significantly larger reduction in the subacromial space when moving the arm from 0° to 60° of abduction, a common arc of motion in tennis. Similar subacromial space changes have also been documented in basketball players [22]. Changes in subacromial space may predispose athletes to subacromial impingement syndrome and subsequent pathologies.

15.9 Surgical Significance of the Anatomy

Arthroscopic subacromial decompression is a safe and efficacious procedure to treat subacromial impingement syndrome. Resection of the anteroinferior acromion, subacromial bursa, and release of the coracoacromial ligament all lead to an increase in volume of the subacromial space. The anatomy of the subacromial bursa can make it difficult for the surgeon to navigate because of the lack of easily identifiable landmarks. Furthermore, its weak ligamentous capsule can lead to fluid extravasation into the muscle and subcutaneous envelope of the shoulder. Therefore, surgical time should be limited and the fluid pump pressure and flow should be kept to a minimum.

For optimal visualization of the subacromial bursa, the arm should be positioned in 20° of abduction and 5° of forward elevation. Less than 15 lb (6.8 kg) of traction are needed to move the greater tuberosity inferiorly and laterally out of the way to open the subacromial space. To perform a diagnostic burscoscopy, the arthroscopic cannula should be placed into the posterior portal aiming for the posterolateral border of the acromion and advanced to the posterior acromial edge. The 30° arthroscope is inserted into the subacromial space and directed to the tip of the cannula. A radiofrequency device can be used to ablate and debride the bursal adhesions and the posterior bursal curtain (posterior "veil of tears"). Visualization can be hindered throughout the procedure without debridement of this posterior bursal curtain. A medial to lateral "sweep" is performed from the medial border of the acromion to the level of the lateral portal to break up any bursal adhesions and create a "room with a view".

In order to delineate the subacromial space and widen it, the anterior and lateral borders of the acromion should be defined and the undersurface of the acromion should be debrided using a radiofrequency device. The remaining bursa can then be debrided using a full-radius motorized shaver (Fig. 15.7). It is the preference of the senior author to debride the entire bursa until the blood vessels overlying the superior rotator cuff can be clearly seen (Fig. 15.8). Use of the motorized shaver medial to the supraspinatus myotendinous junction should be avoided due to bleeding. When releasing the coracoacromial ligament, bleeding from the acromial branch of the thoracoacromial artery should be ablated with a radiofrequency device. The coracoacromial ligament should be performed at the most lateral aspect to avoid this vessel.

To gain a better appreciation of the acromial morphology, especially with a type III or hooked acromion, visualization from the lateral portal is best. A 6.0 mm oval hooded burr is placed through the lateral portal and oriented along the anterior border of the acromion [23] A 4.0 mm burr should be utilized for smaller individuals. The depth of the acromial resection is established by burying the burr to the diameter of the burr. The resection is begun just medial to the acromioclavicular joint to avoid violation of the acromioclavicular joint capsule. Once the extent of the most anterior resection is established, the remaining hook of the acromion is resected until the acromion is flat

The arthroscope is placed underneath the acromion staying in a plane parallel to the acromion. Scraping the trochar of the arthroscope directly under the acromion is avoided as the cannula may end up above the bursa. Conversely, aiming the trochar inferiorly may penetrate the infraspinatus and miss the bursa inferiorly. The cannula is aimed toward the anterior and middle (anterior to posterior) third of the acromion because the subacromial bursa is located in the anterior half of the subacromial space in front of the orientation line drawn at the beginning of the case.

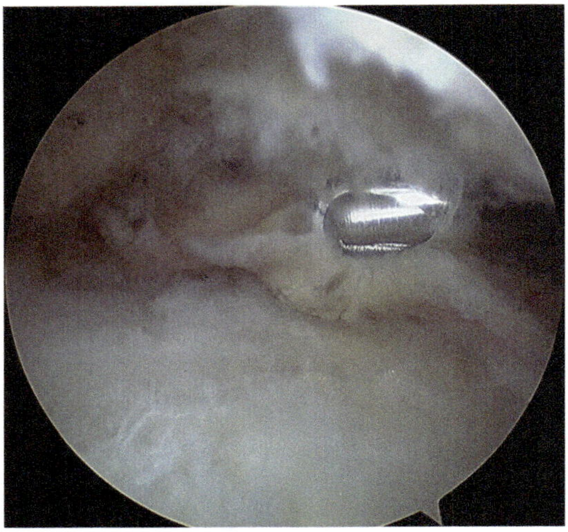

Fig. 15.7 Arthroscopic view using an arthroscopic shaver to completely debride the subacromial bursa (Copyright Kevin D. Plancher)

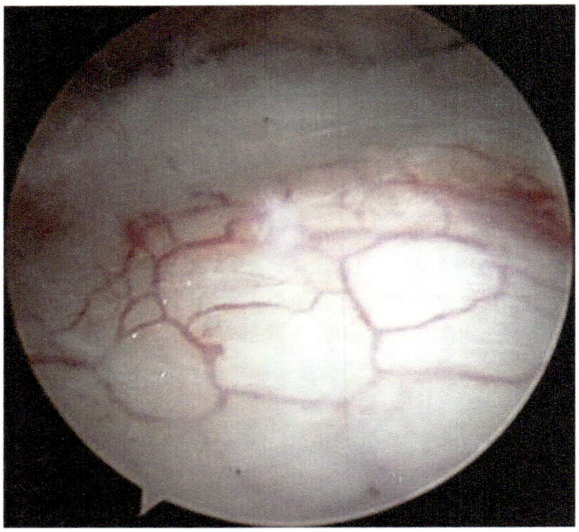

Fig. 15.8 Arthroscopic view showing the blood vessels on the superior aspect of the rotator cuff signifying a complete bursectomy (Copyright Kevin D. Plancher)

If an inside-out portal is desired to be created anteriorly, a long guide rod can be inserted to palpate the coracoacromial ligament. The rod is gently placed underneath the coracoacromial ligament and out through the anterior-superior portal. An outflow cannula is placed over the guide rod back into the bursa in a retrograde manner. The arthroscope and camera are inserted into the trochar and the pump is turned on. The distended bursal space should immediately open up into a room with a view. If muscle or fatty tissue is seen, the instruments are removed and the steps are repeated until a bursal view is achieved. If continued difficulty is encountered, the shaver is placed into the anterior portal and the bursa is carefully removed, aiming the blades superiorly toward the acromion and away from the rotator cuff tendons.

Alternately, an outside-in portal can be made laterally in the middle third of the acromion as previously described. The shaver is introduced and resection of the bursa is completed in a routine fashion [23].

The posterior, "cutting block", approach, popularized by many, is another alternative to the lateral approach for subacromial decompression [24]. The posterior portal is created 1–2 cm superior and slightly lateral to the usual posterior portal for glenohumeral arthroscopy which is too low and could therefore increase the risk of over-resection of the anterior aspect of the acromion A 6.0 mm oval burr, or a 4.0 mm burr in smaller individuals, is placed into the posterior portal and the arthroscope is placed in the lateral portal at the "50 yard line" for adequate visualization. Coplaning of the acromion is initiated at the posterior border of the clavicle and advanced forward to the anterior border of the acromion using the undersurface of the posterior acromion as the "cutting block". Each pass of the burr serves as a guide for each subsequent pass, beginning at the medial acromion moving laterally toward the lateral border. The AC joint capsule should never be violated unless an infraclavicular spur is noted on preoperative x-rays. The hooded portion of the burr can be used as a guide to assess the "flatness" of the acromioplasty. The arthroscope is placed in the posterior portal to check the lateral edge of the acromion for any remaining spurs. A nasal rasp can be used as a reference to ensure the surface of the acromion is flat.

15.10 Subacromial Bursal Examination: 8-Point Anatomy Review

For complete visualization of the anatomy of the shoulder, a methodical evaluation should be conducted for a comprehensive assessment of the shoulder. The arthroscopic 8-point anatomy exam will allow the surgeon to complete the diagnostic burscoscopy. Various procedures may be best carried out during the examination; however, following completion, the 8-point examination should be repeated.

The first five positions of the 8-point subacromial bursal examination are typically viewed from the posterior portal. *Position 1* allows for inspection of the anterior-inferior surface of the acromion and the attachment of the coracoacromial ligament at the anterolateral edge of the acromion before diving anteromedially to attach to the coracoid. Conventionally, the arthroscope is angled superiorly so the acromion is superior and the cuff is inferior. The coracoacromial ligament can extend under the entire anterior half of the acromion, attach solely to the central portion, or extend laterally under the deltoid attachment. The ligament should be smooth and glistening. Evidence of fraying or reactive bursitis should raise suspicion of impingement. The camera is then aimed laterally at the anterolateral edge of the acromion, *Position 2*, for inspection of the lateral subdeltoid shelf. It is important to differentiate the underlying rotator cuff from the plica-like shelf of bursal tissue. The arthroscope is then angled inferolaterally, moving to *Position 3*, where the insertion of the supraspinatus and infraspinatus tendons on the greater tuberosity are in view. For a complete inspection of the entire rotator cuff footprint, the arm may be internally and externally rotated. It is not uncommon for the lateral bursal shelf to obstruct the rotator cuff footprint. In cases such as these, the shelf can be removed for adequate inspection. Fraying of the supraspinatus and infraspinatus tendons is indicative of impingement. The camera is then rotated inferiorly to *Position 4* and the tip of the arthroscope is moved medially to observe the rotator cuff located medial to the tendon-bone interface. This portion of the rotator cuff is poorly vascularized and subsequently is often the first area to fail.

This area should also be inspected for evidence of calcific tendinitis. The arthroscope is then moved medially to *Position 5* to allow for inspection of the subacromial bursa. Normal bursa is smooth and vascular. Inflamed bursal tissue can become hypertrophic with significant vascular fatty tissue. This tissue needs to be removed in order to gain access to the acromioclavicular joint. If this area is not exposed, large osteophytes can be easily overlooked in the area of the medial facet of the acromion and lateral clavicle. The spine of the scapula that divides the supraspinatus and infraspinatus muscle bellies is visualized more posteriorly. The instruments should not be moved medial to the spine of the scapula because the suprascapular nerve lies within this region as it curves around the spinoglenoid notch to innervate the infraspinatus muscle.

Switching sticks are used to move the arthroscope from the posterior to the lateral portal for inspection in the remaining three positions if desired, or if a portal gains easy entry, this can be accomplished without the aid of switching sticks. *Position 6* allows for inspection of the posterior bursal curtain which extends from the posterior border of the acromioclavicular joint to the lateral border of the acromion. This curtain separates the bursa from the posterior subacromial space and is the reason the camera must be inserted into the anterior half of the space to visualize the room with a view. The curtain may become hypertrophic and obstruct the view with significant bursitis. The tip of the arthroscope is then moved laterally and aimed inferomedially to observe the posterior aspect of the infraspinatus tendon attachment on the greater tuberosity in *Position 7*. Once again, the arm can be internally rotated to complete this view. Lastly, in *Position 8* the anterior portion of the rotator cuff, the rotator interval, and the anterior bursal recess can be visualized, as well as the shape of the acromion.

15.11 Pathoanatomy

The cause of true external impingement is a result of the rotator cuff impinging against the anterior edge of the acromion with forced forward flexion. Between the undersurface of the acromion and the superior surface of the rotator cuff lies the subacromial bursa. Inflammation of the subacromial bursa can also lead to a reduction in the subacromial space due to hypertrophy and pain with overhead movements. Additionally, hypertrophy of the coracoacromial ligament can decrease the subacromial space leading to external impingement of the shoulder. Existing subacromial pathology is often correlated with altered scapular kinematics during humeral elevation including decreased upward rotation or posterior tilting. These kinematic changes have the potential to mechanically impinge on subacromial structures and narrow subacromial space [25]. Supraspinatus tendon thickness may be a causative factor of impingement [26]. Patients with subacromial impingement disorder have significantly thicker supraspinatus tendons and greater tendon occupation ratios of the subacromial space.

15.11.1 Effect of Trauma

Impingement syndrome can result from a direct blow to the superolateral aspect of the shoulder as well as the result of an axial load through the upper extremity causing the humeral head to be compressed against the inferior aspect of the acromion. This is commonly seen in sports such as skiing due to a fall or football and hockey as a result of improperly fitted shoulder pads. The resultant inflammation of the subacromial bursa or contusion of the underlying rotator cuff causes discomfort with overhead motion.

15.11.2 Effect of Disease

Patients with full-thickness rotator cuff tears have a narrower subacromial space than patients with impingement or no pathology [27]. Additionally, there is increased superior translation of the humerus in patients with rotator cuff deficiency contributing to subacromial impingement symptoms. Several factors contribute to changes in the subacromial space in these patients including the shape of the acromion (rotator cuff tears more prevalent in patients with hooked acromion), shape of the coracoid, the acromial angle, and the spine-scapula angle [28]. Cuff deficiency and other similar pathologic states increase superior translation of the glenohumeral joint due to altered muscle activation patterns.

Patients with shoulder instability may also present with signs of impingement syndrome. If patients present with persistent posterior shoulder pain, the surgeon must have a high suspicion for internal rather than external impingement. This diagnosis is most commonly restricted to overhead athletes. Up to 30 % of patients with clinical signs of subacromial impingement syndrome also have degenerative changes in the acromioclavicular joint. [29] If associated degenerative osteophytes form inferiorly and project into the subacromial space, the dimensions of this area are reduced (Fig. 15.9). This is more common in people older than 40 years of age.

15.12 Surgical Significance of the Pathoanatomy

In patients with shoulder instability and changes in the subacromial space, a subacromial decompression is indicated in conjunction with the primary superior labral repair [30]. However, the differential diagnosis is important. Subacromial decompression is contraindicated in patients with internal impingement as this could lead to further destabilization and a worsening of symptoms [30].

The role of arthroscopic subacromial decompression in rotator cuff disease may be dependent on the size of the tear. Patients with primary impingement and articular-sided partial supraspinatus tears (e.g. type 1 or 2) demonstrate good results with subacromial decompression alone, without concomitant repair, if the tear size is less than 50 % of tendon thickness [31, 32]. Subacromial decompression alone is typically contraindicated in patients that demonstrate superior migration of the humerus on AP radiographs as the result of insufficient force couples and anterior-superior escape. Removing a portion of the acromion and releasing the coracoacromial ligament in these patients increase the risk of loss of superior containment of the humeral head [30].

Scarring within the bursa following injury or surgery can restrict motion of the shoulder. This can be resected arthroscopically and improve motion (Fig. 15.10).

There are multiple vessels within the subacromial space, which can bleed and interrupt the joy of arthroscopy (Figs. 15.11, 15.12, 15.13, and 15.20). These include the acromial

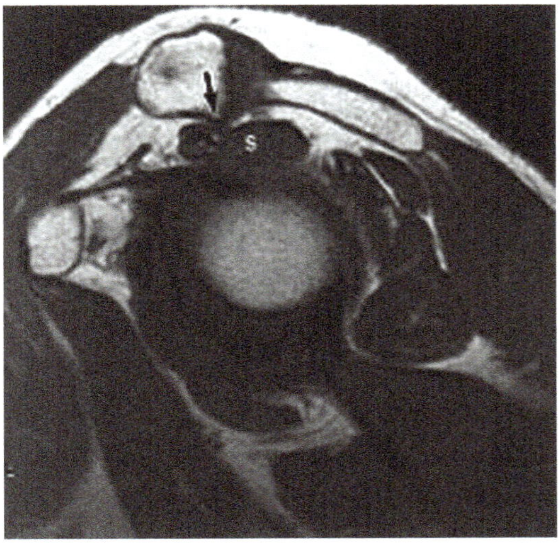

Fig. 15.9 MRI demonstrating inferiorly directed osteophyte of the acromioclavicular joint causing a reduction in the subacromial space (Copyright Kevin D. Plancher)

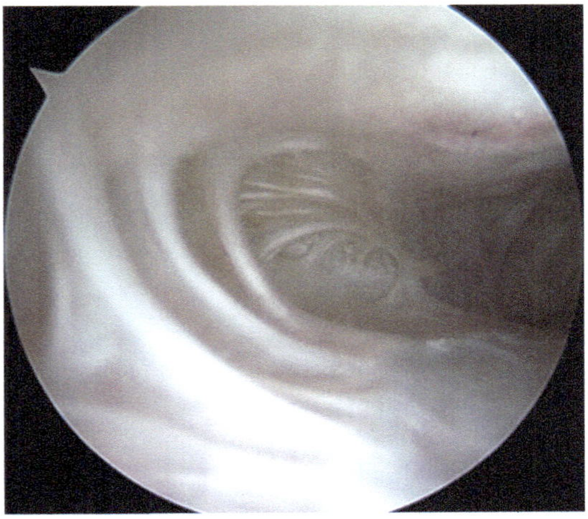

Fig. 15.10 Thickened contracted scarred subacromial bursa following previous surgery (Image Courtesy of Dr. Augustus Mazzocca)

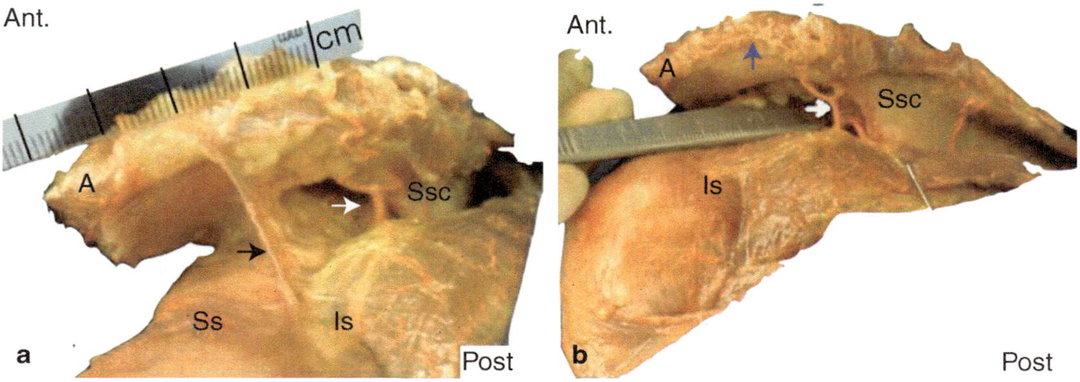

Fig. 15.11 Left shoulder (deltoid removed). (**a**) Lateral view. The black arrow shows the posterior wall of the bursa sac. The white arrow signals the posteromedial acromial artery (a branch of the suprascapular artery). The anterior border of the acromion (*A*), supraspinatus muscle (*Ss*), infraspinatus muscle (*Is*), and spine of the scapula (*Ssc*) were marked. (**b**) Posterior bursa sac removed, showing the spine of the scapula (*Ssc*) and the posteromedial acromial artery (*white arrow*). The artery terminates at the lateral border of the acromion (*A*). The *blue arrow* shows its course on the inferolateral border of the acromion (Used with permission from Yepes et al. [33])

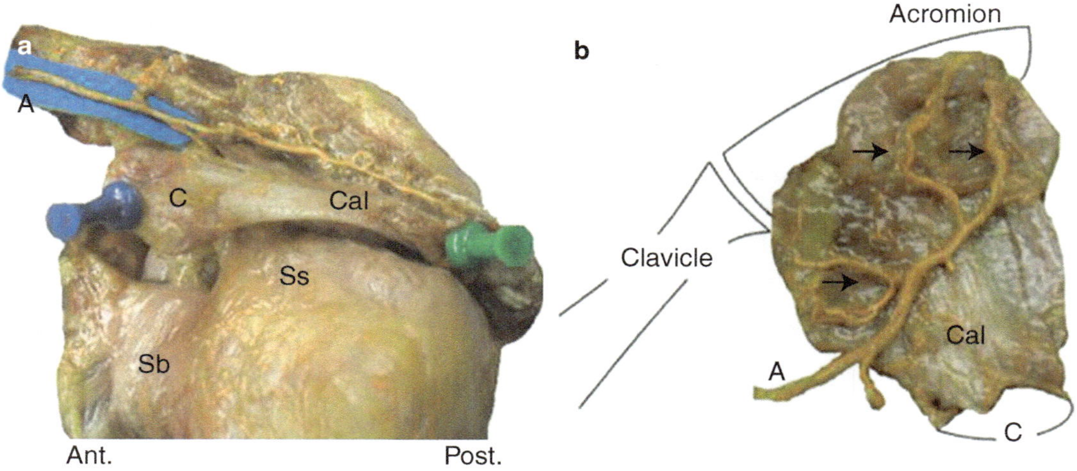

Fig. 15.12 (**a**) Anterolateral view of left shoulder (deltoid removed). The acromial artery (*A*) courses medial to lateral, passing over the coracoid (*C*) and coracoacromial ligament (*Cal*). Supraspinatus tendon (*Ss*) and subscapularis tendon (*Sb*). (**b**) Left coracoacromial ligament and vessels. The acromial artery (*A*) courses from medial/inferior to lateral/superior and divides into coracoacromial arterioles (*arrows*)

vessels, those over the coracoacromial ligament and AC Joint, those adjacent to the coracohumeral ligament and the scapular spine, and those within the deltoid muscle. Avoiding injury of these vessels will make arthroscopy considerably easier.

15.13 Summary

The subacromial space has important implications in shoulder pathology, contributing to subacromial impingement symptoms either in isolation or in conjunction with other pathologies includ-

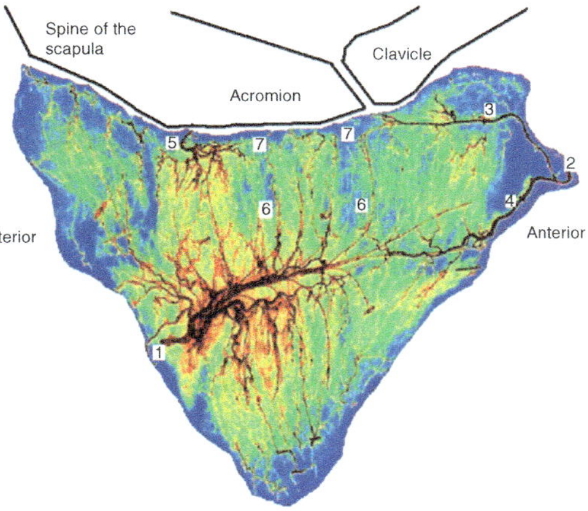

Fig. 15.13 Angiogram of right deltoid muscle. The major vessels are the posterior humeral circumflex artery (*1*) and the thoracoacromial artery (*2*) and its branches, the AC joint (*3*) and intramuscular anastomotic arteries (*4*). There are a series of anastomotic arcades (*6*) between the posterior acromial artery (*5*) and the intramuscular arteries coming from the posterior circumflex complex. Anastomotic vessels between the anterior and posterior acromial arteries (*7*) (Used with permission from Yepes et al. [33])

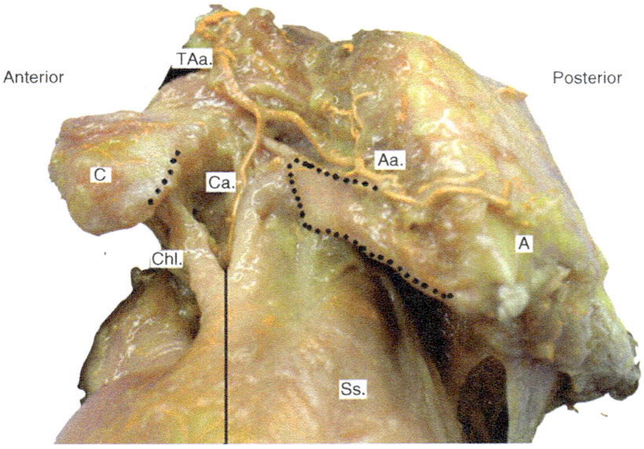

Fig. 15.14 Anterolateral view of the left shoulder (deltoid removed and the cut coracoacromial ligament identified with *dotted lines*). The thoracoacromial artery (*TAa*) and its branches (acromial artery (*Aa*) and anterior capsular artery (*Ca*). Note the anterior capsule artery passes between the supraspinatus (*Ss*), and coracohumeral ligament (*Chl*). This artery will be damaged when releasing the coracohumeral ligament or with an anterior interval slide (*solid line*) (Used with permission from Yepes et al. [33])

ing rotator cuff pathology and shoulder instability. The subacromial space is impacted not only by the bony and soft tissue anatomy but also by the glenohumeral and scapular biomechanics. A thorough understanding of the anatomy, biomechanics, and pathoanatomy is necessary to make an accurate diagnosis and appreciate the surgical implications when managing these patients.

References

1. Neer 2nd CS. Anterior acromioplasty for the chronic impingement syndrome in the shoulder: a preliminary report. J Bone Joint Surg Am. 1972;54(1):41–50.
2. Page P. Shoulder muscle imbalance and subacromial impingement syndrome in overhead athletes. Int J Sports Phys Ther. 2011;6(1):51–8.
3. Umer M, Qadir I, Azam M. Subacromial impingement syndrome. Orthop Rev (Pavia). 2012;4(2):e18.
4. Petersson CJ, Redlund-Johnell I. The subacromial space in normal shoulder radiographs. Acta Orthop Scand. 1984;55(1):57–8.
5. Bigliani L, Morrison D, April E. The morphology of the acromion and its relationship to rotator cuff tears. Orthop Trans. 1986;10:228.
6. Gerber C, Terrier F, Ganz R. The role of the coracoid process in the chronic impingement syndrome. J Bone Joint Surg Br. 1985;67(5):703–8.
7. Matthews LS, Fadale PD. Subacromial anatomy for the arthroscopist. Arthroscopy. 1989;5(1):36–40.
8. Graichen H, et al. Subacromial space width changes during abduction and rotation – a 3-D MR imaging study. Surg Radiol Anat. 1999;21(1):59–64.
9. Hinterwimmer S, et al. Influence of adducting and abducting muscle forces on the subacromial space width. Med Sci Sports Exerc. 2003;35(12):2055–9.
10. Sigholm G, et al. Pressure recording in the subacromial bursa. J Orthop Res. 1988;6(1):123–8.
11. Silva RT, et al. Clinical and ultrasonographic correlation between scapular dyskinesia and subacromial space measurement among junior elite tennis players. Br J Sports Med. 2010;44(6):407–10.
12. Papatheodorou A, et al. US of the shoulder: rotator cuff and non-rotator cuff disorders. Radiographics. 2006;26(1):e23.
13. Park HB, et al. Diagnostic accuracy of clinical tests for the different degrees of subacromial impingement syndrome. J Bone Joint Surg Am. 2005;87(7):1446–55.
14. Kitay GS, et al. Roentgenographic assessment of acromial morphologic condition in rotator cuff impingement syndrome. J Shoulder Elbow Surg. 1995;4(6):441–8.
15. McCreesh KM, Crotty JM, Lewis JS. Acromiohumeral distance measurement in rotator cuff tendinopathy: is there a reliable, clinically applicable method? A systematic review. Br J Sports Med. 2015;49(5):298–305.
16. Cotty P, et al. Rupture of the rotator cuff. Quantification of indirect signs in standard radiology and the Leclercq maneuver. J Radiol. 1988;69(11):633–8.
17. Kim H, et al. Comparative analysis of acromiohumeral distances according to the locations of the arms

and humeral rotation. J Phys Ther Sci. 2014;26(1):97–100.
18. Weiner DS, Macnab I. Superior migration of the humeral head. A radiological aid in the diagnosis of tears of the rotator cuff. J Bone Joint Surg Br. 1970;52(3):524–7.
19. Fehringer EV, et al. The radiographic acromiohumeral interval is affected by arm and radiographic beam position. Skeletal Radiol. 2008;37(6):535–9.
20. Henseler JF, et al. Cranial translation of the humeral head on radiographs in rotator cuff tear patients: the modified active abduction view. Med Biol Eng Comput. 2014;52(3):233–40.
21. Maenhout A, et al. Quantifying acromiohumeral distance in overhead athletes with glenohumeral internal rotation loss and the influence of a stretching program. Am J Sports Med. 2012;40(9):2105–12.
22. Girometti R, et al. Supraspinatus tendon US morphology in basketball players: correlation with main pathologic models of secondary impingement syndrome in young overhead athletes. Preliminary report. Radiol Med. 2006;111(1):42–52.
23. Hawkins RJ, et al. Arthroscopic subacromial decompression. J Shoulder Elbow Surg. 2001;10(3):225–30.
24. Sampson TG, Nisbet JK, Glick JM. Precision acromioplasty in arthroscopic subacromial decompression of the shoulder. Arthroscopy. 1991;7(3):301–7.
25. Wassinger CA, Sole G, Osborne H. Clinical measurement of scapular upward rotation in response to acute subacromial pain. J Orthop Sports Phys Ther. 2013;43(4):199–203.
26. Michener LA, et al. Supraspinatus tendon and subacromial space parameters measured on ultrasonographic imaging in subacromial impingement syndrome. Knee Surg Sports Traumatol Arthrosc. 2015;23(2):363–9.
27. Seitz AL, Michener LA. Ultrasonographic measures of subacromial space in patients with rotator cuff disease: a systematic review. J Clin Ultrasound. 2011;39(3):146–54.
28. Anetzberger H, et al. The architecture of the subacromial space after full thickness supraspinatus tears. Z Orthop Ihre Grenzgeb. 2004;142(2):221–7.
29. Bunker TD, Wallace WA, Austin S. Shoulder arthroscopy. St. Louis: Mosby Year Book; 1991.
30. Ryu RK, et al. Complex topics in arthroscopic subacromial space and rotator cuff surgery. Arthroscopy. 2002;18(2 Suppl 1):51–64.
31. Liem D, et al. Clinical and structural results of partial supraspinatus tears treated by subacromial decompression without repair. Knee Surg Sports Traumatol Arthrosc. 2008;16(10):967–72.
32. Burkhart SS. A stepwise approach to arthroscopic rotator cuff repair based on biomechanical principles. Arthroscopy. 2000;16(1):82–90.
33. Yepes H, et al. Vascular anatomy of the subacromial space. Arthroscopy. 2007;23(9):978–84.

16 Scapulothoracic and Subscapular Bursae

Ronald L. Diercks

16.1 Introduction

There are many bursae around the shoulder (Figs. 16.1 and 16.2). The best known of these is the subacromial bursa (see Chap. 15), but there are many other smaller bursae to protect the soft tissues over the bony prominences. Bursae are synovial-lined spaces between two moving surfaces, which provide "friction-free" motion.

16.2 Supra-acromial Bursa

The supra-acromial bursa lies on top of the acromion, subcutaneously, and does not communicate with any of the other bursae or the joint.

16.3 Subcoracoid Bursa

The subcoracoid bursa [1, 2] is located between the subscapularis tendon and the coracoid process, extending under the conjoined tendon of the coracobrachialis and the short head of the biceps. It extends posteriorly beneath the coracoid process. It facilitates movement by reducing friction between the subscapularis tendon and the tendons of the short head of the biceps and the coracobrachialis during the arc of rotation of the humeral head. In many instances, there is a connection of the glenohumeral joint cavity with this bursa. In a small percentage of patients, a natural communication exists between the subcoracoid bursa and the subacromial bursa.

16.4 Subscapular Recess

The subcoracoid bursa should not be confused with the subscapular recess. The subscapular recess, unlike the subcoracoid bursa, is not a separate bursa but rather an outpouching of the glenohumeral joint that protrudes between the glenohumeral ligaments and lies between the subscapularis muscle and the anterior surface of the scapula. The subscapular recess may occasionally extend superiorly and anteriorly over the subscapularis muscle into the subcoracoid space, close to the subcoracoid bursa, and may be confused with the subcoracoid bursa. Although the subcoracoid bursa may communicate with the subacromial-bursa in healthy patients, the subscapular recess does not. The subcoracoid bursa should not communicate with the glenohumeral joint under normal circumstances, the subscapular recess should always communicate with the joint.

Electronic supplementary material The online version of this chapter 10.1007/978-3-662-45719-1_16 contains supplementary material, which is available to authorized users.

R.L. Diercks
Departmet of Orthopedic Surgery, Sports Medicine Center, University Medical Centre Groningen, University of Groningen, Groningen, The Netherlands
e-mail: r.l.diercks@umcg.nl

G.I. Bain et al. (eds.), *Normal and Pathological Anatomy of the Shoulder*,
DOI 10.1007/978-3-662-45719-1_16, © ISAKOS 2015

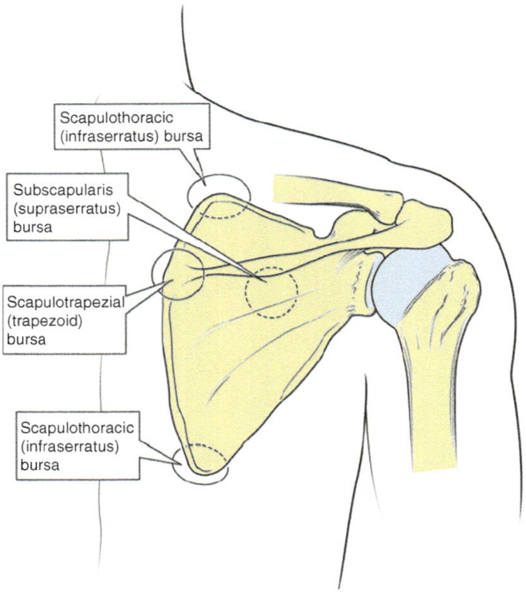

Fig. 16.1 Multiple named bursae around the shoulder, AP view (Hubbard and Poehling [6], p 551, with permission of Springer science + Business Media)

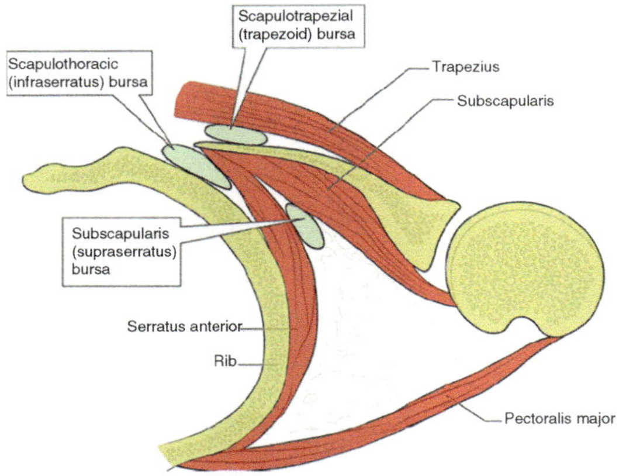

Fig. 16.2 Multiple named bursae around the shoulder, axial view (Hubbard and Poehling [6], p 551, with permission of Springer science + Business Media)

16.5 Coracoclavicular Bursa

The coracoclavicular bursa or supracoracoidal bursa is a small bursa between the conoid and trapezoid (coracoclavicular) ligaments. Sometimes more small bursae are described in this region.

16.6 Scapulothoracic Bursae

There are six bursae (two major, four minor) reported participating in scapulothoracic articulation. Two of these bursae, including the *scapulothoracic (infraserratus) bursa* and *subscapularis (supraserratus) bursa*, are the primary physiologic bursae [3]. The subscapularis bursa, located between the subscapularis and serratus anterior muscles, is on average 5.3 × 5.3 cm when present. The infraserratus bursa is found under the serratus anterior, overlying the posterolateral chest wall, and is on average. The four minor bursae are not consistently found and are often a result of abnormal scapulothoracic articulation. These are typically found along the inferior angle of the spine, at the superomedial border of the scapula, either above or below the serratus anterior, or deep to the trapezius muscle at the medial base of the scapular spine. The bursae of the superomedial border and inferior angle are frequently pathologic and responsible for symptom generation. One superficial bursa has also been described located between the latissimus dorsi and the inferomedial angle of the scapula. A *scapulotrapezial (trapezoid) bursa* has been described located between the superomedial scapula and the trapezius muscles [4].

16.7 Pathology of Subscapular Bursitis

Abnormal motion of the scapula on the underlying thorax is the basis for the development of the snapping scapula syndrome. Patients with scapulothoracic bursitis often present without a history of trauma or injury to the shoulder, although they may report a history of repetitive overhead activity such as swimming or pitching, gymnastics, rowing or weightlifting. An anatomic abnormality is often responsible for symptoms. Excessive anterior angulation of the superior angle of the scapula is an example of skeletal abnormalities. A hook-shaped extension of the superomedial border of the scapula (Luschka Tubercle) may cause irregular scapulothoracic

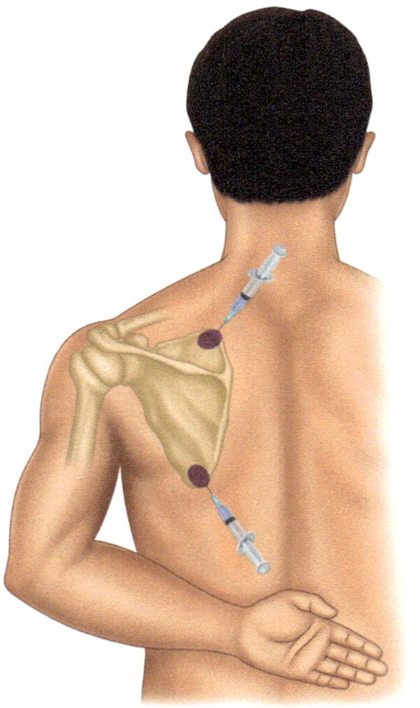

Fig. 16.3 Possible injection sites for painful bursae (Hubbard and Poehling [6], p553, with permission of Springer science + Business Media)

articulation. Changes in the thorax due to kyphosis can also alter the scapulothoracic articulation, as can abnormal posture.

16.8 Treatment of Thoacoacromial Bursitis

Treatment options for the thoracoacromial bursa include cortisone injections and bursoscopy (Fig. 16.3). To safely perform the procedure the surgeon needs to be aware of the adjacent neurological structures. Under endoscopic vision, a bursectomy can be performed, and any prominence of the superior medial scapula can be resected [5]. The suprascapular, spinal accessory and dorsal scapula nerves are all at risk with this procedure and need to be taken into account, to avoid inadvertent injury (Fig. 16.4). The endoscopic visualisation and resection is similar to the principles of performing an acromioplasty, and demonstrated in the video, courtesy of Dr Simon Bell, Melbourne, Australia (Video 16.1).

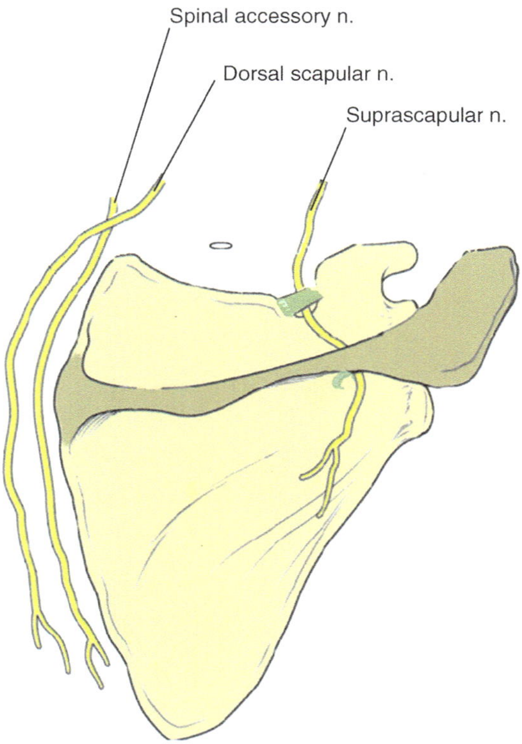

Fig. 16.4 At-risk neurological structures adjacent to the scapula (Hubbard and Poehling, [6], p 556, with kind permission of Springer science + Business Media)

References

1. Colas F, Nevoux J, Gagey O. The subscapular and subcoracoid bursae: descriptive and functional anatomy. J Shoulder Elbow Surg. 2004;13(4):454–8.
2. Schraner AB, Major NM. MR imaging of the subcoracoid bursa. AJR Am J Roentgenol. 1999;172(6):1567–71.
3. Conduah AH, Baker 3rd CL, Baker Jr CL. Clinical management of scapulothoracic bursitis and the snapping scapula. Sports Health. 2010;2(2):147–55.
4. Williams Jr GR, et al. Anatomy of the scapulothoracic articulation. Clin Orthop Relat Res. 1999;359:237–46.
5. Bell SN, van Riet RP. Safe zone for arthroscopic resection of the superomedial scapular border in the treatment of snapping scapula syndrome. J Shoulder Elbow Surg. 2008;17(4):647–9.
6. Hubbard JB, Poehling GG. Chapter 46: Scapulothoracic disorders. In: Milano G, Grasso A, editors. Shoulder arthroscopy principles and practice. London: Springer; 2014.

17 Acromioclavicular Joint

Yon-Sik Yoo

17.1 Introduction

The shoulder complex is composed of four bones—clavicle, scapula, humerus, and posterior aspect of ribs; and four articulations—glenohumeral (GH), acromioclavicular (AC), sternoclavicular (SC), and scapulothoracic. The AC joint (ACJ) and SC joint (SCJ) permit the scapula to move against the chest wall during movements of the arm, allowing the glenoid fossa to follow the head of the humerus. The functional interrelationships between these four joint mechanisms are critical in providing a full range of movements (ROM). For most of the traumatic conditions and disorders involving ACJ there are many controversial treatment options, most of them being supported by level three or four literature backup. Hence it is important to understand the evolution, development, anatomy, and biomechanics of the AC joint to critically evaluate the existing studies and guide us to select an appropriate treatment.

Y.-S. Yoo
Department of Orthopedic Surgery,
Shoulder and Sports trauma Center,
Hallym University Hospital, 7, Keunjaebong-gil,
Hwaseong-si, Gyeonggi-do 445-907, Dongtan, Korea
e-mail: yooo@hallym.ac.kr

17.2 Evolution

As humans evolved to assume orthograde posture, the inherent osseous articular congruity of upper limb joints was sacrificed for soft tissue stability to achieve greater degree of mobility. Over decades these evolutions include development of strong clavicle, a large coracoid and a widened, strong scapula set at 45° to the midline. The scapular index and more profoundly the infraspinatus index decreased in humans allowing the infraspinatus and teres minor muscles to act as more effective depressors and external rotators of humeral head. Acromion has become a massive structure over the humeral head increasing the mechanical advantage of deltoid muscle.

17.3 Development

The clavicle is the first bone in the body to ossify, during the fifth week of gestation and the last bone to fuse. Major portion of clavicle forms by intramembranous ossification from two ossification centers, of which lateral center is more prominent. Cells at the acromial end take on a cartilaginous pattern to form acromioclavicular joint. The clavicle increases in diameter by intramembranous ossification of the periosteum and grows in length through endochondral activity at the cartilaginous ends with major contribution from the medial end. Compared to the medial

G.I. Bain et al. (eds.), *Normal and Pathological Anatomy of the Shoulder*,
DOI 10.1007/978-3-662-45719-1_17, © ISAKOS 2015

epiphysis which begin to ossify at 18 years and fuses with the clavicle at 25 years, the lateral epiphysis is less constant and appear as wafer-like edge of bone. During sixth week, the scapula enlarges and extends from C4 to T7 vertebra. Two ossification centers for acromion appear at 13–16 years of age along with the 3 ossification centers for coracoid and unite at 14–20 years of age.

Very few studies have focused on the development of AC joint unlike that of the glenohumeral joint. AC joint development begins in the fetal period (after 8 weeks of intrauterine life) and the three-layered interzone is not seen. During this period only clavicle and spine of scapula are ossified [1]. Fetal anatomy similar to that of adult AC joint has been confirmed by arthroscopy [2].

17.4 Gross Anatomy

The human torso bears the upper limb by means of suspensory muscles to the scapula, clavicle, and humerus. The clavicle largely exists to assist the scapula in shoulder function maintain optimal scapular position during arm motion and serves the following functions:

1. Forms a strut that braces the glenohumeral joint at a fixed distance from the axial skeleton that permits optimal movement and power
2. Acts as a rigid base for muscular attachments of the shoulder, neck, and chest
3. Provides protection for the neurovascular structures

Clavicle in transverse plane resembles an "italic S" (Fig. 17.1a–c). Anterior concavity of the lateral end has less radius of curvature than the medial end (4.1 and 9.7 cm). The lateral end is flat in cross section. There is an inverse relationship between degree of downward facing and radius of curvature of the lateral end of the clavicle [3]. It has been claimed to be unique in that it does not have a medullary cavity [4, 5]. The inferior surface of the lateral end of the clavicle has conoid tubercle at the posterior border and sagittally oriented trapezoid line located at an average of 25 and 45 mm, respectively, from the lateral end of the clavicle. They provide attachment for the conoid and trapezoid parts of the coracoclavicular ligaments and the relative position of

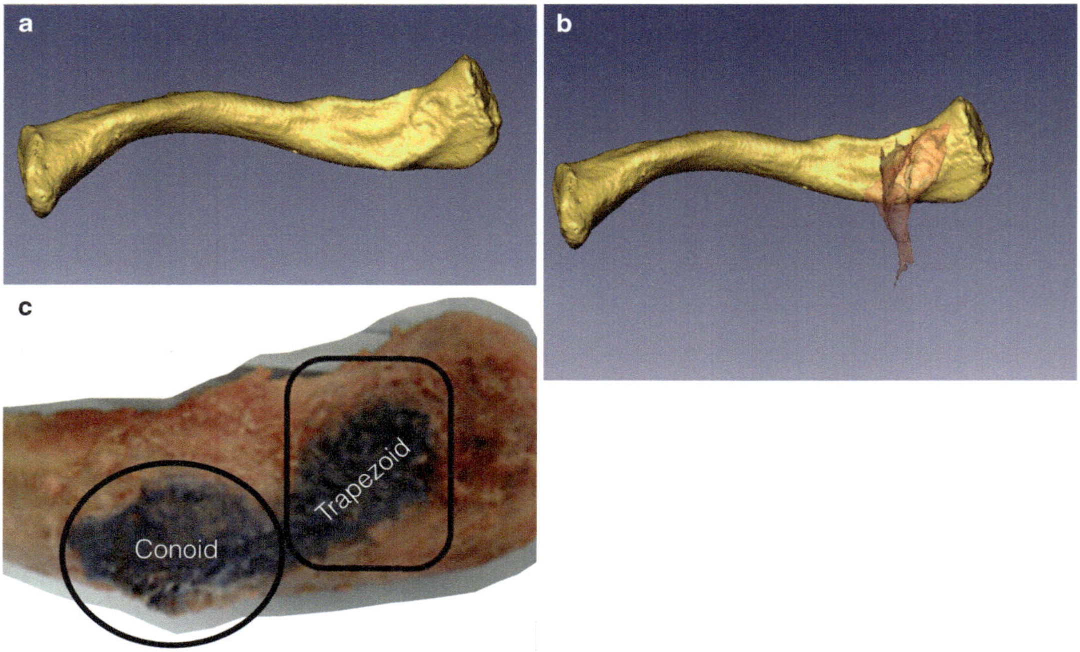

Fig. 17.1 Left clavicle (inferior view). (**a**) Attachments of the conoid and trapezoid ligaments. (**b**) Relative position of the coronoid process. (**c**) Cadaveric attachment sites

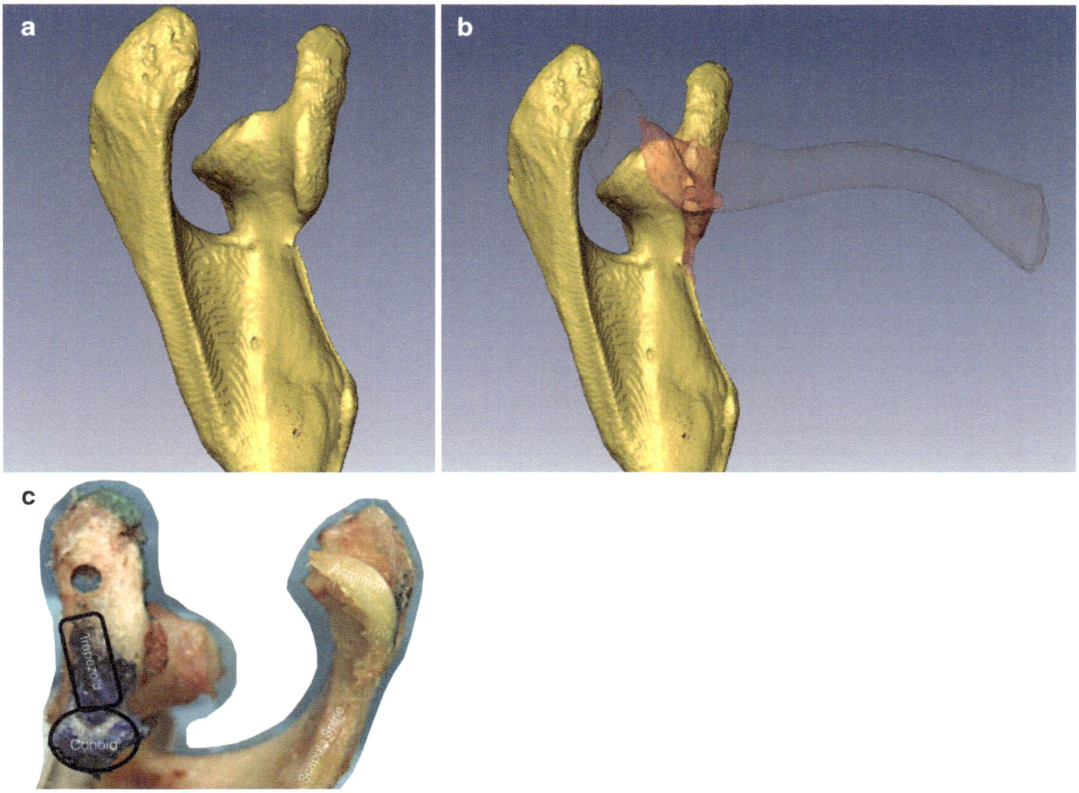

Fig. 17.2 Left scapula with coracoid process (superior view). (**a**) Attachments of the conoid and trapezoid ligaments. (**b**) Relative position of the clavicle. (**c**) Cadaveric attachment sites

these ligament insertions is important in their function [6]. A comparison of the attachment sites on the coracoid process, and how they relate together (Figs. 17.2a–c).

Although 1 % of people have coracoclavicular bar or joint, AC joint is the only articulation between clavicle and scapula [8]. In spite of many anatomical variations, AC joint is a synovial type of planar diarthrodial joint (Fig. 17.3). Although variable the hyaline cartilage covered convex and oval facet on anterior portion of the distal clavicle articulate with the concave small facet-like portion of the anteromedial aspect of the acromion process. The mean size of the joint is 9 × 19 mm in adults. The joint line is oblique and slightly curved permitting the protraction and retraction of the scapula [8] (Figs. 17.4, 17.5, 17.6, and 17.7).

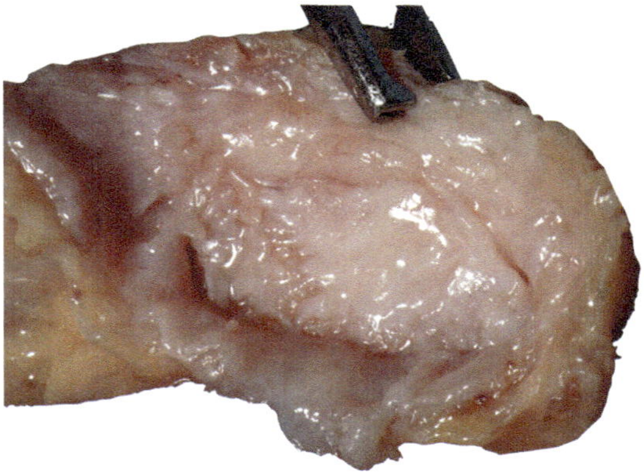

Fig. 17.3 Lateral clavicle articulation (lateral view). Lateral articulation of the clavicle demonstrating that the AC joint is a synovial planar diarthrodial joint, which is covered with variable hyaline cartilage

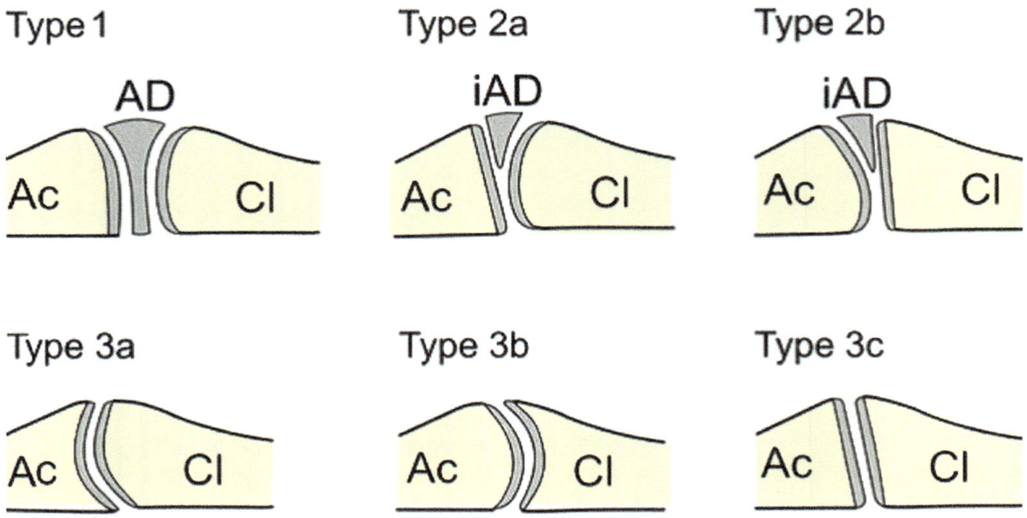

Fig. 17.4 Types of AC joints. Type 1: double ellipsoid joint (4 %). The articular disk (wedge shape) completely divides the articular cavity. It is attached to the articular capsule at its periphery. Both articular surfaces are slightly convex. Type 2: incomplete articular disk incompletely divides the AC joint cavity (25 %). Type 2a: clavicle facet is convex and the acromion flat. Type 2b: clavicle facet is flat and the acromion convex. Although the articular surfaces were incongruent, the incomplete disk compensates. Type 3: absent articular disk (71 %). Type 3a: clavicle facet convex and acromion facet concave- an ellipsoid joint. Type 3b: clavicle facet concave and acromion facet convex. Type 3c: both articular surfaces flat – planar joints

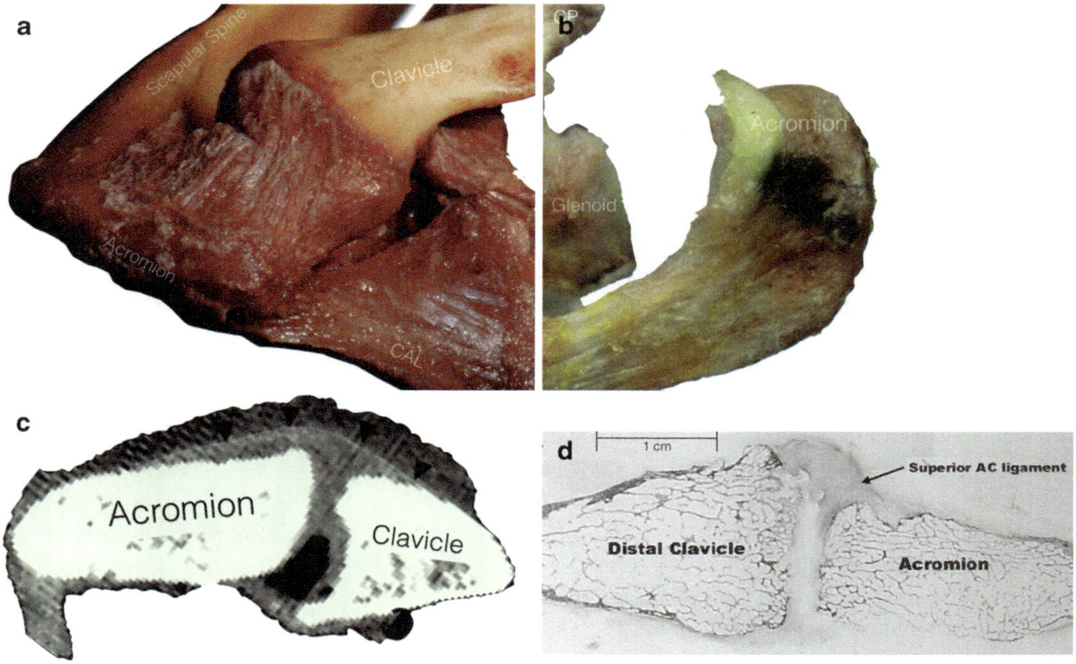

Fig. 17.5 Superior AC ligament. The superior AC capsular ligament attaches close the AC joint articulation. (**a**) Cadaveric view of superior capsule. (**b**) Cadaveric site of insertion of the superior AC joint ligament. (**c**) CT scan. (**d**) Histology

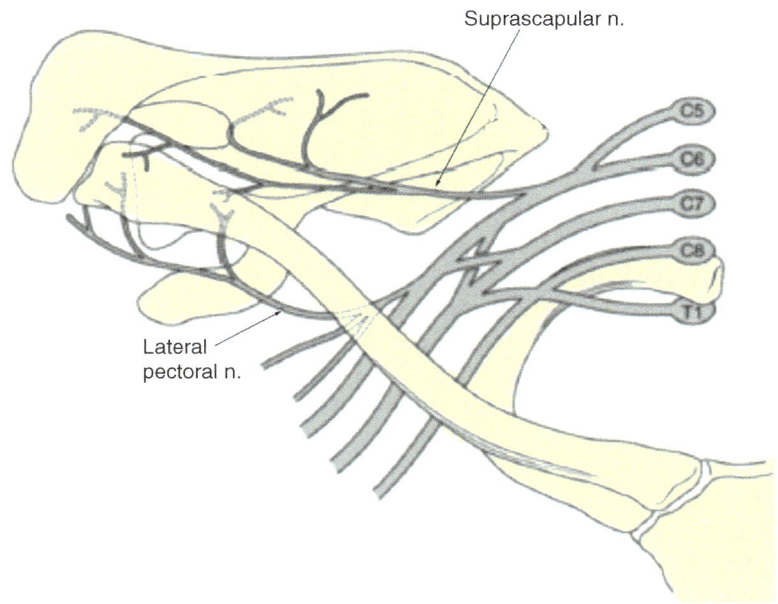

Fig. 17.6 Nerve supply of the AC joint. The AC joint is supplied by the Suprascapular, axillary and lateral pectoral nerves

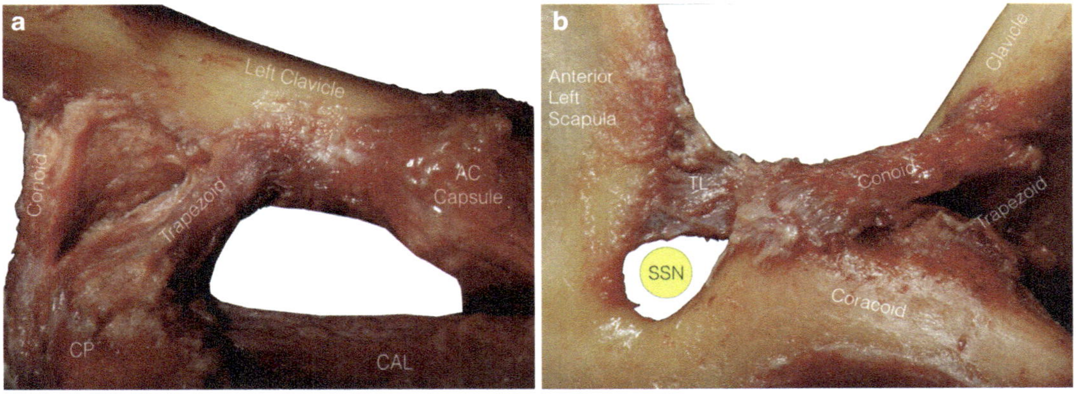

Fig. 17.7 Gross anatomy of coracoclavicular ligaments. (**a**) Anterior view. (**b**) Anterior medial view. *CP* coracoid process, *TL* transverse ligament, *SSN* suprascapular nerve, *CAL* coracoacromial ligament

A fibrocartilaginous disk cushions the joint, corrects for incongruencies, and acts in a load-bearing fashion similar to the meniscus in the knee [9] but others have attributed negligible function to it. This is composed of 75 % water, 20 % collagen (90 % type I), and 5 % proteoglycans, elastin, and other cells [10]. Variable inclinations exist, with being nearly vertical to angled downward and medially accounting for up to 50° [11]. Degeneration of the intra-articular disk, commonly observed in patients over the age of 50 years, begins as early as the timesfibers were confluent with the inferior AC ligament (ACL) at the acromial insertion. The small distance from the medial acromial articular surface to the beginning of the coracoacromial ligament (mean, 3.5 mm) stresses the close proximity of the coracoacromial ligament to the capsular insertions on the anteroinferior acromial surface, which can inadvertently be taken down during distal clavicle resection or co-planing. On the superior side of the AC joint, the trapezius was found to be confluent with the posterosuperior AC ligaments [13].

Barber et al. found no long-term instability after co-planing or hemi-resection (58 patients) of the AC joint or distal clavicle excision (23 patients) in a study in which all patients required resection of the inferior AC capsule and ligament. This illustrates the importance of maintaining the superior and posterior structures to ensure stability [15].

3-D CT scan showed [16] significant variability of the bone shape and size near the AC joint. ACJ subtends mean angle of 51° in the axial plane and 12° in the coronal plane with respect to the clavicular shaft. Hence distal clavicle resections should respect these angles and address the unique morphology of each AC joint to provide a symmetric bone resection without disruption of the AC joint capsule ACL. They proposed this anatomic-based recommendation of 5–7 mm of total resection (combining acromial and distal clavicle resection) should be adequate to provide relief of symptoms and likely to provide a more reliable outcome than larger resections.

Surgically important stabilizer of AC joint is extrinsic coracoclavicular ligaments (CCL) (conoid and trapezoid ligaments). Rather than resisting the traumatic displacements, these ligaments function to control and guide the AC joint movements and provide stability by complex interplay with AC joint capsule, AC ligament and dynamic stabilizers like deltoid, trapezius, and coracoids muscles. The CCL were determined to be greater than 3 times stiffer than the AC ligament. The AC joint compression loads can only occur when the distal clavicle is intact.

CCLs are responsible for suspending the scapula and the upper extremity from the under surface of the clavicle. They are the stronger, more vertically oriented ligaments. They arise from superior surface of coracoid posterior to the pectoralis minor attachment. The mean length of CCL is 19.4 mm. Conoid and trapezoid components are functionally and anatomically distinct and separated by a bursa between them. The conoid ligament thick and triangular, posteromedial in location, with short stout fibers almost vertical and insertion ends approximately 30 mm (females 28.9, males 33.5) from the joint line (Fig. 17.8). The trapezoid ligament is broad, thin, quadrilateral shape, and anterolateral in location. The insertion ends at mid-arc of the lateral curve and trapezoid ridge approximately 16 mm (females 16.1, males 16.7) from the ACJ line [17]. Average vertical height of coracoclavicular space is 1.1–1.3 cm [14]. The CCL strengthens the ACJ and mediate the synchronous scapuloclavicular rotation and scapulohumeral movement.

The manner of attachment provides a mechanism for producing increased external rotation of the scapula. With the elevation of the humerus, scapula rotates to displace the coracoids inferiorly. The resulting tension in the CCL acts on the lateral curve to rotate the clavicle on its long axis. The crank-like phenomenon provided by the coracoclavicular ligaments and the S shape of the clavicle will not restrict the abduction of the arm.

The majority of anteroposterior stability (90 % resistance to anterior translation of scapula on clavicle) and distraction (91 % resistance to distraction) of the ACJ is provided by ACL. Most of the vertical stability (77 % of resistance to inferior translation of the scapula) is conferred by CCL. The conoid ligament is the primary restraint against anterior and superior loading, whereas the trapezoid ligament is the primary restraint against posterior loading. About 75 % of resistance to compression of ACJ is provided by trapezoid ligament [18].

The difference in the contributions by the two ligaments is most likely due to their relative orientations. In situ forces in each ligament are affected by coupled motions that occur during loading. This soft tissue force is redistributed during loading when a greater number of degrees of freedom of motion are allowed [19, 20].

The literature has supported the concept that the AC joint capsule is integral in maintaining normal joint contact and primarily resists motion in the AP (horizontal) plane, and the CCL primarily resist motion in the superoinferior (vertical) direction [21].

Fukuda et al. [18] quantified "the displacement as a function of the ligamentous constraints." From selective ligament section studies they reported that with small displacements the acromioclavicular ligaments are the primary restraints to posterior (89 %) and superior (68 %) translation of the clavicle. With larger displacement, the conoid ligament was found to be the primary restraint (62 %) to superior translation. The trapezoid ligament was found to be the primary

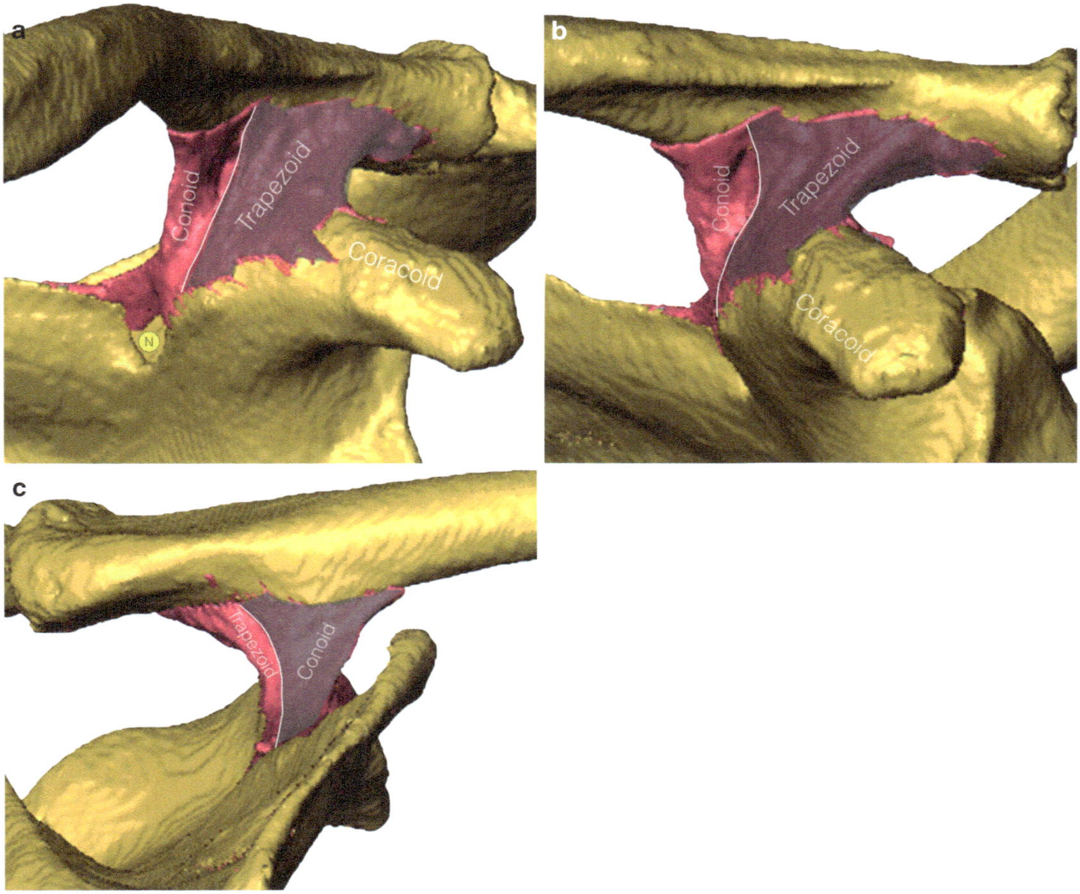

Fig. 17.8 Costoclavicular ligaments. Images demonstrate the relative position of the trapezoid and conoid ligaments: (**a**) Anterior view. (**b**) Anteriolateral view. (**c**) Aosteromedial view

restraint to compression of the acromioclavicular joint at both small and large displacements. Hence the overriding of AC joint is resisted by the trapezoid ligament. Hence small displacement is limited by ACL, but large displacements are resisted by the CCL. Lee and coworkers further determined that the trapezoid ligament was the primary restraint to posterior displacement of the distal clavicle with an intact AC joint [22].

When the arm is elevated through abduction, increased tension on the conoid ligament, causes a backward axial rotation of the clavicle of 50°. This permits the glenoid fossa to continue to elevate and increase the possible degree of arm elevation. When the clavicle is prevented from rotating, the arm can be abducted actively to only 120°. A principal role of the AC joint in the abduction of the arm is to permit continued lateral rotation of the scapula after about 100° of abduction when sternoclavicular movement is restrained by the sternoclavicular ligaments.

Functionally, the two major movements at the acromioclavicular joint are a gliding movement as the shoulder joint flexes and extends and an elevation and depression movement to conform with changes in the relationship between the scapula and the humerus during abduction. The sternoclavicular and acromioclavicular joints play essential and distinct roles in the movements of the shoulder complex [23].

Dynamic stabilization of the joint is provided by the anterior deltoid, trapezius, and serratus anterior as well as the muscles connecting to the coracoids process. They help support the weight of the arm and are force coupled to help stabilize the AC joint [9]. Other muscles around the shoulder also have significant effect on AC joint.

17.5 Movements and Constraint

AC joint movements include gliding [24] and rotations around the vertical, anteroposterior, and horizontal axes [25]. When the arm is elevated, the scapula progressively rotates upward, rotates externally, and tilts posteriorly [26–28].

Rockwood et al. have reported that there is approximately 5–8° of motion detected at the acromioclavicular joint with forward elevation and abduction to 180° [29]. But there are reports that the clavicle rotates 40–50° during full overhead elevation. This motion is combined with scapular rotation rather than through the acromioclavicular joint. This synchronous motion of the clavicle, rotating upward, and the scapula, rotating downward, during abduction and forward elevation was described by Codman as synchronous scapula clavicular rotation [12]. This is coordinated by the CCL [14].

Movements of the clavicle occur with arm movement: elevation/depression in the frontal plane, protraction/retraction in sagittal plane, and anterior or posterior rotation along the long axis in transverse plane. During arm elevation, clavicular rotation of 40–50° occur in synchrony with scapular motion. Only 5–8° of motion actually occur at the AC joint. When high loads are applied to the arm, the static ACL serve to restrain superior and posterior translation of the clavicle. During lower loads, the AC ligaments continue to resist posterior translation while the conoid ligament primarily resists superior translation of the trapezoid ligament serving as the primary compression restraint. As the arm is elevated above 90°, load bearing increases across the CC ligaments as a result of the ligaments trying to help maintain the AC joint articulation during scapulohumeral rhythm. In cases where the clavicle has become shortened >8 mm, either through injury or excessive distal clavicle resection, increased motion and subsequently increased forces can occur at the AC joint, resulting in dysfunctional movement and/or pain to occur [30].

The movement at AC joint is limited by the complex arrangement of ACL and CCL. Three-dimensional kinetic analysis using open MRI of volunteer shoulders demonstrated that anterior axial rotation of the clavicle at AC joint increased linearly with abduction and reached an average of 30° at maximum abduction (Figs. 17.8, 17.9, and 17.10). They also showed that anteroposterior rotation is three times as great as the superinferior rotation [24].

The potential movement present at ACJ and SCJ exceeds that actually attained during the active movement of the shoulder complex. Current data indicate that accurate demonstration of this phasic 3-dimensional movement is a complex problem. During elevation of the extremity, clavicular elevation of about 30° occurs, with maximum at about 1° of elevation. The clavicle also rotates anteriorly of about 10° during the first 40° of elevation. No rotation occurs during the next 90° of elevation, but an additional 15–20° of anterior rotation occurs during the terminal arc of elevation. Flexion of the arm demonstrates an identical pattern of clavicular movements similar to that of during abduction [31].

The axial rotation of the clavicle is an essential and fundamental feature of shoulder movements, particularly arm elevation. If clavicular rotation is not allowed, arm elevation of about 110° only is possible. Superior rotation of the clavicle begins after the arm has attained an arc of about 90° of elevation and then progresses in a rather linier fashion. About 40° of superior rotation at full elevation [32]. Sahara et al. [24] reported that 35° of axial rotation occurred at AC joint. Fixation of the clavicle to the coracoids by a screw or ankylosis due to ectopic ossification causes minimal loss of arm elevation. Hence loss of movements at AC joint is better tolerated.

Resection of the distal clavicle increases posterior translation by about 30 % [20, 33] during posterior loading and increased the in situ force on the CCL three times greater than the intact

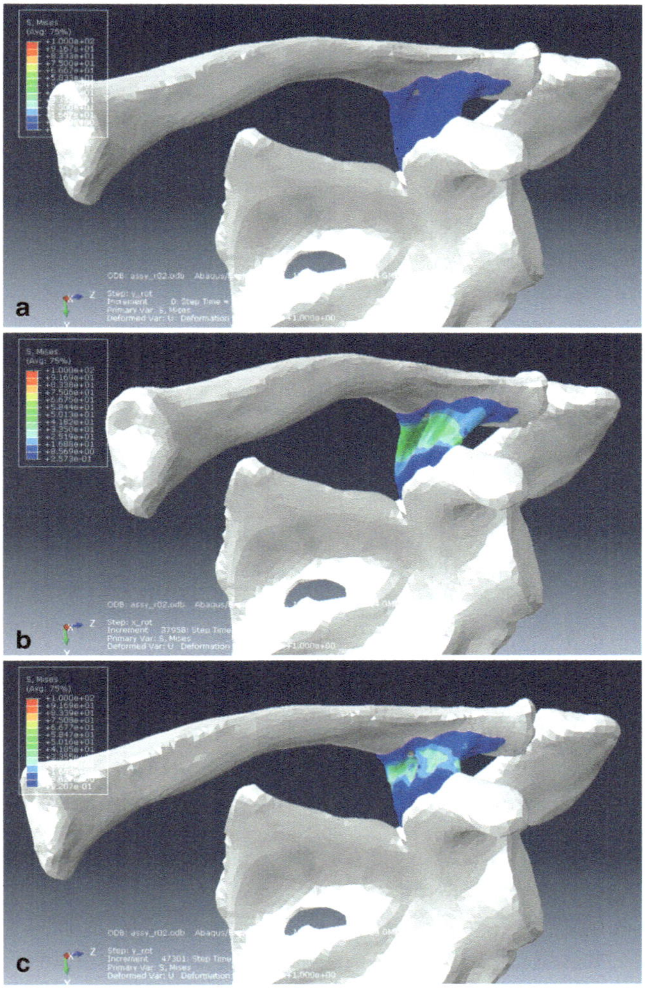

Fig. 17.9 Stress patterns of the coracoacromial ligaments. (**a**) Neutral position. (**b**) Scapular protraction. (**c**) Scapular retraction

shoulder during anterior loading [20]. Thus, the significant effect of DCR on the movement and stability of AC joint should be carefully considered.

17.6 Clinical Implications

The AC joint is part of the six-component superior shoulder suspensory complex (SSSC). Although disruption of one component does not compromise stability, injury to two or more components require surgical repair or reconstruction.

The AC joint is critical coupling clavicular and scapular motion, and thus scapular dyskinesis has been associated with AC joint injury. Due to its peculiar anatomy and biomechanics, articular cartilage is subjected to very high stresses accelerating osteoarthritic changes.

Intact CC ligaments cannot compensate for the loss of the AC joint capsule's function during AP loading, which is typical in type II (or greater) AC joint injuries. The literature clearly suggests that the movement of the AC joint is clinically important. Fusion of AC joint or coracoclavicular screw fixation still allows full forward elevation in abduction resulting in migration and failure of implants over time.

During DCR, resections of greater than 10 mm may result in damage to ACL most patients and CCL in some patients.

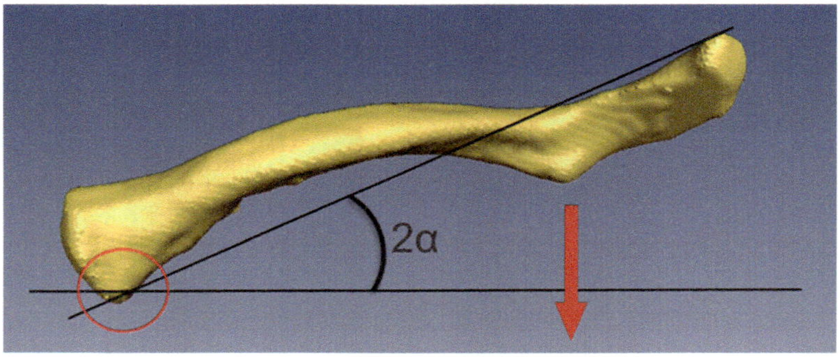

Fig. 17.10 Clavicular rotation. At 30° of clavicle elevation, the motion is restrained by the costoclavicular ligaments, and substituted for backward (external) rotation by the conoid ligament. *Red arrow*: downward force by conoid ligament

The literature demonstrates more than 60 different surgical procedures for AC joint injuries. The large variation in treatment strategies may be due in part to the lack of information on the complex interplay between the ACJ capsule, ACL and CCL. DCR along with reconstruction of CCL for post-traumatic instability of ACJ may increase the risk of posterior instability postoperatively.

References

1. Fealy S, Rodeo SA, Ficado DF, et al. The developmental anatomy of the neonatal glenohumeral joint. J Shoulder Elbow Surg. 2000;9:217–22.
2. Ten AJ, Barrio Asensio C, Puerta Fonolla J, et al. Arthroscopic study of the shoulder joint in fetuses. Arthroscopy. 2005;21:1114–9.
3. Rockwood CA. The shoulder. Philadelphia: Saunders Elsevier; 2008.
4. Moseley HF. The clavicle: its anatomy and function. Clin Orthop. 1968;58:17–27.
5. O'Brien S, Arnoczsky S, Warren R, et al. Developmental anatomy of the shoulder and anatomy of the glenohumeral joint. In: Rockwood C, Matsen F, editors. The shoulder. Philadelphia: WB Saunders; 1990. p. 1–33.
6. Rosse C, Gaddum-Rosse P, Hollinshead WH. Hollinshead's textbook of anatomy. 5th ed. Philadelphia: Lippincott-Raven; 1997.
7. Lewis OJ. The coracoclavicular joint. J Anat. 1959; 93:296–303.
8. Tyurina TV. Age related characteristics of human acromioclavicular joint. Arkh Anat Gistol Embriol. 1985;89:75–81.
9. Mall NA, Foley E, Chalmers PN, Cole BJ, Romeo AA, Bach Jr BR. Degenerative joint disease of the acromioclavicular joint: a review. Am J Sports Med. 2013;41(11):2684–92.
10. DePalama AF. Surgical anatomy of sternoclavicular and acromioclavicular joint. Surg Clin North Am. 1963;43:1541–50.
11. Jobe CM. Anatomy and surgical approaches. In: Jobe F, editor. Operative techniques in upper extremity sports medicine. St. Louis: Mosby; 1996. p. 124–60.
12. Emura K, Arakawa T, Miki A, Terashima T. Anatomical observations of the human acromioclavicular joint. Clin Anat. 2014;27(7):1046–52.
13. Stine IA, Thomas Vangsness Jr C. Analysis of the capsule and ligament insertions about the acromioclavicular joint: a Cadaveric Study. Arthroscopy J Arthrosc Relat Surg. 2009;25(9):968–74.
14. Sellards R. Anatomy and biomechanics of the acromioclavicular joint. Oper Tech Sports Med. 2004; 12(1):2–5.
15. Barber FA. Long-term results of acromioclavicular joint coplaning. Arthroscopy. 2006;22:125–9.
16. Nourissat G, Henon A, Debet-Mejean A, et al. Three-dimensional computed tomographic scan of the external third of the clavicle. Arthroscopy. 2007;23:29–33.
17. Harris RI, Wallace AL, Harper GD, et al. Structural properties of the intact and the reconstructed coracoclavicular ligament complex. Am J Sports Med. 2000;28:103–8.
18. Fukuda K, Craig EV, An KN, et al. Biomechanical study of the ligamentous system of the acromioclavicular joint. J Bone Joint Surg Am. 1986;68:343–440.
19. Debski RE, Parsons IM, Fenwick J, et al. Ligament mechanics during three degree of freedom motion at he acromioclavicular joint. Ann Biomed Eng. 2000; 28:612–8.
20. Debski RE, Parsons IM, Woo SL, et al. Effect of capsular injury on acromioclavicular joint mechanics. J Bone Joint Surg Am. 2003;83:1344–51.
21. Dawson PA, Adamson GJ, Pink MM, Kornswiet M, Lin S, Shankwiler JA, Lee TQ. Relative contribution of acromioclavicular joint capsule and coracoclavicular ligaments to acromioclavicular stability. J Shoulder Elbow Surg. 2009;18:237–44.
22. Lee KW, Debski RE, Chen CH, et al. Functional evaluation of the ligaments at the acromioclavicular joint during anteroposterior and superoinferior translation. Am J Sports Med. 1997;25:858–62.
23. Peat M. Functional anatomy of the shoulder complex. Phys Ther. 1986;66:1855–65.
24. Sahara W, Sugamoto K, Murai M, Tanaka H, Yoshikawa H. 3D kinematic analysis of the acromio-

clavicular joint during arm abduction using vertically open MRI. J Orthop Res. 2006;24:1823–31.

25. Teece RM, Lunden JB, Lloyd AS, Kaiser AP, Cieminski CJ, Ludewig PM. Three-dimensional acromioclavicular joint motions during elevation of the arm. J Orthop Sports Phys Ther. 2008;38: 181–90.
26. McClure PW, Michener LA, Sennett BJ, Karduna AR. Direct 3-dimensional measurement of scapular kinematics during dynamic movements in vivo. J Shoulder Elbow Surg. 2001;10:269–77.
27. Bourne DA, Choo ATM, Regan WD, MacIntyre DL, Oxland TR. Three-dimensional rotation of the scapula during functional movements: an in vivo study in healthy volunteers. J Shoulder Elbow Surg. 2007;16: 150–62.
28. Ludewig PM, Phadke V, Braman JP, Hassett DR, Cieminski CJ, LaPrade RF. Motion of the shoulder complex during multiplanar humeral elevation. J Bone Joint Surg Am. 2009;91:378–89.
29. Rockwood CJ, Williams G, Young D. Disorders of the acromioclavicular joint. In: Rockwood CJ, Matsen F, editors. The shoulder, vol. 1. Philadelphia: Saunders; 1998. p. 483–553.
30. Kibler WB, Sciascia A, Dome D. AC Joint Injuries, ISAKOS Newsletter 2010;14(2):28–30,
31. Laumann U. Kinesiology of the shoulder joint. In: Koelbel R, Helbig B, Blauth W, editors. Shoulder replacement. Berlin: Springer; 1987. p. 23–31.
32. Inman VT, Saunders JR, Abbott LC. Observations on the function of the shoulder joint. J Bone Joint Surg. 1994;26:1–30.
33. Corteen DP, Teitge RA. Stabilisation of the clavicle after distal resection: a biomechanical study. Am J Sports Med. 2005;33:61–7.

Pathoanatomy of Acromioclavicular Joint Instability

18

Joideep Phadnis, Gregory I. Bain, and Klaus Bak

18.1 Introduction

Regarding acromioclavicular (AC) joint instability, Hippocrates famously stated in 460 BC that 'no impediment, small or great, would result from such an injury', despite that it was not possible to return the clavicle to its original position once dislocated [1]. Centuries later, AC joint instability remains one of the most contentious topics in orthopaedic surgery with continued debate regarding pathoanatomy, biomechanics and treatment. The aim of this chapter is to provide the reader with a comprehensive up-to-date review of the pathoanatomy of AC joint instability using relevant historical, contemporary and new concepts.

Electronic supplementary material The online version of this chapter 10.1007/978-3-662-45719-1_18 contains supplementary material, which is available to authorized users.

J. Phadnis, FRCS (Tr&Orth) (✉)
G.I. Bain, PhD, MBBS, FRACS, FA(Ortho)A
Department of Orthopaedic Surgery,
Flinders University of South Australia,
Adelaide, SA, Australia
e-mail: joideep@doctors.org.uk;
greg@gregbain.com.au

K. Bak, MD
Teres Hospital Parken,
Oester Alle 42, 2nd DK-2100,
Copenhagen (OE), Denmark
e-mail: skulderbak@gmail.com

18.2 Biomechanics of Acromioclavicular Joint Instability

Motion of the intact AC joint occurs in three planes: anterior-posterior, superior-inferior and axial rotation. The joint has little inherent stability, relying on the coracoclavicular ligaments; the acromioclavicular ligaments and capsule and the deltotrapezial fascia for stability. The conoid and trapezoid ligaments have distinct insertions on the clavicle suggesting they have different biomechanic roles to each other [2] (Fig. 18.1). The anatomy of these structures is covered in detail in Chap. 17.

The AC joint and its stabilizing ligaments form part of the superior shoulder suspensory complex (Fig. 18.2). This is the concept of a bony and ligamentous ring that stabilizes the shoulder. Using this model, the authors of this chapter propose that rupture of the AC joint stabilizing ligaments uncouples the upper extremity from the axial skeleton as the suspensory ring is disrupted in two places. This disruption is of biomechanical significance because the AC joint transfers the cumulative forces generated by the kinetic chain originating in the lower limbs and axial skeleton when a throwing or lifting motion is initiated. Disruption of this joint will therefore disrupt this coordinated chain of loading and will alter the mechanics, power and efficiency of the desired action.

G.I. Bain et al. (eds.), *Normal and Pathological Anatomy of the Shoulder*,
DOI 10.1007/978-3-662-45719-1_18, © ISAKOS 2015

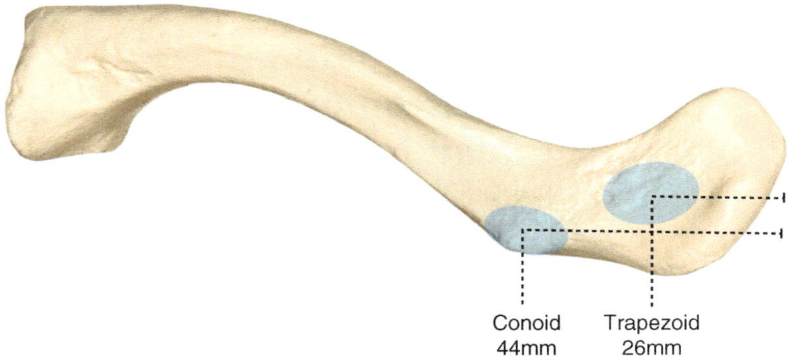

Fig. 18.1 Clavicular attachments of the conoid and trapezoid ligaments. Note the conoid attachment at the most posterior aspect of the clavicle, at the apex of curvature. These anatomic insertion points should be used for tunnel placement

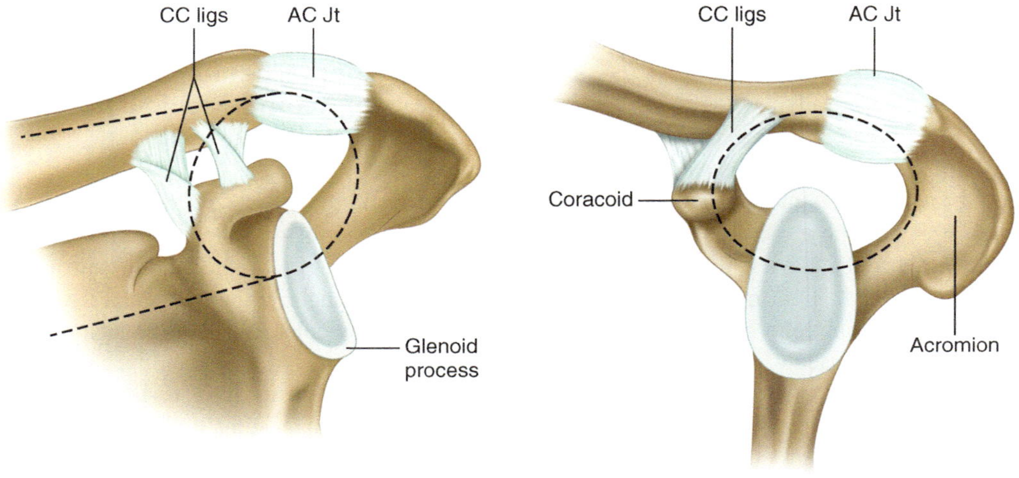

Fig. 18.2 The superior shoulder suspensory complex illustrates why complete disruption of the AC joint unlinks the upper extremity from the axial skeleton [32] (Modified with permission Jeray [36], Copyright Lippincott, Williams and Wilkins)

Fukuda performed a serial ligament sectioning study and reported that each ligament had a specific mode of restraint on the clavicle [3]. The AC ligaments restrain posterior displacement; the conoid ligament restrains superior but also anterior displacement and the trapezoid ligament restrains lateral displacement of the clavicle. Other authors have confirmed these findings [4]. AC joint instability can cause significant functional deficit because of deficiency of these stabilizing structures, which results in the scapula being displaced medially, inferiorly and anteriorly.

18.3 Imaging of the Unstable AC Joint

18.3.1 Plain Radiographs

The recommended radiographs for the diagnosis and quantification of AC joint injuries are a Zanca view (10–15° cephalad-tilted anterior-posterior radiograph) and axillary view to identify posterior displacement of the clavicle. Optimal exposure of the AC joint is achieved with around half the X-ray penetration required for the glenohumeral joint. Another useful view is the Stryker Notch view, which is taken with the patient supine and the arm flexed so that the hand lies on the head. The beam is directed 10° cephalad. This shows the coracoid process in profile, and should be considered if there is an AC joint dislocation with preserved coracoclavicular distance to rule out an associated coracoid fracture.

18.3.2 Stress Radiographs

Stress views can illustrate superior-inferior, anterior-posterior (AP) and medial-lateral instability of the AC joint.

AP stress views with weights strapped to the wrists have been suggested as a method to understand the degree of AC joint instability and distinguish between those dislocations with the deltotrapezial fascia intact and those without. However, these are not frequently used due to inconvenience and because the degree of instability that can be masked if the patient cannot relax the shoulder due to pain [5, 6]. The lateral stress view is a 'scapula Y' view in which the patient thrusts their shoulders forward to accentuate posterior-superior translation of the clavicle. This test adds little in terms of pathoanatomical understanding of the injury as compared to static views. Basamania recently described the medial stress view. This is a cross-arm adduction to demonstrate medial instability of the scapula evident when the scapula passes under the lateral clavicle [7] (Fig. 18.3). This is important from a pathoanatomical viewpoint, as chronic distal clavicle instability is a cause of scapula dyskinesia, pain and dysfunction [8].

18.3.3 Magnetic Resonance Imaging (MRI)

MRI can display the pathoanatomical structures relevant to AC joint instability especially when specific AC joint protocols are used, although interpretation can be difficult due to oedema and haematoma [9]. Moreover, AC joint instability is a dynamic entity, which is not apparent on an MRI scan. There is also evidence that MRI findings of AC joint instability do not correlate with radiographic findings or currently used classification systems [10, 11]. In our experience, we have found that the constellation of ligament injuries seen on MRI and intra-operatively is not consistent with the current classification systems.

18.3.4 4D Computed Tomography Scan (4D CT)

Advances in CT technology now allow motion capture of the skeleton in real time. We have used this to assess the unstable lateral clavicle, especially after previous surgery. It will no doubt be useful for kinematic studies, and assessment of the unstable AC joint (Video 18.1).

18.4 Pathoanatomy and Classification of AC Joint Instability

Cadenat first introduced the concept of a sequential soft tissue injury that led to progressive AC joint instability [12]. Tossy et al. [13] developed a three-stage classification, which was expanded by Rockwood in 1984 to differentiate the complete AC joint dislocations into subtypes [14]. This is the most widely used system today. These classifications were based upon clinical, radiographic and cadaveric dissections but did not include in vivo assessment or new generation imaging. Bannister [15] developed a classification system based upon stress views and intra-operative findings of the injured AC joints. In Type A the dislocation reduced, in Type B the dislocation remained unchanged and in Type C the dislocation increased. Type C were also defined as having >2 cm of coracoclavicular displacement. They documented ligamentous injuries at the time of surgery and found the coracoclavicular ligaments to be disrupted in all 21 patients they explored. The deltotrapezial fascia was sometimes torn in Type B dislocations and sometimes intact in Type C dislocations. This study highlights the discrepancy between findings at surgery and those that might be predicted.

Horn reported that the AC ligaments, the coracoclavicular ligaments and deltotrapezial fascia

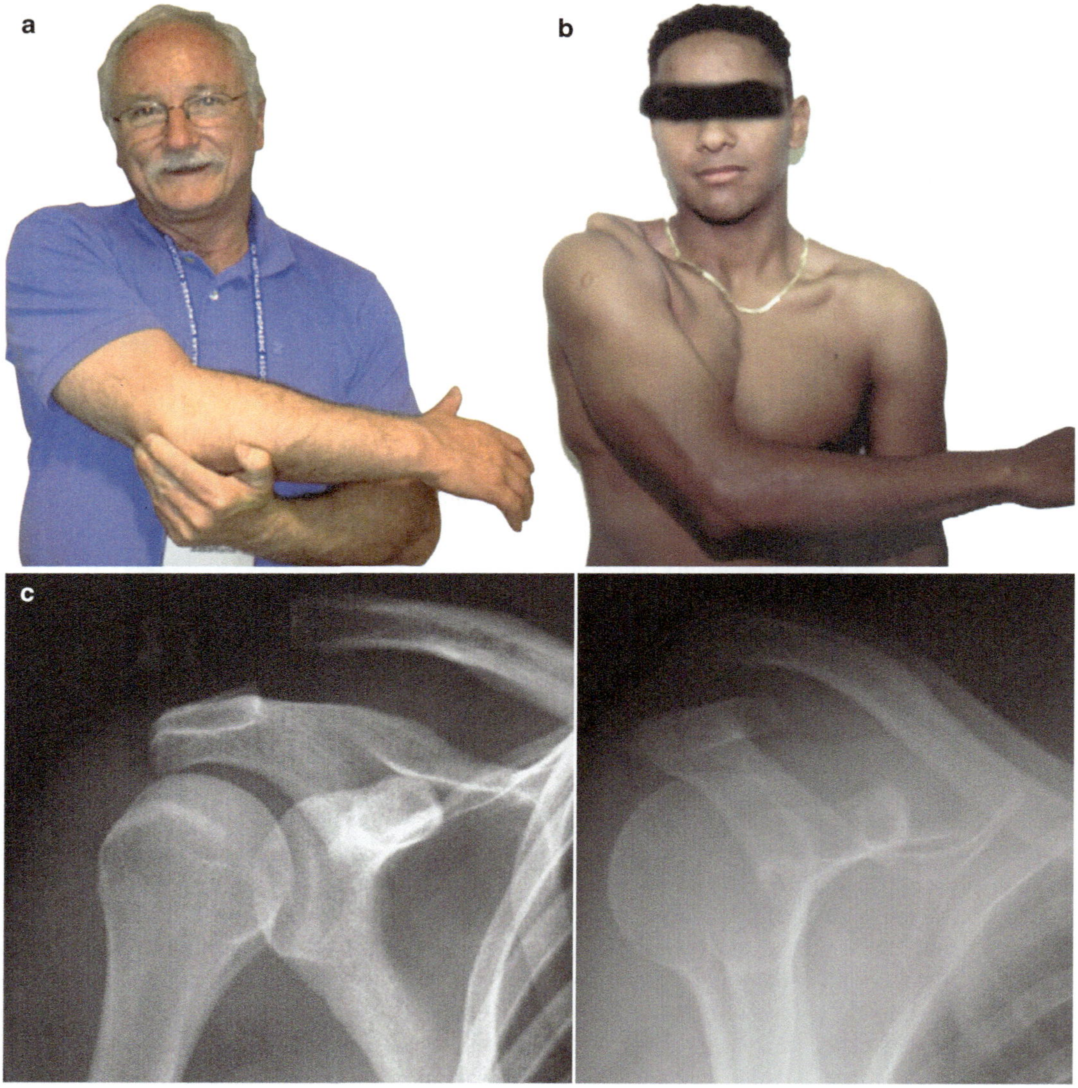

Fig. 18.3 (**a–c**) Cross-arm adduction provocation test. (**a**) The cross-arm adduction provocation test was described and demonstrated here by Carl Basamania. (**b**) With the arm in adduction, the prominent lateral and superior clavicle is seen. (**c**) Plain radiographs demonstrate that the scapula has been medialized in this Rockwood Type 5 AC joint dislocation (Image courtesy of Dr Carl Basamania, Duke University, USA)

were torn in all nine cases. He noted that some deltoid insertion injuries were concealed under intact fascia [16]. They also reported that the meniscus of the AC joint was always avulsed from the clavicle and at least partially attached to the acromion. This study preceded any classification system so that the radiographic degree of clavicle displacement was unknown. Copeland re-emphasized the concept of sequential disruption based on his extensive personal experience and on cadaveric dissections [17]. Although not substantiated by data, he felt the injury started at the superior acromioclavicular ligaments followed by injury to the inferior acromioclavicular ligaments and periosteal stripping of the underside of the clavicle, which led to mid-substance tears of the coracoclavicular ligaments and finally deltotrapezial damage. He also stated that the meniscus always remained attached to the acromion.

Lizaur reported on the operative findings in 46 patients with 'complete' AC joint dislocation [18]. They found the acromioclavicular ligaments and joint capsule to be torn in all patients. The meniscus was avulsed from the clavicle in 38 patients; the coracoclavicular ligaments were torn in 40 patients and the deltotrapezial fascia torn in 43 patients. Interestingly those who had intact coracoclavicular ligaments all had a torn deltotrapezial fascia and vice versa.

All these studies address the coracoclavicular ligaments as one structure; however, they are distinct anatomic structures with different biomechanical roles and thus should not be regarded solely as one unit. Moreover, the established classification systems are not based on advanced imaging or in vivo findings, which are necessary to establish a pathoanatomical understanding of the injury.

18.5 Updated Pathoanatomy of AC Joint Instability

There has been recent recognition of the need to update the classification of AC joint instability by the International Society of Arthroscopy, Knee surgery and Orthopaedic Sports Medicine (ISAKOS) [7]. In a consensus report from ISAKOS it was proposed to subclassify Rockwood grade 3, into 3a (stable) and 3b (unstable), with this differentiation being primarily functional rather than anatomic. The 3b injuries are those with ongoing symptoms of pain and dysfunction despite a period of non-operative management. Radiographs using the Basamania cross-arm view were proposed as a method of confirming a greater degree of instability [7].

Author's Perspective

In order to better understand the pathoanatomy of AC joint injury, we are currently prospectively assessing the advanced imaging characteristics and in vivo operative findings of acute (<4 weeks) AC joint injuries. To date, MRI scan has confirmed that in all grade 2 injuries the coracoclavicular ligaments are intact (Fig. 18.4). None of these patients underwent surgery. In Rockwood grade 3 injuries, MRI demonstrated signal change in the coracoclavicular ligaments but did not sufficiently demonstrate the details of the individual ligaments. Of the AC joint instabilities that underwent surgery, we documented that the trapezoid ligament was avulsed from the coracoid in all patients (Fig. 18.5). The conoid ligament was intact in eight out of nine grade 3 patients but often found to be lengthened (Fig. 18.6). In the grade 5 injuries the conoid was usually torn from the clavicle. The torn proximal conoid ligament often remained attached to the inferior clavicular periosteum, which was stripped medially. Figure 18.7 demonstrates a grade 5 dislocation with the typical pattern of injury. In all cases the articular meniscus remained attached to the acromion with the distal clavicle presented as a bare head with the superior acromioclavicular ligaments avulsed from the clavicle (Fig. 18.8). There was frequently a buttonhole in the deltotrapezial fascia (Fig. 18.9) and in those grade 3 injuries without a buttonhole; there was deep surface stripping from the clavicle (Fig. 18.10).

These findings confirm that instability occurs in a progressive manner with a predictable sequential failure of the stabilizing structures. We also noted on concurrent arthroscopic examination that there was a significant incidence of superior labral tears (Fig. 18.11). This has also been noted by Imhoff who demonstrated a 14 % incidence of SLAP tears in their series with less frequent incidence of other intra-articular pathologies [19]. The reason for this may be the mechanism of injury where typically the patient falls directly onto the point of their shoulder. This drives the scapula medially causing disruption of the AC joint. In this mechanism, the humeral head will also be driven medially causing an axial loading injury to the labrum. Alternatively, it may be that patients who have concurrent labral injuries actually have a different injury mechanism with the arm above the head. This requires further investigation.

We strongly feel that the coracoclavicular ligaments should be considered separately and that they, along with the acromioclavicular ligaments, each have a primary individualized function in AC joint stability. This phenomenon is evident in

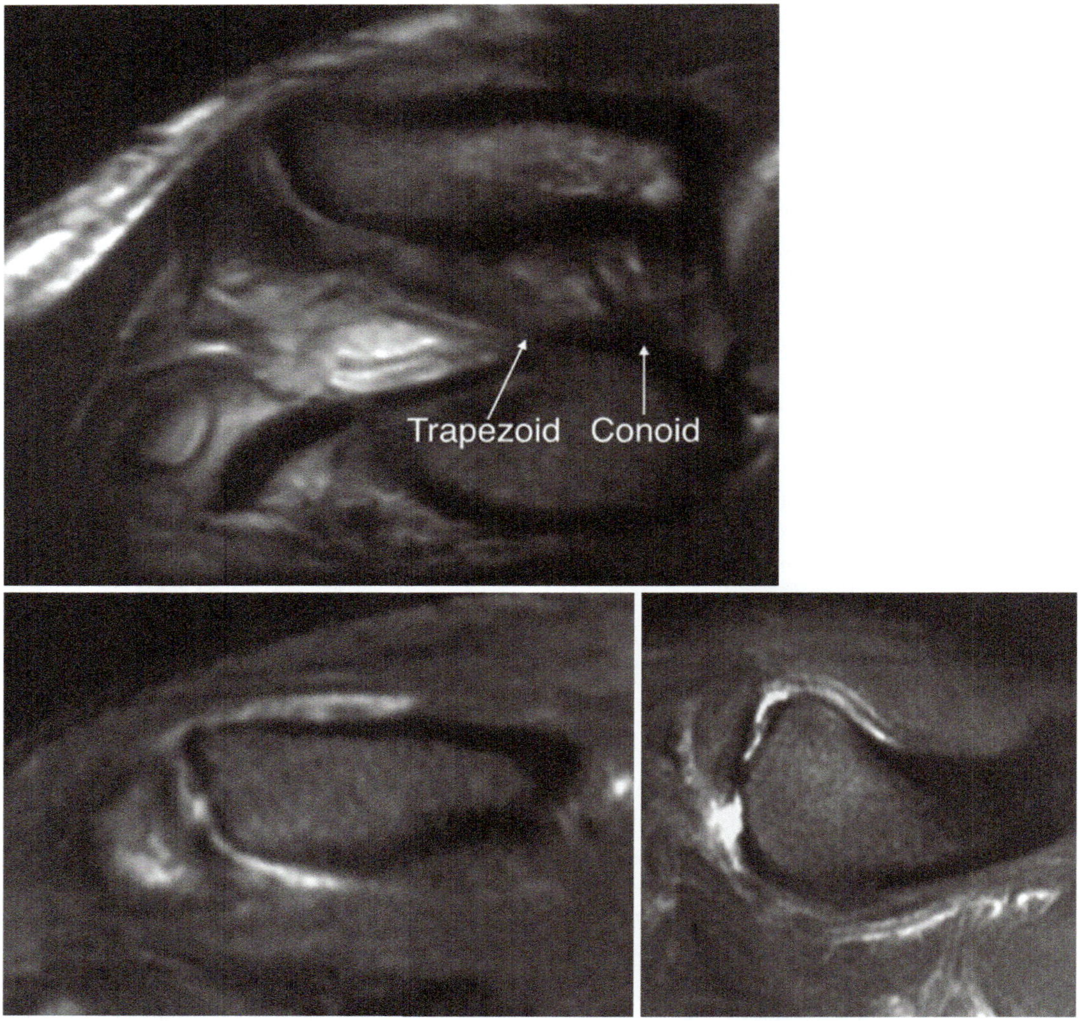

Fig. 18.4 T2 weighted MRI scans of a Rockwood type 2 AC joint injury showing intact coracoclavicular ligaments but high signal around the acromioclavicular ligaments without *posterior translation of the distal clavicle* (Copyright Gregory Bain)

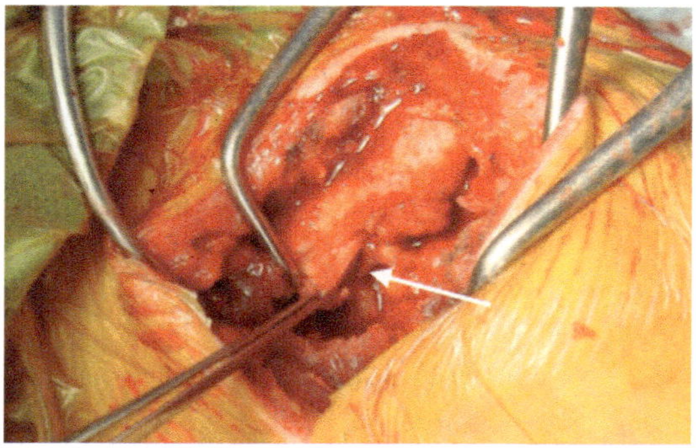

Fig. 18.5 *White arrows* indicate the trapezoid ligament torn from the coracoid in—this is a typical finding (Copyright Gregory Bain)

other joints such as the elbow, where the component parts of the medial and lateral collateral ligaments provide a specific type of restraint [20]. In an AC joint injury the ligaments fail in a predictable sequential manner. They do so because failure of one ligament results in supra-physiologic AC joint motion that further transmits the plane and magnitude of the injury to the remaining ligaments, which fail like dominos as their capacity to absorb the concentrated force is sequentially overcome. The first structures to fail are the acromioclavicular joint ligaments and capsule. The most important of these are the superior and

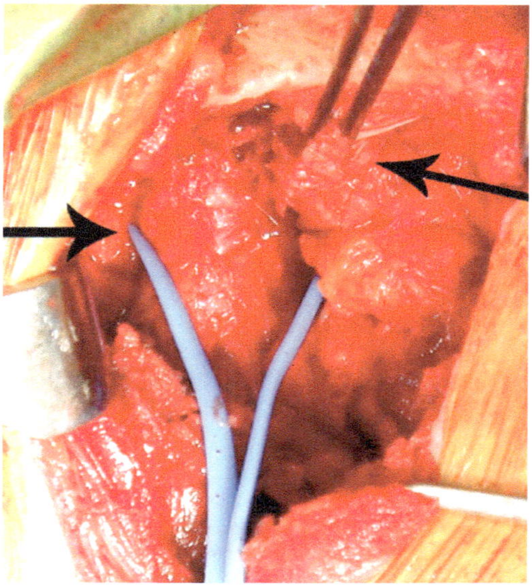

Fig. 18.6 Grade 3 AC joint dislocation with intact conoid identified by vessel tape. Torn trapezoid from the coracoid held in forceps (Copyright Gregory Bain)

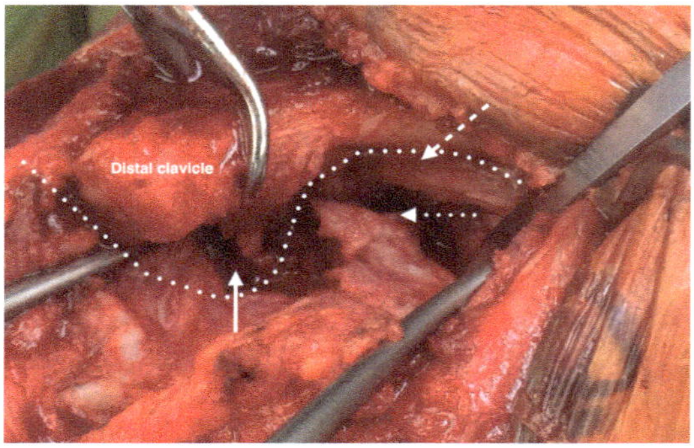

Fig. 18.7 Grade 5 AC joint dislocation. Dotted line represents the propagation of injury. *Solid white arrow* shows trapezoid ligament torn from the coracoid. *Dotted arrow* shows conoid ligament torn from the clavicle. *Dashed arrow* shows the inferior clavicular periosteum stripped medially (Copyright Gregory Bain)

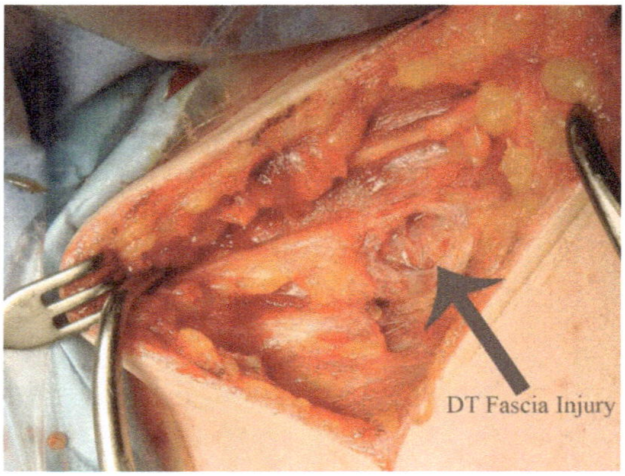

Fig. 18.9 Buttonhole in deltotrapezial fascia (Copyright Gregory Bain)

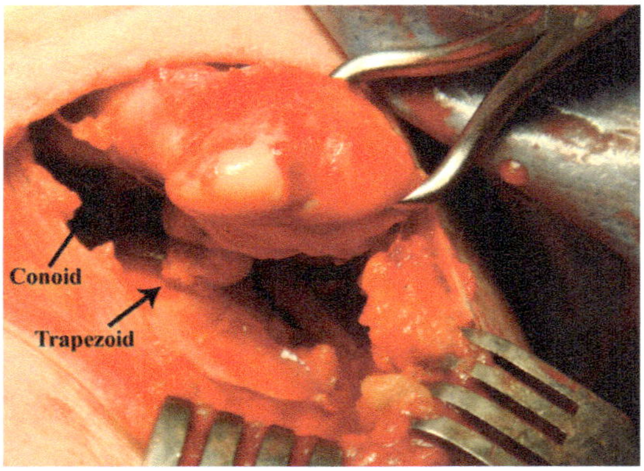

Fig. 18.8 Bare distal clavicle with AC joint meniscus left attached to acromion (Copyright Gregory Bain)

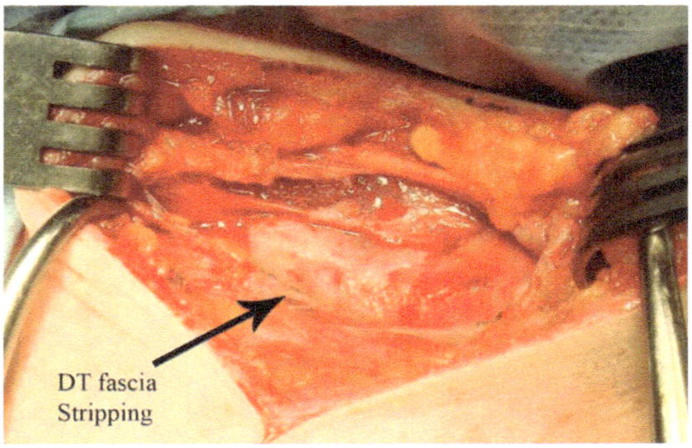

Fig. 18.10 Concealed periosteal stripping. The superficial deltotrapezial fascia was intact (Copyright Gregory Bain)

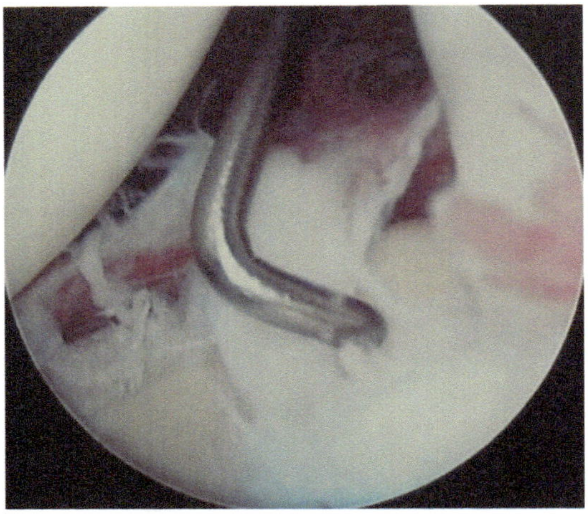

Fig. 18.11 Arthroscopic view of the glenohumeral joint in a patient with Type 3 AC joint dislocation showing the presence of a concurrent superior labral tear. This is known to be a frequent association (Copyright Gregory Bain)

posterior ligaments. The scapula becomes uncoupled from the axial skeleton, medializes and the clavicle externally rotates, placing the trapezoid ligament under maximal strain [3]. Trapezoid failure occurs under tension from the coracoid. There is increased coracoclavicular distance and moderate but complete superior translation of the distal clavicle, which is permitted but also limited by uncoiling and lengthening of the more posteriorly placed conoid ligament as the clavicle externally rotates (Fig. 18.12). This limit to distal clavicle displacement is apparent on radiographs, which illustrate that the degree of lateral clavicle displacement in a Rockwood grade 3 separation is consistently the same.

On the basis of our pathoanatomical findings, we propose a change to the interpretation of the Rockwood classification such that grade 3 injuries display limited but complete superior migration because the conoid ligament remains intact.

18.6 AC Joint Instability Equivalent Injuries

There are a number of injury patterns around the AC joint, which refer to as 'AC joint instability Equivalent Injuries' (similar to the concept 'Monteggia Equivalent injuries'). The injuries occur less commonly, but can be challenging. The lessons learnt from managing AC joint instability can assist in managing these equivalent conditions.

18.6.1 Distal Clavicle Fractures

Distal clavicle fractures are a common variant with similar pathoanatomy to pure AC joint instability and have been classified by several authors [21–23]. The most pertinent system to our understanding of the pathoanatomy has been Craig's modification of the Neer classification [24]. This describes the variation of fractures in relation to the coracoclavicular ligaments, where a Type 2a fracture extends medial to the coracoclavicular ligaments leaving both trapezoid and conoid attached to the lateral fragment. Type 2a fractures tend to be grossly displaced similar to a Rockwood Type 5 AC joint dislocation. In Type 2b fractures, the force is said to travel between the coracoclavicular ligaments to fracture the clavicle. According to Craig, the conoid is also torn in this injury but the trapezoid remains attached to the distal fragment. We remain uncertain of this on the basis that in a 2b injury the clavicle does not displace significantly as in a 2a injury which has no ligament attachment and that in general ligament attachments are spared by fractures everywhere else in the skeleton. We feel the reason the clavicle does not displace significantly is that the conoid is intact as in a Type 3 AC joint dislocation (Figs. 18.13 and 18.14). The principles of distal clavicle fixation are also similar to that of AC joint stabilization. The emphasis should be on restoration of the ligamentous anatomy, which is achieved by realignment of the bone fragments, and stabilization using appropriate implants. In symptomatic, chronic displaced distal clavicle fractures the ligaments should be reconstructed as part of the procedure to produce a load-sharing construct with the bony fixation.

18.6.2 Coracoid Fracture

This variant is a rare injury that has been described in the adult and paediatric population [25, 26]. The hallmark feature on an AP X-ray is

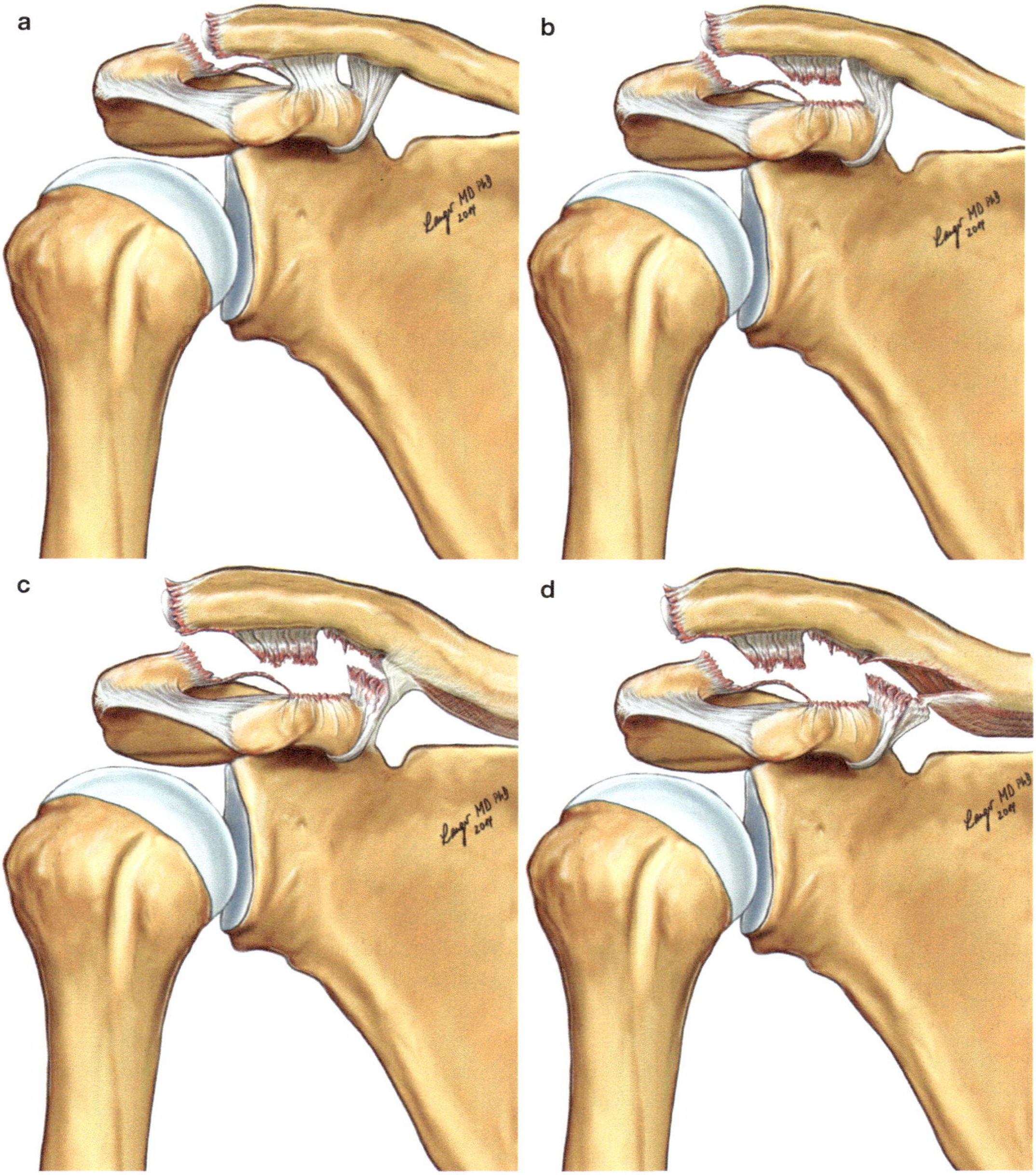

Fig. 18.12 (**a–d**) Author's modified pattern and classification of progressive and sequential AC joint instability and scapula medialization. (**a**) *Stage 1* – Failure of AC ligaments and capsule. (**b**) *Stage 2* – Failure of the trapezoid at the coracoid. (**c**) *Stage 3* – Failure of the conoid at the clavicle. (**d**) *Stage 4* – Stripping of the inferior periosteum from the medial clavicle (Copyright Dr Martin Langer)

an obvious AC joint dislocation without increase in the coracoclavicular distance. The diagnosis is confirmed on Axillary or Stryker Notch views or on CT scan. Even more unusual is the 'triple injury' where the coracoclavicular ligaments are ruptured resulting in AC joint dislocation along with a coracoid fracture [27]. Most of these injuries have been treated non-operatively; hence, the extent of the ligament injury is unknown. The degree of displacement is likely related to

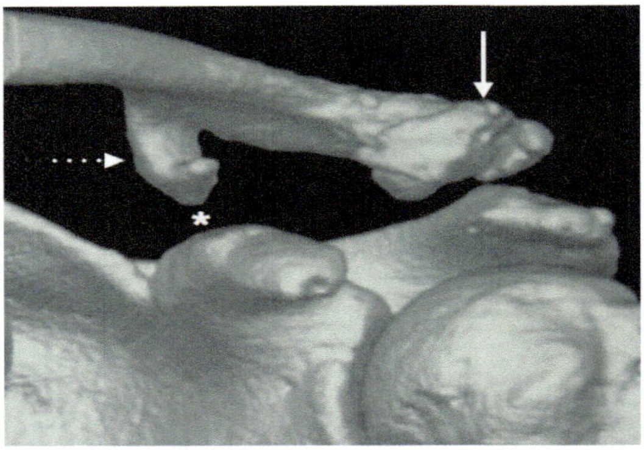

Fig. 18.13 Craig Type 2B distal clavicle fracture. The propagation is thought to pass between the conoid and trapezoid ligaments (Copyright Gregory Bain)

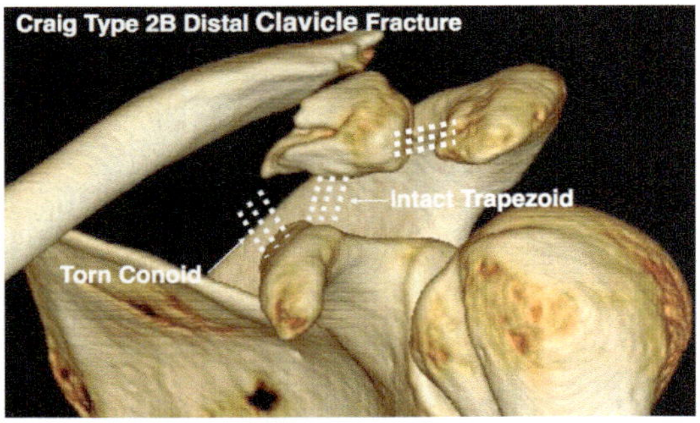

Fig. 18.14 CT scan showing a Craig Type 2B distal clavicle fracture (*white solid arrow*) treated non-operatively. There has been ossification within the periosteal sleeve (*dotted arrow*) stripped as a result of conoid tearing from the clavicle. The *asterisk* indicates where the conoid ligament would remain attached to the coracoid (Copyright Gregory Bain)

the amount of coracoid fractured. A fracture distal to the elbow of the coracoid would spare the coracoclavicular ligaments, which explains why the majority does not result in coracoclavicular diastasis. When coracoid fracture and diastasis does occur, this is either due to a fracture proximal to the coracoid elbow or concurrent ligamentous rupture. Coracoid base fractures may extend into the superior glenoid causing an intra-articular fracture of the glenohumeral joint. Why the coracoid fractures is unclear, however it could be an avulsion due to powerful contraction of the conjoined tendon, or be caused by an inferior dislocation of the AC joint that impacts and fractures the coracoid rather than getting stuck beneath it as in a Rockwood type 6 dislocation.

These injuries pose a challenge to treat because of the difficulty in stabilizing the AC joint in the presence of an unstable coracoid. A hook plate may be most appropriate as it reduces the AC joint and indirectly reduces the coracoid while negating the need for stabilization into or around the coracoid.

18.6.3 Paediatric Injuries

The clavicle has medial and lateral secondary ossification centres. The medial physis is the last in the body to close at around 20–25 years and the distal (lateral) epiphysis only ossifies in the late teenage years. True AC joint dislocation in children is rare and the term *Pseudo-dislocation* is preferred [28]. A displaced distal clavicle usually represents a Salter Harris 1 or 2 fracture through the distal physis. In a Salter Harris 2 fracture, the presence of a Thurston Holland fragment may be visible on plain radiographs. Dameron and Rockwood classified paediatric distal clavicle fractures in the same format as the Rockwood adult AC joint classification [29]. The important pathoantomical point of this classification is that type 3, 4 and 5 injuries all have intact coracoclavicular ligaments. The ligaments remain attached to a thick inferior periosteal sleeve, whereas the superior sleeve is disrupted allowing posterior, superior migration of the distal clavicle. Although the vast majority of these injuries are treated non-operatively, in the rare circumstance when surgery is considered, repair should concentrate on reconstitution of the periosteal tube around the clavicle which results in stability [28, 29].

18.6.4 Iatrogenic Instability

Arthroscopic and open resection of the AC joint is a common procedure performed for debilitating AC joint arthritis. Renfree described the anatomic location of the coracoclavicular and acromioclavicular ligaments in a cadaveric study [2]. A resection of 11 mm in men and women was found to never violate the trapezoid ligament; however, only a 5.2 mm resection in women (7.6 mm in men) risked violation of the superior AC joint ligament. It may be prudent to only resect up to 5 mm from the clavicle and routinely resect up to 5 mm from the acromion to eliminate any risk of posterior impingement. It is appreciated that a small group of patients that undergo lateral clavicle excision develop persistent pain, despite adequate resection. When examined this tends to be because of AP instability suggesting incompetence of the posterior superior AC joint ligaments. This may be related to overzealous resection; however, it may actually be that some patients develop AC joint arthritis because of subtle instability perhaps related to the shape of the AC joint. Resection of the distal clavicle in this group can unmask this AP instability and compound the situation. This entity is subtle and can be difficult to diagnose clinically. We have found 4D CT scanning to be a useful adjunct in this scenario (video). The patient undergoes a dynamic CT scan during which they perform a pre-set series of shoulder movements such as abduction, adduction and throwing motion. A 3D CT film is generated, which can be used to assess for dynamic impingement of the distal clavicle in this scenario. In terms of management, we have had some success in iatrogenic AC joint instability with isolated reconstruction of the AC joint ligament.

18.7 Principles of Acute Repair

The goal of repair is to restore and maintain the osseous anatomy so that the torn ligamentous stabilizers are able to heal. Successful outcome depends upon the initial fixation method, which should be strong enough to withstand cyclic loading until the ligaments heal. Direct repair of the coracoclavicular ligaments as part of the procedure has been suggested; however, it is our experience that these ligaments do not fail in their mid-substance and hence direct repair if performed should be to bone. The acromioclavicular ligaments, acromioclavicular joint capsule and deltotrapezial fascia should be meticulously repaired to recreate the soft tissue envelope around the lateral clavicle.

18.8 Principles of Chronic Reconstruction

Chronic reconstruction may be anatomic or non-anatomic and utilize biologic or synthetic materials. The goal is to reconstitute the bony and ligamentous anatomy in order to restore normal scapulothoracic kinematics, which will restore function and improve pain. A mainstay technique has been the Weaver Dunn procedure and its modifications. This is successful in up to 80 % of patients but has its drawbacks [30]. Firstly, it is non-anatomic and aims only to prevent superior migration of the clavicle rather than to restore kinematics. It also has a lower load to failure as compared to other biologic options and harvests the coracoacromial ligament, which is an important structure in preventing anterior-superior migration of the humeral head. We feel the criteria for an optimal reconstruction are best met by an anatomic and biologic reconstruction. The 'ACCR' reconstruction popularized by Mazzoca has a higher load to failure when used with interference screws [31]. We have modified this technique to use a combined endobutton and interference screw in the coracoid to secure the apex of a two-strand semitendinosus graft. We feel this more accurately recreates the anatomic insertion of the coracoclavicular ligaments on the coracoid. Looping the graft around the coracoid, which is a common technique, places the graft more anterior than the native insertion site and allows pivoting in the line of the graft, which does not truly replicate the ligament-bone interface and may lead to chronic attenuation of the graft. We also recommend routine reconstruction of the acromioclavicular ligaments using the tails of the same graft through acromial tunnels. This gives a robust, durable and biologic reconstruction; however, a salient technical point when doing a combined reconstruction is to reduce the acromioclavicular joint distance prior to securing the coracoclavicular tunnels. This is because it is easy to over reduce the coracoclavicular distance. If this occurs, the acromioclavicular distance is impossible to reduce as the acromion and coracoid are both part of the scapula and the clavicle must be mobile in order to reduce the acromioclavicular distance.

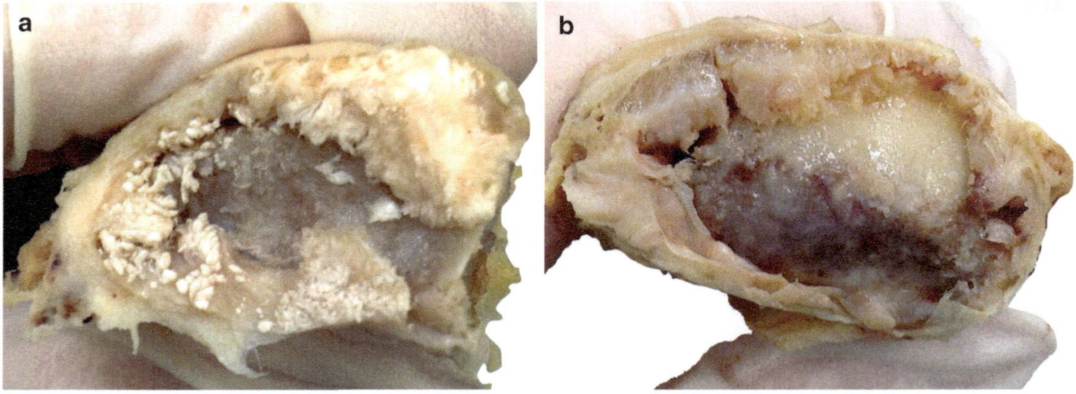

Fig. 18.15 (**a**, **b**) AC joint osteoarthritis. (**a**) Resected cadaveric lateral clavicle. Degenerate AC ligaments and full thickness articular cartilage loss. (**b**) Superior disc with adjacent intact articular cartilage. The inferior articulation is oblique and degenerate (Images courtesy of Dr Simon Bell, Melbourne, Australia)

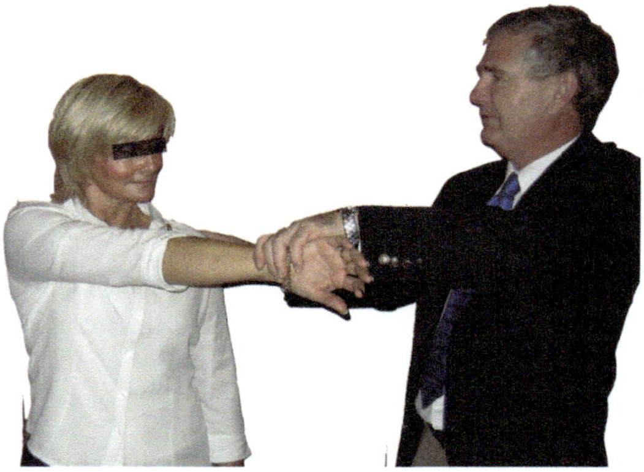

Fig. 18.16 The Bell van Riet (BvR) test is a more sensitive AC joint provocation test. The patient places the arm in forward flexion, *adduction* and internal rotation. The examiner (Simon Bell) resists upward translation of the arm. The patient reports that the pain is localized to the AC joint (Image courtesy of Dr Simon Bell, Melbourne, Australia); Van Riet and Bell [35])

18.9 AC Joint Arthritis

Degenerative arthritis is a common clinical problem. There are many anatomical factors, which predispose to the arthritis the absence of the articular disc and shape of the articulation (Fig. 18.15a, b) [33]. Clinical assessment includes identification of localized tenderness over the joint. We use the 'drop to the floor test'. The examiner palpates the tender AC joint, and that the patient withdraws due to the discomfort. There is an impression that the patient would drop to the floor, if the examiner were to keep pushing. With abduction the patient has 'terminal abduction pain', which discriminates it from the painful arc syndrome, seen in subacromial impingement.

We have previously used the O'Brien sign as an AC joint provocation test, but now use the Bell van Riet test [34, 35]). Localisation of pain to the AC joint during this test is more specific for AC joint pathology (Fig. 18.16).

For those patients with a painful arthritic joint, a cortisone injection can provide symptomatic improvement, and avoid the need for surgery [33]. Arthroscopic resection of the lateral clavicle is usually a successful procedure, but the anatomical attachments of the AC joint capsule need to be respected. See Chap. 19 and iatrogenic AC instability.

18.10 Summary

AC joint stability is reliant on the individualized function of the stabilizing ligaments. These ligaments fail in a sequential and predictable manner. We propose a modification to the classification of AC joint instability where the primary difference between Rockwood grade 3 and 5 injuries is the integrity of the conoid ligament. The concept of the unstable grade 3 injury should be adopted to better guide management and reconstruction should focus on anatomic restoration of the ligaments with biologic, durable grafts.

References

1. Adams F, Hippocrates H. The genuine works of Hippocrates; volume 1 – primary source edition. Nabu Press; 2014.
2. Renfree KJ, Riley MK, Wheeler D, Hentz JG, Wright TW. Ligamentous anatomy of the distal clavicle. J Shoulder Elbow Surg. 2003;12(4):355–9. doi:10.1016/S1058-2746(03)00029-6.
3. Fukuda K, Craig EV, An KN, Cofield RH, Chao EY. Biomechanical study of the ligamentous system of the acromioclavicular joint. J Bone Joint Surg Am. 1986;68(3):434–40.
4. Debski RE, Parsons I, Woo SL, Fu FH. Effect of capsular injury on acromioclavicular joint mechanics. J Bone Joint Surg Am. 2001;83(9):1344–51.
5. Yap JJ, Curl LA, Kvitne RS, McFarland EG. The value of weighted views of the acromioclavicular joint. Results of a survey. Am J Sports Med. 1999;27(6):806–9.
6. Bossart PJ, Joyce SM, Manaster BJ, Packer SM. Lack of efficacy of "weighted" radiographs in diagnosing acute acromioclavicular separation. Ann Emerg Med. 1988;17(1):20–4.
7. Beitzel K, Mazzocca AD, Bak K, et al. ISAKOS upper extremity committee consensus statement on the need for diversification of the Rockwood classification for acromioclavicular joint injuries. Arthroscopy. 2014;30:271–8. doi:10.1016/j.arthro.2013.11.005.
8. Kibler WB, Sciascia A, Wilkes T. Scapular dyskinesis and its relation to shoulder injury. J Am Acad Orthop Surg. 2012;20(6):364–72.
9. Schaefer FK, Schaefer PJ, Brossmann J, Hilgert RE, Heller M, Jahnke T. Experimental and clinical evaluation of acromioclavicular joint structures with new scan orientations in MRI. Eur Radiol. 2006;16(7):1488–93. doi:10.1007/s00330-005-0093-1.
10. Barnes CJ, Higgins LD, Major NM, Basamania CJ. Magnetic resonance imaging of the coracoclavicular ligaments: its role in defining pathoanatomy at the acromioclavicular joint. J Surg Orthop Adv. 2004;13(2):69–75.
11. Nemec U, Oberleitner G, Nemec SF, et al. MRI versus radiography of acromioclavicular joint dislocation. AJR Am J Roentgenol. 2011;197(4):968–73. doi:10.2214/AJR.10.6378.
12. Cadenat FM. The treatment of dislocations and fractures of the outer end of the clavicle. Int Clin. 1917;1:145–69.
13. Tossy JD, Mead NC, Sigmond HM. Acromioclavicular separations: useful and practical classification for treatment. Clin Orthop Relat Res. 1963;28:111–9.
14. Bucholz RW. Rockwood and Green's fractures in adults. Philadelphia: Lippincott Williams & Wilkins; 2012.
15. Bannister GC, Wallace WA, Stableforth PG, Hutson MA. A classification of acute acromioclavicular dislocation: a clinical, radiological and anatomical study. Injury. 1992;23(3):194–6.
16. HORN JS. The traumatic anatomy and treatment of acute acromio-clavicular dislocation. J Bone Joint Surg Br. 1954;36-B(2):194–201.
17. Copeland S, Kessel L. Disruption of the acromioclavicular joint: surgical anatomy and biological reconstruction. Injury. 1980;11(3):208–14.
18. Lizaur A, Marco L, Cebrian R. Acute dislocation of the acromioclavicular joint. Traumatic anatomy and the importance of deltoid and trapezius. J Bone Joint Surg Br. 1994;76(4):602–6.
19. Tischer T, Salzmann GM, El-Azab H, Vogt S, Imhoff AB. Incidence of associated injuries with acute acromioclavicular joint dislocations types III through V. Am J Sports Med. 2009;37(1):136–9. doi:10.1177/0363546508322891.
20. O'Driscoll SW, Jupiter JB, King GJ, Hotchkiss RN, Morrey BF. The unstable elbow. J Bone Joint Surg Am. 2000;82(5):724–38.
21. Allman Jr FL. Fractures and ligamentous injuries of the clavicle and its articulation. J Bone Joint Surg Am. 1967;49(4):774–84.
22. Neer CS. Fractures of the distal third of the clavicle. Clin Orthop Relat Res. 1968;58:43–50.
23. Robinson CM. Fractures of the clavicle in the adult. Epidemiology and classification. J Bone Joint Surg Br. 1998;80(3):476–84.
24. Rockwood Jr CA, Matsen III FA, Lippitt SB. The shoulder. Philadelphia: Elsevier Health Sciences; 2009.
25. Carr AJ, Broughton NS. Acromioclavicular dislocation associated with fracture of the coracoid process. J Trauma Injury Infect Crit Care. 1989;29(1):125–6.
26. Combalía A, Arandes JM, Alemany X, Ramón R. Acromioclavicular dislocation with epiphyseal separation of the coracoid process: report of a case and review of the literature. J Trauma Injury Infect Crit Care. 1995;38(5):812–5.
27. Wilson KM, Colwill JC. Combined acromioclavicular dislocation with coracoclavicular ligament disruption and coracoid process fracture. Am J Sports Med. 1989;17(5):697–8.
28. Falstie-Jensen S, Mikkelsen P. Pseudodislocation of the acromioclavicular joint. J Bone Joint Surg Br. 1982;64(3):368–9.
29. Rockwood CA, Wilkins KE, Beaty JH, Kasser JR. Rockwood and Wilkins' fractures in children. Philadelphia: Lippincott Williams & Wilkins; 2006.
30. Sood A, Wallwork N, Bain GI. Clinical results of coracoacromial ligament transfer in acromioclavicular dis-

locations: a review of published literature. Int J Shoulder Surg. 2008;2(1):13–21. doi:10.4103/0973-6042.39582.
31. Carofino BC, Mazzocca AD. The anatomic coracoclavicular ligament reconstruction: surgical technique and indications. J Shoulder Elbow Surg. 2010; 19(2 Suppl):37–46. doi:10.1016/j.jse.2010.01.004.
32. DeFranco MJ, Patterson BM. The floating shoulder. J Am Acad Orthop Surg. 2006;14(8):499–509.
33. Bain GI, Van Riet RP, Gooi C, Ashwood N. The long-term efficacy of corticosteroid injection into the acromioclavicular joint using a dynamic fluoroscopic method. Int J Shoulder Surg. 2007;1(4): 104–7.
34. O'Brien SJ, Pagnani MJ, Fealy S, McGlynn SR, Wilson JB. The active compression test: a new and effective test for diagnosing labral tears and acromioclavicular joint abnormality. Am J Sports Med. 1998;26:610–3.
35. van Riet RP, Bell SN. Clinical evaluation of acromioclavicular joint pathology: sensitivity of a new test. J Shoulder Elbow Surg. 2011;20(1):73–6. doi:10.1016/j.jse.2010.05.023.
36. Jeray K. Acute midshaft clavicle fractures. JAAOS. 2007;15:239–48.

Sternoclavicular Joint Anatomy and Pathology

19

Michael B. O'Sullivan, Justin Yang, Benjamin Barden, Hardeep Singh, Jessica Divenere, and Augustus D. Mazzocca

19.1 Introduction

The sternoclavicular (SC) joint is a saddle-type joint, which provides the only true articulation between the upper extremity and the axial skeleton [11, 29, 44, 55, 56]. It is a diarthrodial synovial joint, formed by the articulation of the sternal end of the clavicle, the clavicular notch of the manubrium, and the cartilage of the first rib [3, 7, 17, 57]. Much of the stability of the joint is provided from ligamentous structures given the lack of inherent osseous stability. Fortunately, serious pathology of the SC joint is rare, given the vital structures in close proximity.

19.2 Osseous Anatomy

The sternoclavicular joint is subcutaneous, with the medial clavicle being prominent. The clavicle and manubrium articular and the first rib are in close association.

19.2.1 Clavicle

The clavicle provides the only bony connection between the upper extremity and the axial skeleton. The medial half of the clavicle is convex anteriorly, originating from an enlarged and bulbous sternal head [38]. Medially, the clavicle articulates with the manubrium and the synchondrosis of the first rib (Fig. 19.1). The ventral and inferior quadrant of the sternal end of the clavicle is covered in fibrocartilage, while the superior and dorsal crescent serves as the insertion site for the posterior capsule and intra-articular disk (Fig. 19.2) [3, 44, 57]. Only the inferior and medial portion of the clavicular head articulates with the manubrium, while more than half of the clavicular head lies above this articulation (Fig. 19.3) [44, 50]. The medial epiphysis of the clavicle is the last epiphysis in the body to ossify and fuse, occurring between the ages 18–20 and 23–25, respectively [44].

19.2.2 Manubrium

The manubrium is the most cranial of the three bones that constitute the sternum. It is attached to the sternal body by a synchondrosis that ossifies in middle to late adulthood [38]. The manubrium has curved, shallow clavicular notches at its superolateral borders, which articulate with both sternal heads (Fig. 19.3). These clavicular notches are covered in fibrocartilage [44].

Disclosures Dr. Augustus D. Mazzocca receives research support from, and is a consultant for, Arthrex Inc. Naples, FL, USA

M.B. O'Sullivan, MD • J. Yang, MD
B. Barden, MD • H. Singh, MD • J. Divenere, BS
A.D. Mazzocca, MS, MD (✉)
Department of Orthopaedic Surgery,
University of Connecticut Health Center,
Farmington, CT 06032, USA
e-mail: mazzocca@uchc.edu

G.I. Bain et al. (eds.), *Normal and Pathological Anatomy of the Shoulder*,
DOI 10.1007/978-3-662-45719-1_19, © ISAKOS 2015

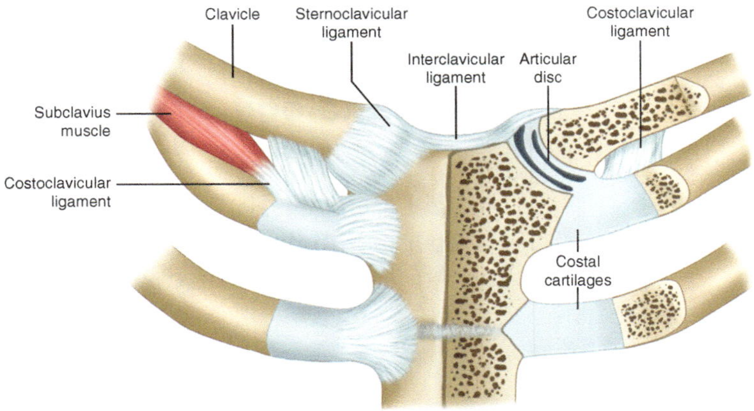

Fig. 19.1 The bony anatomy of the sternoclavicular joint including the medial clavicle, manubrium, and first rib is depicted. In addition, the sternoclavicular ligament, interclavicular ligament, costoclavicular ligament, and intra-articular disk are shown

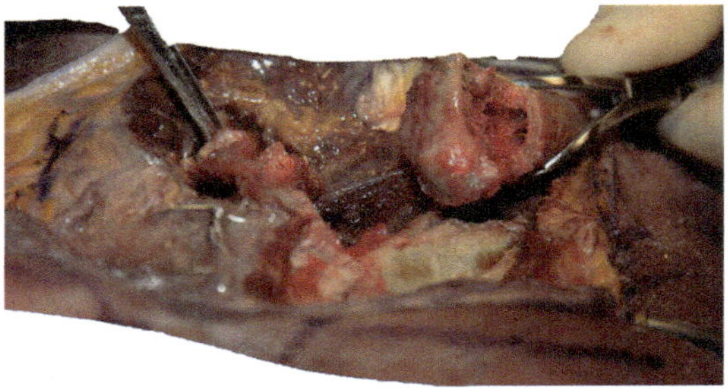

Fig. 19.2 The left medial clavicular head is shown. The medial clavicle is cut to demonstrate the bony anatomy of the medial clavicle. Note also the intra-articular disk, which is reflected from the SC joint

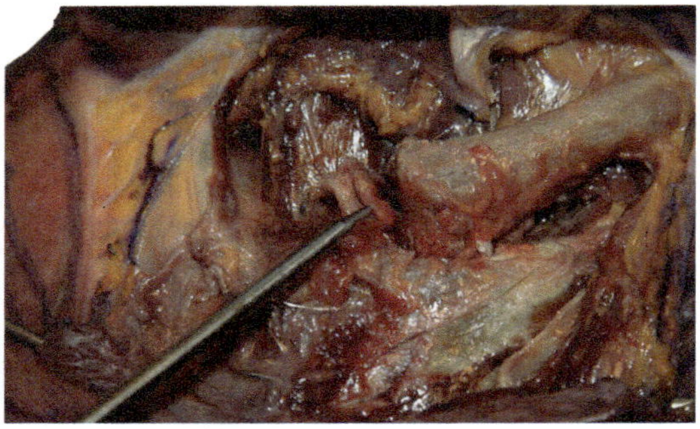

Fig. 19.3 The intra-articular disk of the left sternoclavicular joint is shown. Note that the disk in this specimen is partially degenerated. Approximately half of the medial clavicular head lies cranial to the relatively shallow clavicular notch of the manubrium. Also visualized are the sternothyroid and sternohyoid muscles posterior to the sternoclavicular joint

Both the anterior and posterior sternoclavicular ligaments insert on the manubrium, while the interclavicular ligament courses along the superior portion of the manubrium (Fig. 19.1).

19.2.3 First Rib

The first rib is the broadest and shortest true rib. Anteriorly, it contains costal cartilage, which attaches to the manubrium through a synchondrosis of dense, adherent fibrocartilage [38]. This synchondrosis is located inferolaterally from the clavicular notch of the manubrium (Fig. 19.1) [38]. It represents the most inferior and lateral aspect of the sternoclavicular joint. The synchondrosis of the first rib serves as an attachment site for the costoclavicular ligament. The first ribs, along with the manubrium and first thoracic vertebrae, constitute the thoracic outlet.

19.2.4 Fibrocartilaginous Disk

The sternoclavicular joint contains an intra-articular disk, which attaches to the anterior and posterior sternoclavicular ligaments (Fig. 19.3) [50, 57]. The disk is analogous to the meniscus in the knee and divides the joint into two distinct synovial cavities [50]. It is comprised of a dense, vertically oriented, fibrocartilaginous medial sternoclavicular portion and a thinner more horizontally oriented lateral costoclavicular portion [3]. The intra-articular disk is thought to degenerate with age [50]. Van Tongel et al. [57] demonstrated that the disk ligament was incomplete in 56 % of cadaver specimens evaluated, containing centrally located holes associated with fraying of the disk and degeneration of the clavicular cartilage. The presence of an incomplete intra-articular disk ligament is likely to be related to degenerative changes rather than developmental abnormalities, because this variant was only noted in cadaver specimens greater than 75 years of age [57]. The disk is believed to enhance shock absorption, protect the articular surfaces of the SC joint, aid in rotation of the clavicle, and help prevent medial displacement of the clavicle [17, 14, 44, 55].

19.3 Ligamentous Anatomy

There is limited osseous contact between the clavicle and the manubrium; so, the stability of the sternoclavicular joint is primarily provided by the ligamentous structures [44, 55]. The stabilizing ligaments of the sternoclavicular joint include the costoclavicular ligament, interclavicular ligament, and the anterior and posterior sternoclavicular (SC) ligaments (Fig. 19.1) [3, 7, 17, 44, 50, 55, 56].

19.3.1 Costoclavicular Ligament

The costoclavicular ligament contains an anterior and a posterior fasciculus, which arises from the chondral surface of the first rib and inserts on the undersurface of the medial clavicle (Fig. 19.4) [44]. The anterior fasciculus originates on the anterior medial aspect of the chondral surface and its fibers run superolaterally, while the posterior fasciculus originates on the posterior lateral aspect and its fibers run superomedially [44]. The anterior fasciculus of the costoclavicular ligament limits both lateral translation and upward rotation of the clavicle, while the posterior fasciculus limits medial translation and downward rotation [5, 44, 55].

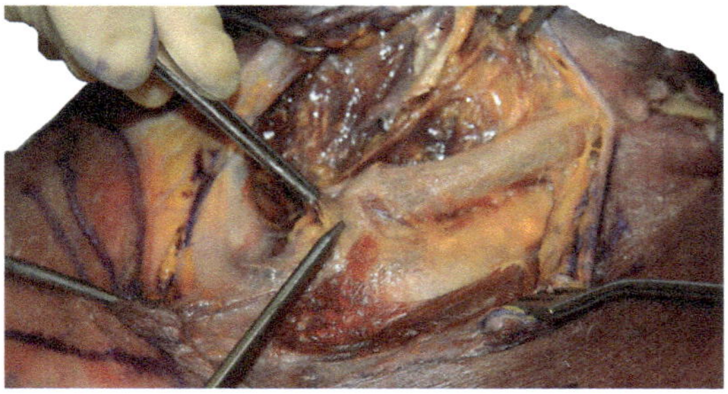

Fig. 19.4 The interclavicular ligament is shown with the use of the forceps. The *probe* indicates the position of the sternoclavicular joint. The fibers of the anterior sternoclavicular ligament are well visualized at the level of the joint. The anterior and posterior fasciculi of the costoclavicular ligament are seen lateral to the anterior SC ligament. Also note the internal jugular vein superolaterally from the sternoclavicular joint

19.3.2 Interclavicular Ligament

The interclavicular ligament links the superomedial aspects of each clavicle and runs along the superior margin of the manubrium (Fig. 19.4). The ligament functions as a secondary stabilizer of the SC joint, preventing excessive depression of the clavicle [44].

19.3.3 Capsular Ligaments

The anterior and posterior SC ligaments represent capsular thickenings, which function as the primary stabilizers of the SC joint, with the posterior ligament being the stronger of the two (Fig. 19.4) [52]. The anterior SC ligament fibers run obliquely from the medial clavicle to the sternum in a downward and medial direction, serving to prevent anterior displacement of the clavicle. The posterior SC ligament fibers cover the posterior aspect of the SC joint, preventing both anterior and posterior displacements of the clavicle [52].

19.3.4 Intra-articular Ligament

The presence of an intra-articular disk ligament is a matter of debate. Some authors argue that the intra-articular disk ligament complex passes from the dorsal and caudal clavicular head, through the SC joint, to the synchondral junction of the first rib and manubrium [3, 44, 55]. Yet others argue that the disk only inserts on the anterior and posterior capsular ligaments, which then insert onto the bone, and thus is not a ligament itself [50, 57]. Regardless, by virtue of its inferior lateral and superomedial insertions, the disk acts as a tether, preventing medial translation of the clavicular head [44].

19.4 Anatomic Structures in Close Proximity

The sternoclavicular joint is easily palpated at the base of the anterior neck. The clavicle has numerous muscular attachments including the sternohyoid, sternocleidomastoid, pectoralis major, subclavius, deltoid, and trapezius. The manubrium has attachments including the sternocleidomastoid, sternothyroid, and sternohyoid. At the sternoclavicular joint, most superficially, are the inferior fibers of the platysma and the deep cervical fascia. Deep to this lie the sternocleidomastoid muscle and the sternoclavicular joint. Pectoralis major lies inferior to the sternoclavicular joint. Immediately posterior and medial to the SC joint are the sternothyroid and sternohyoid muscles (Figs. 19.3 and 19.4). The anterior jugular vein courses anteriorly along this musculature in the region of the sternoclavicular joint. The internal jugular vein courses laterally to this musculature in the region of the SC joint, where it joins the subclavian vein to form the brachiocephalic vein (Figs. 19.4 and 19.5). The trachea exists deep and medial to these muscles. Deep and medial to the brachiocephalic vein on the right is the vagus nerve and brachiocephalic artery and on the left is the vagus nerve and common carotid artery at the level of the sternoclavicular joint (Fig. 19.6a–c). The medial supraclavicular nerve and the nerve to the subclavius innervate the SC joint. The joint receives its arterial blood supply from the branches of the internal thoracic artery and suprascapular

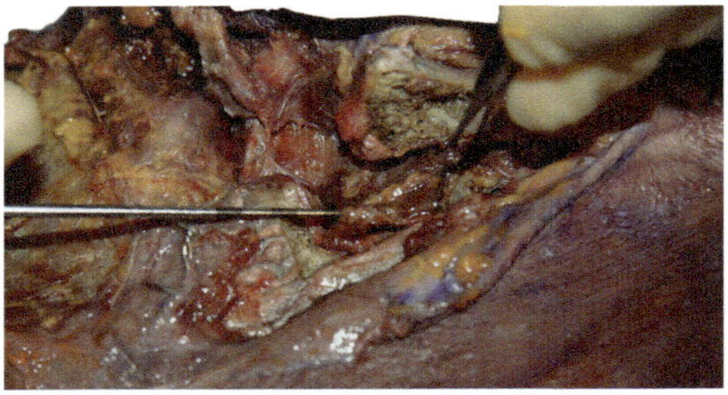

Fig. 19.5 The left brachiocephalic vein is shown posterior to the retracted clavicle at the level of the sternoclavicular joint. Also visualized is the left subclavian vein as it joins the internal jugular vein to form the brachiocephalic vein

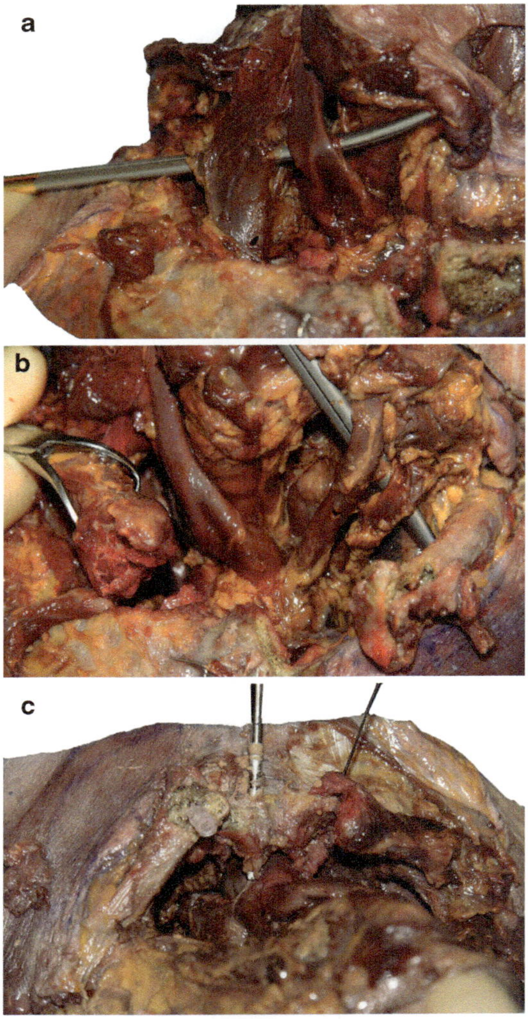

Fig. 19.6 (**a–c**) The arterial structures in proximity to the sternoclavicular joint are depicted. The right common carotid artery (**a**) is shown deep and medial to the right internal jugular vein. The right brachiocephalic artery, left common carotid artery, and left subclavian artery (**b**) are shown as they branch from the arch of the aorta with the clavicles retracted and dissection of the sternothyroid and sternohyoid muscles. The trachea is also visualized deep to these structures. A drill-tip piercing the manubrium and abutting the vital arterial structures (**c**) is shown to emphasize the proximity of these structures to the sternoclavicular joint

arteries. The subclavian and external jugular veins receive venous drainage from the joint [39, 53].

19.5 Biomechanics

The sternoclavicular joint is a diarthrodial joint that has been described both as a saddle joint and as a ball-and-socket joint [42]. The osteology of the sternoclavicular joint is reciprocally concave and convex. This allows motion in coronal and sagittal planes. However, there is also a rotational component in the normal shoulder motion, which translates to the sternoclavicular joint, making it biomechanically similar to a ball-and-socket joint [28, 44]. Rockwood and others propose that the clavicle acts as a crankshaft, allowing the scapula to rotate in a 60° arc around the sternoclavicular joint [20, 45]. During the normal motion of the shoulder, the sternoclavicular joint is also capable of 30–35° of elevation and 35° of flexion and extension [45]. Motion of the sternoclavicular joint occurs mostly in the first 90° of arm elevation, with a 4° of sternoclavicular motion for every 10° of shoulder elevation. Almost no sternoclavicular motion occurs at high degrees of shoulder elevation [29].

The sternoclavicular joint is the only true diarthrodial articulation between the upper extremity and the axial skeleton in most adults. In 2.5 % of people, an articulation exists between the clavicle and the first rib [11]. However, less than half of the medial clavicle (inferior pole) articulates with the upper angle of the sternum, making this an inherently unstable joint [44].

To compensate for the lack of inherent osseous instability, the capsule and ligaments surrounding the SC joint are some of the strongest in the human body [20]. In ligament sectioning experiments using cadaveric specimens, Spencer et al. [52] found significant increases in posterior and anterior translation (107 and 42 %, respectively), which resulted from cutting the posterior capsule. Cutting the anterior capsule only produced increases in anterior translation, but to a lesser degree (26 %) than the sectioning of the posterior capsule. Cutting the costoclavicular and interclavicular ligaments had little effect on sternoclavicular joint translation. On the basis of these experiments, Spencer et al. [52] concluded that the posterior capsule is the most important restraint for posterior and anterior translation of the medial clavicle. The authors also performed a load-to-failure test on native specimens, and interestingly at a maximum load of 552 N, the failure occurred at the bone–cement interface of the testing apparatus [52]. No studies exist on what the true load-to-failure load is of the native anterior or posterior capsule.

Dynamic muscular stabilization of the sternoclavicular joint is poorly understood. Clearly, muscular stabilization plays some role, as medial excision of the clavicle in chronic instability has been reported with good clinical results [1]. In cases of atraumatic anterior subluxation, trapezius weakness has been thought to be a contributing factor. As the superior portion of the trapezius elevates the lateral clavicle, the medial clavicle is depressed, improving the stability of the sternoclavicular joint [61].

19.6 Pathoanatomy: Atraumatic Conditions

Atraumatic conditions of the sternoclavicular joint are rare, but always raise concerns, because they may be sinister, and their surgical management can theoretically be life threatening.

19.6.1 Sternocostoclavicular Hyperostosis

Sternocostoclavicular hyperostosis (SCCH) is a rare chronic inflammatory disorder of the anterior chest wall. The disease typically begins with inflammation and calcification of the sternoclavicular ligaments and sternoclavicular perichondritis, which progresses to erosive arthritis [10, 22]. With time, progressive hyperosteotic changes are seen extending to the medial clavicle, sternoclavicular joint, manubrium, first rib, and soft tissue [10, 22]. Occasionally, changes are seen in ribs two through seven [22]. Complete fusion of the sternoclavicular joints can occur after years of chronic inflammation. Biopsy specimens demonstrate nonspecific osteosclerosis with the presence of a round cell infiltrate and granulation tissue [22]. SCCH can be associated with extrasternal manifestations such as sclerosis of the axial skeleton (vertebrae, pelvis, sacroiliac joint), peripheral arthritis, and, most commonly, palmoplantar pustulosis [10, 22, 51]. The etiology of this rare disorder and the overlap between this and others that present with multifocal nonsuppurative periosteitis and hyperostosis, namely, SAPHO (synovitis, acne, pustulosis, hyperostosis, and osteitis) syndrome and chronic recurrent multifocal osteomyelitis (CRMO), is poorly understood [4]

19.6.2 Condensing Osteitis

Condensing osteitis is a rare disorder involving sclerosis and enlargement of the inferomedial clavicle, which spares the sternoclavicular joint. The condition was first described in 1974 [8], and only 40 cases have been documented in the literature [27]. The disease typically presents unilaterally in women of childbearing age without a history of trauma. Patients experience an insidious onset of pain, which may radiate to the supraclavicular fossa, and a fusiform swelling over the medial clavicle [8, 34]. Pain is exacerbated with shoulder abduction and forward flexion. The etiology is currently unknown. Some authors argue the disorder is a response to mechanical stress, while others support an infectious etiology [8, 12, 30, 34]. Radiographs and CT scans demonstrate sclerosis, minor expansion of the inferomedial clavicle, and loss of marrow space [25]. Hypointensity in the affected region of the clavicle on T1-weighted SE images and low to intermediate signal on T2-weighted SE images is noted on MRI [41]. Bone scans with technetium-99 m methylene diphosphonate or pyrophosphate demonstrates focal increased uptake in the ipsilateral medial clavicle, while both indium and gallium scans show no focal accumulation of white blood cells at the lesion, arguing against infectious etiology [34]. Imaging typically demonstrates no involvement of the sternoclavicular joint, manubrium, or first rib in this disorder. Biopsy and gross pathology show increase and thickening of the cancellous bone, periosteal reaction, inferomedial osteophyte formation, and enlargement of the clavicular head without evidence of necrosis, bony destruction, or soft tissue involvement [8, 34].

19.6.3 Friedrich's Disease

Friedrich's disease, or avascular necrosis of the medial clavicle, was first described in 1924 [21]. Like most conditions involving the sternoclavicular joint, Friedrich's disease is a rare disorder with sparse literature describing the condition. The etiology is currently unknown. While the disease has been described in men, most case reports involve female patients with unilateral disease [31, 35]. Patients typically present with insidious onset of localized pain and swelling at the sternoclavicular joint. This pain tends to increase with shoulder abduction and is not associated with a direct history of trauma. Radiographs and CT scans show sclerosis that is predominantly located in the inferomedial clavicle, but may involve the entire medial clavicular head, and an irregular sternoclavicular joint with bony destruction [9, 25]. Histological evaluation demonstrates characteristic findings of avascular necrosis, namely, Haversian systems with empty lacunae and fibrotic bone marrow [19, 35]. Some argue that given the overlap between Friedrich's disease and condensing osteitis, specifically the presence of inferomedial clavicular sclerosis, these disorders may represent the same disease with age-related radiographic differences [31].

19.6.4 Septic Arthritis

Sternoclavicular septic arthritis accounts for 1 % of all cases of septic arthritis in normal healthy individuals but 17 % of septic arthritis cases in intravenous drug users [46]. Intravenous drug users may be preferentially affected from the spread of the infection through the wall of the subclavian vein to the sternoclavicular joint following injection into the upper extremity [23]. In a review of 168 patients with septic arthritis of the sternoclavicular joint, 77 % of the patients were shown to have risk factors for infection, with intravenous drug use (21 %), infection at a distant site (15 %), diabetes mellitus (13 %), history of trauma (12 %), infected central venous catheter (9 %), chronic renal failure (8 %), and HIV infection (4 %) among the most prevalent [46]. The most common causative organisms are *Staphylococcus aureus* (49 %), *Pseudomonas aeruginosa* (10 %), *Brucella melitensis* (7 %), and *Escherichia coli* (5 %) [46]. In a review of 27 healthy adult patients, the most common causative organisms were *S. aureus* (50 %), Mycobacteria (12 %), *Streptococcus* species (8 %), anaerobes (8 %), and Pasteurella (8 %) [2]. While some patients may present with symptoms consistent with sepsis, others may present with vague shoulder pain or an anterior chest wall mass [6]. Radiographs and CT scans are frequently inconclusive in the acute setting [6]. MRI is the most sensitive imaging modality to evaluate for osseous destruction, fluid accumulation, edema, and abscess [6].

19.6.5 Osteoarthritis

Osteoarthritis is a common cause of sternoclavicular pathology. Osteoarthritis is rare in patients under 40, but seen in 53 % of patients over 60 years of age, typically with bilateral involvement [33]. In a study of 25 sternoclavicular cadaver specimens, van Tongel et al. [57] noted at least moderate cartilage damage in all specimens. Patients at risk for osteoarthritis of the sternoclavicular joint include postmenopausal women, patients with a history of manual labor, and patients with a history of surgical neck dissection [26, 62]. Patients are typically asymptomatic, but may present with pain, which is worse with shoulder abduction, swelling, and crepitus at the sternoclavicular joint [26, 62]. Radiographs and CT scans demonstrate joint space narrowing, subchondral sclerosis and cysts, and osteophyte formation (Fig. 19.7). The inferomedial portion of the clavicle, which is responsible for articulating with the manubrium, most clearly demonstrates these arthritic changes [33].

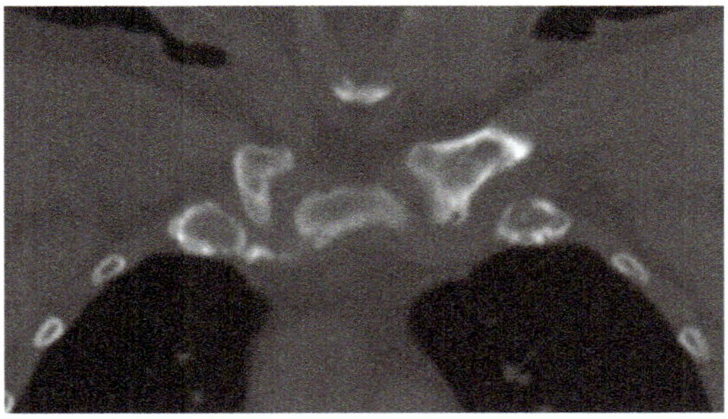

Fig. 19.7 A coronal CT image of a 53-year-old male with a history of manual labor demonstrates osteoarthritic changes in the bilateral sternoclavicular joints

19.6.6 Tietze's Syndrome

Tietze's syndrome is a diagnosis of exclusion. The disorder is defined as a nonsuppurative inflammation of the anterior chest wall including the costosternal and sternoclavicular joints, which is both benign and self-limiting in nature [27, 49, 58]. The second and third costochondral joints on the left side are most frequently affected; however, sternoclavicular joint involvement is common [49]. When the sternoclavicular joint is involved, patients present with insidious onset of localized pain and swelling. The etiology of Tietze's syndrome is currently unknown, but some believe it is related to microtrauma, as patients often report a history of cough [32, 49]. Radiographs are typically normal. CT scans may demonstrate thickening of the costal cartilage of the effected region, ventral angulation of the effected region, or normal anatomy [16]. MRI findings in Tietze's syndrome demonstrate thickening of the cartilage of the effected region with focal or widespread increased signal intensity in the cartilage on T2-weigthed imaging and possible subchondral bone marrow edema [58]. Biopsy has demonstrated inconsistent findings, including normal cartilage, metaplasia of the cartilage, increased vascularity, swelling of the perichondrium, and low-grade inflammation [32]. Synovitis may represent the pathological lesion in this disorder, but biopsy data in the literature has not included synovial or ligamentous findings [32].

19.6.7 Atraumatic Instability

Spontaneous atraumatic anterior subluxation of the sternoclavicular joint is rare. Only 14 % of 273 patients presenting with complaints involving the sternoclavicular joint were diagnosed with this disorder [43]. Patients are typically adolescents to middle age when they first present for evaluation. Rockwood and Odor [43] reported equal distribution among gender in their series of 37 patients, while Sadr and Swann [47] reported a female predominance in their series of 22 patients. The dominant and nondominant arms are affected in equal proportion, and the disorder is occasionally bilateral. Generalized ligamentous laxity is seen in approximately 80 % of patients [43]. Subluxation occurs when the patient's ipsilateral arm is in an overhead position and is typically associated with a "pop" and a visible prominence at the sternoclavicular joint. Spontaneous reduction occurs when the arm is brought down to the patient's side. While the index subluxation may cause some discomfort, subsequent episodes are typically painless [43, 47]. Radiographs may show elevation of the medial clavicle on serendipity view and anterior subluxation of the medial clavicle on a stress view without evidence of bony abnormalities [15]. We were unable to identify any studies in the English language describing the MRI or histological findings of the stabilizing ligaments in patients with spontaneous atraumatic anterior subluxation of the sternoclavicular joint. Given the association with generalized ligamentous laxity, we believe this disorder is likely caused by repetitive microtrauma leading to ligamentous and capsular attenuation in genetically predisposed individuals.

19.7 Pathoanatomy: Traumatic Conditions

Traumatic conditions of the sternoclavicular joint are seen in major trauma, sportsman, and can occur almost spontaneously. Because of the close proximity of the vital structures, they are always taken seriously and require professional management.

19.7.1 Posterior Dislocation

Since initially described by Sir Astley Cooper in 1824, only 140 cases of posterior sternoclavicular dislocation have been reported in the literature [54, 36]. The relative rarity of this injury combined with the subtle physical findings can make diagnosis difficult. It is not uncommon for these injuries to be missed on initial evaluation [37]. Unfortunately, the consequences of this injury can be devastating given the potential for significant neurologic and vascular compromise of the mediastinal structures that can occur with posterior dislocation of the sternoclavicular joint. The most common mechanism of sternoclavicular injury is from motor vehicle accidents, but several sport-related collision injuries have been reported in youth, collegiate, and professional rugby and American football players [36, 37, 60]. Two distinct patterns of posterior sternoclavicular dislocation exist: (1) a direct posterior force to the medial clavicle with the arm in an adducted and internally rotated position and (2) an indirect anterior force to the posterolateral shoulder, usually as a result of a fall. Radiographs may be difficult to interpret given the degree of bony overlap at the SC joint. However, a CT scan will readily reveal the diagnosis (Fig. 19.8). Posterior dislocations should be urgently evaluated and a closed reduction should be attempted initially. An urgent open reduction with reconstruction of the sternoclavicular joint is recommended if a closed reduction attempt fails or if the patient develops neurovascular or airway compromise [24, 54].

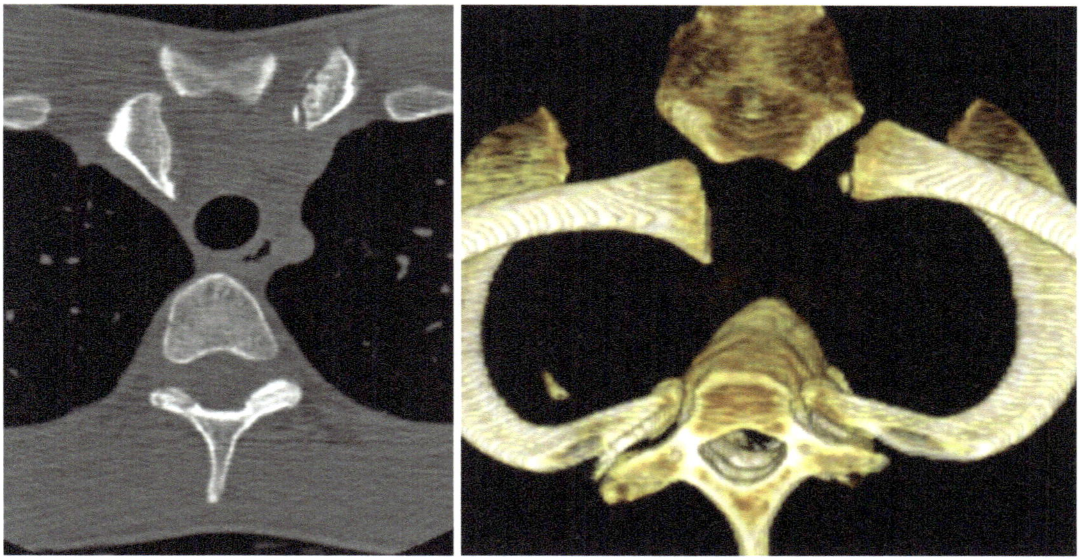

Fig. 19.8 An axial CT image demonstrates a right-sided posterior sternoclavicular joint dislocation in a 17-year-old female who experienced a blunt force to the medial clavicle during athletic competition

19.7.2 Anterior Dislocation

Anterior dislocation occurs more frequently than posterior dislocations; although, a majority of these anterior dislocations are atraumatic in nature. Traumatic anterior dislocations are typically caused indirectly by a posterior force to the anterolateral shoulder, such as from a seat belt in a motor vehicle accident [24]. Closed reduction can be attempted, although the success rates that have been reported range from 21 to 100 % [18, 40]. Long-term outcomes with conservative management are good, with the main complaint being cosmesis [13, 24, 48]. Surgical reconstruction following anterior dislocations is controversial, with the risks possibly outweighing the benefits [24].

19.7.3 Physeal Injury

The epiphysis at the medial end of the clavicle is the last to appear (at age 18–20), and the physis is one of the last to close (at age 23–25) [59]. One postmortem evaluation of the medial clavicular physis revealed that a complete closure of the medial physis may not occur until 31 years of age [59]. Furthermore, the physis is much weaker biomechanically than the strong capsular ligaments of the sternoclavicular joint. As a result, many "dislocations" in teenage and adolescent patients are actually physeal injuries. Closed reductions can be performed for both anterior and posterior displaced physeal fractures. However, except in the case of irreducible posteriorly displaced fractures, open reduction is not recommended due to the remodeling potential [24].

Conclusion

The sternoclavicular joint represents the only true articulation between the axial skeleton and the upper extremity. Ligamentous restraint plays a pivotal role in the stability of the joint given the lack of osseous constraint on the joint. Symptomatic pathology of the joint is rare, so literature on the joint is sparse. Most conditions regarding the joint can be managed conservatively. Given the potential for morbidity and mortality with a posterior dislocation of the sternoclavicular joint, urgent reduction and possible operative stabilization is recommended.

References

1. Acus 3rd RW, Bell RH, Fisher DL. Proximal clavicle excision: an analysis of results. J Shoulder Elbow Surg. 1995;4(3):182–7.
2. Bar-Natan M, Salai M, Sidi Y, Gur H. Sternoclavicular infectious arthritis in previously healthy adults. Semin Arthritis Rheum. 2002;32(3):189–95.
3. Barbaix E, Lapierre M, Van Roy P, Clarijs JP. The sternoclavicular joint: variants of the discus articularis. Clin Biomech (Bristol, Avon). 2000;15 Suppl 1:S3–7.
4. Beals RK, Sauser DD. Nontraumatic disorders of the clavicle. J Am Acad Orthop Surg. 2006;14(4):205–14.
5. Bearn JG. Direct observations on the function of the capsule of the sternoclavicular joint in clavicular support. J Anat. 1967;101(Pt 1):159–70.
6. Bodker T, Tottrup M, Kjaer K, Jurik AG. Diagnostics of septic arthritis in the sternoclavicular region: 10 consecutive patients and literature review. Acta Radiol. 2013;54:67–74. doi:10.1258/ar.2012.120363.
7. Brossmann J, Stabler A, Preidler KW, Trudell D, Resnick D. Sternoclavicular joint: MR imaging–anatomic correlation. Radiology. 1996;198(1):193–8.
8. Brower AC, Sweet DE, Keats TE. Condensing osteitis of the clavicle: a new entity. Am J Roentgenol Radium Ther Nucl Med. 1974;121(1):17–21.
9. Carrera EF, Carvalho RL, Netto NA, Nicolao FA, Pedro FJ. Avascular necrosis of the sternal end of the clavicle: a case report. J Shoulder Elbow Surg. 2008;17(4):e31–3. doi:10.1016/j.jse.2007.08.008.
10. Carroll MB. Sternocostoclavicular hyperostosis: a review. Ther Adv Musculoskel Dis. 2011;3(2):101–10. doi:10.1177/1759720X11398333.
11. Cave AJ. The nature and morphology of the costoclavicular ligament. J Anat. 1961;95:170–9.
12. Cone RO, Resnick D, Goergen TG, Robinson C, Vint V, Haghigi P. Condensing osteitis of the clavicle. AJR Am J Roentgenol. 1983;141(2):387–8.
13. de Jong KP, Sukul DM. Anterior sternoclavicular dislocation: a long-term follow-up study. J Orthop Trauma. 1990;4(4):420–3.
14. DePalma AF. The role of the disks of the sternoclavicular and the acromioclavicular joints. Clin Orthop. 1959;13:222–33.
15. Echlin PS, Michaelson JE. Adolescent butterfly swimmer with bilateral subluxing sternoclavicular joints. Br J Sports Med. 2006;40(4):e12. doi:10.1136/bjsm.2005.020115.
16. Edelstein G, Levitt RG, Slaker DP, Murphy WA. Computer tomography of Tietze syndrome. J Comput Assist Tomogr. 1984;8(1):20–3.
17. Emura K, Arakawa T, Terashima T, Miki A. Macroscopic and histological observations on the human sternoclavicular joint disc. Anat Sci Int. 2009;84(3):182–8. doi:10.1007/s12565-009-0014-5.
18. Eskola A. Sternoclavicular dislocation. A plea for open treatment. Acta Orthop Scand. 1986;57(3):227–8.
19. Fischel RE, Bernstein D. Friedrich's disease. Br J Radiol. 1975;48(568):318–9.
20. Flatow EL. The biomechanics of the acromioclavicular, sternoclavicular, and scapulothoracic joints. Instr Course Lect. 1993;42:237–45.
21. Friedrich H. Uber ein noch nicht beschriebenes, der Perthess-chen Erkralkung, Krankheitsbild des Sternlen Clavikelendes. Dtsch Z Chir. 1924;187:385–98.
22. Fritz P, Baldauf G, Wilke HJ, Reitter I. Sternocostoclavicular hyperostosis: its progression and radiologic features. A study of 12 cases. Ann Rheum Dis. 1992;51(5):658–64.
23. George S, Wagner M. Septic arthritis of the sternoclavicular joint. Clin Infect Dis. 1995;21(6):1525–6.
24. Groh GI, Wirth MA. Management of traumatic sternoclavicular joint injuries. J Am Acad Orthop Surg. 2011;19(1):1–7.
25. Harden SP, Argent JD, Blaquiere RM. Painful sclerosis of the medial end of the clavicle. Clin Radiol. 2004;59(11):992–9.
26. Higginbotham TO, Kuhn JE. Atraumatic disorders of the sternoclavicular joint. J Am Acad Orthop Surg. 2005;13(2):138–45.
27. Hiramuro-Shoji F, Wirth MA, Rockwood Jr CA. Atraumatic conditions of the sternoclavicular joint. J Shoulder Elbow Surg. 2003;12(1):79–88.
28. Hobbs DW. Sternoclavicular joint: a new axial radiographic view. Radiology. 1968;90(4):801.
29. Inman VT, Saunders JB, Abbott LC. Observations on the function of the shoulder joint. J Bone Joint Surg Am. 1944;26:1–30.
30. Jones MW, Carty H, Taylor JF, Ibrahim SK. Condensing osteitis of the clavicle: does it exist? J Bone Joint Surg Br. 1990;72(3):464–7.
31. Jurik AG. Noninflammatory sclerosis of the sternal end of the clavicle: a follow-up study and review of the literature. Skeletal Radiol. 1994;23(5):373–8.
32. Jurik AG, Graudal H. Sternocostal joint swelling—clinical Tietze's syndrome. Report of sixteen cases and review of the literature. Scand J Rheumatol. 1988;17(1):33–42.
33. Kier R, Wain SL, Apple J, Martinez S. Osteoarthritis of the sternoclavicular joint. Radiographic features and pathologic correlation. Invest Radiol. 1986;21(3):227–33.
34. Kruger GD, Rock MG, Munro TG. Condensing osteitis of the clavicle. A review of the literature and report of three cases. J Bone Joint Surg Am. 1987;69(4):550–7.
35. Levy M, Goldberg I, Fischel RE, Frisch E, Maor P. Friedrich's disease. Aseptic necrosis of the sternal end of the clavicle. J Bone Joint Surg Br. 1981;63B(4):539–41.
36. Marker LB, Klareskov B. Posterior sternoclavicular dislocation: an American football injury. Br J Sports Med. 1996;30(1):71–2.
37. Mirza AH, Alam K, Ali A. Posterior sternoclavicular dislocation in a rugby player as a cause of silent vascular compromise: a case report. Br J Sports Med. 2005;39(5):e28.
38. Moore KL, Dalley AF, Agur AMR, editors. Clinically oriented anatomy. 6th ed. Baltimore: Lippincott Williams & Wilkins; 2010. p. 71–83 and 673–5.

39. Netter FH. Atlas of human anatomy. 4th ed. Philadelphia: Saunders; 2006. p. 26–33 and 188–91.
40. Nettles JL, Linscheid RL. Sternoclavicular dislocations. J Trauma. 1968;8(2):158–64.
41. Rand T, Schweitzer M, Rafii M, Nguyen K, Garcia M, Resnick D. Condensing osteitis of the clavicle: MRI. J Comput Assist Tomogr. 1998;22(4):621–4.
42. Renfree KJ, Wright TW. Anatomy and biomechanics of the acromioclavicular and sternoclavicular joints. Clin Sports Med. 2003;22(2):219–37.
43. Rockwood Jr CA, Odor JM. Spontaneous anterior subluxation of the sternoclavicular joint. J Bone Joint Surg Am. 1989;71(9):1280–8.
44. Rockwood Jr CA, Wirth MA. Disorders of sternoclavicular joint. In: Rockwood Jr CA, Matsen III FA, editors. The shoulder. 2nd ed. Philadelphia: Saunders; 1990. p. 555–609.
45. Rockwood Jr CA, Williams GR, Young DC. Disorders of the acromioclavicular joint. In: Rockwood Jr CA, Mattson III FA, editors. The shoulder. 2nd ed. Philadelphia: Saunders; 1990. p. 483–553.
46. Ross JJ, Shamsuddin H. Sternoclavicular septic arthritis: review of 180 cases. Medicine (Baltimore). 2004;83(3):139–48.
47. Sadr B, Swann M. Spontaneous dislocation of the sterno-clavicular joint. Acta Orthop Scand. 1979; 50(3):269–74.
48. Salvatore JE. Sternoclavicular joint dislocation. Clin Orthop Relat Res. 1968;58:51–5.
49. Semble EL, Wise CM. Chest pain: a rheumatologist's perspective. South Med J. 1988;81(1):64–8.
50. Sewell MD, Al-Hadithy N, Le Leu A, Lambert SM. Instability of the sternoclavicular joint: current concepts in classification, treatment and outcomes. J Bone Joint Surg Br. 2013;95-B(6):721–31. doi:10.1302/0301-620X.95B6.31064.
51. Sonozaki H, Mitsui H, Miyanaga Y, Okitsu K, Igarashi M, Hayashi Y, Matsuura M, Azuma A, Okai K, Kawashima M. Clinical features of 53 cases with pustulotic arthro-osteitis. Ann Rheum Dis. 1981;40(6): 547–53.
52. Spencer EE, Kuhn JE, Huston LJ, Carpenter JE, Hughes RE. Ligamentous restraints to anterior and posterior translation of the sternoclavicular joint. J Shoulder Elbow Surg. 2002;11(1):43–7.
53. Standring S, Borley NR, Crossman AR, Gatzoulis MA, Healy JC, Johnson D, Mahadevan V, Newell RLM, Wigley CB. Pectoral girdle, shoulder region and axilla. In: Standring S, editor. Gray's anatomy: the anatomical basis of clinical practice. 40th ed. Edinburgh/New York: Churchill Livingstone; 2008. p. 791–822.
54. Tepolt F, Carry PM, Heyn PC, Miller NH. Posterior sternoclavicular joint injuries in the adolescent population: a meta-analysis. Am J Sports Med. 2014. doi:10.1177/0363546514523386.
55. Terry GC, Chopp TM. Functional anatomy of the shoulder. J Athl Train. 2000;35(3):248–55.
56. Tubbs RS, Loukas M, Slappey JB, McEvoy WC, Linganna S, Shoja MM, Oakes WJ. Surgical and clinical anatomy of the interclavicular ligament. Surg Radiol Anat. 2007;29(5):357–60.
57. van Tongel A, MacDonald P, Leiter J, Pouliart N, Peeler J. A cadaveric study of the structural anatomy of the sternoclavicular joint. Clin Anat. 2012;25(7):903–10. doi:10.1002/ca.22021.
58. Volterrani L, Mazzei MA, Giordano N, Nuti R, Galeazzi M, Fioravanti A. Magnetic resonance imaging in Tietze's syndrome. Clin Exp Rheumatol. 2008;26(5):848–53.
59. Webb PA, Suchey JM. Epiphyseal union of the anterior iliac crest and medial clavicle in a modern multiracial sample of American males and females. Am J Phys Anthropol. 1985;68(4):457–66.
60. Yang JS, Bogunovic L, Brophy RH, Wright RW, Scott R, Matava MJ. A case of posterior sternoclavicular dislocation in a professional American football player. Sports Health. 2013. doi:10.1177/1941738113502153.
61. Yeh GL, Williams Jr GR. Conservative management of sternoclavicular injuries. Orthop Clin North Am. 2000;31(2):189–203.
62. Yood RA, Goldenberg DL. Sternoclavicular joint arthritis. Arthritis Rheum. 1980;23(2):232–9.

Part V

Musculo-Tendinous Structures

20 Rotator Cuff

Akimoto Nimura, Keiichi Akita, and Hiroyuki Sugaya

20.1 Introduction

Quadrapods have four muscles that mobilize the humeral head, but it is only the bipodal animals that have the shoulder tendons linked to form the common tendon, which we call the rotator cuff [9].

20.2 The Supraspinatus and Infraspinatus Muscles

20.2.1 Humeral Insertions

Most anatomy textbooks and authors of several anatomic studies have stated that the supraspinatus inserts into the highest impression of the greater tuberosity of and the infraspinatus inserts into the middle impression of the greater tuberosity [5]. However, Clark and Harryman [4] indicated the difficulty of separating these tendons with their integrated fibers. Minagawa et al. [7] reported overlapping areas of these two tendons on the footprint and claimed the footprint of the supraspinatus to have a wider area than had been previously described.

The supraspinatus muscle originates from the supraspinatus fossa and the superior surface of the spine of the scapula, and it passes laterally. The infraspinatus muscle originated from both the infraspinatus fossa and the inferior surface of the spine of the scapula, and it passes superolaterally (Fig. 20.1). Recently, Mochizuki et al. [8] reported the new findings about the humeral insertions of the supraspinatus and infraspinatus. The supraspinatus and infraspinatus appear to mingle into one structure at their insertions on the humerus (Fig. 20.1). However, after removal of the coracohumeral ligament and the loose connective tissues overlying the supraspinatus and infraspinatus, the anterior border of the infraspinatus can be clearly traced and the border between two muscles became more apparent. The anterior margin of the infraspinatus is slightly protuberant compared with the posterior margin of the supraspinatus. The anterior part of the infraspinatus partially covers the posterolateral part of the supraspinatus (Fig. 20.2).

The upper surface of the greater tuberosity has been generally described as being marked by three impressions: the highest, the middle, and the lowest. However, the humeral insertion of the infraspinatus actually occupies about half of the highest impression and all of the middle impression (Fig. 20.3). The anteriormost region

A. Nimura, MD, PhD (✉) • K. Akita, MD, PhD
Department of Clinical Anatomy,
Graduate School of Medical and Dental Sciences,
Tokyo Medical and Dental University, Tokyo, Japan
e-mail: nimura.orj@tmd.ac.jp; akita.fana@tmd.ac.jp

H. Sugaya, MD, PhD
Shoulder and Elbow Service,
Funabashi Orthopaedic Sports Medicine Center,
Funabashi, Japan
e-mail: Hsugaya@nifty.com

G.I. Bain et al. (eds.), *Normal and Pathological Anatomy of the Shoulder*,
DOI 10.1007/978-3-662-45719-1_20, © ISAKOS 2015

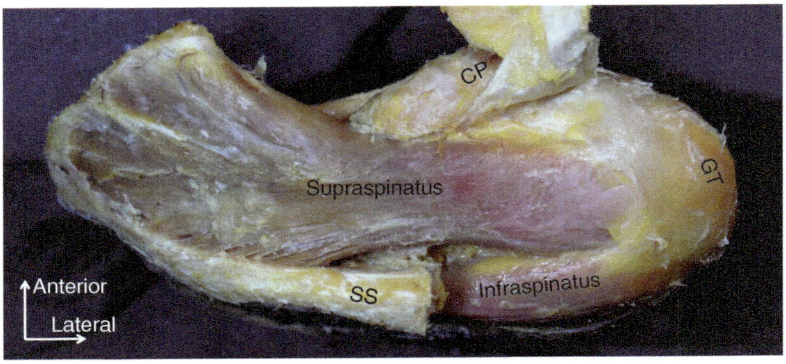

Fig. 20.1 The superior view of the supraspinatus and infraspinatus (right shoulder, the acromion has been removed and reflected to anterior). Both tendons appear to mingle into one structure at the greater tuberosity (*GT*). *SS* scapular spine, *CP* coracoid process

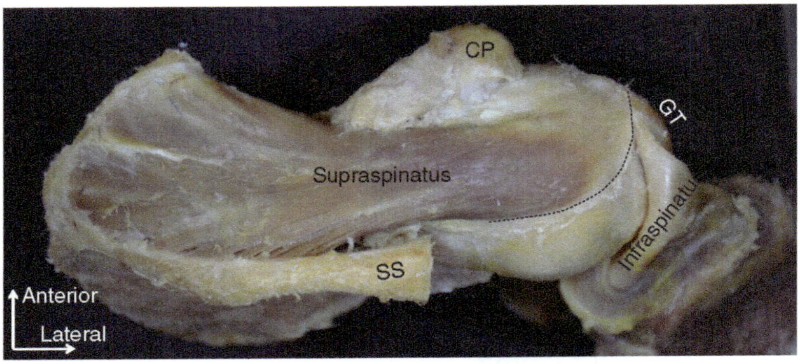

Fig. 20.2 The superior view of the border between the supraspinatus and the infraspinatus (shown as the *black dotted line*). The infraspinatus has been detached from the scapula and the articular capsule and reflected to lateral. *SS* scapular spine, *CP* coracoid process, *GT* greater tuberosity

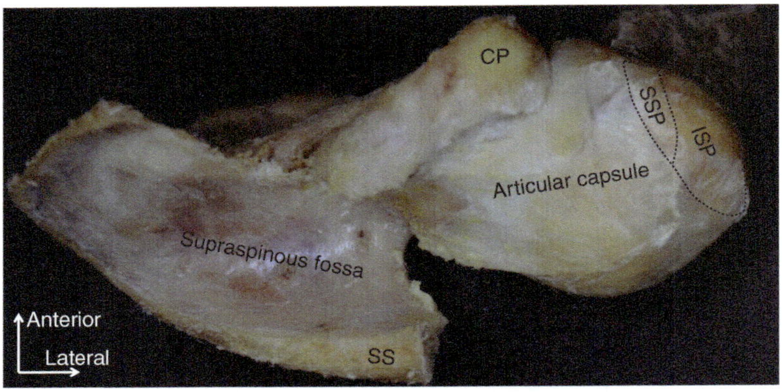

Fig. 20.3 The humeral side insertions of the supraspinatus and infraspinatus (the supraspinatus, *SSP*; the infraspinatus, *ISP*). Note that the articular capsule of the shoulder joint is completely separated from the supraspinatus and infraspinatus and preserved. *SSP* supraspinatus, *ISP* infraspinatus

of the humeral insertion of the infraspinatus almost reaches the anterior margin of the highest impression of the greater tuberosity. The supraspinatus inserts into the anteromedial area of the highest impression of the greater tuberosity (Fig. 20.3). The footprint of the supraspinatus is in the shape of a right triangle, with the base lying along the articular surface. In addition to the greater tuberosity, the supraspinatus also inserts in the lesser tuberosity in one-fifth of specimens. In these specimens, the anteriormost portion of the supraspinatus tendon covers the superior part of the bicipital groove.

20.2.2 Muscular and Tendinous Portions

Most of the muscle fibers of the supraspinatus, especially those of its superficial layer, run anterolaterally toward the anterior tendinous portion, while the rest of the fibers from the deep layer runs laterally toward the medial margin of the highest impression on the greater tuberosity. The supraspinatus tendon is composed of two portions: the anterior half is long and thick, and the posterior half is short and thin.

The superoanterior two-thirds of the infraspinatus is composed of a thick and long tendinous portion. A thin and short tendinous portion which occupied the rest of the infraspinatus muscle joined the thin and short tendinous portion of the teres minor.

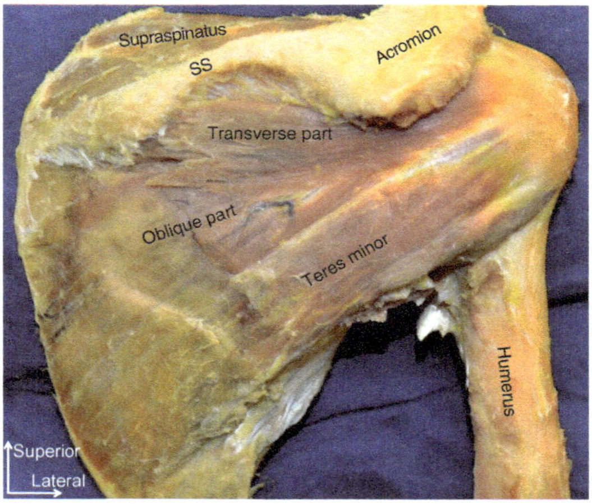

Fig. 20.4 The posterior view of the right shoulder. The transverse part of the infraspinatus is shown to attach to the oblique part, *SS* scapular spine

20.2.3 Oblique and Transverse Part of the Infraspinatus

The infraspinatus is identified to be composed of oblique and transverse parts according to the direction of muscle fibers (Fig. 20.4) [6]. The oblique part is a fan-shaped muscle bundle and originates from the infraspinatus fossa running superolaterally. The transverse part originates from the inferior surface of the spine and passes laterally; it then attached to the oblique part of the middle portion of the tendinous part. Both parts are connected to each other at the superior area of the muscular portions; however, in the distal tendinous portions, they can be clearly separated. Although the oblique part is to the greater tuberosity, the transverse part does not reach the tuberosity (Fig. 20.5). The transverse part adjoined the posterior surface of the middle area of the tendinous portion of the oblique part. It is suggested that significant strength from the oblique part of the infraspinatus can focus more anteriorly and contribute to the shoulder abduction; on the other hand, the transverse part may play only a supportive role in the infraspinatus function and stabilize the tendinous portion of the oblique part during the shoulder motion from above.

Origins of the innervating branch of the suprascapular nerve to the transverse part of the infraspinatus are variable. Branches arise from the branch to the supraspinatus muscle and/or from the main trunk of the suprascapular nerve after branching off the branches to the supraspinatus

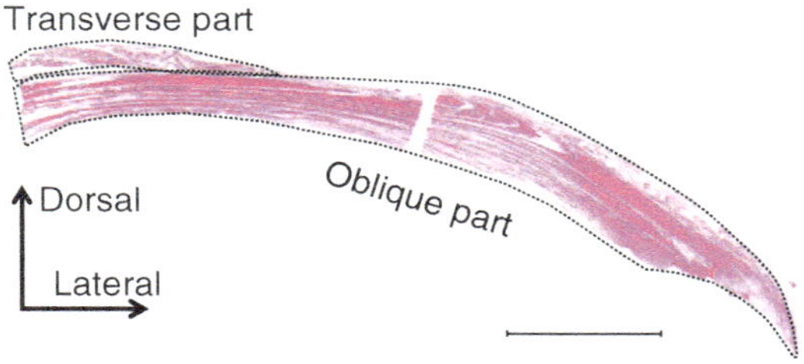

Fig. 20.5 Histological section of the distal part of the infraspinatus stained by hematoxylin-eosin stain. The longitudinal section through the distal part of the infraspinatus is shown. The transverse part is shown as the dorsal *dotted area*. The oblique part is shown as ventral *dotted area*. *Scale bar* 10 mm

muscle (Fig. 20.6). No branch is found to pierce the transverse part to innervate the oblique part and vice versa. Although the transverse part is a part of the infraspinatus, according to its innervation, the transverse part might be closely more related to the supraspinatus.

20.2.4 Variations

The supraspinatus is very constant in its form and attachments. However, rare variations include connections with the pectoralis major or minor and an additional slip from the superior transverse scapular ligament. The infraspinatus is

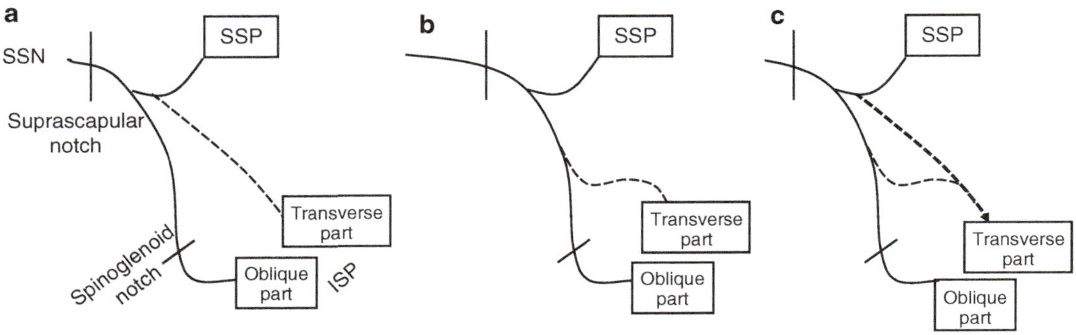

Fig. 20.6 Schematic illustrations represented origins of the branch to the transverse part of the infraspinatus. (**a**) Branches arise from branches to the supraspinatus muscle. (**b**) Branches arise from branches to the infraspinatus muscle. (**c**) Branches arise from branches to both muscles. *SSN* suprascapular nerve, *SSP* supraspinatus, *ISP* infraspinatus

sometimes inseparably united with the teres minor. A slip connecting with the posterior border of the deltoid has been described [3].

20.3 Subscapularis Muscle

20.3.1 Rotator Cuff Structure

The subscapularis muscle insertion is composed of the superior two-third tendinous insertion and the inferior one-third insertion where the muscle attaches to the humerus almost directly by way of a thin membranous structure [2]. The superior-most insertion of the subscapularis tendon is wide in the uppermost margin of the lesser tuberosity, whereas the rest of the subscapularis tendon inserts into the anteromedial portion of the lesser tuberosity (Fig. 20.7). Moreover, the superior-most insertion of the subscapularis tendon extends a thin tendinous slip, which attaches to the fovea capitis of the humerus (Fig. 20.8).

By removing muscular tissues, several intramuscular tendons can be observed. Those tendons aggregate laterally and form a tendinous insertion. The superior-most insertion of the subscapularis tendon is derived from the cranial part of intramuscular tendons. The superior-most insertion, the lateral portion of the cranial part of the intramuscular tendons, and the tendinous slip comprised a structure that is in direct contact with the inferior side of the corner

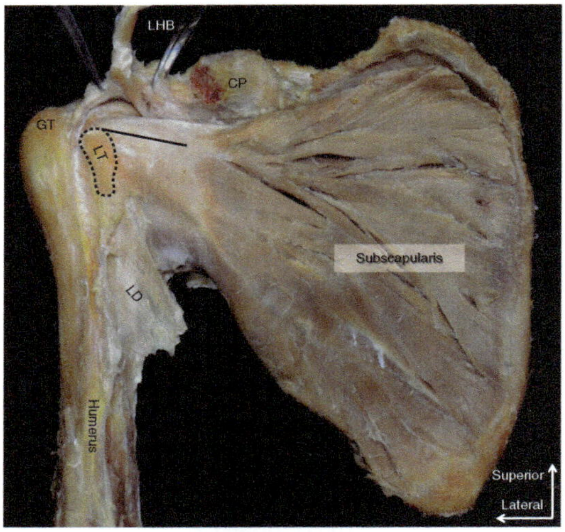

Fig. 20.7 Anterior view of the right shoulder. The long head of the biceps tendon (*LHB*) is reflected. The coracoid process (*CP*) is partially resected. The cranial part of the subscapularis tendon inserts superior to the uppermost margin (*black line*) of the lesser tuberosity (*LT*, *dotted area*). *GT* greater tuberosity, *LD* latissimus dorsi

portion of the long head of the biceps tendon. This structure continued the pathway of the long head of the biceps tendon proximally from the osseous medial wall of the intertubercular groove.

The medial portion of the superior glenohumeral ligament (SGHL) appears at the anterosuperior part of the internal wall of the joint cavity [1]. It passes spinally and finally attaches

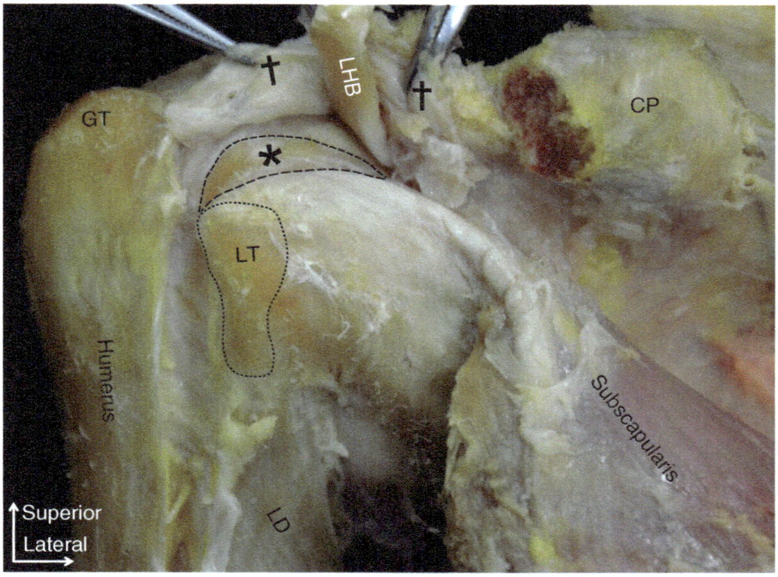

Fig. 20.8 Superior-most insertion of the subscapularis tendon. The long head of the biceps tendon (*LHB*) is reflected. The coracohumeral ligament is also detached from the subscapularis tendon and reflected with forceps (*cross*). The coracoid process (*CP*) is partially resected. The subscapularis muscle is detached from the scapular origin and reflected to anterior. The superior-most insertion of the subscapularis tendon extends a thin tendinous slip, which attaches to the fovea capitis of the humerus (*dotted area* marked with the *asterisk*). *GT* greater tuberosity, *LD* latissimus dorsi, *LT* lesser tuberosity

to the tendinous slip of the subscapularis insertion. The SGHL supports the long head of the biceps from the anteroinferior side. In other words, the running course of the long head of the biceps is formed from the SGHL to the intertubercular groove. Just above the intertubercular groove, the SGHL attached to the surface of the tendinous slip of the subscapularis insertion.

20.3.2 Rotator Cuff Variations

The major variation is an additional muscle called the subscapularis minor or secundus, which arises from the upper part of the axillary border of the scapula and is inserted into the joint capsule, the crest of the lesser tuberosity of the humerus, or distal to the lesser tuberosity. Rarer variations are slips arising from the subscapularis tendon and passing to the axillary fascia, the pectoralis major, or the short head of the biceps muscle [3].

20.4 Teres Minor Muscle

20.4.1 Description of Structure

The teres minor muscle locates inferior to the infraspinatus, and originates from the lateral edge of the dorsal scapula. The teres minor muscle inserts to the lowest impression of the greater tuberosity of the humerus and additionally inserts to the posterior side of the surgical neck of the humerus (Figs. 20.9 and 20.10). The border between the infraspinatus and the teres minor is separated by the tendinous fascia, which is sometimes unclear and disappears at their insertion.

At the musculotendinous junction of the teres minor muscle, it can be separated into the superior and inferior bundle. The superior bundle at the insertion originates from the lateral edge of the dorsal scapula and inserts to the lowest impression as an oval footprint (Fig. 20.11). On the other hand, the inferior bundle at the insertion mainly originates from the tendinous fascia, which forms a septum between the infraspinatus and the teres minor, and

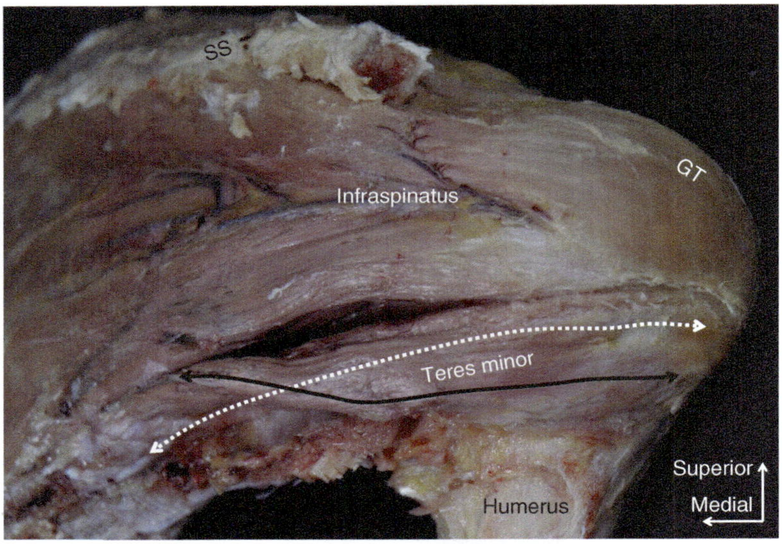

Fig. 20.9 Posterior view of the right shoulder. The acromion is resected. The running course of the superior bundle of the teres minor at the insertion is indicated as the *white double-headed arrow* with a *dotted line*. The running course of the inferior bundle at the insertion is indicated as the *black double-headed arrow*. *GT* greater tuberosity, *SS* scapular spine

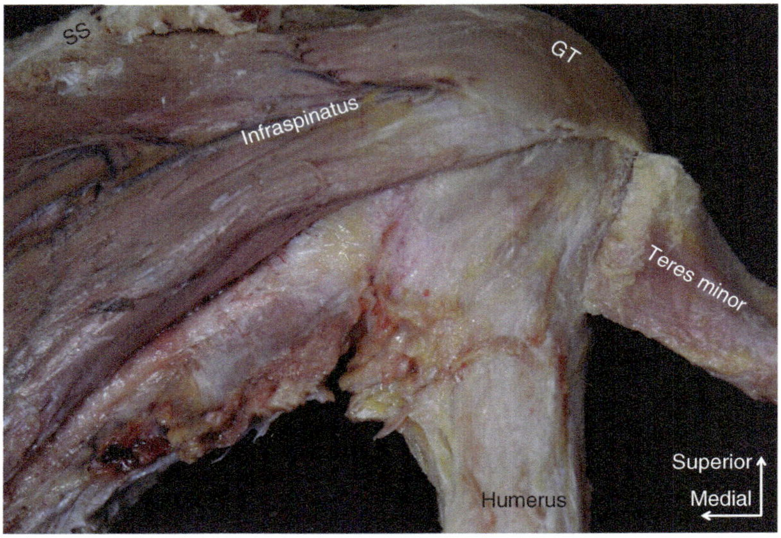

Fig. 20.10 Detached and reflected muscle of the teres minor. The teres minor muscle is detached from the scapular origin and reflected to lateral. *GT* greater tuberosity, *SS* scapular spine

partially originates from the lateral edge of the dorsal scapula. The inferior bundle of the teres minor runs dorsal to the superior bundle and inserts into the distal to the lowest impression as a linear shape (Fig. 20.11). At the origin of the teres minor, there is no structure which separates the two bundles. Both bundles are innervated by the branch of the axillary nerve which supply from the dorsal or inferior side of the teres minor muscle, not from ventral side of it.

20.4.2 Variations

Absence of the teres minor has been reported. The origin of the muscle may be extended so that it entirely covers the infraspinatus and replaces the fascial sheet covering that muscle. Various slips, which appear to be related with the deltoid muscle, are suggesting an associated development of these two muscles [3].

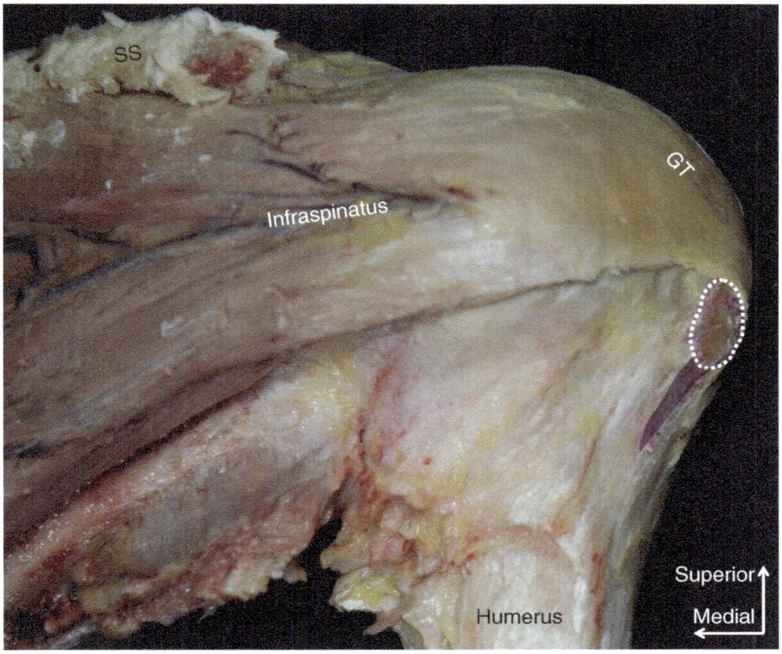

Fig. 20.11 Insertion of the teres minor muscle. The teres minor muscle is detached from the humeral insertion. The insertion of the superior bundle of the teres minor is shown as the white dotted area. The insertion of the inferior bundle of the teres minor is shown as the black area with the shape of an arrowhead. *GT* greater tuberosity, *SS* scapular spine

References

1. Arai R, Mochizuki T, Yamaguchi K, Sugaya H, Kobayashi M, Nakamura T, Akita K. Functional anatomy of the superior glenohumeral and coracohumeral ligaments and the subscapularis tendon in view of stabilization of the long head of the biceps tendon. J Shoulder Elbow Surg. 2010;19(1):58–64. doi:10.1016/j.jse.2009.04.001. S1058-2746(09)00194-3 [pii].
2. Arai R, Sugaya H, Mochizuki T, Nimura A, Moriishi J, Akita K. Subscapularis tendon tear: an anatomic and clinical investigation. Arthroscopy. 2008;24(9):997–1004. doi:10.1016/j.arthro.2008.04.076. S0749-8063(08)00359-9 [pii].
3. Bergman AR, Thompson SA, Afifi KA, Saadeh AF. Thorax, shoulder, and arm. In: Compendium of human anatomic variation text, atlas, and world literature. Baltimore: Urban & Schwarzenberg; 1988. p. 7–18.
4. Clark JM, Harryman 2nd DT. Tendons, ligaments, and capsule of the rotator cuff. Gross and microscopic anatomy. J Bone Joint Surg Am. 1992;74(5):713–25.
5. Clemente C. Osteology, and muscles and fasciae of the upper limb. In: Gray's anatomy of the human body. 30th ed. Philadelphia: Lea & Febiger; 1985. p. 233–4.
6. Kato A, Nimura A, Yamaguchi K, Mochizuki T, Sugaya H, Akita K. An anatomical study of the transverse part of the infraspinatus muscle that is closely related with the supraspinatus muscle. Surg Radiol Anat. 2012;34(3):257–65. doi:10.1007/s00276-011-0872-0.
7. Minagawa H, Itoi E, Konno N, Kido T, Sano A, Urayama M, Sato K. Humeral attachment of the supraspinatus and infraspinatus tendons: an anatomic study. Arthroscopy. 1998;14(3):302–6.
8. Mochizuki T, Sugaya H, Uomizu M, Maeda K, Matsuki K, Sekiya I, Muneta T, Akita K. Humeral insertion of the supraspinatus and infraspinatus. New anatomical findings regarding the footprint of the rotator cuff. J Bone Joint Surg Am. 2008;90(5):962–9.
9. Sonnabend DH, Young AA. Comparative anatomy of the rotator cuff. J Bone Joint Surg Br. 2009;91(12):1632–7.

Ultrastructure and Pathoanatomy of the Rotator Cuff

21

Matthias A. Zumstein, Nandoun Abeysekera, Pietro Pellegrino, Beat K. Moor, and Michael O. Schär

21.1 Introduction

The rotator cuff is a complex musculotendinous unit, which plays a major role in glenohumeral joint stability and mobilization. Tears of the rotator cuff are common, and the incidence increases with age. Several structures such as the muscle, tendon, and bone may contribute to the development of a tear as well as on the outcome following a rotator cuff repair.

The goal of this chapter is to discuss the evidence, which exists with regard to the pathophysiological changes in the muscle, tendon, and bone that lead to a rotator cuff rupture as well as the changes that occur in these structures after a tear has occurred.

21.2 Muscle

21.2.1 Ultrastructure Anatomy and Physiology

Four muscles from the scapula converge to form a common tendon, the rotator cuff. The supraspinatus, infraspinatus, and teres minor are inseparable, while the subscapularis muscle is separated from the remainder by the rotator interval [1]. These muscles are composed of muscle fibers that contain sarcomeres. Several factors are responsible for the generation of force within these muscles.

21.2.2 Force Generation

21.2.2.1 Cross Sectional Area and Moment Arm of the Muscle

Force generation in these muscles is predominantly determined by the muscle architecture. This includes the cross-sectional area, muscle fiber arrangement, and the moment arm of the muscle. When comparing the cross-sectional area between the different muscles of the rotator cuff, the subscapularis shows the greatest force-producing capacity (102 ± 12 g), followed by the infraspinatus (78 ± 8 g), supraspinatus (34 ± 4 g), and teres minor (21 ± 2 g) (Table 21.1) [2].

The average muscle fiber length has been reported to be between 4.5 and 6.6 cm, with the supraspinatus muscle showing the smallest fiber length (Table 21.1) [2]. To compensate for the small fiber length, the supraspinatus muscle has been shown to operate over the widest range of sarcomere lengths [2].

M.A. Zumstein (✉) • N. Abeysekera • P. Pellegrino
B.K. Moor • M.O. Schär
Shoulder, Elbow and Orthopaedic Sports Medicine, Department of Orthopaedic Surgery and Traumatology, Inselspital, University of Bern, Bern, Switzerland
e-mail: matthias.zumstein@insel.ch

G.I. Bain et al. (eds.), *Normal and Pathological Anatomy of the Shoulder*,
DOI 10.1007/978-3-662-45719-1_21,

Table 21.1 Muscle architecture properties

Muscle	Mass (g)	Muscle length (cm)	L_f (cm)	L_f/L_m	L_s (μm)	S_n	Pennation angle (°)	PCSA (cm^2)
Supraspinatus	34.0 ± 4.3[a]	8.5 ± 0.4[a]	4.50 ± 0.32[a]	0.53 ± 0.03[b]	3.23 ± 0.05[b,c]	16,655 ± 1182[a]	5.1 ± 0.8[a]	6.65 ± 0.56[a]
Infraspinatus	78 ± 7.5[a]	2.1 ± 0.5[c,d]	6.57 ± 0.33[d]	0.55 ± 0.02[b]	3.18 ± 0.06[§ b]	24,332 ± 1203[d]	1.4 ± 0.4[† II]	10.71 ± 0.95[a]
Teres minor	21.2 ± 2[a]	10.8 ± 0.6[d,e]	6.09 ± 0.35[d]	0.57 ± 0.03[b]	2.80 ± 0.07[d,e]	22,569 ± 1299[d]	0.6 ± 0.3[d]	3.18 ± 0.30[a]
Subscapularis	101.8 ± 11.5[a]	13 ± 0.6[d]	6 ± 0.47[d]	0.45 ± 0.02[a]	2.52 ± 0.09[d, e]	22,069 ± 1735[d]	0 ± 0[d,e]	15.53 ± 1.41[a]

Reproduced from Ward et al. [2]

Lf normalized fiber length, *Lf/Lm* normalized fiber length/muscle length ratio, *Ls* resting sarcomere length, *Sn* sarcomere number, *PCSA* physiologic cross-sectional area

Values are mean ± standard error of 10 specimens

[a]Significantly different than supraspinatus

[b]Significantly different than infraspinatus

[c]Significantly different than teres minor

[d]Significantly different than subscapularis

[e]Significantly different than all other muscles

21.2.2.2 Pennation Angle

The pennation angle is defined as the angle between the muscle fibers and the intramuscular central tendon. For mechanical and geometrical reasons, pennated muscles work better when the pennation angle is smaller.

The pennation angle in the intact supraspinatus muscle varies from 10° for the medial fibers to 85° for the lateral fibers [3–5]. This provides greater contraction ability on the medial side, with a subsequent increased shear stress on the tendon when compared to the lateral side. Meyer et al. [6] reported an increase in the pennation angle after rotator cuff tears. Others found a positive correlation between the pennation angle of the supraspinatus muscle and the tear size of the supraspinatus enthesis [7].

21.3 Pathophysiology of the Rotator Cuff Muscle

Although rotator cuff tears are a common pathology, the lesion rarely involves directly the muscular belly or the muscle-tendon junction. Nevertheless, muscles are deeply and substantially involved through three mechanisms: retraction, atrophy, and fatty infiltration. These pathologic changes may lead to worse outcomes and may guide surgical indications.

21.3.1 Retraction and Atrophy

A rotator cuff muscle without stimulation and with a torn tendon will undergo atrophy and retraction. Ruptures lead to severe changes in the muscle [8–10]. These changes are more accentuated in slow contracting type 1 muscle fibers when compared to fast contracting type 2A and 2B fibers [6, 11]. Fibrous tissue is stored intramuscularly [11], and a short time after the occurrence of the tear, the muscle starts to retract and loses its ability to build up tension [12]. In a supraspinatus tenotomy model in sheep, the muscle retracted 29 mm on average 16 weeks after tenotomy. This retraction corresponds with the physiological range of movement of the muscle. The pennation angle increased from 30° to 55°, and the cross-sectional area decreased by 57 % compared to the healthy opposite side [12]. This retraction leads to shortening of the muscle fibers by the breakdown of serially arranged sarcomeres up to 50 %. Due to this loss of sarcomeres, the muscle gets shorter instead of thinner.

Over the years, the cross-sectional area can decrease. With increasing retraction, the pennation angle of the muscle increases. As discussed above, this leads to the incorporation of fatty infiltration in between the muscle fibers. In this model, retraction and atrophy are therefore caused by shortening of "healthy" muscle tissue and not muscle degeneration in the proper sense [6]. Steinbacher et al. [13] confirmed this in a study where they reported that the cause for atrophy in rotator cuff tears greater than or equal to 3 cm (Bateman grade III and IV) is found in the decrease of the absolute myofibril volume and not in the death of fibers [13].

21.3.1.1 Changes in Gene Expression with Atrophy

Gene expressions of several genes that induce muscle atrophy are altered after rotator cuff tears. The two key regulators for the induction of Muscle RING-finger protein-1 (MuRF1) und Atrogin-1 are upregulated shortly after the rotator cuff tear [14, 15]. In massive rotator cuff tears, genes that are involved in sustaining muscle atrophy (e.g., Forkhead box protein O1A (FOXO1A), Calpain, Ubiquitin-conjugating enzyme-E2B (UBE2B) and -E3A (UBE3A), and Cathepsin B (CTSB)) are greatly upregulated when compared to smaller tears. This could explain the fact that rotator cuff reconstructions show a better outcome after small tears compared to massive tears.

21.3.1.2 Generation of Force in Atrophic Muscle

An increased pennation angle results in an inefficient force application from the muscle fibers to the tendon due to the fact that the muscle fibers are not pulling in the direction of the tendon [9].

Meyer et al. [16] reported in a study, in which they performed a tenotomy of the infraspinatus tendon in sheep, that the fatty infiltration negatively correlates with the loss of strength and that atrophy correlates with the contractile amplitude. The larger the muscle density and the smaller the fatty infiltration measured in computer tomography [16], the greater is the maximal force of the muscle. The maximum contraction force reached by the muscle decreased significantly with increasing atrophy. This finding is crucial for the outcome of a reconstruction. Even if a rotator cuff tendon is adapted without traction to the footprint, its force may be diminished. Furthermore, passive tension is increased, which may lead to a limitation of the range of motion.

21.3.2 Fatty Infiltration

Deposition of fat into the muscle is termed fatty infiltration (Fig. 21.1). Fatty infiltration is a common finding not only in cuff tears but also in neurologic lesions of the rotator cuff. This infiltration can occur within different sites of the muscle, for example, in the interstitial space [13] where its accumulation leads to a limitation of its mechanics [6, 17]. Its accumulation not only occurs in between the muscle fibers but also in type I muscle fibers in the sarcoplasm [13]. On the other hand, fatty infiltration is also found in the extramuscular space (epimysium of the muscle belly) as well as in the torn tendon [10].

Over the years, different theories were established to explain the origin of the fatty infiltration.

21.3.2.1 Changes in Muscle Architecture Lead to Fatty Infiltration

Rotator cuff tears and the associated loss of tension seem to favor processes that induce fatty infiltration [18]. After a rotator cuff tear, muscle fibers shorten and the pennation angle increases [6]. This leads to a newly created space, which is filled with fat and connective tissue (Fig. 21.2). This fatty infiltration in between the muscle fibers decreases the elasticity of the muscle, leading to decreased mechanical properties of the muscle. This theory is supported by the fact that the fatty infiltration correlates with the size of the rotator cuff tear [19, 20]. Furthermore, the progression of the fatty infiltration is inversely proportionate to the strength development [21].

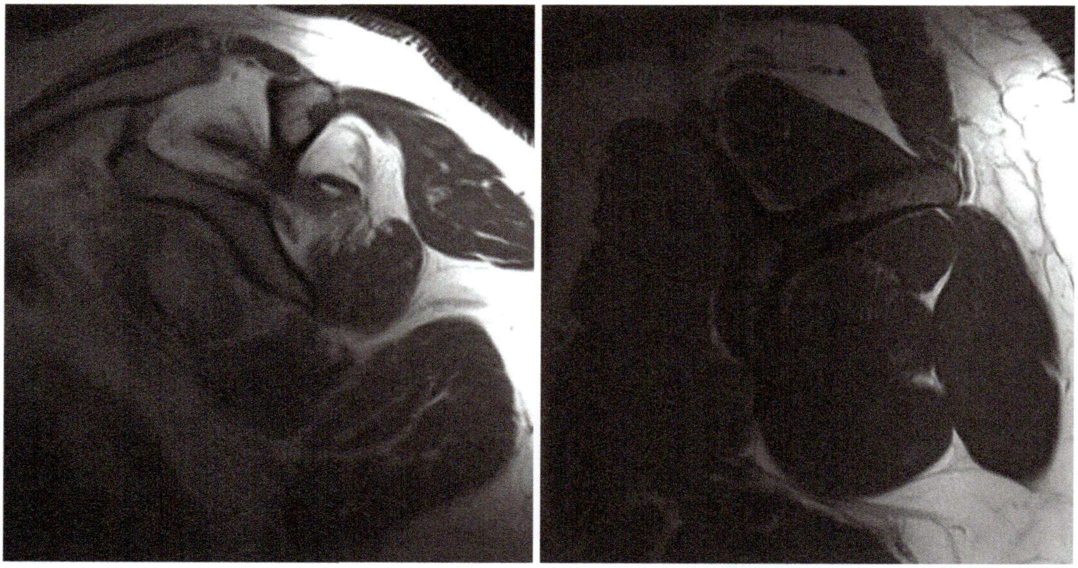

Fig. 21.1 (a) Sagittal MR scans of the shoulder showing fatty infiltration Goutallier stage 4 and (b) sagittal MR scan of the shoulder showing fatty infiltration Goutallier stage 1

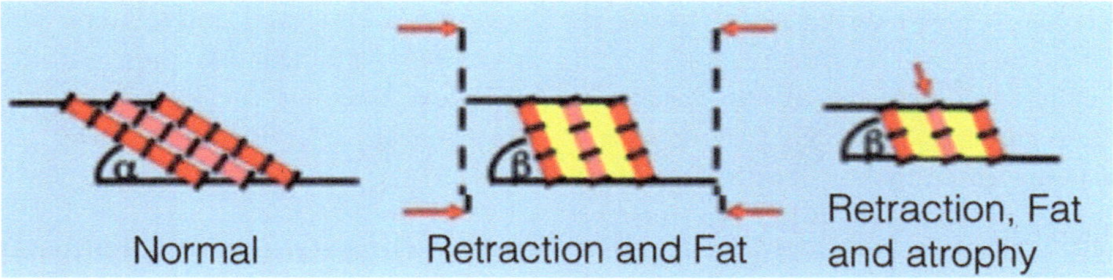

Fig. 21.2 (**a**) Normal muscle. (**b**) Retraction of muscle (decrease of *x*) leads to an increased pennation angle (*beta*) and to newly created space, which is filled with fat (*yellow*) (**c**) Atrophy (decrease of *y*) leads to a decrease in cross sectional area

This indicates that the loss of strength is not only caused by the atrophy but also by the fatty infiltration [9].

21.3.2.2 Changes in Gene Expression Patterns Lead to Fatty Infiltration

Frey et al. [22] were able to show that after tenotomy of the infraspinatus muscle in sheep, several transcription factors that are important for myogenic differentiation are upregulated. This suggests that the body intends to solve the problem by increasing the amount of muscle. On the other hand, CAAT/enhancer binding protein β (C/EBPβ) and peroxisome proliferator-activated receptor γ (PPARγ), two factors involved in adipogenesis, were upregulated. The concentration of these factors significantly decreased after refixation of the rotator cuff.

21.3.2.3 Traumatic Partial Denervation May Lead to Fatty Infiltration

Another possible cause for the development of fatty infiltration may be related to neurogenic changes. This is supported by the fact that 7 of 28 patients with a complete rupture of the rotator cuff showed signs of a peripheral neuropathy in electromyography [23]. Albritton et al. [24] reported a partial denervation after complete tear of the supraspinatus tendon. This may lead to atrophy not only of the supraspinatus but also of the infraspinatus muscle [24].

The fact that neurogenic injuries lead to a fat distribution pattern that is more diffuse and irregular when compared to the one seen after rotator cuff tears makes it unlikely that denervation is the main factor contributing to fatty infiltration.

21.3.2.4 Increased Neovascularization and Amount of Mitochondria Correlate with Fatty Infiltration

Gerber et al. [17] were able to show that an increased neovascularization of the supraspinatus muscle and an increase in the amount of mitochondria occur after rotator cuff tears in the human. This correlated with the fatty infiltration and atrophy [17].

In another study [25], an increase in two angiogenetic factors (hypoxia-inducible factor (HIF) and vascular endothelial growth factor (VEGF)) were shown after rotator cuff tears.

21.4 Tendon and Enthesis

21.4.1 Ultrastructure and Physiology of the Tendon

Tendons of the rotator cuff are essentially made of two components: an extracellular matrix (ECM) and a cellular component. The extracellular matrix, which is the most important part of tendons in terms of mechanical properties, is made up of collagen (65–80 %), elastin (1–2 %), and a mixture of water, proteoglycans, glycosaminoglycans (GAGs), and glycoproteins. In rotator cuff tendons, there are more proteoglycans than in other purely tensional tendons. The most common proteoglycans are called decorin and biglycan (containing one or two GAG chains).

The cellular component of the tendon is divided into tenoblasts and tenocytes, representing 95 % of the whole amount of cells. Tenoblasts are precursors of tenocytes [26], while tenocytes are mature cells producing collagen and other components of the ECM .

The collagen fibers of the tendon are directly responsible for its mechanical properties. Lake et al. [27, 28] have extensively investigated the ultrastructure and mechanical properties of the rotator cuff tendons, especially of the supraspinatus tendon in recent works. The shoulder joint has a great range of motion and subsequently experiences multiaxial stresses. This is reflected in the collagen pattern having different distributions in tendon tissue. The stress-strain curve of the supraspinatus tendon, the most complex and commonly involved in damage, shows particular mechanical properties. These properties are different in the anterior, posterior, and bursal joint sides and can be attributed to the varying distribution of collagen. While axial loading shows higher stiffness in medial and anterolateral parts, resistance to transverse loading results in higher stiffness in posterolateral and anterolateral sides. This may be explained by the multidirectional complex loadings near the osteotendinous insertion.

21.4.2 Ultrastructure and Physiology of the Enthesis

The enthesis is the intermediary between the tendon and bone, and it has a remarkable difference in mechanical properties. The enthesis exists as a means of reducing the concentration of stress between the soft tendinous tissue and the hard bony tissue. As such, this site is most prone to rupture because of the great stresses it has to experience. The healthy enthesis employs different strategies to increase the strength of this crucial zone such as functional grading and change in microstructure through a transitional tissue, a reduced angle of fiber direction at the attachment, and interdigitation of tissue with bone [29].

According to literature, the enthesis is commonly described as consisting of four distinct zones (Fig. 21.3). More recent literature suggests that in reality these four zones are not distinctly

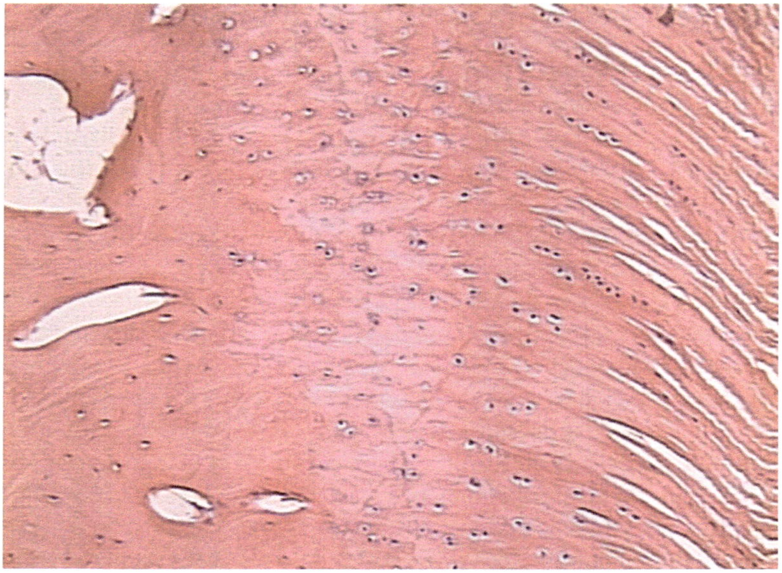

Fig. 21.3 Section of an intact enthesis, showing the four zones (bone, mineralized cartilage, unmineralized cartilage, and tendon) (From Schär and Rodeo [73])

segregated; rather, it exists as a fibrocartilaginous zone and has a graded transition from tendon to bone [30]. These four zones have been described as follows:

- Zone I: This zone constitutes the tendon and is composed of type I collagen fibers that have been well-aligned, and is also contributed by a small percentage of decorin.
- Zone II: This zone constitutes fibrocartilage, and it is mainly composed of type II and III collagen, with a small percentage being contributed from types I, IX, and X collagen, as well as some aggrecan and decorin. This zone represents the start of the shift from tendinous to bony tissue.
- Zone III: In this zone, mineralized fibrocartilage is predominant. Here, primarily type II collagen, and a high percentage of type X collagen and aggrecan are found. This zone represents the shift toward bony material.
- Zone IV: This zone is composed of bone. This zone has a high mineral content and is composed of type I collagen.

In fibrocartilaginous tissue, the main cell population is constituted by chondrocytes that are arranged in rows, which follow the tensile strains of the tendon [31, 32]. The mineral content gradient seems to be one of the main characteristics of the enthesis. It has been shown that mineral clusters and collagens fibrils seem to interdigitate until they reach the fully mineralized region.

21.4.3 Pathology of the Enthesis and Tendon

21.4.3.1 Pathological Changes That May Lead to Rotator Cuff Tear

Rotator cuff tears are one of the most common orthopedic injuries and are the leading cause of shoulder pain and disability. In most cases, a chronic degeneration leads to rupture, although an acute traumatic event may cause tears even in younger patients. The etiology of this degeneration is multifactorial, and can be attributed to both intrinsic and extrinsic factors.

Although detachment could occur in the middle of the tendon or intramuscular zone inside the rotator cuff in traumatic cases, most detachments occur where the mechanical forces change dramatically: the enthesis [33].

It has been suggested by Ogata and Uhthoff that tendon degeneration is the primary cause for partial tears of the rotator cuff [34]. Different pathological changes haven been suggested to lead to degeneration of the rotator cuff tendon, eventually increasing the likelihood of a rotator cuff tear.

Increased Collagen Type III Gene Expression

Neviaser et al. [35] showed in an experimental model that gene expression for Col-III increases in tendons after cyclic fatigue loading. This finding is supported by an altered expression of TGF-β1 (which regulates collagen production in cuff), both in overloaded and torn tendons [36]. Such changes in collagen properties could lead to macroarchitectural changes.

Type I collagen, the most common type of collagen inside rotator cuff tendons, is frequently replaced by type III collagen (usually present in scar tissue) during degeneration. This has less cross-links in between, and consequently worse mechanical properties [37].

Type II collagen, instead, is most frequently seen on the chondral surface of the attachment to bone (zone II and zone III). There is also a shift toward type III collagen during degeneration [38].

These changes could lead to a reduction in strength and a subsequent rupture at the enthesis.

Increased Apoptosis, Oxidative Stress and Autophagy

Tenocytes, in addition to collagen, are of great interest in tendon degeneration. Their number initially decreases, and then a structural change occurs with rounding of the nuclei and apoptosis. This may lead to a change in the structural properties of the tendon, impairing the capability of tenocytes to produce normal healthy collagen. It has also been documented that degenerative tendons have a higher rate of cells undergoing apoptosis compared to histologically normal tendons [39]. Apoptosis may be caused by oxidative stress, with an increase of oxygen-reactive species and JNK (Map-K) expression [40, 41]. Their presence in both torn and unhealthy tendons is not surprising and could lead to a decreased cellular response to damage. As in the muscle belly, it has been observed that autophagic cell death is present as well, suggesting common degenerative mechanisms but with unclear pathways [42] (Table 21.2).

Increased Differentiation into Myofibroblasts

The other remarkable finding is the differentiation of tendon cells into myofibroblasts. It has been observed that these cells have a contractile capability in torn tendon samples although this phenomenon is not present in healthy ones. It has been shown that they increase when the structures are submitted to higher strain, suggesting that they intend to maintain the integrity of the tendon representing a response to a tendinopathic tendon that loses its normal mechanical properties [42].

Changes in Vascularization

There is some evidence in the literature that suggests that the alteration of vascularization in rotator cuff tears may have caused rotator cuff tears. Though hypervascularization and hypovascularization patterns have been described previously [43, 44], in a recent study, no alterations have been shown in between torn and healthy tendons, although a regional hypovascular age-related pattern was reported [45]. Nevertheless, a major expression of bursal-spreading vessels and thin unorganized hypervascular patterns seems to be present in rotator cuff tears [46].

Changes in MMP and TIMP Expression

Recently, it has been described that metalloproteases (MMPs) as well as tissue inhibitors of MMPs (TIMPs) are altered not only in torn but also in tendinopathic tendons of a partially damaged rotator cuff [47, 48]. Metalloproteases (MMPs) are endogenous enzymes involved in collagen degradation. In healthy tendons, MMPs and TIMPs are in balance and are involved in normal tendon development and remodeling. Studies suggest that MMPs are involved in tendon degradation following immobilization and also in oxidative stress due to overload [49, 50].

21.4.4 Pathological Changes of the Torn Tendon

21.4.4.1 Scar Tissue Formation in Enthesis

The four zones of the enthesis are not restored properly. Instead, scar tissue is formed (Fig. 21.4). This loss of organization at the enthesis during reparative processes could explain the reduction in local strength after a rotator cuff tear.

The normal pattern of tendon healing in torn rotator cuffs is an extrinsic process. This means

Table 21.2 Cellular characteristics in the extracellular matrix

Cellular characteristics	Extracellular matrix grade			
	0	1	2	3
Autophagic cell death (%)	3.9 ± 3.6 (51 fields)	42.9 ± 1.8 (209 fields)	1.9 ± 1.5 (371 fields)	46.0 ± 1.8 (269 fields)
Apoptotic cell death (%)	21.4 ± 2.2 (54 fields)	26.0 ± 1.4 (237 fields)	31.0 ± 1.2 (363 fields)	34.8 ± 1.6 (246 fields)
Myofibroblasts (%)	6.7 ± 1.0 (50 fields)	13.8 ± 0.9 (229 fields)	16.9 ± 1.0 (358 fields)	19.8 ± 1.3 (263 fields)
Cell density (cells/mm^2)	555 ± 41 (57 fields)	674 ± 27 (246 fields)	529 ± 17 (358 fields)	395 ± 17 (239 fields)

Reproduced from Wu et al. [42]
All the data were presented as mean ± SEM
Percentage of autophagic cell death, apoptosis, myofibroblasts, and the cell density in the ECM graded 0–3, respectively

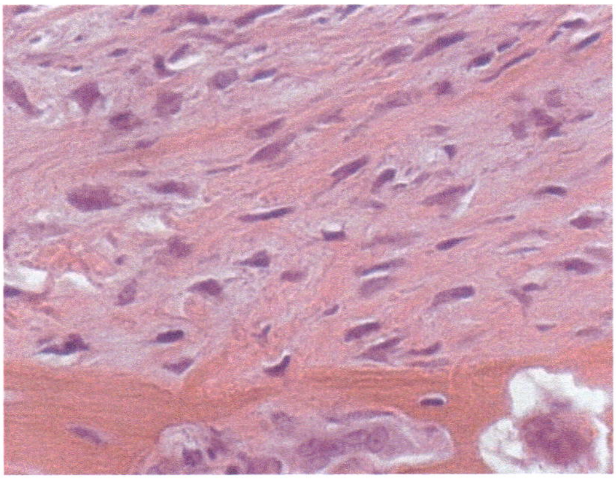

Fig. 21.4 Scar tissue formation after rotator cuff reconstruction (From Schär and Rodeo [73])

that cells from the surrounding tissues migrate to the tendon for healing. To a lesser extent, intrinsic factors may also increase healing. This is carried out by tenocytes and tenocyte-like stem cells from the epitenon and endotenon.

Tendons and the enthesis heal in three steps:

1. *Inflammatory phase* (first days): An inflammatory response shows neutrophil and macrophagic infiltration with phagocytosis of the necrotic tissue. At the same time, a release of vasoactive and chemotactic factors increase angiogenesis and stimulate tenocyte growth, which start producing type III collagen [51].
2. *Proliferative phase* (first days up to 6 weeks): In this phase, cells proliferate and collagen type III production peaks, with high concentration of GAGs and water.
3. *Remodeling phase* (6 weeks up to 10 weeks): In this phase, there is a change from cellular to fibrous tissue with a continuing tenocyte activity and an organization of collagen into stress-directed fibers. At the same time, a large amount of type I collagen starts to be produced [52]. After 10 weeks, the tenocyte metabolism decreases and the fibrous tissue develops into a scar-like tendon. Typically, this process lasts for 1 year.

21.4.4.2 Tendon Shortening

Though it has been widely stated that the main retraction is attributed to the muscle belly, there is evidence that the retraction can also be partly attributed to the tendon, especially during early stages of FI (Fatty infiltration) (Goutailler stage 1) or late stages of FI (Goutailler stage 4). According to Wolff's law of tendon retraction, the late retraction can be due to the intrinsic elastic properties of the tendon, while the initial shortening is attributed to a remaining tendon substance that is left on the greater tuberosity after the initial tear [53].

21.4.5 Clinical Application

All these findings are important to assess the restoration power of a surgical procedure on rotator cuff. Surprisingly, a single stage procedure with restoration of a normal muscle length impairs muscle properties, with progressive degeneration and atrophy, probably due to excessive stress and subsequent cellular damage [54]. Nevertheless, it has been demonstrated that a continued tensile force, though unable to restore normal anatomy to the muscles, could arrest and partially revert ultrastructural changes [21] (Fig. 21.5). Stem cells may have a role to play in the future, as they have demonstrated their potential in decreasing fatty infiltration [55].

21.5 Bone

21.5.1 Ultrastructure Anatomy Changes Among Humeral Head

Several authors have investigated the bony structures at the level of the humeral head. To our knowledge, there are no predisposing factors of the ultrastructure of the bone that leads to the pathophysiology of a rotator cuff tear. Conversely, the ultrastructure has a technical as well as a biological influence on repair. Recent works suggest that computed tomography, especially high-resolution peripheral quantitative computed tomography (HR-pQCT) could assess volumetric bone mineral density (vBMD) changes in the humeral head [56, 57]. The most significant

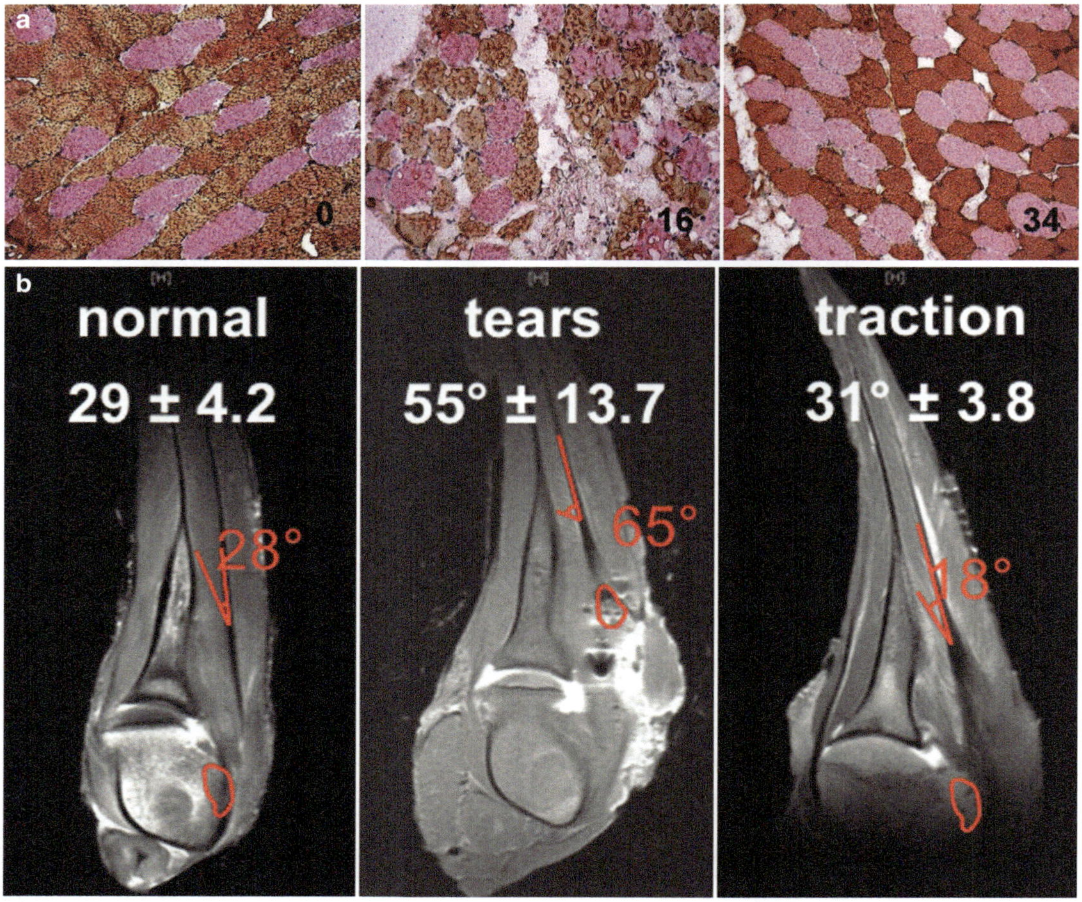

Fig. 21.5 Changes in (**a**) fatty infiltration and (**b**) the pennation angle from normal to torn and to tracted muscle (From Gerber et al. [21])

differences have been highlighted between the humeral head, greater tuberosity, and lesser tuberosity. In fact, factors such as increasing age have been shown to be associated with decreased bone density and mineralization, especially at greater tuberosity [58] (Fig. 21.6).

Other factors such as vitamin D, biphosphonates, and estrogen levels have been demonstrated to have a positive effect both on osseous quality and tendon strength, although only in animal models [59, 60].

Moreover, it has been shown that full thickness rotator cuff tears in humans are associated with loss of bone density at greater tuberosity [61]. The distribution of bone density among tuberosities could be very important for rotator cuff repair (such as anchors or knots restraint). The rotator cuff tear could play a role in bone loss, probably because of the loss of mechanical stimulation over osseous structures and osteoclasts' higher activity [62, 63].

This could explain why structural changes have been observed in rotator cuffs presenting with only impingement without complete tears [64].

Additionally, there is no real consensus about whether the most osteoporotic zone is located on the medial or lateral aspect of the tuberosities [65]. Nevertheless, many studies have explained

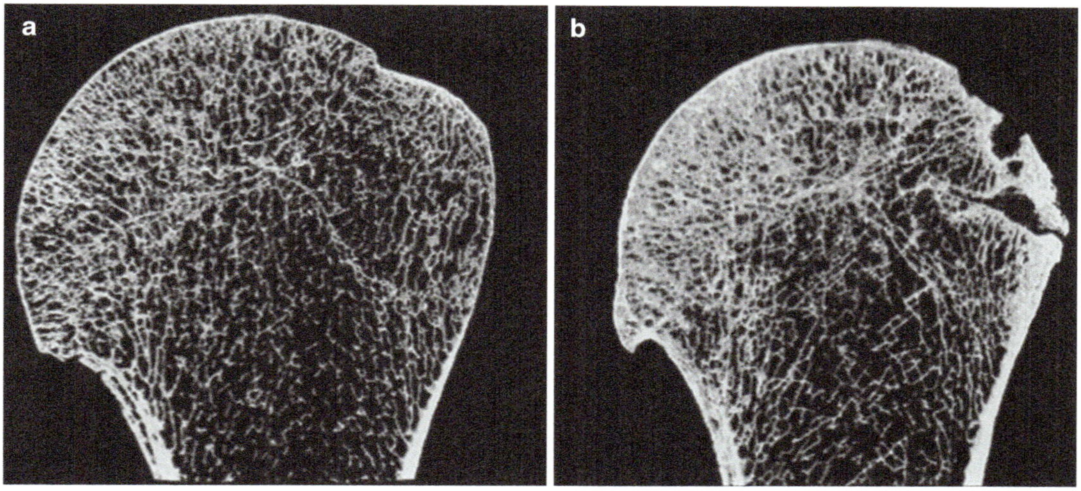

Fig. 21.6 Micro-CT scan demonstrating the osseous structure of the proximal humerus. (a) normal left head, (b) right head with rotator cuff tear

that the best place for fixation is on the medial side of both the greater and lesser tuberosities [66, 67].

21.5.2 Skeletal Morphology Acromion-Type CSA, Lateral Acromial Index

Skeletal morphology, especially of the acromion and glenoid, has been fully investigated as an important cause of rotator cuff tear. Acromion type has probably been historically the first one. Bigliani et al. [68] described acromion type (hooked, curved, flat) is strictly associated with rotator cuff tears and they developed the concept of "impingement" as a cause of rotator cuff tear. More recently, the concept that acromion type is more a consequence than a cause of rotator cuff tears has been developed too. The acromion could affect lever brace of deltoid with a consequentially greater displacement force upward. Other studies such as acromial index [69] or glenoid inclination angle [70] have been suggested as predictors of a cuff tear or glenohumeral arthritis. Both of them, nevertheless, have not been validated. More recently, the critical shoulder angle [71, 72] (Fig. 21.7), a union of both indexes, has been fully investigated and validated as a predictor of rotator cuff tear and also more weakly correlated with concentric glenohumeral arthritis.

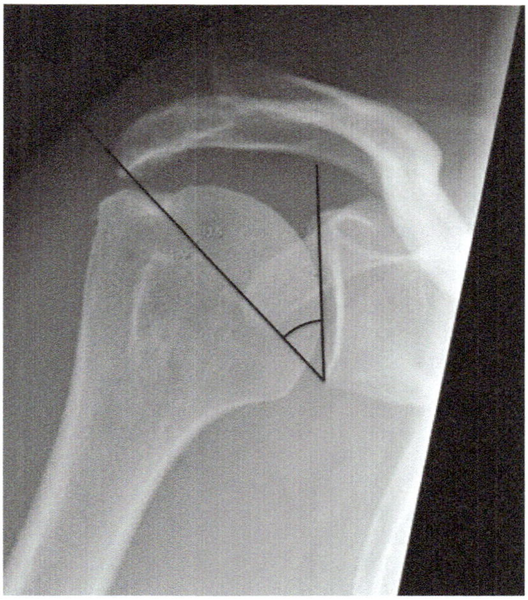

Fig. 21.7 Radiograph of the shoulder containing the Critical Shoulder Angle (CSA)

It has been evaluated, in fact, that an angle of 38° or more is strictly associated with a rotator cuff tear, due to greater shear forces that may only be mitigated by a major effort by the supraspinatus and a subsequent increased risk of tear.

Conclusion

In rotator cuff pathology, all the three structures of the rotator cuff have shown to play a crucial role in glenohumeral joint stability as well as mobilization. Several structures such as the muscle, tendon, and bone may contribute to the development of a tear as well as on the outcome following a rotator cuff repair. Further studies are needed to further our understanding of the physiological as well as the pathophysiological properties. Knowledge of these structures may help to improve rotator cuff healing after rotator cuff tear.

References

1. Jost B, Koch PP, Gerber C. Anatomy and functional aspects of the rotator interval. J Shoulder Elbow Surg. 2000;9(4):336–41.
2. Ward SR, Hentzen ER, Smallwood LH, et al. Rotator cuff muscle architecture: implications for glenohumeral stability. Clin Orthop Relat Res. 2006;448:157–63.
3. Kim SY, Boynton EL, Ravichandiran K, Fung LY, Bleakney R, Agur AM. Three-dimensional study of the musculotendinous architecture of supraspinatus and its functional correlations. Clin Anat. 2007;20:648–55.
4. Kim S, Bleakney R, Boynton E, et al. Investigation of the static and dynamic musculotendinous architecture of supraspinatus. Clin Anat. 2010;23(1):48–55.
5. Roh MS, Wang VM, April EW, Pollock RG, Bigliani LU, Flatow EL. Anterior and posterior musculotendinous anatomy of the supraspinatus. J Shoulder Elbow Surg. 2000;9:436–40.
6. Meyer DC, Hoppeler H, von Rechenberg B, Gerber C. A pathomechanical concept explains muscle loss and fatty muscular changes following surgical tendon release. J Orthop Res: Off Publ Orthop Res Soc. 2004;22(5):1004–7.
7. Zuo J, Sano H, Itoi E. Changes in pennation angle in rotator cuff muscles with torn tendons. J Orthop Sci: Off J Jpn Orthop Assoc. 2012;17(1):58–63.
8. Gladstone JN, Bishop JY, Lo IK, Flatow EL. Fatty infiltration and atrophy of the rotator cuff do not improve after rotator cuff repair and correlate with poor functional outcome. Am J Sports Med. 2007;35(5):719–28.
9. Gerber C, Schneeberger AG, Hoppeler H, Meyer DC. Correlation of atrophy and fatty infiltration on strength and integrity of rotator cuff repairs: a study in thirteen patients. J Shoulder Elbow Surg. 2007;16(6):691–6.
10. Gerber C, Fuchs B, Hodler J. The results of repair of massive tears of the rotator cuff. J Bone Joint Surg Am. 2000;82-A(4):505–15.
11. Józsa L, Kannus P, Thöring J, Reffy A, Järvinen M, Kvist M. The effect of tenotomy and immobilisation on intramuscular connective tissue. A morphometric and microscopic study in rat calf muscles. J Bone Joint Surg. 1990;72:293–7.
12. Crawford GN. Some effects of tenotomy on adult striated muscles. J Anat. 1977;123(Pt 2):389–96.
13. Steinbacher P, Tauber M, Kogler S, Stoiber W, Resch H, Sanger AM. Effects of rotator cuff ruptures on the cellular and intracellular composition of the human supraspinatus muscle. Tissue Cell. 2010;42(1):37–41.
14. Schmutz S, Fuchs T, Regenfelder F, Steinmann P, Zumstein M, Fuchs B. Expression of atrophy mRNA relates to tendon tear size in supraspinatus muscle. Clin Orthop Relat Res. 2009;467(2):457–64.
15. Sacheck JM, Hyatt JP, Raffaello A, et al. Rapid disuse and denervation atrophy involve transcriptional changes similar to those of muscle wasting during systemic diseases. Faseb J. 2007;21(1):140–55.
16. Meyer DC, Gerber C, Von Rechenberg B, Wirth SH, Farshad M. Amplitude and strength of muscle contraction are reduced in experimental tears of the rotator cuff. Am J Sports Med. 2011;39(7):1456–61.
17. Gerber C, Meyer DC, Schneeberger AG, Hoppeler H, von Rechenberg B. Effect of tendon release and delayed repair on the structure of the muscles of the rotator cuff: an experimental study in sheep. J Bone Joint Surg Am. 2004;86-A(9):1973–82.
18. Coleman SH, Fealy S, Ehteshami JR, et al. Chronic rotator cuff injury and repair model in sheep. J Bone Joint Surg Am. 2003;85-A(12):2391–402.
19. Melis B, Nemoz C, Walch G. Muscle fatty infiltration in rotator cuff tears: descriptive analysis of 1688 cases. Orthop Traumatol Surg Res: OTSR. 2009;95(5):319–24.
20. Melis B, Wall B, Walch G. Natural history of infraspinatus fatty infiltration in rotator cuff tears. J Shoulder Elbow Surg. 2010;19(5):757–63.
21. Gerber C, Meyer DC, Frey E, et al. Neer Award 2007: Reversion of structural muscle changes caused by chronic rotator cuff tears using continuous musculotendinous traction. An experimental study in sheep. J Shoulder Elbow Surg. 2009;18:163–71.
22. Frey E, Regenfelder F, Sussmann P, et al. Adipogenic and myogenic gene expression in rotator cuff muscle of the sheep after tendon tear. J Orthop Res: Off Publ Orthop Res Soc. 2009;27(4):504–9.
23. Vad VB, Southern D, Warren RF, Altchek DW, Dines D. Prevalence of peripheral neurologic injuries in rotator cuff tears with atrophy. J Shoulder Elbow Surg. 2003;12(4):333–6.

24. Albritton MJ, Graham RD, Richards RS, Basamania CJ. An anatomic study of the effects on the suprascapular nerve due to retraction of the supraspinatus muscle after a rotator cuff tear. J Shoulder Elbow Surg. 2003;12(5):497–500.
25. Lakemeier S, Reichelt JJ, Patzer T, Fuchs-Winkelmann S, Paletta JR, Schofer MD. The association between retraction of the torn rotator cuff and increasing expression of hypoxia inducible factor 1alpha and vascular endothelial growth factor expression: an immunohistological study. BMC Musculoskelet Disord. 2010;11:230.
26. Kannus P, Jozsa L, Jarvinnen M. Basic science of tendons. In: Garrett Jr WE, Speer KP, Kirkendall DT, editors. Principles and practice of orthopaedic sports medicine. Philadelphia: Lippincott Williams & Wilkins; 2000. p. 21–37.
27. Lake SP, Miller KS, Elliott DM, Soslowsky LJ. Effect of fiber distribution and realignment on the nonlinear and inhomogeneous mechanical properties of human supraspinatus tendon under longitudinal tensile loading. J Orthop Res. 2009;27: 1596–602.
28. Lake SP, Miller KS, Elliott DM, Soslowsky LJ. Tensile properties and fiber alignment of human supraspinatus tendon in the transverse direction demonstrate inhomogeneity, nonlinearity, and regional isotropy. J Biomech. 2010;43:727–32.
29. Killian ML, Cavinatto L, Galatz LM, Thomopoulos S. The role of mechanobiology in tendon healing. J Shoulder Elbow Surg. 2012;21:228–37.
30. Thomopoulos S, Genin GM, Galatz LM. The development and morphogenesis of the tendon-to-bone insertion – what development can teach us about healing. J Musculoskelet Neuronal Interact. 2010;10:35–45.
31. Benjamin M, Evans EJ, Copp L. The histology of tendon attachments to bone in man. J Anat. 1986;149:89.
32. Koike Y, Trudel G, Uhthoff H. Formation of a new enthesis after attachment of the supraspinatus tendon: a quantitative histologic study in rabbits. J Orthop Res. 2005;23:1433–40.
33. Zhao S, Peng L, Xie G, Li D, Zhao J, Ning C. Effect of the interposition of calcium phosphate materials on tendon-bone healing during repair of chronic rotator cuff tear. Am J Sports Med. 2014;0363546514532781.
34. Ogata S, Uhthoff HK. Acromial enthesopathy and rotator cuff tear. A radiologic and histologic postmortem investigation of the coracoacromial arch. Clin Orthop Relat Res. 1990;254:39–48.
35. Neviaser A, Andarawis-Puri N, Flatow E. Basic mechanisms of tendon fatigue damage. J Shoulder Elbow Surg. 2012;21:158–63.
36. Soslowsky LJ, Thomopoulos S, Tun S, et al. Neer Award 1999. Overuse activity injures the supraspinatus tendon in an animal model: a histologic and biomechanical study. J Shoulder Elbow Surg. 2000;9:79–84.
37. Kader D, Saxena A, Movin T, Maffulli N. Achilles tendinopathy: some aspects of basic science and clinical management. Br J Sports Med. 2002;36:239–49.
38. Kumagai J, Sarkar K, Uhthoff HK. The collagen types in the attachment zone of rotator cuff tendons in the elderly: an immunohistochemical study. J Rheumatol. 1994;21:2096–100.
39. Yuan J, Murrell GAC, Wei A-Q, Wang M-X. Apoptosis in rotator cuff tendonopathy. J Orthop Res: Off Publ Orthop Res Soc. 2002;20:1372–9.
40. Nho SJ, Yadav H, Shindle MK, MacGillivray JD. Rotator cuff degeneration: etiology and pathogenesis. Am J Sports Med. 2008;36:987–93.
41. Morikawa D, Itoigawa Y, Nojiri H, et al. Contribution of oxidative stress to the degeneration of rotator cuff entheses. J Shoulder Elbow Surg. 2014;23:628–35.
42. Wu B, Chen J, Rosa TD, et al. Cellular response and extracellular matrix breakdown in rotator cuff tendon rupture. Arch Orthop Trauma Surg. 2011;131: 405–11.
43. Fukuda H, Hamada K, Yamanaka K. Pathology and pathogenesis of bursal-side rotator cuff tears viewed from en bloc histologic sections. Clin Orthop Relat Res. 1990;75–80.
44. Goodmurphy CW, Osborn J, Akesson EJ, Johnson S, Stanescu V, Regan WD. An immunocytochemical analysis of torn rotator cuff tendon taken at the time of repair. J Shoulder Elbow Surg. 2003;12:368–74.
45. Funakoshi T, Iwasaki N, Kamishima T, et al. In vivo visualization of vascular patterns of rotator cuff tears using contrast-enhanced ultrasound. Am J Sports Med. 2010;38:2464–71.
46. Hashimoto T, Nobuhara K, Hamada T. Pathologic evidence of degeneration as a primary cause of rotator cuff tear. Clin Orthop Relat Res. 2003;415:111–20.
47. Castagna A, Cesari E, Garofalo R, et al. Matrix metalloproteases and their inhibitors are altered in torn rotator cuff tendons, but also in the macroscopically and histologically intact portion of those tendons. Muscles Ligaments Tendons J. 2013;3:132–8.
48. Dalton S, Cawston TE, Riley GP, Bayley IJ, Hazleman BL. Human shoulder tendon biopsy samples in organ culture produce procollagenase and tissue inhibitor of metalloproteinases. Ann Rheum Dis. 1995;54:571–7.
49. Gardner K, Arnoczky SP, Caballero O, Lavagnino M. The effect of stress-deprivation and cyclic loading on the TIMP/MMP ratio in tendon cells: an in vitro experimental study. Disabil Rehabil. 2008;30: 1523–9.
50. Han Z, Boyle DL, Chang L, et al. c-Jun N-terminal kinase is required for metalloproteinase expression and joint destruction in inflammatory arthritis. J Clin Invest. 2001;108:73–81.
51. Murphy PG, Loitz BJ, Frank CB, Hart DA. Influence of exogenous growth factors on the synthesis and secretion of collagen types I and III by explants of normal and healing rabbit ligaments. Biochem Cell Biol. 1994;72:403–9.
52. Abrahamsson SO. Matrix metabolism and healing in the flexor tendon. Experimental studies on rabbit tendon. Scand J Plast Reconstr Surg Hand Surg Suppl. 1991;23:1–51.

53. Meyer DC, Wieser K, Farshad M, Gerber C. Retraction of supraspinatus muscle and tendon as predictors of success of rotator cuff repair. Am J Sports Med. 2012;40:2242–7.
54. Stauber WT, Miller GR, Grimmett JG, Knack KK. Adaptation of rat soleus muscles to 4 wk of intermittent strain. J Appl Physiol (1985). 1994;77:58–62.
55. Oh JH, Chung SW, Kim SH, Chung JY, Kim JY. 2013 Neer Award: effect of the adipose-derived stem cell for the improvement of fatty degeneration and rotator cuff healing in rabbit model. J Shoulder Elbow Surg. 2014;23:445–55.
56. Kirchhoff C, Braunstein V, Milz S, et al. Age and gender as determinants of the bone quality of the greater tuberosity: a HR-pQCT Cadaver study. BMC Musculoskelet Disord. 2012;13:221.
57. Lee YS, Mihata T, Oh JH. Anatomically reproducible assessment of volumetric bone mineral density — based on clinical computed tomography. J Biomech. 2013;46:767–72.
58. Clavert P, Bouchaïb J, Sommaire C, Flurin P-H, Hardy P. Does bone density of the greater tuberosity change in patients over 70? Orthop Traumatol Surg Res. 2014;100:109–11.
59. Angeline ME, Ma R, Pascual-Garrido C, et al. Effect of diet-induced vitamin D deficiency on rotator cuff healing in a Rat model. Am J Sports Med. 2014;42:27–34.
60. Cadet ER, Vorys GC, Rahman RK, et al. Improving bone density at the rotator cuff footprint increases supraspinatus tendon failure stress in a rat model. J Orthop Res. 2010;28(3):308–14.
61. Jiang Y, Zhao J, van Holsbeeck MT, Flynn MJ, Ouyang X, Genant HK. Trabecular microstructure and surface changes in the greater tuberosity in rotator cuff tears. Skeletal Radiol. 2002;31:522–8.
62. Cadet ER, Hsu JW, Levine WN, Bigliani LU, Ahmad CS. The relationship between greater tuberosity osteopenia and the chronicity of rotator cuff tears. J Shoulder Elbow Surg. 2008;17:73–7.
63. Thomopoulos S, Kim HM, Silva MJ, et al. Effect of bone morphogenetic protein 2 on tendon-to-bone healing in a canine flexor tendon model. J Orthop Res. 2012;30(11):1702–9.
64. Waldorff EI, Lindner J, Kijek TG, et al. Bone density of the greater tuberosity is decreased in rotator cuff disease with and without full-thickness tears. J Shoulder Elbow Surg. 2011;20:904–8.
65. Kirchhoff C, Braunstein V, Milz S, et al. Assessment of bone quality within the tuberosities of the osteoporotic humeral head: relevance for anchor positioning in rotator cuff repair. Am J Sports Med. 2010; 38:564–9.
66. Grimberg J, Kany J. Latissimus dorsi tendon transfer for irreparable postero-superior cuff tears: current concepts, indications, and recent advances. Curr Rev Musculoskelet Med. 2014;7(1):22–32.
67. Castagna A, Garofalo R, Conti M, Borroni M, Snyder SJ. Arthroscopic rotator cuff repair using a triple-loaded suture anchor and a modified Mason-Allen technique (Alex stitch). Arthroscopy: J Arthrosc Relat Surg: Off Publ Arthroscopy Assoc North Am Int Arthrosc Assoc. 2007;23:440.e441–444.
68. Bigliani LU, Morrison DS, April EW. The morphology of the acromion and its relationship to rotator cuff tears. Orthop Trans. 1986;10:228
69. Nyffeler RW, Werner CM, Sukthankar A, Schmid MR, Gerber C. Association of a large lateral extension of the acromion with rotator cuff tears. J Bone Joint Surg Am. 2006;88(4):800–5.
70. Hughes RE, Bryant CR, Hall JM, et al. Glenoid inclination is associated with full-thickness rotator cuff tears. Clin Orthop Relat Res. 2003;407:86–91.
71. Moor BK, Bouaicha S, Rothenfluh DA, Sukthankar A, Gerber C. Is there an association between the individual anatomy of the scapula and the development of rotator cuff tears or osteoarthritis of the glenohumeral joint?: a radiological study of the critical shoulder angle. Bone Joint J. 2013;95-B(7):935–41.
72. Moor BK, Wieser K, Slankamenac K, Gerber C, Bouaicha S. Relationship of individual scapular anatomy and degenerative rotator cuff tears. J Shoulder Elbow Surg. 2014;23(4):536–41.
73. Schär MO, Rodeo SA. Shoulder Arthroscopy, Principles and Practice. Biology of Injury and Repair of Soft Tissues of the Shoulder. Milano G and Grasso A, editors. Springer London. Part 1, 2014:p. 59–72. doi:10.1007/978-1-4471-5427-3_5

22 Kinematics of the Rotator Cuff

Matthew T. Provencher, Stephen A. Parada, Daniel J. Gross, and Petar Golijanin

22.1 Introduction

The biomechanics of the rotator cuff is complex and reliant on the osseous platform and articulations on which the rotator cuff operates. In this chapter, we will discuss the anatomical, functional, and clinical aspects of rotator cuff kinematics.

22.2 Osseous Anatomy

22.2.1 Acromion

The acromion emerges from the lateral aspect of scapular spine as it continues beyond the scapular border and reflects back on itself to meet the distal clavicle, forming the acromioclavicular joint. The acromial angle is the angle of reflection between the scapular spine and the acromion, and forms a mean angle of 78° (64–99°).

The acromial coverage index is the distance between the plane of the glenoid face and the lateral border of the acromion, divided by the distance between the plane of the glenoid face and the lateral aspect of the humeral head [1]. A glenohumeral joint with a higher acromial coverage index is associated with an increased risk of a rotator cuff tear [2].

The critical shoulder angle (CSA) combines the measurements of glenoid inclination and lateral extension of the acromion (the acromial coverage index) [3]. The angle is measured between a line connecting the inferior border to the superior border of the glenoid fossa, and a second line connecting the inferior border of the glenoid to the most inferolateral point of the acromion [4]. The CSA has been used to predict the presence of degenerative rotator cuff tears demonstrating what appears to be a clear causal relationship of the biomechanics of the anatomy affecting the pathology (Fig. 22.1).

22.2.2 Scapula

In the resting position, the scapula extends from the 2nd rib to between the 7th or 9th rib at the inferior angle, and is anteriorly rotated 30° in the axial plane to accommodate the thoracic rib cage [5–7]. In the coronal plane, the scapula is rotated upward 3–10°, and when viewed in the sagittal plane, it is anteflexed 10–20° [8].

22.2.3 Glenoid

The orientation of the glenoid may be angled in both the axial and coronal planes. In the axial

M.T. Provencher, MD (✉) • D.J. Gross, MD
P. Golijanin, BS
Department of Sports Medicine and Surgery, Massachusetts General Hospital, Boston, MA, USA
e-mail: mattprovencher@gmail.com; danielgross23@gmail.com; golijaninp@gmail.com

S.A. Parada, MD
Department of Orthopaedics, Eisenhower Army Medical Center, Fort Gordon, GA, USA
e-mail: stephen.a.parada@gmail.com

G.I. Bain et al. (eds.), *Normal and Pathological Anatomy of the Shoulder*,
DOI 10.1007/978-3-662-45719-1_22, © ISAKOS 2015

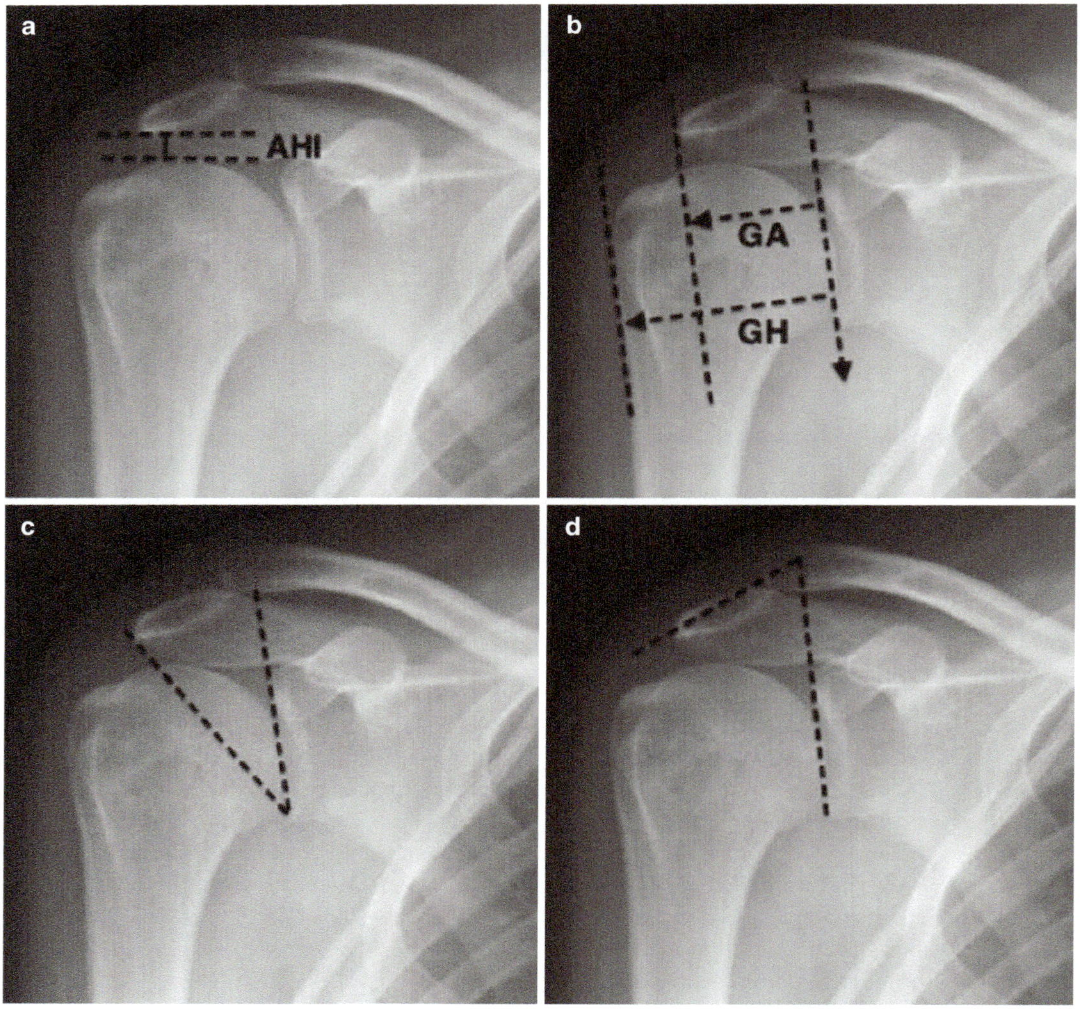

Fig. 22.1 Overview of the assessed radiologic parameters. (**a**) The acromiohumeral interval – distance from superior humerus to inferior acromion. (**b**) The acromion index – ratio GA/GH. *GA* the distance from the glenoid plane to the acromion, and *GH* the distance from the glenoid plane to the lateral aspect of the humeral head. (**c**) The critical shoulder angle – angle between a line connecting the inferior and superior borders of the glenoid fossa and a second line connecting the inferior border of the glenoid to the most inferolateral point of the acromion. (**d**) The lateral acromion angle – angle between a line drawn parallel to the sclerotic line of the acromion undersurface and a second line connecting the superior to the inferior border of the glenoid fossa

plane, anterior or posterior angulation is referred to as version. Glenoid version can range from retroversion to anteversion, with one study finding the mean version to be 1.23° of retroversion, with a range of 9.5° of anteversion to 10.5° retroversion [9]. Other sources have found the mean glenoid version to be closer to 7° of retroversion [5]. In the coronal plane, superior or inferior angulation is referred to as inclination. Churchill et al. found the mean glenoid inclination to be 4.2° of superior inclination, with a range of 7° of inferior inclination to 15.8° of superior inclination [9].

22.2.4 Humeral Head

The humeral head and shaft lie roughly in the plane of the scapula, with the articular surface of

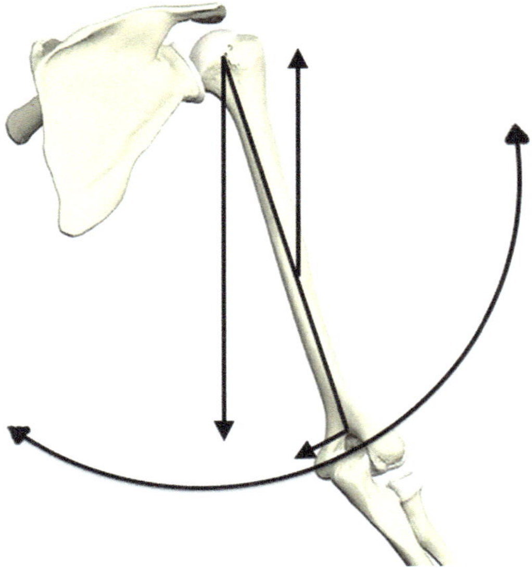

Fig. 22.2 Forces acting on glenohumeral joint. Rotator cuff muscles are predominantly stabilizing glenohumeral joint, but they also significantly contribute to the joint movement as well. They mainly contribute to abduction, external rotation, and internal rotation, and also play important role in joint compression

the humeral head comprising 1/3 of a sphere. The humeral head is oriented superiorly, forming a neck-shaft angle of 45°, and it rests in approximately 30° of retroversion, which is complimented by the anteversion of the scapula [5, 6].

22.3 Tendon Anatomy

22.3.1 Gross Anatomy

The subscapularis muscle alone inserts on the lesser tuberosity of the humerus, and is responsible for internal rotation of the arm (Figs. 22.3 and 22.4). The supraspinatus, infraspinatus, and teres minor all insert on the greater tuberosity (Figs. 22.5, 22.6, 22.7, and 22.8). The supraspinatus tendon passes through the subacromial space beneath the subacromial bursa to its insertion site on the superior facet of the greater tuberosity. This insertion site facilitates abduction of the upper extremity, but also is a component to the frequent impingement of this tendon. Beyond

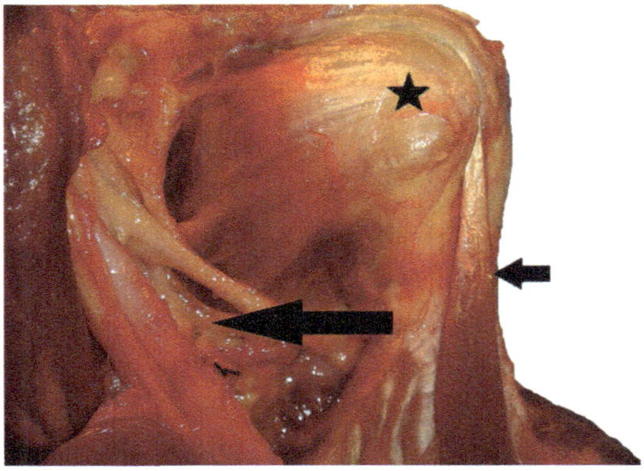

Fig. 22.3 Anterior rotator cuff musculature. Photograph of the anterior aspect of a cadaveric shoulder specimen demonstrating the subscapularis (*star*), the intact biceps tendon (*short arrow*), and the nearby presence of the neurovascular plexus (*long arrow*)

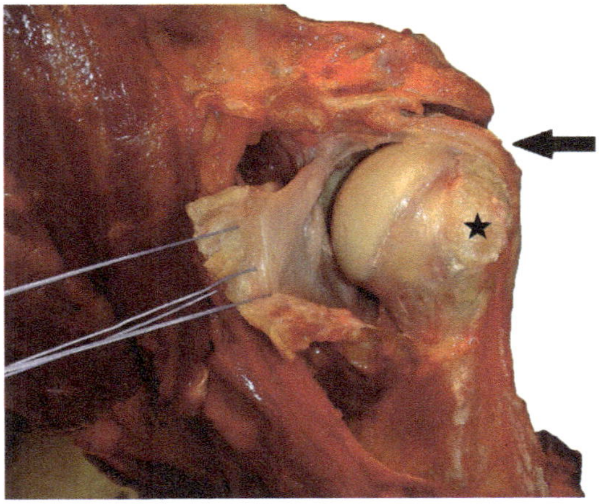

Fig. 22.4 Photograph of the anterior aspect of a cadaveric shoulder specimen after the subscapularis and capsule have been reflected from the lesser tuberosity (*star*). The supraspinatus remains intact on the greater tuberosity (*arrow*)

70° of abduction, the greater tuberosity has a tendency to come into contact with the acromion, resulting in the impingement of both the tendon and the subacromial bursa [10, 11]. The infraspinatus and teres minor tendons insert on the posteroinferior facet of the greater tuberosity and working conjunction to facilitate external rotation.

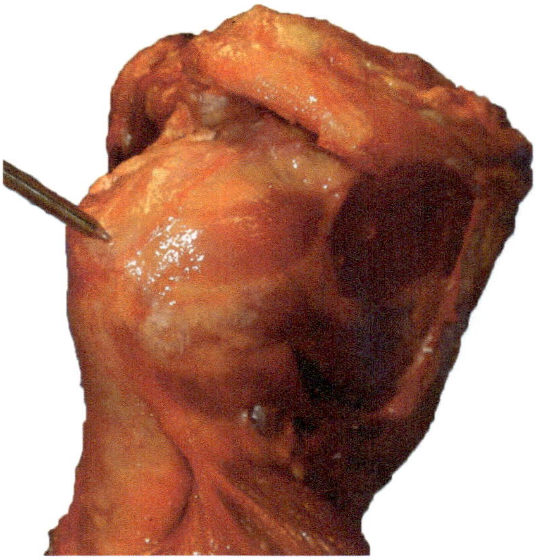

Fig. 22.5 Photograph of the posterior-superior aspect of a cadaveric shoulder specimen with the posterior cuff intact and the pointer resting on the lateral aspect of the greater tuberosity

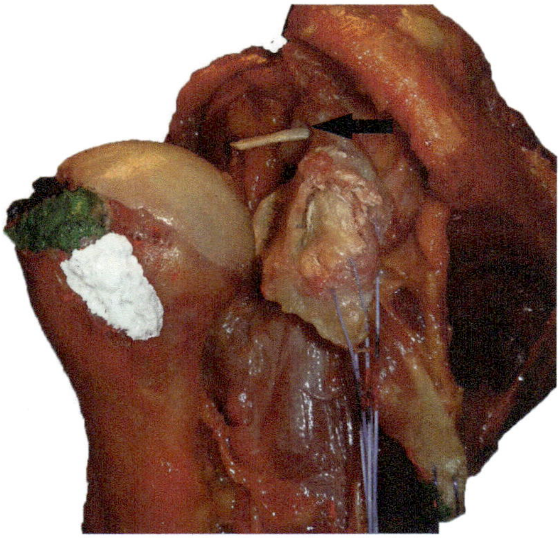

Fig. 22.7 Photograph of the posterior aspect of a cadaveric shoulder specimen with sutures in the reflected posterior rotator cuff. The attachment of the supraspinatus has been painted *black* (barely visible), the infraspinatus has been painted *green*, and the teres *minor white*. The biceps has been tenotomized, and its proximal segment is visible at the superior glenoid tubercle

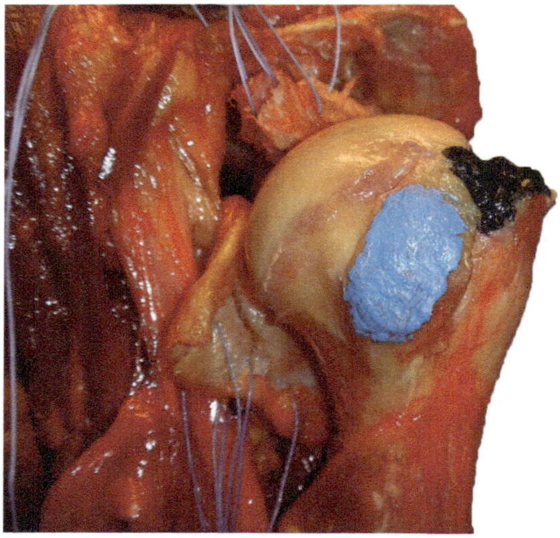

Fig. 22.6 Photograph of the anterior aspect of a cadaveric shoulder specimen after the subscapularis and supraspinatus have been reflected from the lesser tuberosity (*painted blue*) and the greater tuberosity (*painted black*)

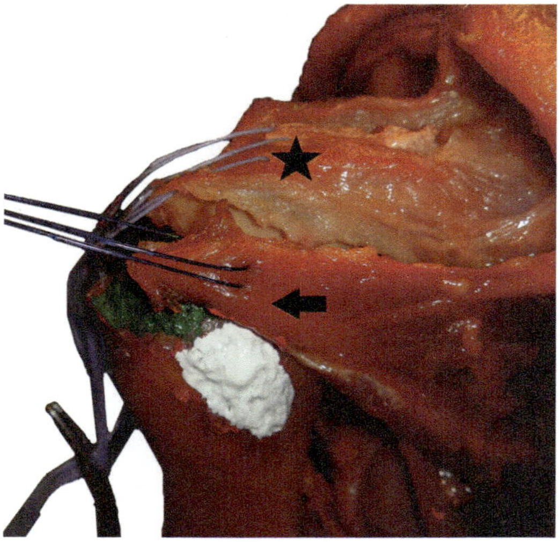

Fig. 22.8 Photograph of the posterior aspect of a cadaveric shoulder specimen with sutures in the posterior rotator cuff, which have been pulled back in-line toward their attachments. The attachment of the supraspinatus (*star*) has been painted *black* and the infraspinatus (*arrow*) has been painted *green*. The attachment of the teres minor has been painted *white* and remains reflected

22.3.2 Tendon Composition

Tendons are primarily composed of type I collagen, which comprises approximately 85 % of their dry weight. There is also a high concentration of proteoglycans and elastin. Proteoglycans are responsible for the viscoelasticity found in tendons [12–14].

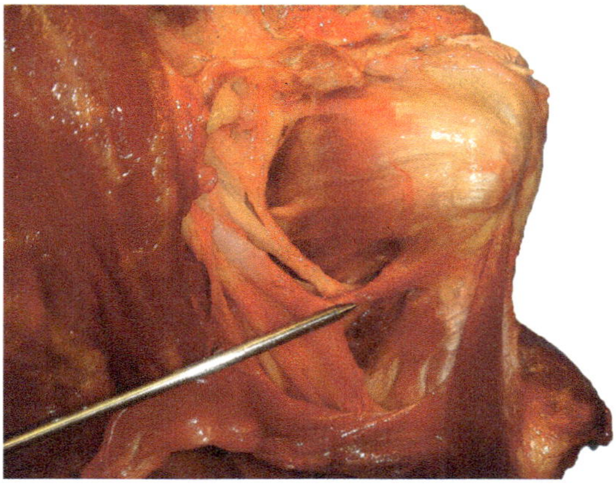

Fig. 22.9 Photograph of the anterior aspect of a cadaveric shoulder specimen with the pointer on the branch of the axillary artery providing the blood supply to the anterior aspect of the rotator cuff

22.3.3 Vascularity

The muscles of the rotator cuff derive their blood supply from the branches of the axillary artery (Fig. 22.9). The axillary artery is generally divided in to three segments, based on the borders of the pectoralis minor. Proximally, the thoracoacromial artery emerges from the axillary artery at the level of the upper border of the pectoralis minor. The artery pierces the clavipectoral fascia, and then divides into four branches that supply the muscles of the shoulder and proximal humerus. Of these four branches, the deltoid (or humeral) branch and acromial branch are the primary blood suppliers to the scapulohumeral muscles.

At the lateral border of the pectoralis minor, the subscapular artery emerges to pass between the radial and median nerves and descends caudally to supply the subscapularis muscle. Eventually, the subscapular artery will give rise to the circumflex scapular artery whose branches form an anastomosis with the suprascapular and dorsal scapular arteries. Distally, the anterior and posterior circumflex arteries emerge and encircle the humerus, and provide blood supply to the glenohumeral joint capsule and rotator cuff tendons via its terminal branches [15, 16].

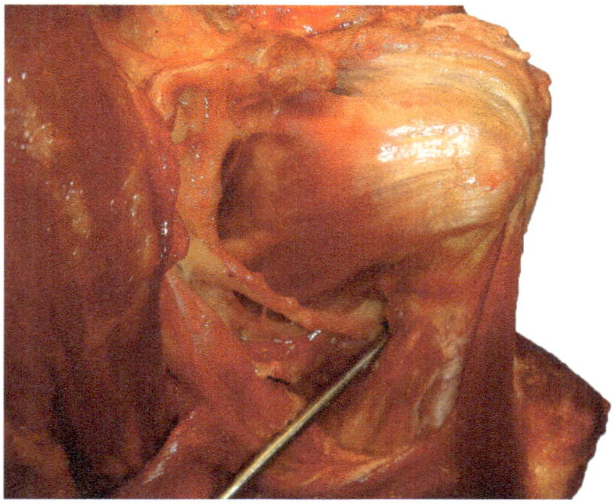

Fig. 22.10 Photograph of the anterior aspect of a cadaveric shoulder specimen with the pointer on the axillary nerve as it travels from anterior to posterior to exit posterior to the humerus (see next figure)

22.3.4 Nerve Supply

The muscles of the rotator cuff are innervated by the brachial plexus, which is formed from the branches of spinal roots C5–T1. In terms of surgical anatomy, the most important nerves are the suprascapular nerve and the axillary nerve. The suprascapular nerve arises from the superior trunk of the brachial plexus and passes through the suprascapular notch to enter into the supraspinatous fossa where it gives off two motor branches to innervate the supraspinatus muscle. It is as it passes under the superior transverse ligament that the nerve becomes most susceptible to injury via compression and shearing forces [5]. The nerve then continues to travel around the lateral border of the scapular spine and into the infraspinatous fossa via the spinoglenoid notch. The subscapular nerve arises from the posterior cord of the brachial plexus, and divides into an upper and lower nerve. The upper nerves insert directly into the subscapularis muscle, while the lower nerve continues to innervate the inferior portion of the subscapularis. The remaining innervation of the teres minor comes from the axillary nerve (Figs. 22.10 and 22.11) [17].

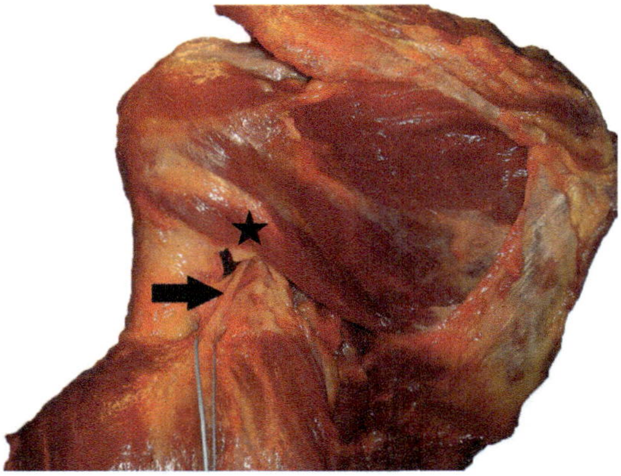

Fig. 22.11 Photograph of the posterior aspect of a cadaveric shoulder specimen with the axillary nerve (*arrow*), which is tagged with a suture as it exits the quadrilateral space after supplying the teres minor (*star*) with its innervation

22.4 Shoulder Function

22.4.1 Shoulder Motion

The motion of the shoulder is complex and takes into account motion at the acromioclavicular and sternoclavicular joints, the glenohumeral joint, and the scapulothoracic articulation. There have been numerous methods developed to describe the motion that occurs at each joint. For research purposes, the motion at the glenohumeral joint is often referred to in regard to rotation around three axes to give forward elevation, elevation in the plane of the scapula (or abduction), and humeral rotation. The term "scapulothoracic rhythm" was first used by Codman to describe the complex, coordinated movements of both the scapulothoracic articulation and the glenohumeral joint during arm motion. Further proving that this motion is coupled, pathology of the glenohumeral joint, such as the pain of a full-thickness RC tear, affects the kinematics of the scapulothoracic articulation [18].

During humeral elevation, a greater degree of motion occurs at the glenohumeral joint during the early arc of motion (first 30°). Scapular and clavicular motion are minimal with the arm in less than 90° of elevation. Beyond 90° of humeral elevation, there is upward rotation, posterior tilting, and external rotation of the scapula to allow for the full arc of motion in the upper extremity [19–21]. During this, the clavicle retracts, elevates, and rotates backward to accommodate this motion. The humerus must externally rotate to avoid impingement of the greater tuberosity under the acromion [22].

There is a lack of linearity to the ratio of glenohumeral to scapulothoracic motion with a ratio of 4:1 during the initial 25° of motion, which almost equates at 5:4 after that to average approximately 2:1 for the entire arc of elevation [23–26]. This coupled motion has been further elucidated with dual fluoroscopic imaging systems in an in vitro model [27].

22.4.2 Center of Rotation

When simplified to a single plane, the center of rotation lies within 6 mm of the geometric center of the humeral head [28]. This tight focus regarding the center of rotation is largely due to the small amounts of translation that occur during initial elevation. The center of rotation is also affected by the integrity of the rotator cuff as well as the long head of the biceps tendon [29, 30].

22.4.3 Biomechanics

Forces at the glenohumeral joint depend on multiple factors including the overall condition of the muscle, the size of the muscle as measured by the cross-sectional area, and the position of the joint. In general, a muscle is strongest at the midpoint of its excursion compared to when it is fully contracted or extended.

If the muscle is atrophic due to muscular or neural pathophysiology, it will not function as otherwise expected based on the biomechanics alone. Cross-sectional area of the muscle has been measured to determine a muscle's volume and subsequent forces, which it is able to generate [31]. The orientation of the muscles surrounding the shoulder to the glenoid creates a joint reaction force with a large component perpendicular to the glenoid, further aiding humeral head compression [32]. The location of the muscle and tendon unit with regard to the joint also becomes important when setting up the testing apparatus during biomechanical testing. Tears of the RC affect the overall amount of force that can be produced in the abducted arm. When a tear was created that involved 1/3 or 2/3 of the supraspinatus, the force decreased only 5 %, and a complete, retracted tear of the supraspinatus caused only a loss of torque of 58 % [33].

The position of the arm also effects the direction of pull of the muscle, with the most obvious example being the supraspinatus, which can perform abduction or external rotation based on the arm's position (Fig. 22.2). The position of the arm also affects the morphology of the rotator cuff, as was shown in an MRI study evaluating supraspinatus tendon during different positions of rotation and abduction of the humerus. Abduction over 30° was found to shorten the tendon, while external and internal rotations caused elongation of the anterior and posterior portions of the tendon, respectively [34]. The size of the individual rotator cuff muscle can be compensated for by positioning that creates a more effective moment arm. The subscapularis and infraspinatus muscles generate forces of two to three times more than the supraspinatus; however, a more effective moment arm causes the supraspinatus to be a much more effective abductor [35].

22.4.4 Function

The rotator cuff complex functions overall in three broad categories: rotation of the humerus about the scapula, compression of the humeral head into the glenoid, and providing muscle balance to the glenohumeral joint. The subscapularis functions to internally rotate the humerus, while the infraspinatus and teres minor are external rotators. The supraspinatus functions to abduct as well as provide weak external rotation with the arm in adduction [36].

Compression of the humeral head into the glenoid has been researched by determining reaction force testing. This is accomplished by using a dynamic shoulder testing apparatus that has determined that joint forces increase throughout abduction and peak at approximately 90° of motion. Firing of the supraspinatus increases the joint forces, and simulated paralysis of the supraspinatus results in a significant decrease in compression [37].

There remains controversy on how much muscle balance and stability to the glenohumeral joint is provided by the rotator cuff complex. Different in vivo studies have been performed in an effort to gain more realistic information on the stabilizing effects of the rotator cuff. Loads to the rotator cuff were given, and the strain of the inferior glenohumeral ligament was measured, showing that the infraspinatus and teres minor were most responsible for aiding to the stability of the GH joint [38]. It was also shown that the subscapularis stabilized anteriorly with the arm in abduction; however, with external rotation, the humerus became less important as the posterior musculature became of primary importance. The rotator cuff musculature has been shown to have a greatest effect on glenohumeral stability in the midrange of motion when the capsule-labral complex is lax [39].

The moment arm and orientations of the rotator cuff muscles change with the abduction angle of the arm. Different authors have measured this through radiographic studies [40, 41]. Not surprisingly, the anterior and middle deltoids as well as the supraspinatus have the largest moment arm.

Much of the work done to corroborate the biomechanical studies of the rotator cuff has utilized electromyography and selective nerve blocks to include or exclude certain muscles and record their activity. This type of research has led to a greater understanding of the proportion of individual rotator cuff muscle involvement with certain motions. The percentage of involvement of the supraspinatus with external rotation was quantified with this pattern of research, which has led to an increased recognition of physical exam tests to detect RC tears [42].

22.5 Pathology

22.5.1 Vascularity

It has been proposed that the rotator cuff tendon tears are related to hypoperfusion. A study utilizing Doppler flowmetry to analyze blood flow to the rotator cuff failed to identify any "critical" zone of hypoperfusion in a normal rotator cuff. Blood flow was found to be highest at the edges of torn rotator cuff tendons and lowest in tendons suffering from chronic impingement [15].

22.5.2 Effect of Trauma

In cadaveric studies utilizing electromagnetic tracking devices, a 2 cm rotator cuff tear was created to determine the effect on kinematics in vitro. EMG data was used to apply force through cables sewn into the muscles. The defect resulted in posterior angulation of the plane of elevation, most notably throughout the midpoint of abduction [43].

The actions of the rotator cuff muscles are coupled in a manner that produces increased strain on the surrounding tendons when a tear in one tendon occurs. In this way, a tear of one tendon could potentially lead to an increased risk for injury in the remaining muscles [44].

The surrounding shoulder musculature has also shown to be involved in a pattern that may differentiate asymptomatic from symptomatic patients with a rotator cuff tear. Compensation from other shoulder girdle muscles may prevent more prominent symptoms in patients with a rotator cuff tear [45]. This knowledge may lead to better identification of patients who would benefit from surgical repair of their rotator cuff tear.

22.5.3 Effect of Degeneration

The supraspinatus tendon itself has a spacer affect that limits superior humeral elevation even when no tension is placed across the tendon. This elevation results in decreased acromiohumeral distance. This distance can be further affected by weakness, fatigue, or shoulder dysfunction. Healthy subjects were studied to determine the effect of fatigue on superior glenohumeral migration. Exercises of the deltoid and rotator cuff caused the humeral head to become inferiorly positioned when in the resting position and superiorly migrated during arm elevation in both static and dynamic evaluations [46, 47].

The primary cause of degeneration in rotator cuff tendons is aging. The connective tissue that makes up the tendon undergoes degeneration and weakness as the rest of the connective tissue in the body and becomes susceptible to injury with decreasing amounts of force required to cause an injury.

22.6 Applied Anatomy: Sports

22.6.1 Forces

The forces transmitted through the rotator cuff in normal daily activities is 140–200 N, while the ultimate tensile load of the supraspinatus measured in cadaveric specimens is 600–800 N [48].

Throwing motion kinematics has been studied extensively as rotator cuff tendon tears are not uncommon in baseball pitchers [49] (Fig. 22.12). The effect of the supraspinatus and infraspinatus on superior-inferior translation has been examined by performing a selective nerve of the suprascapular nerve and then obtaining an MRI with the shoulder in different functional positions (30° and 60° of abduction) [50]. This did not provoke any detectable superior migration of the humeral head, despite paralysis of the supraspinatus and infraspinatus. This led to further research, with the design of sophisticated cadaveric models with robotic systems and actuators to reproduce the pitching motion to better define the kinematics involved [51].

A cadaver study examining the effect of isolated supraspinatus tears and repairs on joint kinematics has demonstrated that 1- and 3-cm tears do not significantly alter glenohumeral translation [52]. It was shown, however, that an overtightening of the tendon can occur with a transosseous equivalent repair. This shifts the glenohumeral joint center of rotation posteriorly, which results in a significant decrease in translation, especially in external rotation greater than 90°.

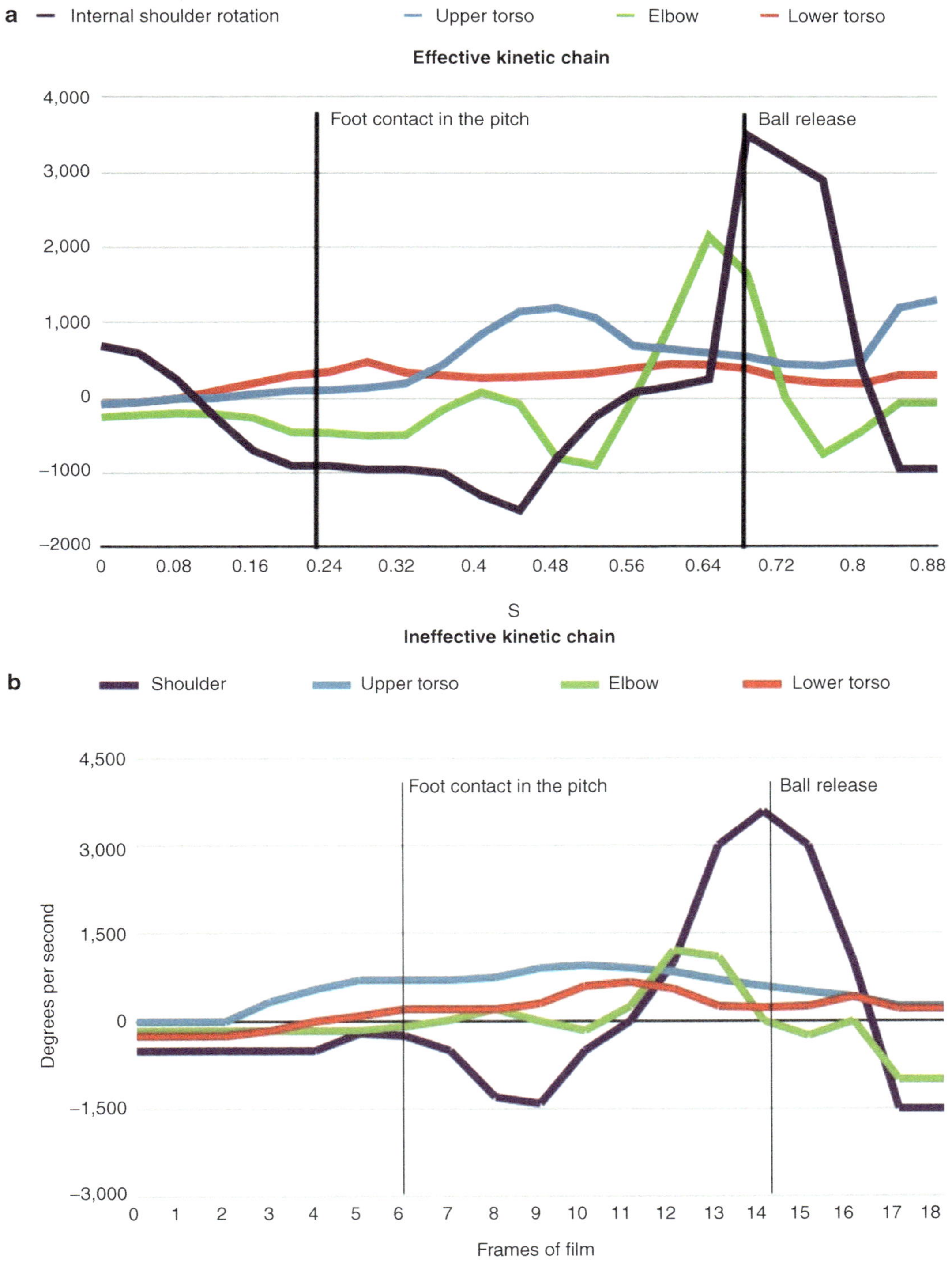

Fig. 22.12 Kinetic chain with throwing. (**a**) This is a graphical illustration of an efficient pitching motion. Power is created in the lower body using speed and the contact off the front foot. The energy is efficiently transferred from the lower body to the arm, creating arm speed with minimal stress to the shoulder and elbow. (**b**) In this example, there is inefficient contribution of power and stability in a pitching motion. Speed is poorly transferred from the lower body, and the arm speed is created without the contribution of the big muscles of the lower body resulting in high stress being placed on the shoulder

22.7 Applied Anatomy: Surgery

22.7.1 Rotator Cuff Tendon Tears

Rotator cuff tendon tears are often asymptomatic, but have also been shown to lead to proximal migration leading to mechanical impingement at the acromion, which can be symptomatic [53]. Other studies have demonstrated abnormal glenohumeral kinematics with superior migration in patients who remain nonsymptomatic [54]. In symptomatic patients with impingement syndrome, MRI has demonstrated increased tendon signal intensity and encroachment of the rotator cuff on the acromion in abduction compared to asymptomatic controls [55].

As previously discussed, certain individuals can also have bony anatomy such as a downsloping acromion, increased upward tilt of the glenoid, excessive anteversion or retroversion of the glenoid, increased lateral acromial extension, or an increased CSA, which can also predispose to having a mechanical impingement [56, 57]. In these situations where the bony anatomy of the acromion is also a contributing factor, surgery to address the rotator cuff should also include addressing the acromion. Specifically, a larger CSA has recently been shown to increase the supraspinatus tendon load at low degrees of abduction compared to a control with a normal CSA in a simulation model [58]. Acromial pathology was once thought to only account for bursal-sided rotator cuff tears; however, other data suggests that a bursal insult can cause degeneration and subsequent tearing on the bursal side, the articular side, or even within the tendon [59]. Certainly, the etiology of rotator cuff tears is often multifactorial, and not a sole intrinsic or extrinsic factor is responsible for the tear [60].

References

1. Ames JB, Horan MP, Van der Meijden OAJ, Leake MJ, Millett PJ. Association between acromial index and outcomes following arthroscopic repair of full-thickness rotator cuff tears. J Bone Joint Surg Am. 2012;94:1862–9. doi:10.2106/JBJS.K.01500.
2. Torrens C, López J-M, Puente I, Cáceres E. The influence of the acromial coverage index in rotator cuff tears. J Shoulder Elbow Surg. 2007;16:347–51. doi:10.1016/j.jse.2006.07.006.
3. Moor BK, Bouaicha S, Rothenfluh DA, Sukthankar A, Gerber C. Is there an association between the individual anatomy of the scapula and the development of rotator cuff tears or osteoarthritis of the glenohumeral joint?: A radiological study of the critical shoulder angle. Bone Joint J. 2013;95-B:935–41. doi:10.1302/0301-620X.95B7.31028.
4. Moor BK, Wieser K, Slankamenac K, Gerber C, Bouaicha S. Relationship of individual scapular anatomy and degenerative rotator cuff tears. J Shoulder Elbow Surg. 2014;23:536–41. doi:10.1016/j.jse.2013.11.008.
5. Rockwood Jr CA, Matsen III FA, Wirth MA, Lippitt SB. The shoulder. Philadelphia: Elsevier Health Sciences; 2009.
6. Llusá M, Merí A, Ruano D, American Academy of Orthopaedic Surgeons. Surgical atlas of the musculoskeletal system. Rosemont: American Academy of Orthopaedic Surgeons; 2008. ix, 422 p.
7. Lin H-T, Hsu A-T, Chang G-L, Chang Chien J-R, An K-N, Su FC. Determining the resting position of the glenohumeral joint in subjects who are healthy. Phys Ther. 2007;87:1669–82. doi:10.2522/ptj.20050391.
8. Ludewig PM, Cook TM, Nawoczenski DA. Three-Dimensional Scapular Orientation and Muscle Activity at Selected Positions of Humeral Elevation. J Orthop Sports Phys Ther. 1996;24:57–65. doi:10.2519/jospt.1996.24.2.57.
9. Churchill RS, Brems JJ, Kotschi H. Glenoid size, inclination, and version: an anatomic study. J Shoulder Elbow Surg. 2001;10:327–32. doi:10.1067/mse.2001.115269.
10. Volk AG, Vangsness CT. An anatomic study of the supraspinatus muscle and tendon. Clin Orthop Relat Res. 2001;384:280–5.
11. DePalma AF. The classic. Surgical anatomy of the rotator cuff and the natural history of degenerative periarthritis. Surg Clin North Am. 1963;43:1507–1520. Clin Orthop Relat Res. 2008;466:543–51. doi:10.1007/s11999-007-0103-5.
12. Sharma P, Maffulli N. Tendon injury and tendinopathy: healing and repair. J Bone Joint Surg Am. 2005;87:187. doi:10.2106/JBJS.D.01850.
13. Wang JH-C. Mechanobiology of tendon. J Biomech. 2006;39:1563–82. doi:10.1016/j.jbiomech.2005.05.011.
14. Wang JH-C, Guo Q, Li B. Tendon biomechanics and mechanobiology–a minireview of basic concepts and recent advancements. J Hand Ther. 2012;25:133–40. doi:10.1016/j.jht.2011.07.004; quiz 141.
15. Levy O, Relwani J, Zaman T, Even T, Venkateswaran B, Copeland S. Measurement of blood flow in the rotator cuff using laser Doppler flowmetry. J Bone Joint Surg Br. 2008;90(7):893–8.
16. Llusá M, Merí A, Ruano D, Surgical atlas of the musculoskeletal system. Rosemont, IL: American Academy of Orthopaedic Surgeons; 2008. ix, 422 p.

17. Tubbs RS, Loukas M, Shahid K, Judge T, Pinyard J, Shoja MM, Slappey JB, McEvoy WC, Oakes WJ. Anatomy and quantitation of the subscapular nerves. Clin Anat. 2007;20:656–9. doi:10.1002/ca.20478.
18. Scibek JS, Mell AG, Downie BK, Carpenter JE, Hughes RE. Shoulder kinematics in patients with full-thickness rotator cuff tears after a subacromial injection. J Shoulder Elbow Surg. 2008;17:172–81. doi:10.1016/j.jse.2007.05.010.
19. Ludewig PM, Phadke V, Braman JP, Hassett DR, Cieminski CJ, LaPrade RF. Motion of the shoulder complex during multiplanar humeral elevation. J Bone Joint Surg Am. 2009;91:378–89. doi:10.2106/JBJS.G.01483.
20. Provencher CMT, Makani A, McNeil JW, Pomerantz ML, Golijanin P, Gross D. The Role of the Scapula in Throwing Disorders. Sports Med Arthrosc. 2014;22:80–7. doi:10.1097/JSA.0000000000000023.
21. Burkhart SS, Morgan CD, Kibler WB. The disabled throwing shoulder: spectrum of pathology Part III: the SICK scapula, scapular dyskinesis, the kinetic chain, and rehabilitation. Arthroscopy. 2003;19:641–61.
22. Talkhani IS, Kelly CP. Movement analysis of asymptomatic normal shoulders: a preliminary study. J Shoulder Elbow Surg. 2001;10:580–4. doi:10.1067/mse.2001.118481.
23. McClure PW, Michener LA, Sennett BJ, Karduna AR. Direct 3-dimensional measurement of scapular kinematics during dynamic movements in vivo. J Shoulder Elbow Surg. 2001;10:269–77. doi:10.1067/mse.2001.112954.
24. Bourne DA, Choo AMT, Regan WD, MacIntyre DL, Oxland TR. Three-dimensional rotation of the scapula during functional movements: an in vivo study in healthy volunteers. J Shoulder Elbow Surg. 2007;16:150–62. doi:10.1016/j.jse.2006.06.011.
25. Fung M, Kato S, Barrance PJ, Elias JJ, McFarland EG, Nobuhara K, Chao EY. Scapular and clavicular kinematics during humeral elevation: a study with cadavers. J Shoulder Elbow Surg. 2001;10:278–85. doi:10.1067/mse.2001.114496.
26. Graichen H, Stammberger T, Bonel H, Haubner M, Englmeier KH, Reiser M, Eckstein F. Magnetic resonance-based motion analysis of the shoulder during elevation. Clin Orthop Relat Res. 2000;370:154–63.
27. Massimini DF, Warner JJP, Li G. Non-invasive determination of coupled motion of the scapula and humerus–an in-vitro validation. J Biomech. 2011;44:408–12. doi:10.1016/j.jbiomech.2010.10.003.
28. Poppen NK, Walker PS. Normal and abnormal motion of the shoulder. J Bone Joint Surg Am. 1976;58:195–201.
29. Terrier A, Reist A, Vogel A, Farron A. Effect of supraspinatus deficiency on humerus translation and glenohumeral contact force during abduction. Clin Biomech (Bristol, Avon). 2007;22:645–51. doi:10.1016/j.clinbiomech.2007.01.015.
30. Warner JJ, McMahon PJ. The role of the long head of the biceps brachii in superior stability of the glenohumeral joint. J Bone Joint Surg Am. 1995;77:366–72.
31. Bassett RW, Browne AO, Morrey BF, An KN. Glenohumeral muscle force and moment mechanics in a position of shoulder instability. J Biomech. 1990;23:405–15.
32. Yu J, McGarry MH, Lee Y-S, Duong LV, Lee TQ. Biomechanical effects of supraspinatus repair on the glenohumeral joint. J Shoulder Elbow Surg. 2005;14:65S–71. doi:10.1016/j.jse.2004.09.019.
33. Halder AM, O'Driscoll SW, Heers G, Mura N, Zobitz ME, An KN, Kreusch-Brinker R. Biomechanical comparison of effects of supraspinatus tendon detachments, tendon defects, and muscle retractions. J Bone Joint Surg Am. 2002;84-A:780–5.
34. Nakajima T, Hughes RE, An KN. Effects of glenohumeral rotations and translations on supraspinatus tendon morphology. Clin Biomech (Bristol, Avon). 2004;19:579–85. doi:10.1016/j.clinbiomech.2004.02.007.
35. Escamilla RF, Yamashiro K, Paulos L, Andrews JR. Shoulder muscle activity and function in common shoulder rehabilitation exercises. Sports Med. 2009;39:663–85. doi:10.2165/00007256-200939080-00004.
36. Kuechle DK, Newman SR, Itoi E, Niebur GL, Morrey BF, An KN. The relevance of the moment arm of shoulder muscles with respect to axial rotation of the glenohumeral joint in four positions. Clin Biomech (Bristol, Avon). 2000;15:322–9.
37. Apreleva M, Parsons IM, Warner JJ, Fu FH, Woo SL. Experimental investigation of reaction forces at the glenohumeral joint during active abduction. J Shoulder Elbow Surg. 2000;9:409–17. doi:10.1067/mse.2000.106321.
38. Cain PR, Mutschler TA, Fu FH, Lee SK. Anterior stability of the glenohumeral joint. A dynamic model. Am J Sports Med. 1987;15:144–8.
39. Labriola JE, Lee TQ, Debski RE, McMahon PJ. Stability and instability of the glenohumeral joint: the role of shoulder muscles. J Shoulder Elbow Surg. 2005;14:32S–8. doi:10.1016/j.jse.2004.09.014.
40. Howell SM, Galinat BJ, Renzi AJ, Marone PJ. Normal and abnormal mechanics of the glenohumeral joint in the horizontal plane. J Bone Joint Surg Am. 1988;70:227–32.
41. Poppen NK, Walker PS. Forces at the glenohumeral joint in abduction. Clin Orthop Relat Res. 1978;135:165–70.
42. Blonna D, Cecchetti S, Tellini A, Bonasia DE, Rossi R, Southgate R, Castoldi F. Contribution of the supraspinatus to the external rotator lag sign: kinematic and electromyographic pattern in an in vivo model. J Shoulder Elbow Surg. 2010;19:392–8. doi:10.1016/j.jse.2009.10.007.
43. Kedgley AE, Shore BJ, Athwal GS, Johnson JA, Faber KJ. An in-vitro study of rotator cuff tear and repair kinematics using single- and double-row suture anchor fixation. Int J Shoulder Surg. 2013;7:46–51. doi:10.4103/0973-6042.114224.
44. Andarawis-Puri N, Kuntz AF, Kim S-Y, Soslowsky LJ. Effect of anterior supraspinatus tendon

partial-thickness tears on infraspinatus tendon strain through a range of joint rotation angles. J Shoulder Elbow Surg. 2010;19:617–23. doi:10.1016/j.jse.2009.10.003.
45. Kelly BT, Williams RJ, Cordasco FA, Backus SI, Otis JC, Weiland DE, Altchek DW, Craig EV, Wickiewicz TL, Warren RF. Differential patterns of muscle activation in patients with symptomatic and asymptomatic rotator cuff tears. J Shoulder Elbow Surg. 2005;14:165–71. doi:10.1016/j.jse.2004.06.010.
46. Chen SK, Simonian PT, Wickiewicz TL, Otis JC, Warren RF. Radiographic evaluation of glenohumeral kinematics: a muscle fatigue model. J Shoulder Elbow Surg. 1999;8:49–52.
47. Teyhen DS, Miller JM, Middag TR, Kane EJ. Rotator cuff fatigue and glenohumeral kinematics in participants without shoulder dysfunction. J Athl Train. 2008;43:352–8. doi:10.4085/1062-6050-43.4.352.
48. Itoi E, Berglund LJ, Grabowski JJ, Schultz FM, Growney ES, Morrey BF, An KN. Tensile properties of the supraspinatus tendon. J Orthop Res. 1995;13:578–84. doi:10.1002/jor.1100130413.
49. Mazoue CG, Andrews JR. Repair of full-thickness rotator cuff tears in professional baseball players. Am J Sports Med. 2006;34(2):182–9.
50. Werner CM, Weishaupt D, Blumenthal S, Curt A, Favre P, Gerber C. Effect of experimental suprascapular nerve block on active glenohumeral translations in vivo. J Orthop Res. 2006;24(3):491–500.
51. Entezari V, Trechsel BL, Dow WA, Stanton SK, Rosso C, Müller A, McKenzie B, Vartanians V, Cereatti A, Della Croce U, Deangelis JP, Ramappa AJ, Nazarian A. Design and manufacture of a novel system to simulate the biomechanics of basic and pitching shoulder motion. Bone Joint Res. 2012;1(5):78–85.
52. Mueller AM, Rosso C, Entezari V, McKenzie B, Hasebroock A, Della Croce U, Nazarian A, Ramappa AJ, DeAngelis JP. The effect of supraspinatus tears on glenohumeral translations in passive pitching motion. Am J Sports Med. 2014;42(10):2455–62.
53. Jensen KL, Williams GR, Russell IJ, Rockwood CA. Rotator cuff tear arthropathy. J Bone Joint Surg Am. 1999;81:1312–24.
54. Yamaguchi K, Sher JS, Andersen WK, Garretson R, Uribe JW, Hechtman K, Neviaser RJ. Glenohumeral motion in patients with rotator cuff tears: a comparison of asymptomatic and symptomatic shoulders. J Shoulder Elbow Surg. 2000;9:6–11.
55. Shibuta H, Tamai K, Tabuchi K. Magnetic resonance imaging of the shoulder in abduction. Clin Orthop Relat Res. 1998;348:109–13.
56. Bigliani LU. The morphology of the acromion and its relationship to rotator cuff tears. Orthop Trans. 1986;10:228.
57. Balke M, Schmidt C, Dedy N, Banerjee M, Bouillon B, Liem D. Correlation of acromial morphology with impingement syndrome and rotator cuff tears. Acta Orthop. 2013;84:178–83. doi:10.3109/17453674.2013.773413.
58. Gerber C, Snedeker JG, Baumgartner D, Viehöfer AF. Supraspinatus tendon load during abduction is dependent on the size of the critical shoulder angle: a biomechanical analysis. J Orthop Res. 2014;32:952–7. doi:10.1002/jor.22621.
59. Luo ZP, Hsu HC, Grabowski JJ, Morrey BF, An KN. Mechanical environment associated with rotator cuff tears. J Shoulder Elbow Surg. 1998;7:616–20.
60. Seitz AL, McClure PW, Finucane S, Boardman ND, Michener LA. Mechanisms of rotator cuff tendinopathy: intrinsic, extrinsic, or both? Clin Biomech (Bristol, Avon). 2011;26:1–12. doi:10.1016/j.clinbiomech.2010.08.001.

Imaging of the Normal Rotator Cuff

23

Eiji Itoi, Shin Hitachi, and Nobuyuki Yamamoto

23.1 Imaging Modalities

The x-rays do not depict the cuff tendons except when there is a calcified deposit in the tendon. CT is good at visualizing the bony structures, but not the soft tissues. Among various imaging modalities, MRI and ultrasound are the most appropriate modalities to visualize the rotator cuff tendons and muscles. A recent meta-analysis showed that the diagnostic accuracy of ultrasound was equivalent to that of MRI or MR arthrography in making a diagnosis of full-thickness rotator cuff tears [2]. For full-thickness tears, the summary sensitivity and specificity were 94 and 93 %, respectively, for MRI (7 studies, 368 shoulders); 94 and 92 %, respectively, for MR arthrography (3 studies, 183 shoulders); and 92 and 93 %, respectively, for ultrasound (10 studies, 729 shoulders). However, for partial-thickness tears, both MRI and ultrasound had poor sensitivity. The sensitivity of ultrasound was much lower than that of MRI. High-resolution MRI using a microscopy coil is one solution to improve the diagnostic accuracy of MRI without using an invasive procedure [1]. It showed higher sensitivity than conventional MRI and equivalent sensitivity to MR arthrography for diagnosing partial-thickness tears.

23.2 Rotator Cuff MRI

Since the scapula, from which all the rotator cuff muscles originate, is located oblique to the coronal plane, MR images in the oblique coronal, oblique sagittal, and axial slices are useful for visualization of the rotator cuff tendons and muscles. The oblique coronal images are along the orientation of the longitudinal axis of the tendons, whereas the sagittal oblique images are perpendicular to the oblique coronal images. Thus, a combination of the oblique coronal and oblique sagittal images is suitable for imaging the superior portions of the rotator cuff such as the supraspinatus tendon and the superior portion of the infraspinatus tendon, whereas a combination of the axial and sagittal oblique images is suitable for imaging the anterior and posterior portions of the rotator cuff such as the inferior portion of the infraspinatus tendon, teres minor, and the subscapularis tendons.

The most anterior coronal oblique image depicts the intramuscular tendons of the subscapularis, which run horizontally in the muscle belly and converge to the subscapularis tendon, and the conjoint tendon, which runs vertically from the tip of the coracoid process (Fig. 23.1). The next slice depicts the anterior portion of the supraspinatus tendon with a thick single intramuscular tendon

E. Itoi, MD, PhD (✉) • N. Yamamoto, MD, PhD
Department of Orthopaedic Surgery, Tohoku University School of Medicine, Sendai, Japan
e-mail: itoi-eiji@med.tohoku.ac.jp

S. Hitachi, MD, PhD
Department of Diagnostic Radiology, Tohoku University School of Medicine, Sendai, Japan

G.I. Bain et al. (eds.), *Normal and Pathological Anatomy of the Shoulder*,
DOI 10.1007/978-3-662-45719-1_23, © ISAKOS 2015

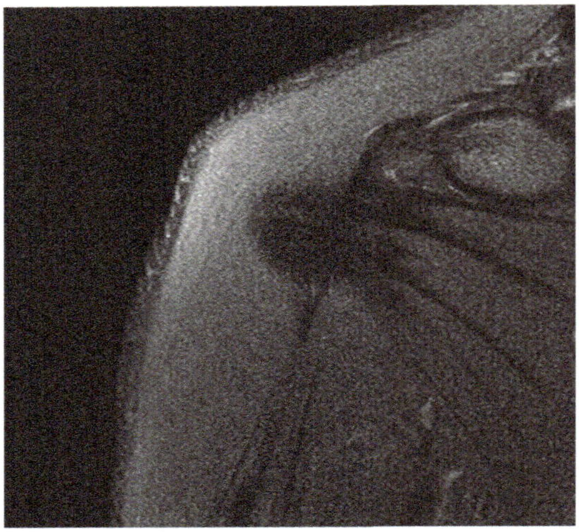

Fig. 23.1 Fast spin-echo T2-weighted oblique coronal image 1. This anterior slice depicts the subscapularis tendon with its several intramuscular tendons spanning into the muscle belly. Above the subscapularis tendon is the coracoid process with the conjoint tendon running vertically

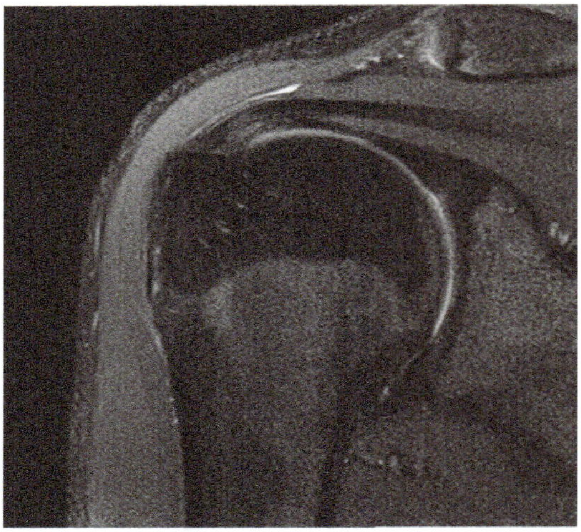

Fig. 23.2 Fast spin-echo T2-weighted oblique coronal image 2. This middle slice depicts the anterior portion of the supraspinatus tendon with a thick single intramuscular tendon (dark band). It attaches to the superior facet of the greater tuberosity, which looks horizontal on this image. There is a small amount of fluid in the glenohumeral joint, acromioclavicular joint, and the subacromial bursa. The mid portion of the glenoid is also depicted on this image

(dark band) (Fig. 23.2). It attaches to the superior facet of the greater tuberosity, which appears as a horizontal facet on top of the greater tuberosity on

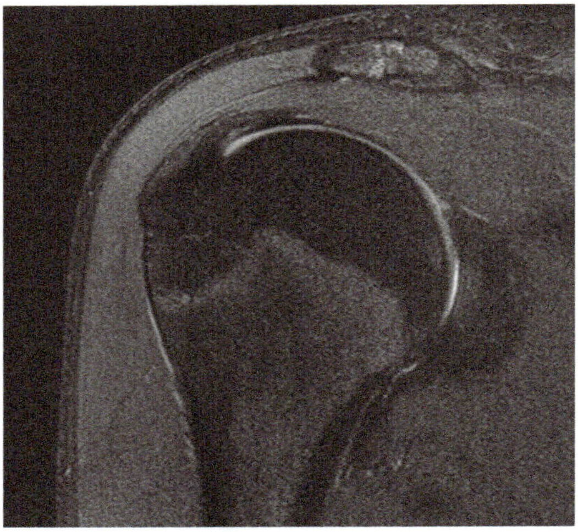

Fig. 23.3 Fast spin-echo T2-weighted oblique coronal image 3. This middle posterior image depicts the superior portion of the infraspinatus tendon that attaches to the middle facet of the greater tuberosity, which looks slightly inclined on this image. The intramuscular tendon (dark band) is observed on the articular side of the tendon. The posterior portion of the glenoid is also depicted on this image

this image. The next slice reveals the superior portion of the infraspinatus tendon, which attaches to the middle facet of the greater tuberosity that appears as a slightly inclined facet on this image (Fig. 23.3). The intramuscular tendon (dark band) is observed on the articular side of the tendon. The most posterior slice shows the posterior portion of the humeral head with the superior portion of the infraspinatus tendon superiorly and the teres minor tendon inferiorly (Fig. 23.4).

The most lateral slice of the sagittal oblique view shows the greater tuberosity with distal cross section of the supraspinatus tendon superiorly, infraspinatus tendon posterosuperiorly, and teres minor tendon posteriorly (Fig. 23.5). More medially, the lesser tuberosity with the cross section of the subscapularis tendon is observed (Fig. 23.6). On this image, the muscle bellies of the supraspinatus, infraspinatus, and teres minor are observed with their intramuscular tendons as dark signal areas inside the muscle belly. The medial slice at the level of the glenoid surface depicts the cross sections of all the rotator cuff muscles with their intramuscular tendons located in the middle of the muscle bellies (Fig. 23.7).

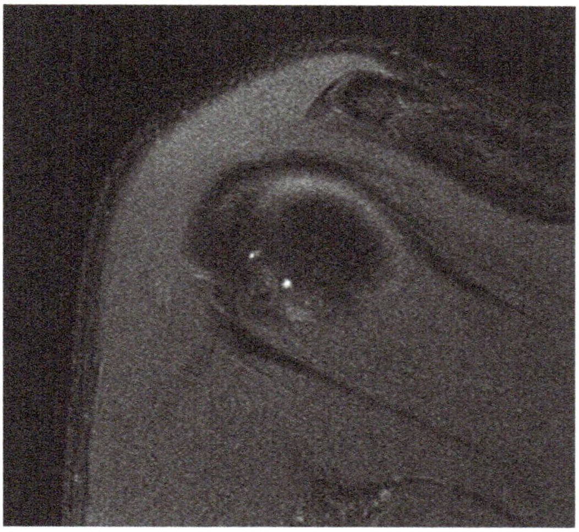

Fig. 23.4 Fast spin-echo T2-weighted oblique coronal image 4. This posterior slice shows the posterior portion of the humeral head with the superior portion of the infraspinatus tendon superiorly and the teres minor tendon inferiorly. There are two small bony cysts (high signal) on the posterior aspect of the humeral head

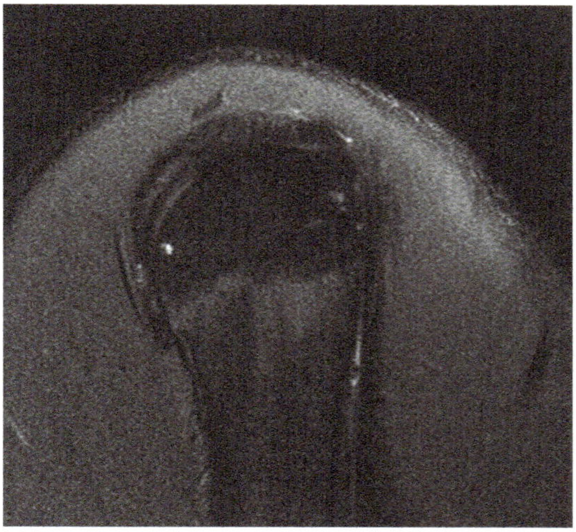

Fig. 23.5 Fast spin-echo T2-weighted oblique sagittal image 1. This lateral slice shows the sagittal view of the greater tuberosity with distal cross sections of the supraspinatus tendon superiorly, the infraspinatus tendon posterosuperiorly, and the teres minor tendon posteriorly. Anteriorly there runs the tendon of the long head of the biceps vertically along the bicipital groove. A small bony cyst is depicted close to the surface of the posterior aspect of the humeral head

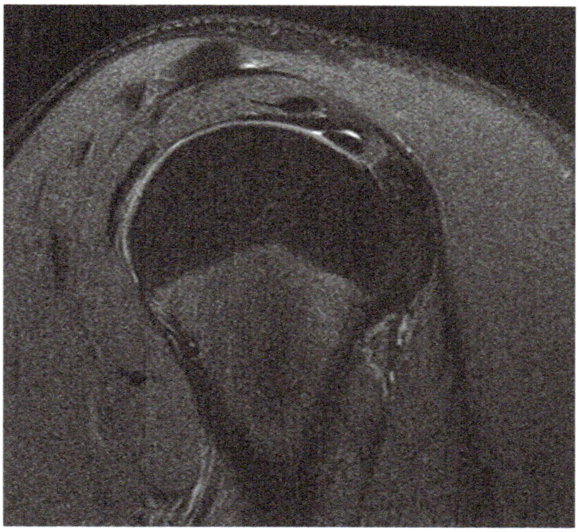

Fig. 23.6 Fast spin-echo T2-weighted oblique sagittal image 2. More medially, the lesser tuberosity with the cross section of the subscapularis tendon is observed. On this image, the long head of the biceps tendon is located in the rotator interval with the coracohumeral ligament. The muscle bellies of the supraspinatus, infraspinatus, and teres minor are observed with their intramuscular tendons as dark signal areas in the muscle belly

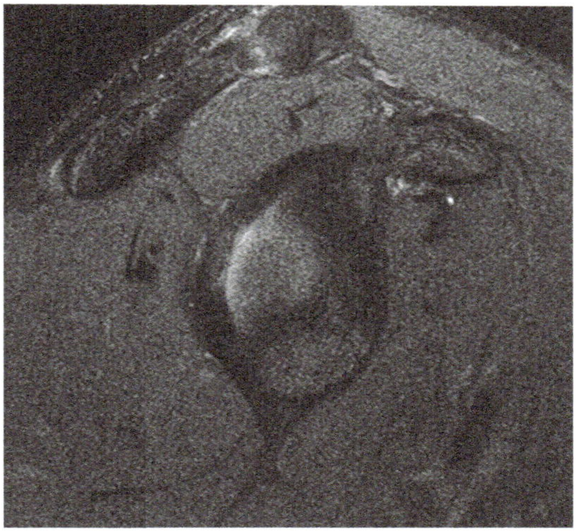

Fig. 23.7 Fast spin-echo T2-weighted oblique sagittal image 3. This medial slice is at the level of the glenoid surface with the surrounding labrum. The hooked coracoid process is observed anterosuperiorly. All the rotator cuff muscles with their intramuscular tendons are clearly depicted on this image

This axial slice at the level of the superior glenoid depicts the subscapularis tendon attaching to the lesser tuberosity anteriorly, and the infraspinatus tendon to the greater tuberosity posteriorly (Fig. 23.8). The intramuscular tendons of the posterior deltoid, which run perpendicular to this plane, are also visible on this image.

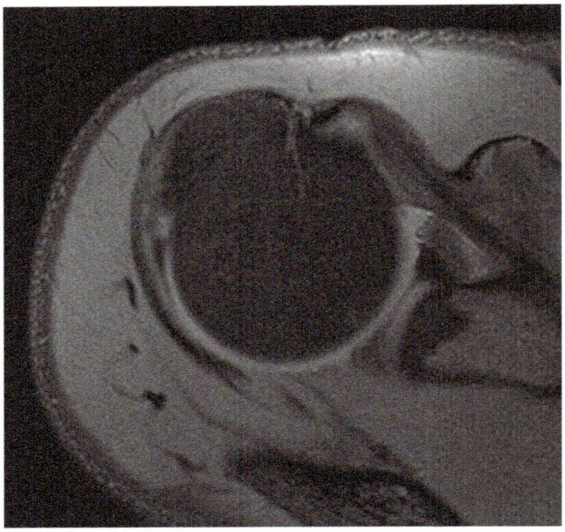

Fig. 23.8 Fast spin-echo proton density-weighted axial image. This slice is at the level of the superior glenoid with the labrum attached. The subscapularis tendon attaches to the lesser tuberosity anteriorly, and the infraspinatus tendon attaches to the greater tuberosity posteriorly. The subscapularis tendon runs between the glenoid and the coracoid process. The intramuscular tendons of the posterior deltoid are clearly visible on this image

23.3 Rotator Cuff Ultrasound

These days, many orthopedic doctors use ultrasound at the outpatient clinic. It is very convenient and useful if you have an ultrasound device in your clinic and are able to handle the device by yourself because (1) you do not need to make a reservation for the examination by an ultrasonographer and (2) a real-time, dynamic monitoring makes it easy not only for the doctors but also for the patients to better understand the normal and pathologic anatomy. The normal tendon shows the fibrillar structure, observed as multiple, closely spaced echogenic parallel lines on longitudinal scanning (Figs. 23.9, 23.10, 23.11, and 23.12). Under pathologic conditions, loss of normal fibrillar structure is commonly observed with increased spacing between the hyperechoic lines and reduced echogenicity, followed by a swelling or disruption of the tendon.

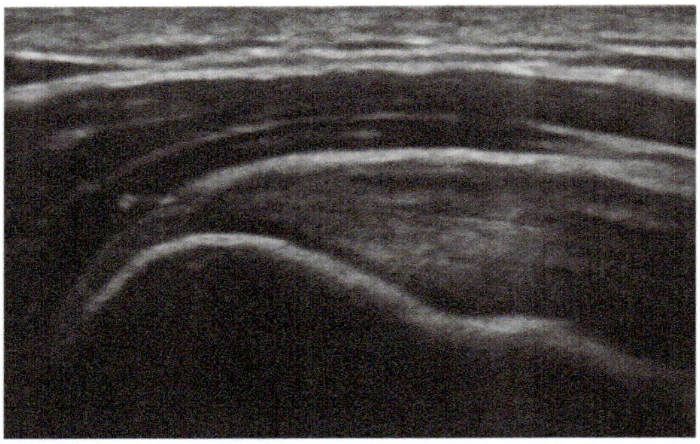

Fig. 23.9 Ultrasound long axis view of the supraspinatus tendon. The supraspinatus tendon appears as a homogeneous band with fibrillar structure running parallel. The tendon attaches to the superior facet of the greater tuberosity. The tendon is covered by the subacromial bursa and the deltoid muscle

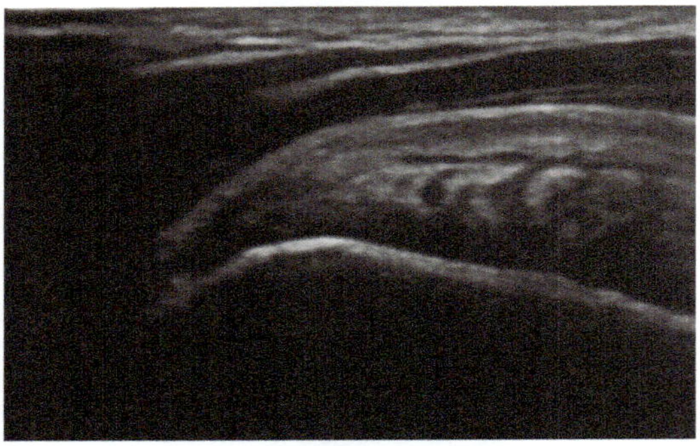

Fig. 23.10 Ultrasound long axis view of the infraspinatus tendon. The infraspinatus tendon attaches to the middle facet of the greater tuberosity. The superficial tendon shows the fibrillar structure, whereas the deep tendon shows an interrupted pattern. This is because the superficial tendon fibers run parallel to the longitudinal axis, but the deep tendon fibers run obliquely at this portion of the infraspinatus tendon. The tendon is covered by the posterior deltoid muscle

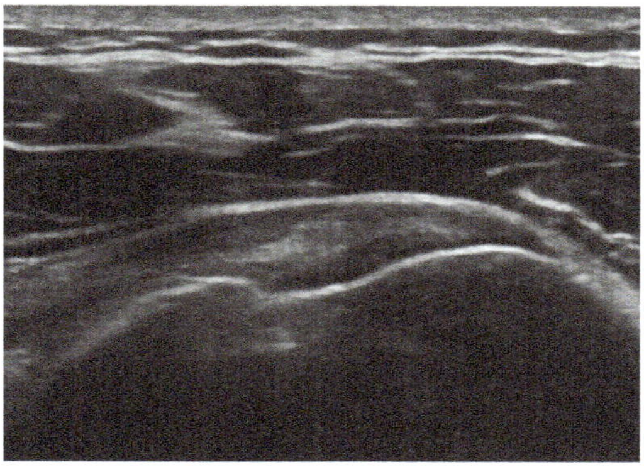

Fig. 23.11 Ultrasound long axis view of the teres minor tendon. The teres minor tendon, a little thinner than the infraspinatus tendon, attaches to the inferior facet of the greater tuberosity. The fibrillar structure is clearly visible

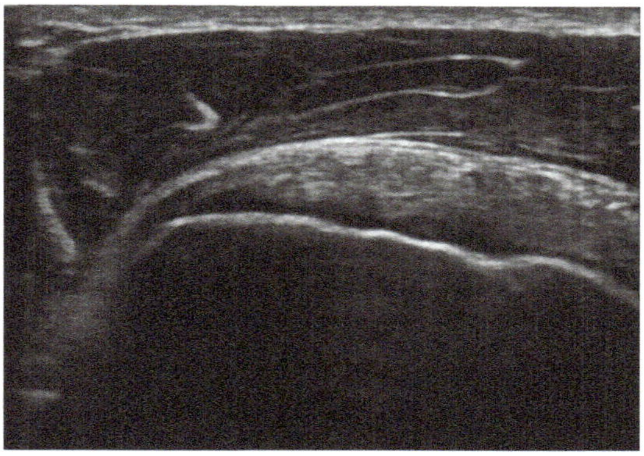

Fig. 23.12 Ultrasound long axis view of the subscapularis tendon. The subscapularis tendon is located anterior to the shoulder joint and attaches to the lesser tuberosity. Here again, the fibrillar structure is clearly visible

References

1. Hitachi S, Takase K, Tanaka M, et al. High-resolution magnetic resonance imaging of rotator cuff tears using a microscopy coil: noninvasive detection without intraarticular contrast material. Jpn J Radiol. 2011;29(7):466–74.
2. Lenza M, Buchbinder R, Takwoingi Y, Johnston RV, Hanchard NC, Faloppa F. Magnetic resonance imaging, magnetic resonance arthrography and ultrasonography for assessing rotator cuff tears in people with shoulder pain for whom surgery is being considered. Cochrane Database Syst Rev. 2013;(9):CD009020.

Rotator Cuff Pathology: A Comparison of Magnetic Resonance Imaging and Arthroscopic Findings

24

Brian B. Gilmer and Dan Guttmann

24.1 Osseous Anatomy

While evidence is limited, osseous impingement may be implicated in the pathogenesis of rotator cuff tears.

24.1.1 Subcoracoid Impingement

Subacromial impingement occurs when a prominent lateral extension of the coracoid process abuts the lesser tuberosity in internal rotation, theoretically resulting in insertional tears of the subscapularis tendon [8]. Burkhart et al. have described subcoracoid stenosis as a coracohumeral distance of 6 mm or less, which can be measured intraoperatively or on axial magnetic resonance imaging (MRI) (Fig. 24.1). The example illustrates a reduced coracohumeral distance and a corresponding tear of the subscapularis noted at time of arthroscopy.

B.B. Gilmer, MD
Orthopedics and Sports Medicine, Mammoth Orthopedic Institute, Mammoth Lakes, CA, USA
e-mail: bbgilmer@gmail.com

D. Guttmann, MD (✉)
Department of Orthopaedic Surgery, Taos Orthopaedic Institute, Taos, NM, USA
e-mail: drg@taosortho.com

24.1.2 Subacromial Impingement

Subacromial impingement has likewise been implicated in the pathogenesis of tears of the supraspinatus tendon. The Bigliani classification, previously described in this text, demonstrates the variations in morphology most commonly evaluated on a scapular-Y radiograph [1].

24.2 Rotator Cuff Tears

Well-designed, reproducible, validated classification systems that are clinically relevant are useful for understanding the spectrum of pathology encountered in the treatment of rotator cuff tears. It should be noted that some common shoulder classification systems do not demonstrate high inter-rater agreement, and that classifications do not capture all tear patterns or components [7]. Nonetheless, the use of existing systems is helpful for the purposes of understanding anatomy and severity.

Rotator cuff tears are often described by their corresponding muscle belly. It is important to remember, however, that the rotator cuff is a confluence of its four constituent tendons and tears frequently extend beyond the margins of a single tendon. Frequently, sagittal images near the cuff insertion tendon on can demonstrate the segment involved in a given tear pattern.

G.I. Bain et al. (eds.), *Normal and Pathological Anatomy of the Shoulder*,
DOI 10.1007/978-3-662-45719-1_24, © ISAKOS 2015

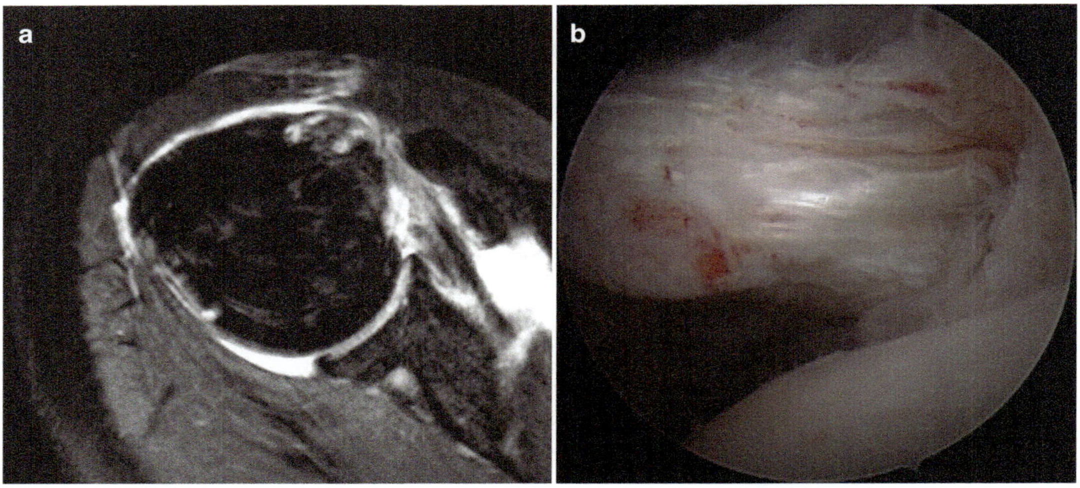

Fig. 24.1 (**a**) An axial non-contrast MR image demonstrates a prominent lateral extension of the coracoid. (**b**) Intraoperative image showing tearing of the subscapularis tendon in the same patient

Table 24.1 Fox and Romeo classification of subscapularis tears

Type 1: partial thickness tears
Type 2: complete tear of upper 25 % of tendon
Type 3: complete tear of upper 50 % of tendon
Type 4: complete rupture of tendon

24.2.1 Subscapularis

Subacromial impingement tendon pathology is easily overlooked unless careful arthroscopic examination is performed and a high index of suspicion maintained. Though no validated classification system for these tears currently exists in the literature, Fox and Romeo described a system for describing these lesions (Table. 24.1) [4].

Tears of the subscapularis frequently occur at its leading, superior border. Therefore, when evaluating these lesions on MRI it is important to closely analyze the axial images immediately below the level of the coracoid as this often reveals subtle tears of the superior margin of the tendon. Figure 24.2 illustrates axial MR images and the corresponding intraoperative arthroscopic images in the same patient.

24.2.2 Partial-Thickness Tears

Partial-thickness of the rotator cuff may involve the articular or bursal side of the tendon. The diagnosis on is frequently more difficult than for full-thickness tears. While all imaging sequences should be closely analyzed for evidence of partial-thickness lesions, the coronal images, viewed in series, are often the most useful. The Ellman system (Fig. 24.3) of partial-thickness tears divides lesions into articular or bursal sided and determines grade by depth of tear and exposed footprint [3]. Grade 1 lesions are less than 3 mm deep, grade 2 lesions are 3–6 mm deep, and grade 3 lesions are greater than 6 mm deep. Figures 24.4 and 24.5 demonstrate articular-sided and bursal-sided partial-thickness tears, respectively, on non-contrast MRI with the associated intraoperative arthroscopic images from the same patient.

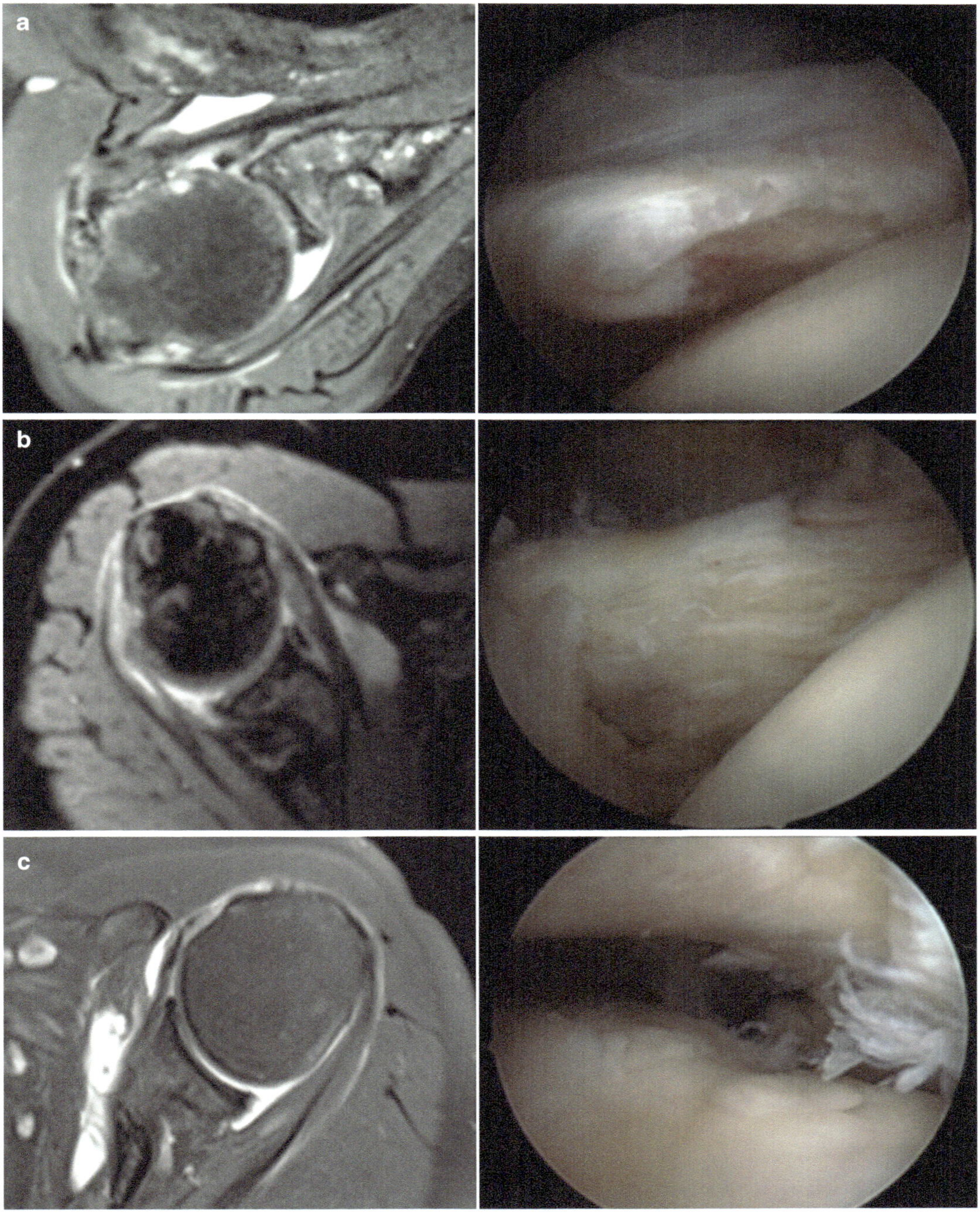

Fig. 24.2 Axial non-contrast MR image with corresponding intraoperative arthroscopic image in the same patient. (**a**) Type 1 lesions represent partial-thickness tears without complete tearing at any level. (**b**) Type 2 lesions represent complete tearing of upper 25 % of tendon. (**c**) Type 4 lesions represent complete rupture of the tendon

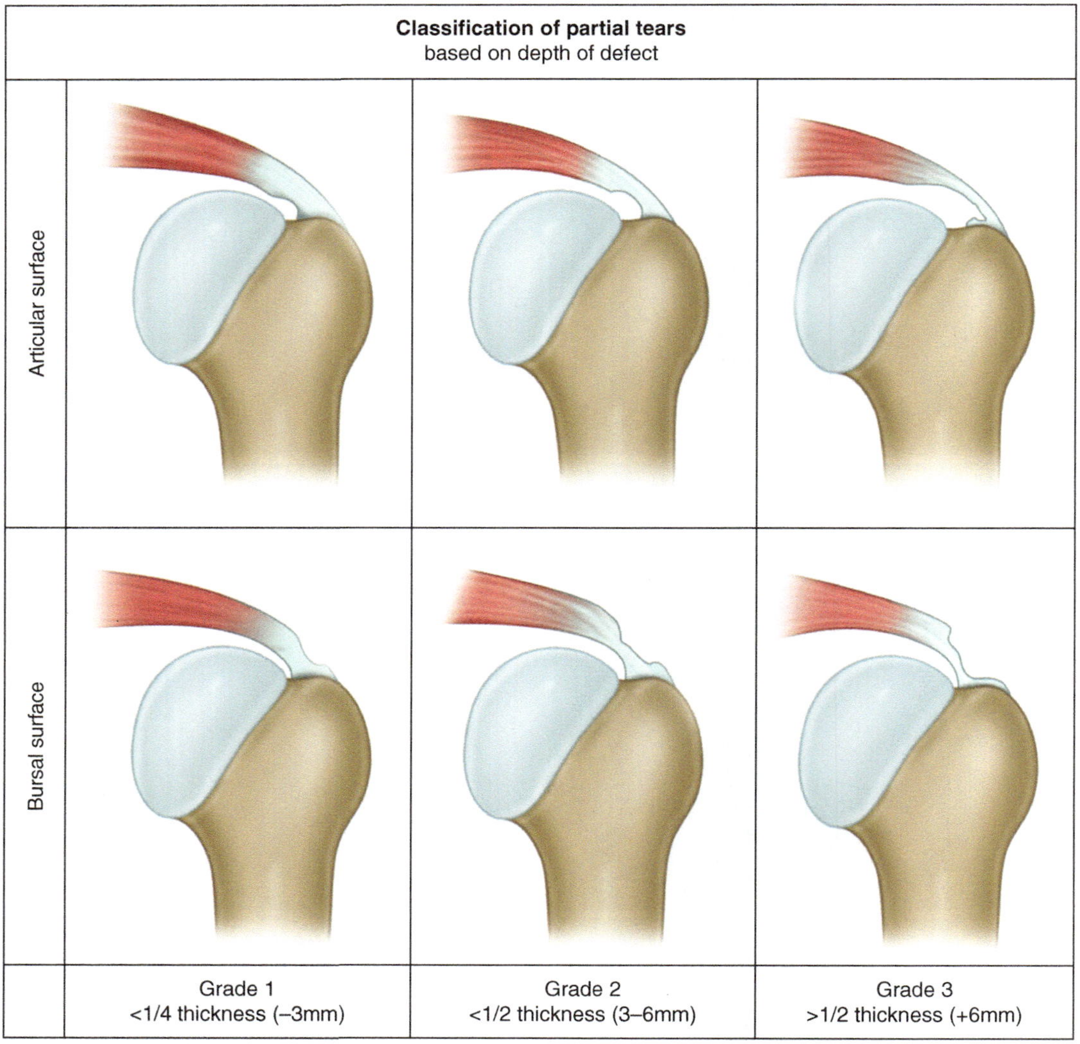

Fig. 24.3 The Ellman classification of partial-thickness rotator cuff tears [3]

24.2.3 Full-Thickness Tears

Full-thickness are perhaps the most common pathology encountered surgically. Symptomatic full-thickness tears are a frequent indication for operative intervention. The pattern of the tear has important implications when planning rotator cuff repairs arthroscopically. The Ellman and Gartsman (Fig. 24.6) classification of full-thickness rotator cuff tears describes the commonly encountered patterns [2]. In most cases, multiple images and sequences are required to fully determine tear pattern. The representative MR images illustrate these patterns and their corresponding intraoperative arthroscopic images (Fig. 24.7).

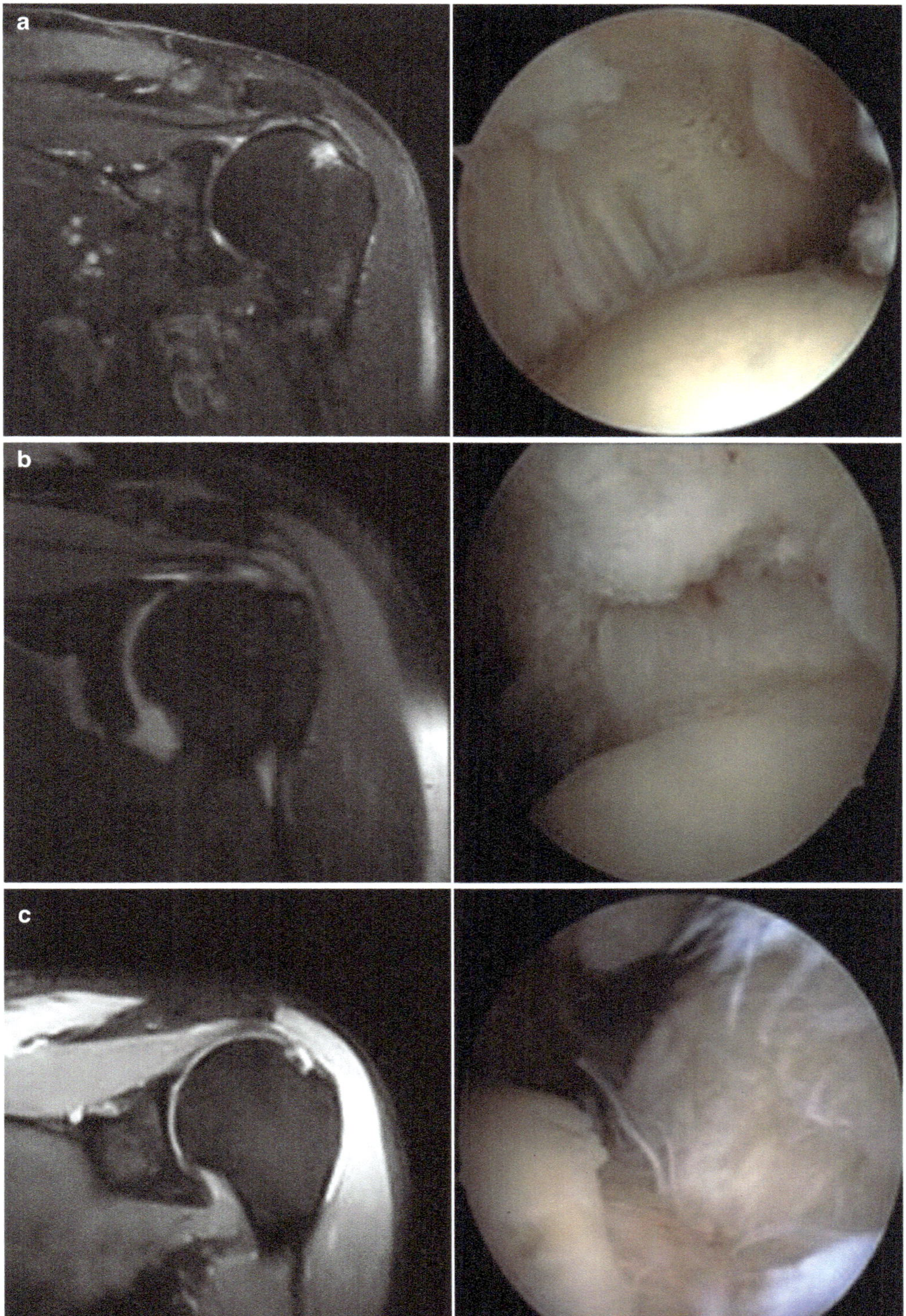

Fig. 24.4 Coronal non-contrast MR image with corresponding intraoperative arthroscopic image in the same patient. (**a**) Grade 1 articular-sided lesion is less than 3 mm deep. (**b**) Grade 2 articular-sided lesion is 3–6 mm deep. (**c**) Grade 3 articular-sided lesion is greater than 6 mm deep

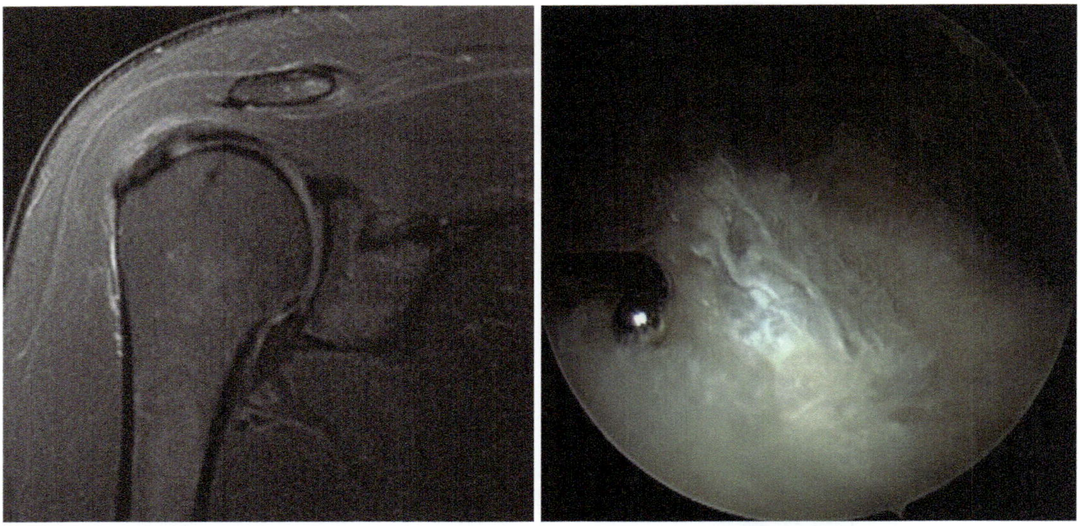

Fig. 24.5 Non-contrast MR image of a grade 1 bursal-sided lesion that is less than 3 mm deep and corresponding intraoperative arthroscopic image in the same patient

24.3 Other Factors Effecting Prognosis

24.3.1 Retraction

For any given pattern, the degree of retraction has important implications for the technical ease of repair and prognosis for healing. More retracted tears often represent more severe or in some cases chronic pathology and as a result may be under more tension after operative fixation.

Several classification systems have been developed to grade the retraction of full-thickness rotator cuff tears. One simple, practical, and clinically relevant classification that yields useful information about the reparability is the Patte classification [10]. In this system, retraction is evaluated on the coronal view and is staged as follows: Stage 1 – proximal stump is close to bony insertion, Stage 2 – proximal stump at level of the humeral head, and Stage 3 – proximal stump at level of the glenoid. Figure 24.8 illustrates the three stages of the Patte classification of retraction with representative MR images and corresponding arthroscopic images in the same patient.

24.3.2 Fatty Atrophy

Fatty and atrophy occur with increasing chronicity of a rotator cuff tear. The Goutallier classification of fatty infiltration and atrophy is commonly used in clinical practice. While initially described based on CT imaging, the classification system has later been validated for MRI [5, 6, 11]. Fatty and atrophy are assessed on the sagittal view and are graded 0–4 based on severity of fatty infiltration and atrophy of the muscle. Fatty and atrophy are prognostic since Goutallier stage 3 and 4 changes have been shown to have inferior clinical outcomes [9]. Figure 24.9 illustrates examples of each stage on sagittal.

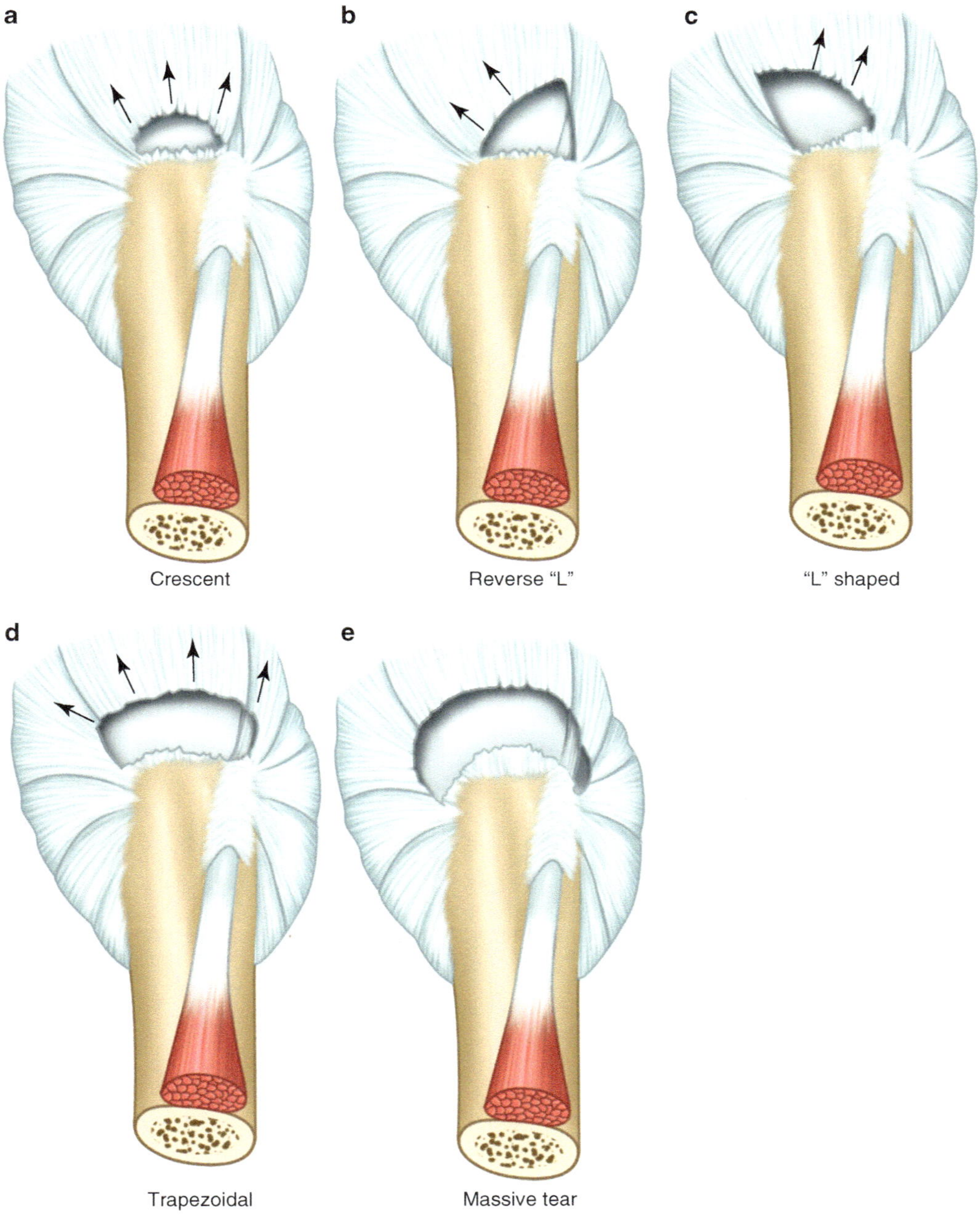

Fig. 24.6 The Ellman and Gartsman classification of full-thickness rotator cuff tears [2]

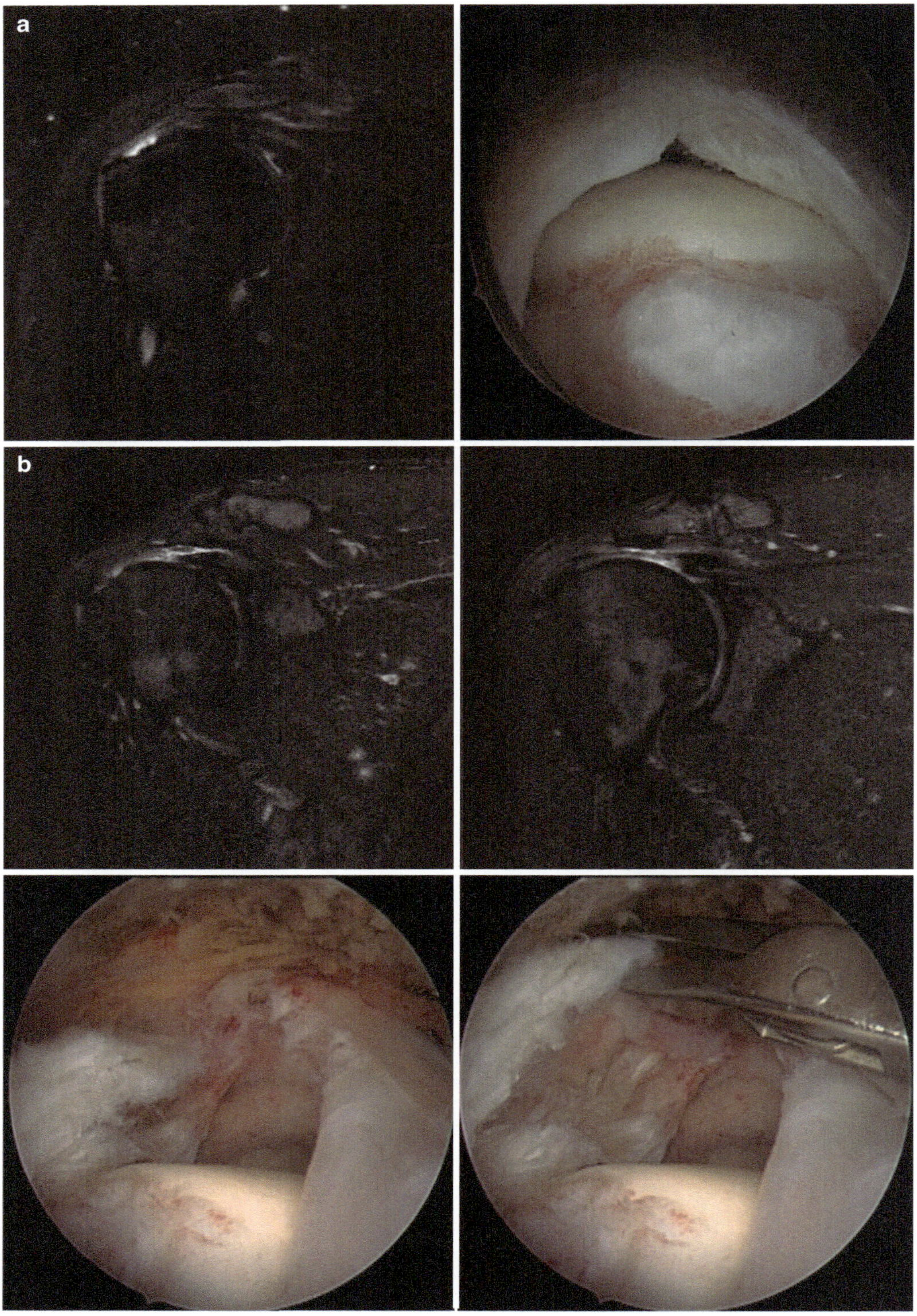
a
b

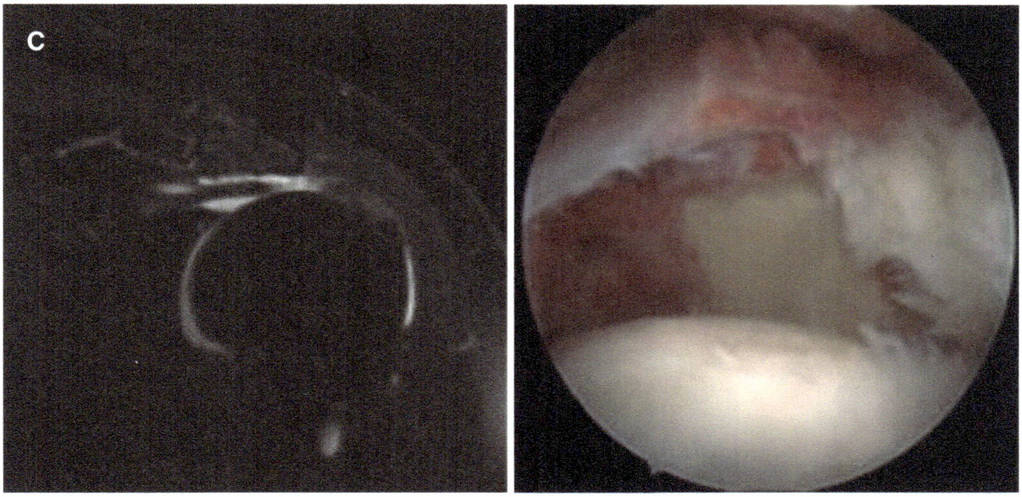

Fig. 24.7 (continued)

Fig. 24.7 Non-contrast MR image with corresponding intraoperative arthroscopic image in the same patient: (**a**) Crescent tear and (**b**) Reverse "L" tear. Note that the tear appears larger on the first, more anterior MR image, then appears smaller in the second, more posterior image. The arthroscopic images demonstrate that the primary fibers of the anterior cuff are retracted posteriorly as indicated by examination with an arthroscopic grasper (**c**) L-shaped tear. Tear appears larger posteriorly and smaller in more anterior images. Arthroscopically, the anterior fibers are more mobile and reduce to the intact posterior fibers. (**d**) Trapezoidal tear. Two coronal images demonstrate relatively even tearing of anterior and posterior fibers. Sagittal image demonstrates tearing of anterior and posterior fibers. Arthroscopic image confirms trapezoidal pattern. (**e**) Massive tear

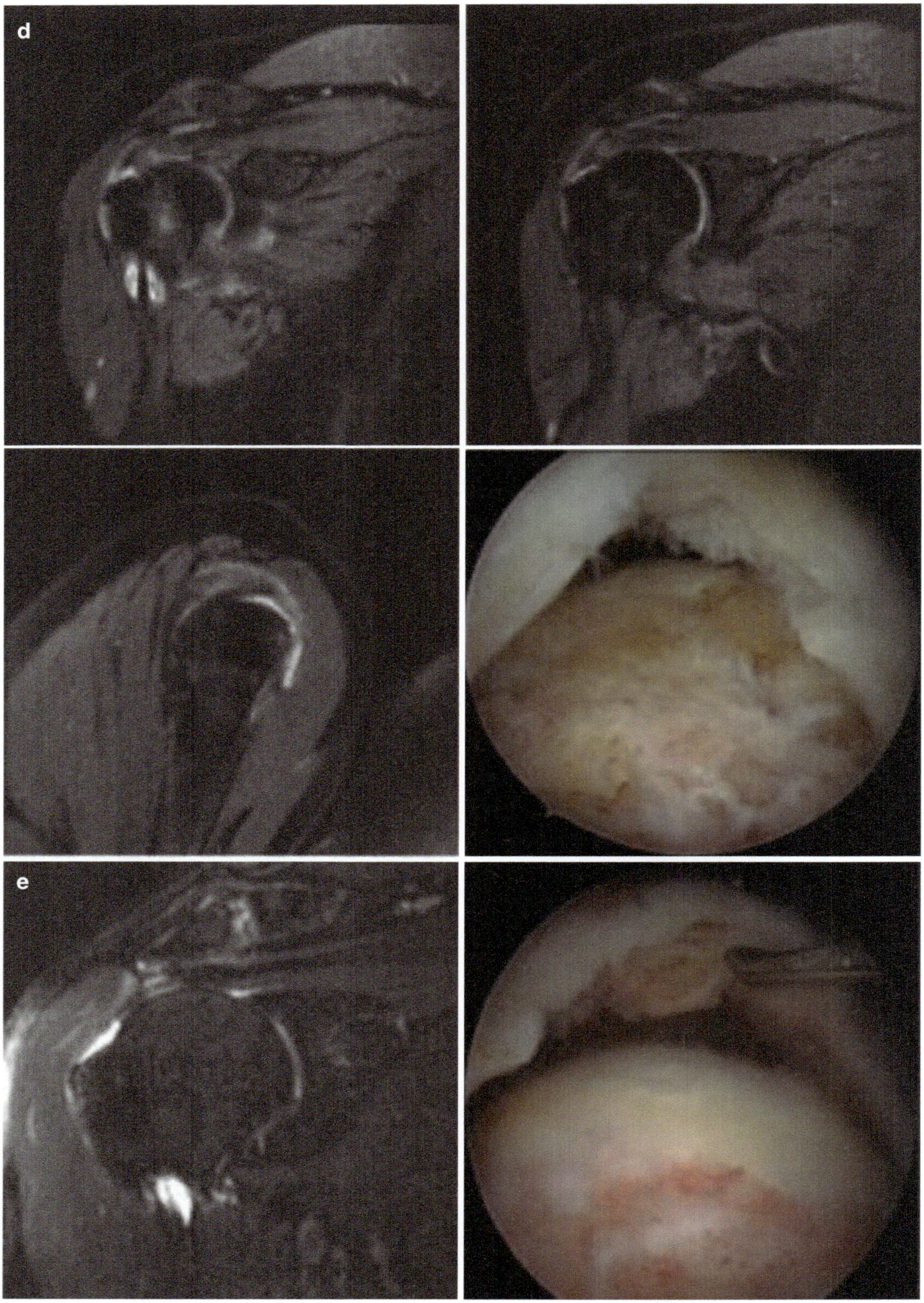

Fig. 24.7 (continued)

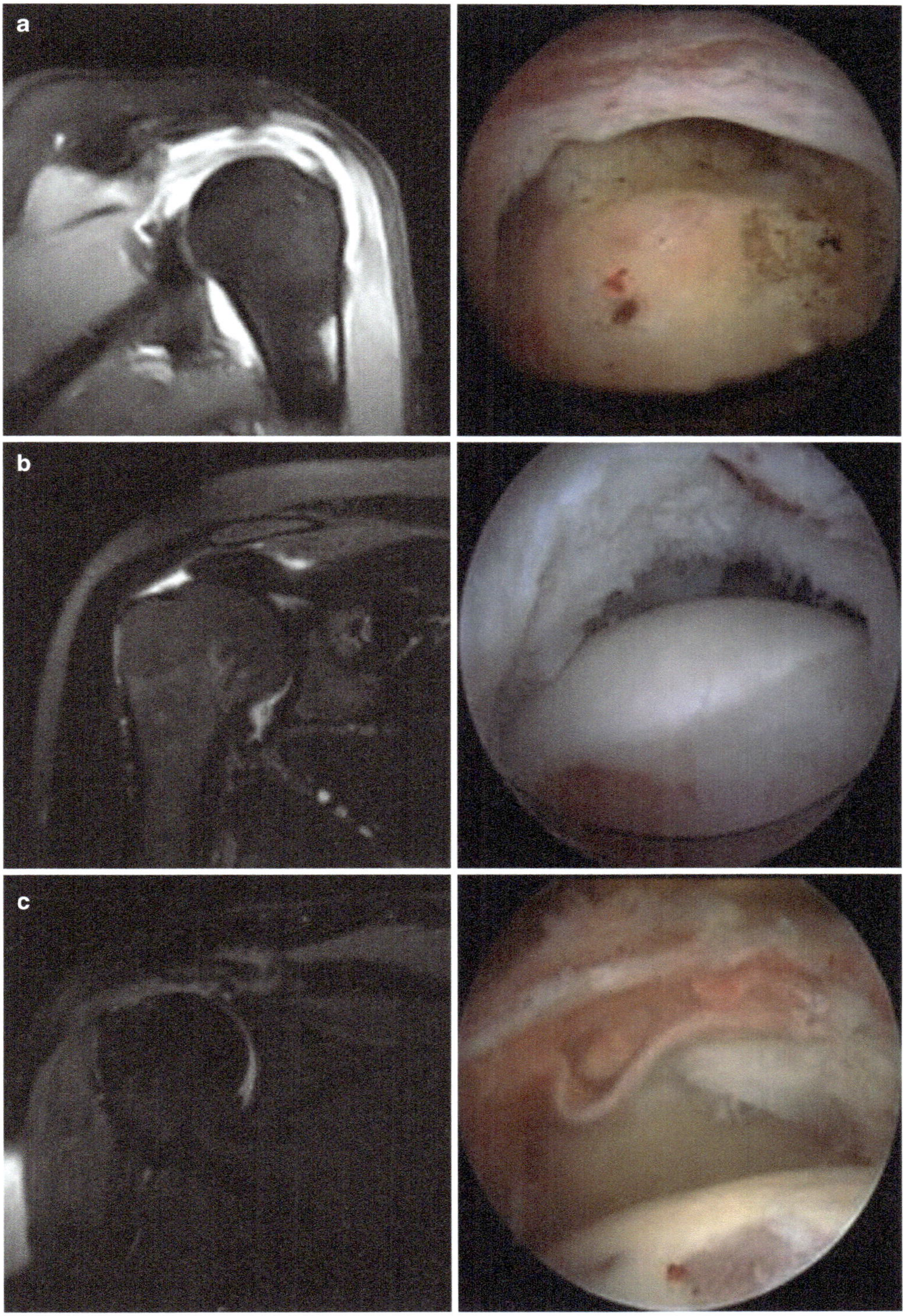

Fig. 24.8 Non-contrast coronal MR image and corresponding intraoperative arthroscopic image in the same patient. (**a**) Stage 1 – proximal stump of tendon is close to bony insertion on the greater tuberosity. (**b**) Stage 2 – proximal stump is retracted to the level of the humeral head. (**c**) Stage 3 – proximal stump is retracted to the level of the glenoid

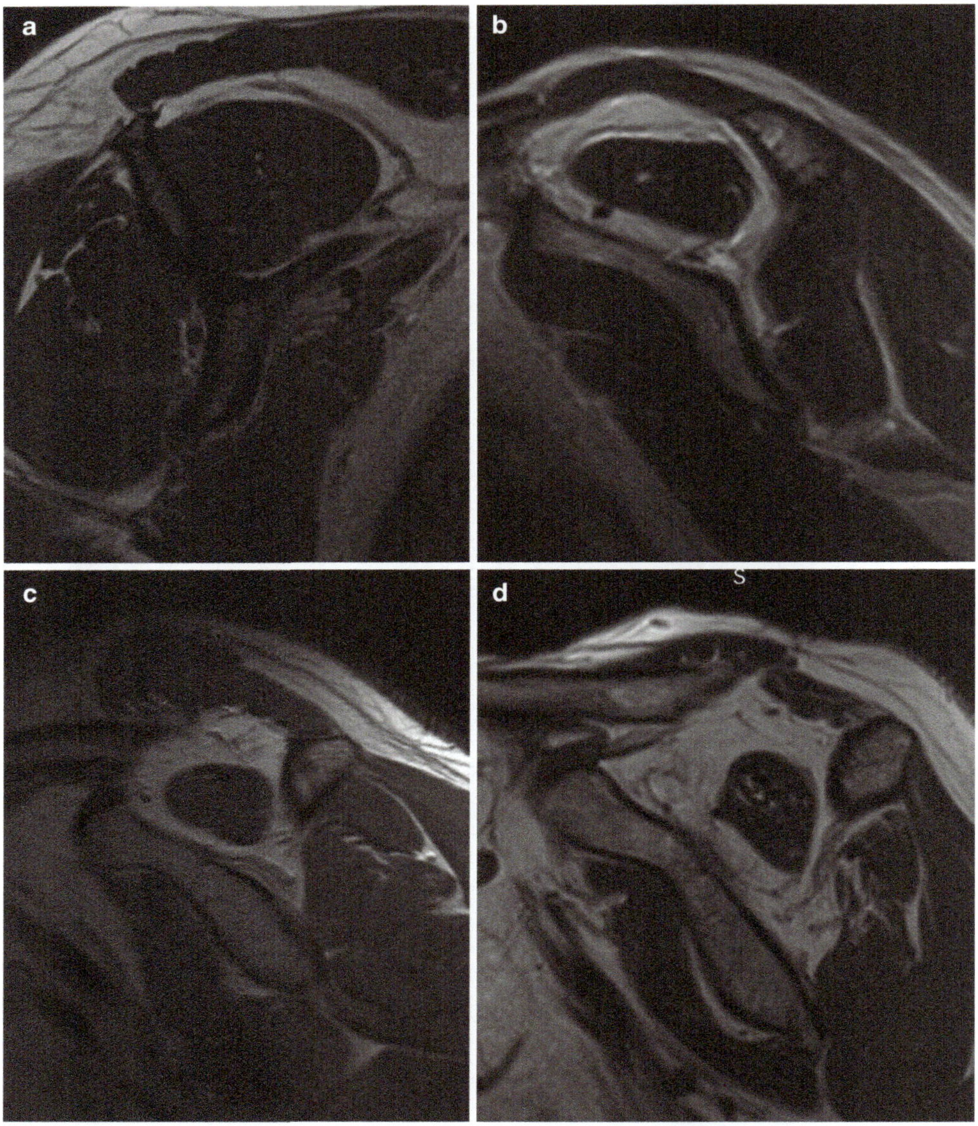

Fig. 24.9 Sagittal non-contrast MR images corresponding to grades of atrophy in the Goutallier system. (**a**) Stage 0 – normal muscle. (**b**) Stage 2 – The fatty streaks are important but there is less than 50 % fatty muscle atrophy. (**c**) Stage 3 – There is 50 % fatty muscle atrophy. (**d**) Stage 4 – There is greater than 50 % fatty muscle atrophy. Note: Stage 1 – the muscle that contains some fatty streaks is not illustrated [6]

Conclusion

Preoperative assessment of bony anatomy, soft tissue tear pattern, retraction, atrophy and fatty infiltration is critical to planning for successful rotator cuff repair. As seen in the above examples all series of images are necessary to understand tear morphology and characteristics. Continually comparing preoperative to intraoperative findings improves the surgeons' ability to plan a surgical approach and provide the patient with important prognostic information.

References

1. Bigliani LU. The morphology of the acromion and its relationship to rotator cuff tears. Orthop Trans. 1986;10:228.
2. Ellman H, Gartsman GM, editors. Open repair of full-thickness rotator cuff tears. Philadelphia/Baltimore/Hong Kong/London/Munich/Sydney/Tokyo: Lea and Febiger; 1993. p. 181–202.
3. Ellman H, Gartsman GM, editors. Treatment of partial-thickness rotator cuff tears: arthroscopic and mini-open. Philadelphia/Baltimore/Hong Kong/London/Munich/Sydney/Tokyo: Lea and Febiger; 1993. p. 155–80.
4. Fox JA, Noerdlinger MA, Romeo AA. Arthroscopic subscapularis repair. Tech Shoulder Elbow Surg. 2003;4(4):154.
5. Fuchs B, et al. Fatty degeneration of the muscles of the rotator cuff: assessment by computed tomography versus magnetic resonance imaging. J Shoulder Elbow Surg. 1999;8(6):599–605.
6. Goutallier D, et al. Fatty muscle degeneration in cuff ruptures. Pre- and postoperative evaluation by CT scan. Clin Orthop Relat Res. 1994; (304):78–83.
7. Lippe J, et al. Inter-rater agreement of the Goutallier, Patte, and Warner classification scores using preoperative magnetic resonance imaging in patients with rotator cuff tears. Arthroscopy J Arthrosc Relat Surg Off Publ Arthrosc Assoc North Am Int Arthrosc Assoc. 2012;28(2):154–9.
8. Lo IKY, Burkhart SS. The etiology and assessment of subscapularis tendon tears: a case for subcoracoid impingement, the roller-wringer effect, and TUFF lesions of the subscapularis. Arthroscopy J Arthrosc Relat Surg Off Publ Arthrosc Assoc North Am Int Arthrosc Assoc. 2003;19(10):1142–50.
9. Oh JH, et al. Prognostic factors affecting anatomic outcome of rotator cuff repair and correlation with functional outcome. Arthroscopy J Arthrosc Relat Surg Off Publ Arthrosc Assoc North Am Int Arthrosc Assoc. 2009;25(1):30–9.
10. Patte D. Classification of rotator cuff lesions. Clin Orthop Relat Res. 1990;254:81–6.
11. Slabaugh MA, et al. Interobserver and intraobserver reliability of the Goutallier classification using magnetic resonance imaging: proposal of a simplified classification system to increase reliability. Am J Sports Med. 2012;40(8):1728–34.

25 Pathoanatomy of Rotator Cuff Tears

Robert U. Hartzler, Richard L. Angelo, and Stephen S. Burkhart

25.1 Introduction

> Since the success of treatment depends in most conditions on our knowledge of the exact pathology, a surgeon treating shoulder lesions should try in each individual case to picture in his mind the relative proportions of the elements of the joint which may be involved. E.A. Codman, The Shoulder, 1934

Shoulder pathology is oftentimes puzzling, and this is nowhere more true than in the pathology of rotator cuff tears. However, we agree wholeheartedly with Dr. Codman: a surgeon will only be successful in treating a rotator cuff tear if he endeavors to understand the pathoanatomy of the tear and associated lesions. Our goal in this chapter is to aid surgeons in recognizing patterns of rotator cuff pathology in order that they might develop rational treatment strategies for their patients.

Shoulder arthroscopy has been a huge advance in the endeavor to understand rotator cuff pathoanatomy. With the arthroscope, specialized instrumentation, and some simple surgical techniques, visualization of rotator cuff pathology is virtually unlimited. Furthermore, as the complexity of the pathology increases, the importance of the arthroscope as a surgical diagnostic and treatment tool increases. Thus, in this chapter, our focus will be on the pathoanatomy of rotator cuff tears from the arthroscopic perspective.

25.2 Etiology of Rotator Cuff Tears and Pathoanatomy of External Impingement

A complete discussion of the various theories of rotator cuff tear pathogenesis is outside the scope of this chapter. Rotator cuff tearing is, generally speaking, a multifactorial process, with contributions from intrinsic tendon degeneration, lack of healing, wear from extrinsic impingement or friction, and trauma. Previous authors have placed varying emphasis on these factors [1–5]. In our opinion, it is important to recognize that the surgeon's perspective on the etiology of rotator cuff tears greatly influences surgical decision making.

The typical rotator cuff tear patterns seldom occur in the absence of degenerative tendinosis, the prevalence of which increases with age. Currently in the orthopedic literature, "intrinsic" factors, such as tendon hypovascularity and

Electronic supplementary material The online version of this chapter 10.1007/978-3-662-45719-1_25 contains supplementary material, which is available to authorized users.

R.U. Hartzler, MD, MS • S.S. Burkhart, MD (✉)
Orthopaedic Surgery, The San Antonio Orthopaedic Group, San Antonio, TX, USA
e-mail: rhartzler@tsaog.com; ssburkhart@msn.com

R.L. Angelo, MD
Department of Orthopedics, University of Washington, Seattle, WA, USA
e-mail: rlamdortho@comcast.net

age-related weakening of the connective tissues, are emphasized over "extrinsic" factors, such as bony impingement, in the development of tendinosis. This theory is supported by the higher prevalence of partial articular-sided and intratendinous tears compared with bursal-sided tears in cadaver studies [1]. On the other hand, it should be noted that the evidence is now conflicting as to whether a "critical" zone of hypovascularity in the supraspinatus tendon actually exists [2–4].

Although the pathogenesis of rotator cuff tears is multifactorial, only one of those factors may, at present, be directly influenced by the surgeon at the time of operation: lesions secondary to extrinsic impingement. In our experience, a critical component to success in treating cuff tears is recognizing and treating external impingement lesions. Thus, we urge the diligent and systematic examination of the subcoracoid and subacromial spaces for evidence of impingement by the CA ligament, coracoid tip, AC joint, and anterior and lateral acromion against the rotator cuff and tuberosities (Fig. 25.1, Video 25.1).

Typically, extrinsic impingement results in abrasive lesions between the opposing structures (Fig. 25.1b, d). However, we also believe that extrinsic impingement creates high tensile forces in the articular tendon fibers (the roller-ringer effect) and can initiate either articular-sided (Fig. 25.1a) or bursal-sided tears [5]. In addition to improving visualization and working space, a thorough subacromial decompression in conjunction with rotator cuff repair has been associated with a lower rate of reoperation and re-repair even at short-term follow-up in high-level studies [6, 7]. Even in massive rotator cuff tears, where we preserve the coracoacromial ligament to prevent anterosuperior escape of the humeral head if the cuff fails to heal, we still take care to expose and treat all other potential sources of extrinsic impingement.

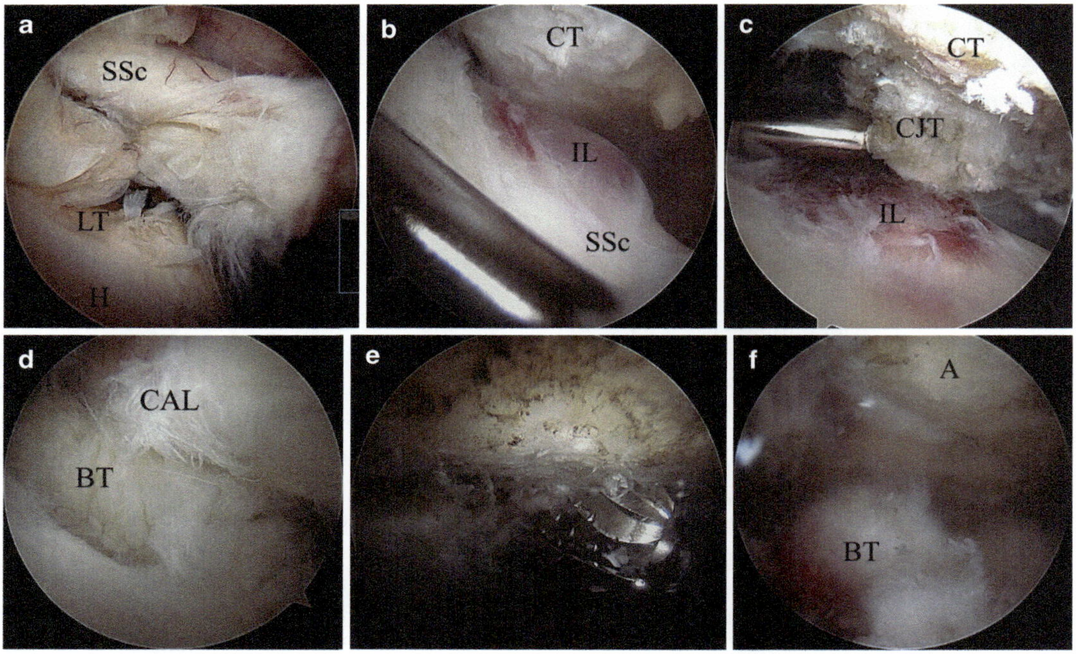

Fig. 25.1 *Top row*: Left shoulder, posterior viewing portal, 70° arthroscopic view of glenohumeral joint. (**a**) The subscapularis tendon (*SSc*) is torn from the lesser tuberosity (*LT*), but not retracted. (**b**) The hemorrhagic and edematous bursal surface of the tendon (*IL*, impingement lesion) shows evidence of extrinsic impingement at the coracoid tip (*CT*), which is also affected. (**c**) After coracoplasty and debridement, there is an adequate subcoracoid space. Left shoulder, posterior viewing portal, 30° arthroscopic view of subacromial space. (**d**) The sharp and worn edge of the coracoacromial ligament (*CAL*) and lateral acromion against a high-grade bursal-sided supraspinatus tear (*BT*). (**e**) Burr against a down sloping lateral acromial osteophyte. (**f**) Adequate subacromial space after decompression including beveling of lateral acromion (*A*)

25.3 Full-Thickness Rotator Cuff Tears

Full-thickness rotator cuff tears present at operation with varying degrees of retraction, scarring, and delamination. The surgeon must differentiate rotator cuff tissue from "bursal leaders," which are thickened, synovialized bands of bursal scar tissue that have an appearance similar to a chronically torn rotator cuff edge (Fig. 25.2) [8]. Bursal leaders insert into the internal deltoid fascia, whereas the intact cuff edge inserts onto the tuberosity. The surgeon must develop the plane between these two tissue edges (Videos 25.2 and 25.3) in order to correctly identify and repair the torn cuff. When cuff delamination occurs, typically only in large and massive tears, each layer should be assessed individually to determine the best repair construct. Not uncommonly, the superficial layer can be repaired with

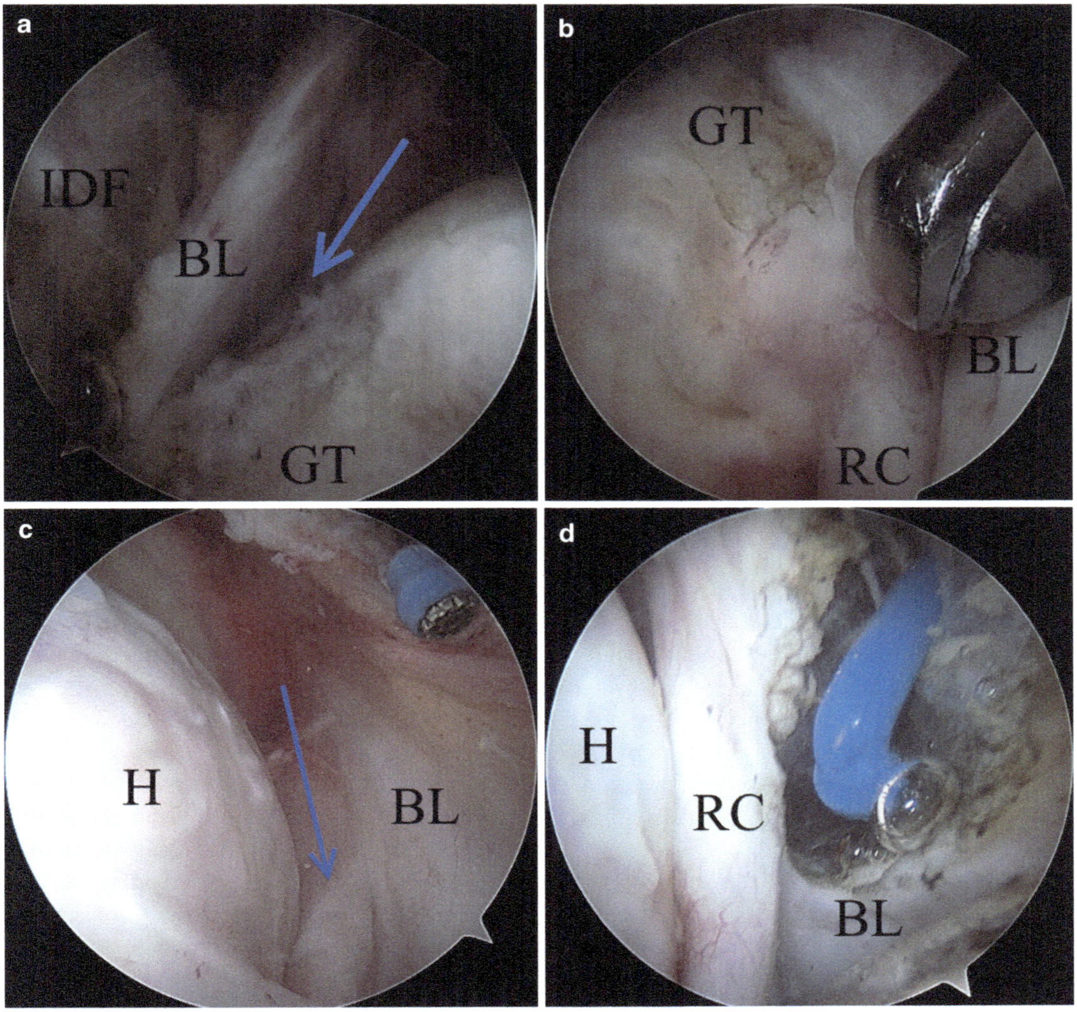

Fig. 25.2 *Top row*: Right shoulder. Bursal leaders as seen in a massive posterosuperior rotator cuff tear. (**a**) Lateral portal, 70° arthroscopic view. A deceptively thick bursal leader (*BL*) travels past the greater tuberosity (*GT*) to insert into the internal deltoid fascia (*IDF*), while the *blue arrow* identifies the intact teres minor tendon. (**b**) Switching the 70° scope to the posterior portal gives an enhanced view of the plane between intact rotator cuff (*RC*) and BL. *Bottom row*: Left shoulder. (**c**) Lateral portal, 70° arthroscopic view. A sheet-like bursal leader inserts into the internal deltoid fascia, while the rotator cuff (*blue arrow*) inserts into the tuberosity. (**d**) The bursal leader has been debrided, and the plane between the deltoid and cuff has been re-established (eg *H* humeral head)

Table 25.1 The geometric classification of full-thickness, posterosuperior rotator cuff tears

Type	Description	Preoperative MRI	Intraoperative mobility	Treatment strategies
1	Crescent	Short and wide	Primarily medial-to-lateral mobility	End-to-bone repair
2	Longitudinal (L, rev-L, U)	Long and narrow	Primarily anterior-to-posterior mobility	Margin convergence, margin-to-bone repair
3	Massive contracted	Long and wide (>2 × 2 cm)	Relatively immobile in any direction	Interval slides, partial repair, load-sharing rip stop
4	Cuff tear arthropathy	GH arthrosis, proximal humeral migration	Completely immobile	Arthroplasty

Adapted from Davidson and Burkhart [9]

a double-row construct, while the deep layer may only be amenable to single row repair in order to avoid over tensioning.

Full-thickness, posterosuperior rotator cuff tears retract in several consistent patterns. The senior author (SSB) developed the geometric classification system for these supraspinatus, infraspinatus, and teres minor cuff tears based on these patterns (Table 25.1) [9]. The surgeon assesses both the size and mobility of the tear edges at the time of surgery using a tendon grasping instrument. The classification system is useful for both diagnosis and enables the creation of a treatment algorithm (Fig. 25.3). We cannot overemphasize the importance of a thorough bursal debridement during the intraoperative assessment of rotator cuff tear pathoanatomy. Our routine exposure proceeds from the spine of the scapula medially to the lateral edge of the muscle tendon units. Additionally, the geometric tear types as determined by preoperative MRI characteristics and the intraoperative assessment are highly correlated, thereby facilitating preoperative planning [10, 11].

Type 1 (crescent) tears have smaller length (ML dimension) than width (AP dimension); however, width varies greatly in size from small to massive. Crescent tears also have medial to lateral mobility that is sufficient for repair of the tendon directly lateral onto the bone bed under minimal tension (Fig. 25.4). These tears typically do not have significant medial retraction. Lastly, medial to lateral mobility should be equal along the length of the tear margin (Video 25.4).

Type 2 (longitudinal) tears have larger length (ML dimension) than width (AP dimension), and this width is typically <2 cm. The subclassification of longitudinal tears as L-, reverse L-, and U-shaped is based on intraoperative assessment of the mobility of each leaf of the tear. L-shaped tears have a "corner" that is located along the anterolateral aspect of the posterior leaf of the tear (Fig. 25.5), and this corner will often have a "surprising" amount of posterior to anterior mobility that allows reduction to the anterolateral aspect of the bone bed (Video 25.5). In contrast, reverse L-shaped tears have a "corner" that is located along the posterolateral aspect of the anterior leaf (Fig. 25.6) and requires reduction in a posterolateral direction (Video 25.6). We have found that the corner of a reverse-L tear does not typically have the dramatic amount of anterior to posterior mobility that can be seen with L tears (opposite direction of reduction). In contrast, U-shaped tears have roughly equal mobility of the anterior and posterior tear margins without a clear "corner" to reduce (Fig. 25.7) (Videos 25.7 and 25.8). Recognizing longitudinal tear variations allows the surgeon to perform tension-free repairs using margin convergence sutures and/or suture anchors with margin-to-bone stitch configurations.

Type 3 (massive contracted) tears are both long and wide, typically >2 cm × 2 cm (Fig. 25.8). These tears require advanced mobilization techniques for repair, because tendon mobility is poor in all directions (Video 25.9) [12]. Usually only partial repairs or single row repairs are possible

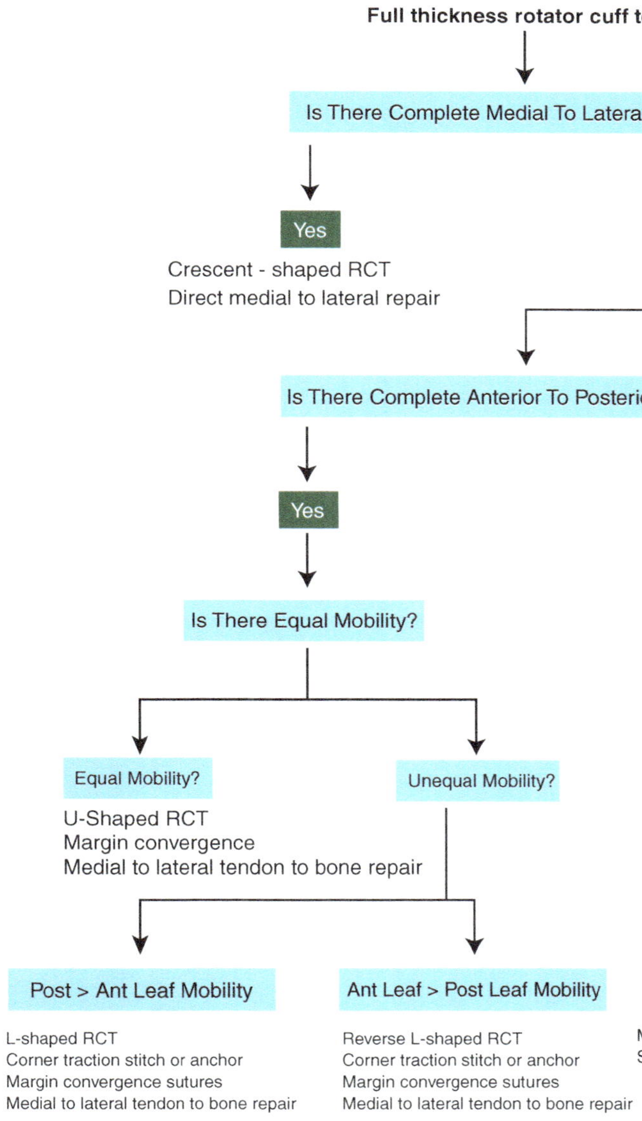

Fig. 25.3 Geometric classification system with treatment considerations. Intraoperative assessment of morphology and mobility of full-thickness rotator cuff tears allow the surgeon to make treatment decisions

for these tears. When assessing tear mobility with a tendon grasper, scarring and fibrosis may be based anteriorly, posteriorly, or in both locations. The anterior interval slide in continuity may be used to release adhesions that tether the cuff to the base of the coracoid via the superior glenohumeral ligament [13]. The posterior interval slide may be used to address posterior retraction and fibrosis [12]. Fibrosis of the joint capsule in a retracted tear may also be a cause of tear immobility; however, we have found that only a few millimeters of added lateral excursion is gained by a capsular release. In contrast, interval slides may result in several centimeters of added excursion.

Type 4 tears (rotator cuff tear arthropathy) are rarely observed during arthroscopy. The cuff tendons appear similarly to those of type 3 tears.

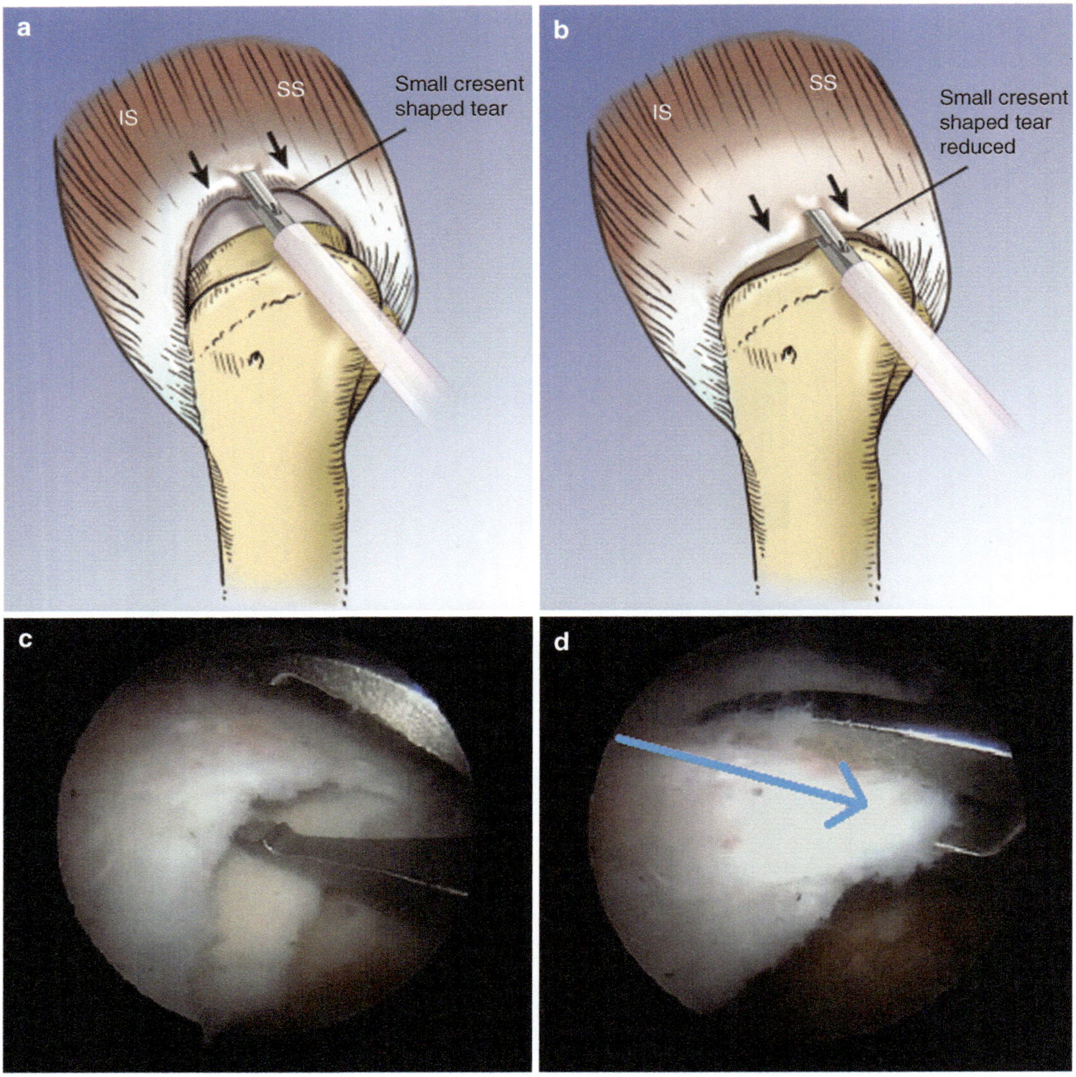

Fig. 25.4 Right shoulder. Crescent tear illustrations (**a**, **b**) and photographs (**c**, **d**) demonstrating the characteristics of width (AP dimension) greater than length (ML dimension), minimal retraction, and direct lateral mobility (*blue arrow*) of the tendon edge. *SS* supraspinatus, *IS* infraspinatus

In contrast, type 4 tears have little mobility, even with advanced mobilization techniques, and have associated advanced degenerative changes of the glenohumeral joint such as full-thickness cartilage loss, eburnation of the joint surfaces, and marginal osteophytes.

In treating full-thickness rotator cuff tears, we always strive to assess, both using preoperative imaging and intraoperative assessment, whether the rotator cable insertions (see Chap. 20) are involved in the tear. Tears that involve the cable insertions, particularly that of the anterior cable, tend to progress in size, retraction, and fatty degeneration, while those that are contained within the crescent, do not [14, 15]. In addition to proper tear pattern recognition and reduction, we will often reinforce the fixation at the cable attachments using extra sutures.

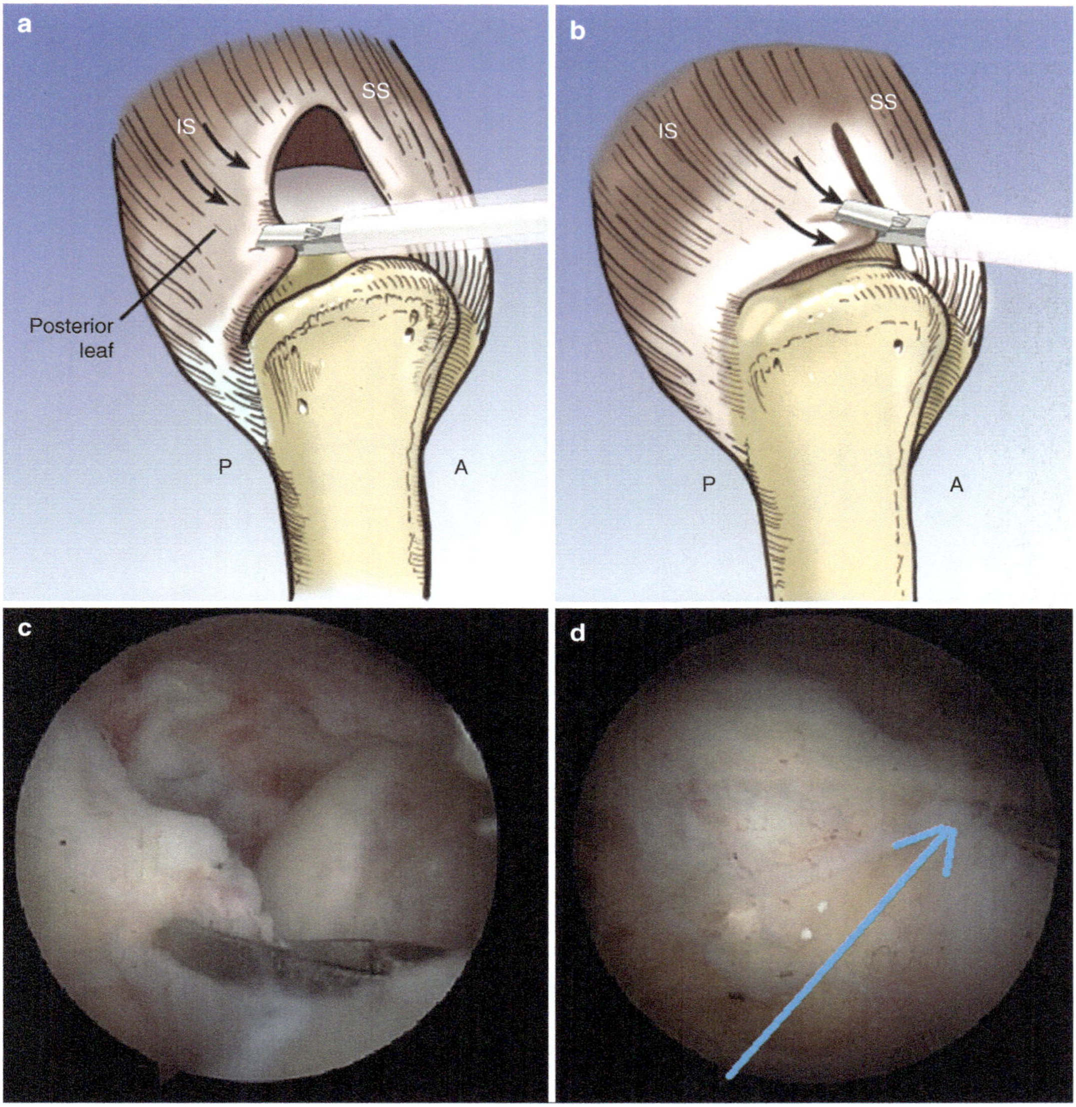

Fig. 25.5 Right shoulder. L-shaped longitudinal tear illustrations (**a**, **b**) and photographs (**c**, **d**) demonstrating characteristics of length (ML dimension) greater than width (AP dimension) and a "corner" that can be reduced to the anterolateral aspect of the bone bed (*blue arrow*). *P* posterior, *A* anterior, *SS* supraspinatus, *IS* infraspinatus

We classify full-thickness subscapularis tears by their proximal to distal length and whether they are retracted or non-retracted. Recognizing the pathoanatomy of retracted subscapularis tears is critical to localizing and repairing the tendon. As seen in Fig. 25.9, when the subscapularis tendon tears and retracts, the superolateral corner of the tendon remains attached to fibers of rotator interval tissue (primarily medial biceps tendon sling, superior glenohumeral and coracohumeral ligaments) that are oriented perpendicular to the subscapularis tendon fibers. This interval tissue is shaped like an arc, and so we described it as the "comma sign"[16]. The retracted subscapularis tendon may appear hidden, but it can be routinely located by recognizing its consistent attachment to the comma tissue and the consistent location of the upper border at the mid-glenoid notch

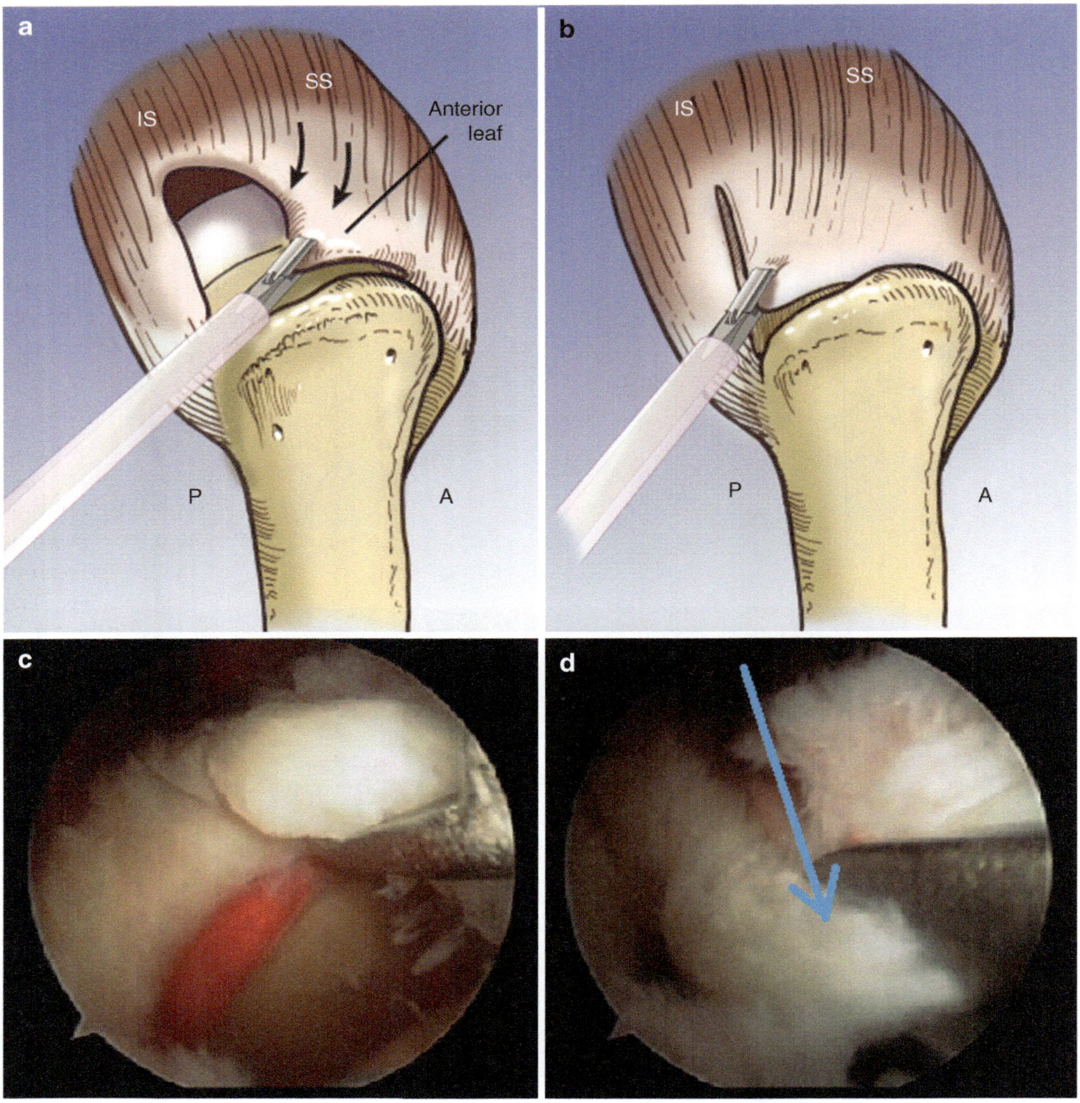

Fig. 25.6 Right shoulder. Reverse L-shaped longitudinal tear illustrations (**a**, **b**) and photographs (**c**, **d**) demonstrating characteristics of length greater than width and a "corner" that can be reduced to the posterolateral aspect of the bone bed (*blue arrow*). *P* posterior, *A* anterior, *SS* supraspinatus, *IS* infraspinatus

(Videos 25.10 and 25.11). In addition to serving as a landmark for the retracted subscapularis tendon, the comma tissue also attaches the subscapularis tendon to the anterior supraspinatus tendon (Video 25.12). Therefore, the comma should be identified and preserved during rotator cuff repair.

Over the years, our ability to repair the rotator cuff has improved as our ability to recognize tear patterns has become more accurate. Our philosophy is that the tear pattern determines the repair pattern. This has allowed us to repair cuff tears under less tension and with better mechanical integrity. Lastly, repairing the cuff according to the tear pattern has allowed us to achieve countless tendon-to-bone repairs which seemed impossible on first inspection. Those tears would not have been possible with only medial-to-lateral tendon reduction and mobilization.

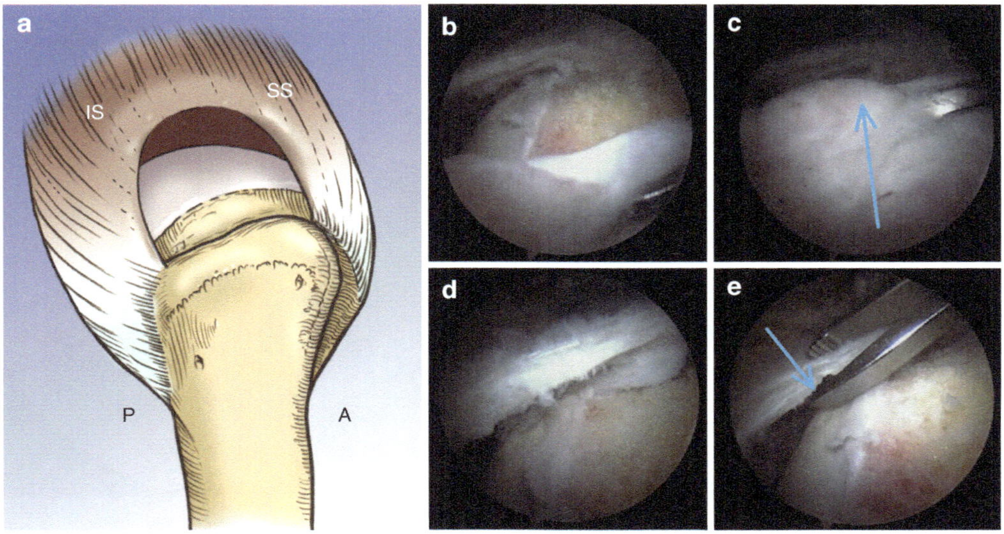

Fig. 25.7 Right shoulder. U-shaped longitudinal tear illustration (**a**). Photographs **b**, **c** show the direction of anterior mobility of the posterior margin, and photos **d**, **e** show the direction of posterior mobility of the anterior margin. *P* sterior, *A* anterior, *SS* supraspinatus, *IS* infraspinatus

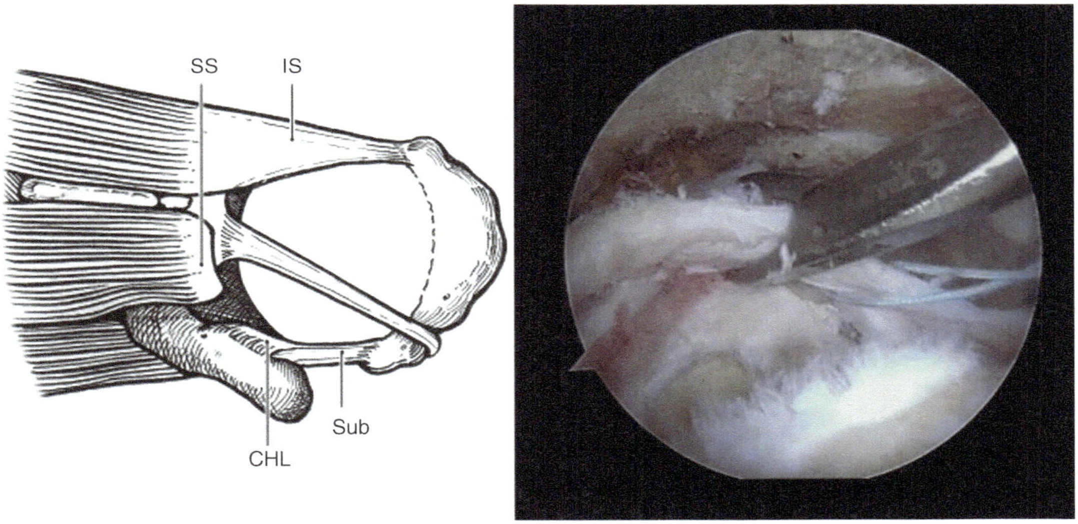

Fig. 25.8 Right shoulder. Type 3 (massive contracted) rotator cuff tears are both long and wide. Tendon mobility is poor, and advanced mobilization techniques are required to repair these tears. *SS* supraspinatus, *IS* infraspinatus, *CHL* coracohumeral ligament, *Sub* subscapularis

25.4 Partial-Thickness Tears and Internal Impingement Lesions of the Rotator Cuff

We classify partial-thickness posterosuperior rotator cuff tears according to location (Fig. 25.10) and size as originally described by Ellman [17]. As tear width is commonly used to make treatment decisions intraoperatively, the dimensions of the tear should be measured with a calibrated probe or an instrument of known size.

Partial articular tears of the posterosuperior cuff are usually readily visible from a standard posterior glenohumeral viewing portal. Debridement

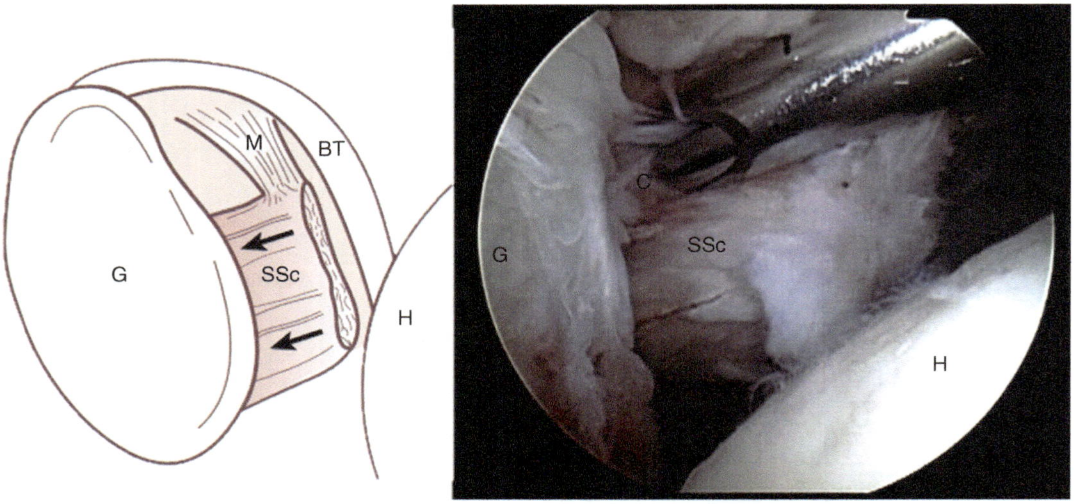

Fig. 25.9 Pathoanatomy of retracted subscapularis tears demonstrating how pulling the comma tissue (located behind the shaver) laterally brings the retracted subscapularis tendon (*SSc*) into clear view. *M* medial biceps sling, *BT* biceps tendon, *H* humeral head, *G* glenoid, *C* coracoid tip

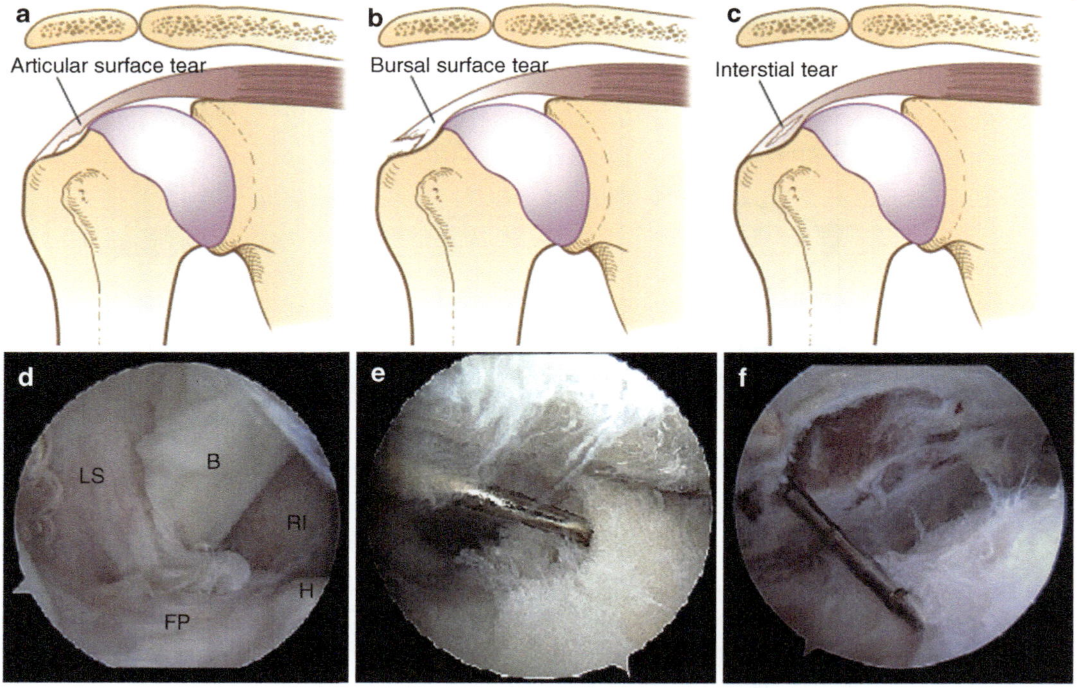

Fig. 25.10 Illustrations (**a–c**) of partial tears of the posterosuperior rotator cuff with corresponding arthroscopic photos (**d–f**). A hooked probe is a useful tool for inspecting the cuff for interstitial tears (**f**), as the probe will often "fall into" the defect. *LS* lateral sling, *B* biceps tendon, *RI* rotator interval, *FP* footprint, *H* humeral head

of the tear edges is necessary to accurately define the extent of the lesion. Placing a spinal needle through the tear while viewing intraarticularly is a useful technique to mark the bursal location of the tear for later inspection from the subacromial space (Video 25.13). Partial articular tears of the infraspinatus, typically seen in the throwing athlete, result from excessive, repetitive torsional loading of the articular surface tendon fibers and internal impingement of the cuff against the posterosuperior glenoid [18]. During arthroscopic examination, the surgeon can recreate the internal impingement mechanism, which typically occurs at high degrees of external rotation in abduction (Video 25.14).

Following a bursectomy, partial bursal tears of the posterosuperior cuff are easily visible in the subacromial space while viewing from a standard posterior portal with either a 30° or 70° arthroscope. Sources of abrasion or impingement should be diligently sought after by the surgeon in the case of a partial bursal tear. Following subacromial debridement, tear size should be measured using appropriate instruments (Video 25.15).

A high index of suspicion based on MRI and physical exam and a systematic approach to diagnosis will aid the surgeon in diagnosing interstitial tears. When articular and bursal fibers are intact, but a lesion is suspected, we employ several arthroscopic examination tests (Videos 25.16 and 25.17). First, we palpate the tendon with a hook probe and assess for the presence of sliding layers. We also push into the tendon with a probe, which will then often "drop into" an interstitial tear defect. Finally, we will perform the "bubble test" by attempting to inject sterile normal saline into the tendon substance with an 18-gauge spinal needle. A positive sign, the easy flow of fluid into the tendon creating a tendon "bubble," indicates an interstitial tear (Video 25.17).

Subscapularis tear classifications, including partial tear sub-types, have been proposed by several authors [19–22]. We prefer a descriptive classification for partial subscapularis tendon tears based on morphology (Fig. 25.11). Linear longitudinal tears in the tendon may occur without frank fiber disruption from the lesser tuberosity (Fig. 25.11a). These linear tears are commonly attributable to external subcoracoid impingement via the roller-wringer effect

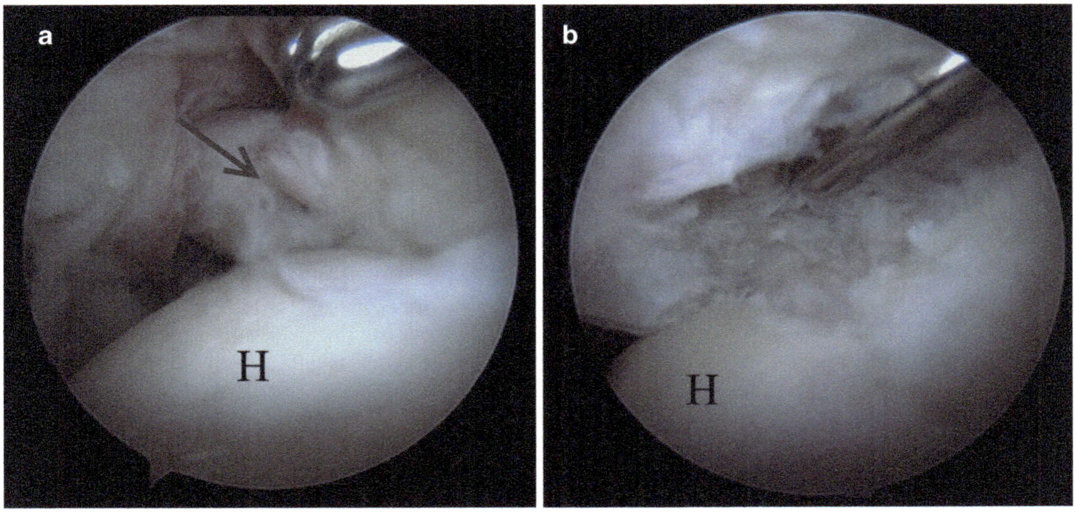

Fig. 25.11 Right shoulders, posterior portal, 70° arthroscopic views. (**a**) Linear longitudinal tears in the subscapularis tendon without fiber disruption from the insertion. (**b**) Arthroscopic shaver lifts the tendon away from a partial articular tear of the upper tendon. Bottom row: occult interstitial subscapularis tear. (**c**) Abrasion and tearing of the biceps tendon and medial sidewall of the groove are demonstrated (*arrows*). (**d**) After biceps tenotomy, debridement of the tear allows the shaver to "drop into" the interstitial tear. *H* humeral head, *TL* transverse humeral ligament, *BT* long head of biceps tendon, *SSc* subscapularis tendon, *C* comma tissue, *GT* greater tuberosity

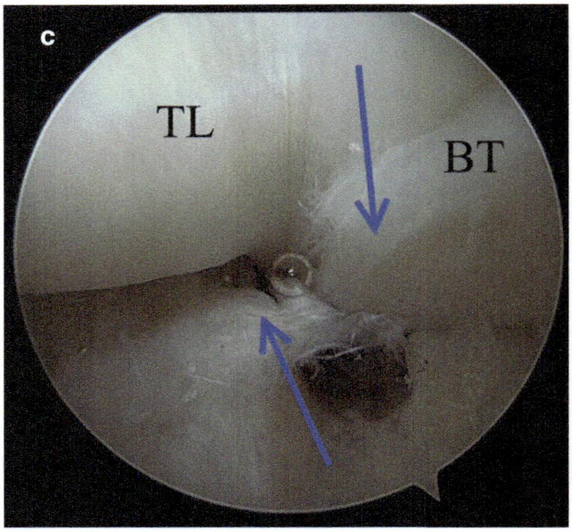

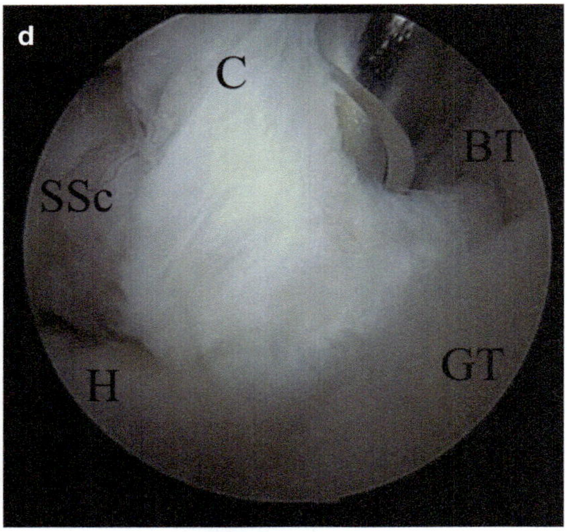

Fig. 25.11 (continued)

(Fig. 25.1a, b) [5]. The most common type of partial subscapularis tear is an articular-sided tear of the tendinous superior boarder (Fig. 25.11b). Partial superior subscapularis tears are also usually a result of subcoracoid impingement, but we are increasingly recognizing internal impingement from a thick middle glenohumeral ligament (Video 25.18) as a potential cause. Debridement of partial articular tears often reveals a tear that is larger than anticipated on casual inspection (Video 25.19). These tears are also commonly associated with linear longitudinal splits in the subscapularis tendon. Partial bursal-sided tears and interstitial tears of the subscapularis are less common, but may be revealed with careful inspection of the bicipital groove with a 70° arthroscope (Fig. 25.11c, d). Abrasion of the biceps tendon and medial sling (Fig. 25.11c) or a tear in the medial side wall of the biceps groove (Fig. 25.11d) should alert the surgeon to the possibility of a partial interstitial or bursal tear (Video 25.20). Rarely, occult subscapularis tendon tears are hidden by the capsuloligamentous tissues of the medial biceps sling and can only be visualized after take down of the sling [23–25].

References

1. Loehr JF, Uhthoff HK. The pathogenesis of degenerative rotator cuff tears. Orthop Trans. 1987;11:237.
2. Rathbun JB, Macnab I. The microvascular pattern of the rotator cuff. J Bone Joint Surg Br. 1970;52(3):540–53. PubMed PMID: 5455089.
3. Moseley HF, Goldie I. The arterial pattern of the rotator cuff of the shoulder. J Bone Joint Surg Br. 1963;45:780–9. PubMed PMID: 14074332.
4. Levy O, Relwani J, Zaman T, Even T, Venkateswaran B, Copeland S. Measurement of blood flow in the rotator cuff using laser Doppler flowmetry. J Bone Joint Surg Br. 2008;90(7):893–8. doi:10.1302/0301-620X.90B7.19918. PubMed PMID: 18591599.
5. Lo IK, Burkhart SS. The etiology and assessment of subscapularis tendon tears: a case for subcoracoid impingement, the roller-wringer effect, and TUFF lesions of the subscapularis. Arthroscopy. 2003;19(10):1142–50. doi:10.1016/j.arthro.2003.10.024. PubMed PMID: 14673459.
6. Abrams GD, Gupta AK, Hussey KE, Tetteh ES, Karas V, Bach BR, et al. Arthroscopic repair of full-thickness rotator cuff tears with and without acromioplasty: randomized prospective trial with 2-year follow-up. Am J Sports Med. 2014;42(6):1296–303. doi:10.1177/0363546514529091. PubMed PMID: 24733157.
7. MacDonald P, McRae S, Leiter J, Mascarenhas R, Lapner P. Arthroscopic rotator cuff repair with and without acromioplasty in the treatment of full-thickness rotator cuff tears: a multicenter, randomized controlled trial. J Bone Joint Surg Am. 2011;93(21):1953–60. doi:10.2106/JBJS.K.00488. PubMed PMID: 22048089.

8. Lo IK, Burkhart SS. Arthroscopic revision of failed rotator cuff repairs: technique and results. Arthroscopy. 2004;20(3):250–67. doi:10.1016/j.arthro.2004.01.006. PubMed PMID: 15007314.
9. Davidson J, Burkhart SS. The geometric classification of rotator cuff tears: a system linking tear pattern to treatment and prognosis. Arthroscopy. 2010;26(3):417–24. doi:10.1016/j.arthro.2009.07.009. PubMed PMID: 20206053.
10. Sela Y, Eshed I, Shapira S, Oran A, Vogel G, Herman A, et al. Rotator cuff tears: correlation between geometric tear patterns on MRI and arthroscopy and pre- and postoperative clinical findings. Acta Radiol. 2014. doi:10.1177/0284185114520861. PubMed PMID: 24445094.
11. Davidson JF, Burkhart SS, Richards DP, Campbell SE. Use of preoperative magnetic resonance imaging to predict rotator cuff tear pattern and method of repair. Arthroscopy. 2005;21(12):1428. doi:10.1016/j.arthro.2005.09.015. PubMed PMID: 16376230.
12. Lo IK, Burkhart SS. Arthroscopic repair of massive, contracted, immobile rotator cuff tears using single and double interval slides: technique and preliminary results. Arthroscopy. 2004;20(1):22–33. doi:10.1016/j.arthro.2003.11.013. PubMed PMID: 14716275.
13. Lo IK, Burkhart SS. The interval slide in continuity: a method of mobilizing the anterosuperior rotator cuff without disrupting the tear margins. Arthroscopy. 2004;20(4):435–41. doi:10.1016/j.arthro.2004.01.016. PubMed PMID: 15067287.
14. Burkhart SS, Esch JC, Jolson RS. The rotator crescent and rotator cable: an anatomic description of the shoulders "suspension bridge". Arthroscopy. 1993;9(6):611–6. PubMed PMID: 8305096.
15. Kim HM, Dahiya N, Teefey SA, Keener JD, Galatz LM, Yamaguchi K. Relationship of tear size and location to fatty degeneration of the rotator cuff. J Bone Joint Surg Am. 2010;92(4):829–39. doi:10.2106/JBJS.H.01746. PubMed PMID: 20360505; PubMed Central PMCID: PMCPMC2842942.
16. Lo IK, Burkhart SS. The comma sign: an arthroscopic guide to the torn subscapularis tendon. Arthroscopy. 2003;19(3):334–7. doi:10.1053/jars.2003.50080. PubMed PMID: 12627163.
17. Ellman H. Diagnosis and treatment of incomplete rotator cuff tears. Clin Orthop Relat Res. 1990;254:64–74. PubMed PMID: 2182260.
18. Burkhart SS, Morgan CD, Kibler WB. The disabled throwing shoulder: spectrum of pathology part I: pathoanatomy and biomechanics. Arthroscopy. 2003;19(4):404–20. doi:10.1053/jars.2003.50128. PubMed PMID: 12671624.
19. Lafosse L, Jost B, Reiland Y, Audebert S, Toussaint B, Gobezie R. Structural integrity and clinical outcomes after arthroscopic repair of isolated subscapularis tears. J Bone Joint Surg Am. 2007;89(6):1184–93. doi:10.2106/JBJS.F.00007. PubMed PMID: 17545420.
20. Garavaglia G, Ufenast H, Taverna E. The frequency of subscapularis tears in arthroscopic rotator cuff repairs: a retrospective study comparing magnetic resonance imaging and arthroscopic findings. Int J Shoulder Surg. 2011;5(4):90–4. doi:10.4103/0973-6042.91000. PubMed PMID: 22223958; PubMed Central PMCID: PMCPMC3249929.
21. Toussaint B, Barth J, Charousset C, Godeneche A, Joudet T, Lefebvre Y, et al. New endoscopic classification for subscapularis lesions. Orthop Traumatol Surg Res. 2012;98(8 Suppl):S186–92. doi:10.1016/j.otsr.2012.10.003. PubMed PMID: 23149162.
22. Fox JA, Noerdlinger MA, Romeo AA. Arthroscopic subscapularis repair. Techn Shoulder Elbow Surg. 2003;4(4):154–68.
23. Koo SS, Burkhart SS. Subscapularis tendon tears: identifying mid to distal footprint disruptions. Arthroscopy. 2010;26(8):1130–4. doi:10.1016/j.arthro.2010.06.017. PubMed PMID: 20678713.
24. Burkhart SSL, Ian KY, Brady PC, Denard PJ. The Cowboy's companion: a trail guide for the arthroscopic shoulder surgeon. Philadelphia: Wolters Kluwer Health/Lippincott Williams & Wilkins; 2014.
25. Hartzler RU, Burkhart SS. Technical note: medial biceps sling takedown may be necessary to expose an occult subscapularis tendon tear. Arthrosc Techn. 2014;3(6):e719–22.

Deltoid Muscle

26

Yoshimasa Sakoma and Eiji Itoi

26.1 Muscular Architecture

The deltoid is a large and triangular muscle, which covers the complex of the humeral head and the rotator cuff. Typically, the deltoid is divided into three anatomical portions: the anterior, middle, and posterior portions (Fig. 26.1). The anterior portion originates from the anterior border and upper surface of the lateral one-third of the clavicle. The middle portion originates from the lateral margin and upper surface of the acromion. The posterior portion originates from the posterior border of the scapular spine. These three portions converge and insert to the deltoid tubercle on the lateral aspect of the humeral shaft. The arrangement of the muscle fibers differs among these three portions. The anterior and posterior portions have the parallel fibers with long excursion, whereas the middle portion has the oblique fibers in a multipennate fashion with short excursion. The superficial fascia of the proximal deltoid continues to the fascia of the trapezius muscle and forms the deltotrapezial fascia.

The origin and insertion of the deltoid have two attachment patterns: the tendinous attachment and the direct attachment of the muscle fibers. The muscle fibers of the anterior portion attach directly to the periosteum of the anterior surface of the distal clavicle. The muscle fibers of

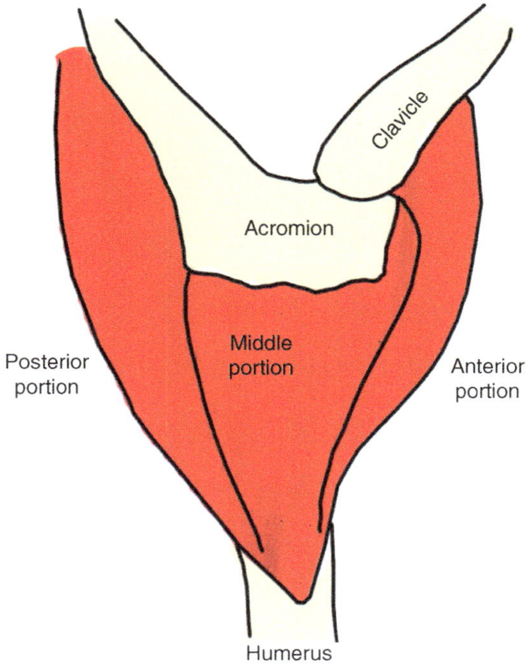

Fig. 26.1 Three parts of the deltoid. The anterior, middle, and posterior portions of the deltoid converge and run distally to the deltoid tuberosity on the humeral shaft

Y. Sakoma, MD, PhD
Department of Orthopaedic Surgery, Onomichi Municipal Hospital, 3-1170-177, Shintakayama, Onomichi 722-8503, Japan
e-mail: ysakoma@gmail.com

E. Itoi, MD, PhD (✉)
Department of Orthopaedic Surgery, Tohoku University School of Medicine, 1-1, Seiryo-machi, Aoba-ku, Sendai 980-8574, Japan
e-mail: itoi-eiji@med.tohoku.ac.jp

G.I. Bain et al. (eds.), *Normal and Pathological Anatomy of the Shoulder*,
DOI 10.1007/978-3-662-45719-1_26, © ISAKOS 2015

the posterior portion also attach directly to the periosteum of the scapular spine. In the middle portion, four intramuscular tendons originate from the lateral aspect of the acromion (Fig. 26.2a) [21, 28]. The muscle fibers arise from these intramuscular tendons and run downward to the intramuscular tendons of the insertion site. Tendinous insertion forms three discrete lines [14] or M-shaped insertion (Fig. 26.2b) [26, 28]. The anterior portion attaches to the anterior tendinous insertion and the posterior portion attaches to the posterior tendinous insertion. However, the middle portion diverges into three portions and each portion attaches to the anterior, middle, and posterior tendinous insertions, respectively. In other words, three origins and three insertions do not match with each other. The anterior insertion has three (A1, A2, and A3), the middle insertion has one (M1), and the posterior insertion has three (P1, P2, and P3) intramuscular tendons [19, 28].

According to the attachment pattern of the muscle fibers to the intramuscular tendons of the insertion, the deltoid can be divided into seven segments: i.e., A1, A2, A3, M1, P1, P2, and P3 segments (Fig. 26.2c) [28]. A1, A2, and A3 segments, M1 segments, and P1, P2, and P3 segments attach to the anterior, middle, and posterior insertions, respectively. In the classical division, A1 segment corresponds to the anterior portion, A2, A3, M1, and P1 segments correspond to the

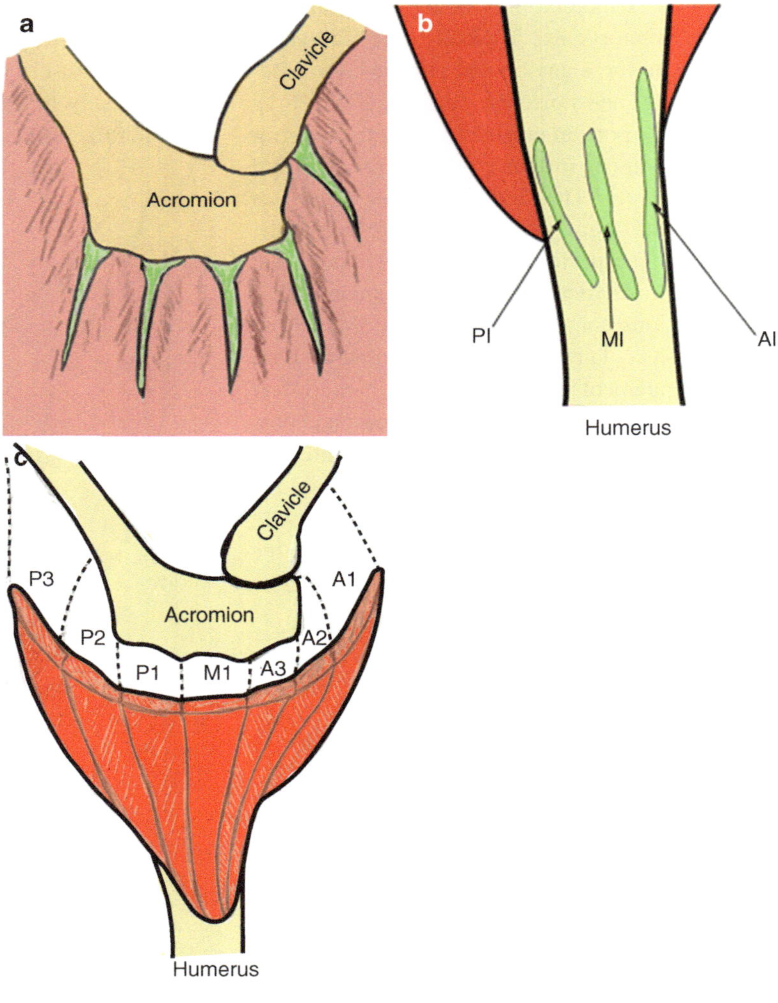

Fig. 26.2 Segments of the deltoid. The deltoid has several intramuscular tendons at its origin (**a**). The tendinous insertion of the deltoid forms three discrete lines (**b**). *AI* anterior insertion, *MI* middle insertion, *PI* posterior insertion (Modified from Rispoli et al. [26]). Based on the origins and insertions, the deltoid muscle can be divided into seven segments (**c**). The anterior segments (*A1*, *A2*, and *A3*) converge and attach to the anterior insertion. The middle segment (*M1*) attaches to the middle insertion. The posterior segments (*P1*, *P2*, and *P3*) converge and attach to the posterior insertion

middle portion, and P2 and P3 segments correspond to the posterior portion. As the intramuscular tendons are clearly depicted on T2-weighted transverse magnetic resonance (MR) images with fat suppression, these seven segments can be differentiated on MR images [28, 37].

The tendon-muscle-tendon unit is known as the basic functional unit of the muscle, and thus, these anatomical segments based on the intramuscular tendons should be taken into consideration when the function of the deltoid is discussed.

26.2 Innervation

Innervation of the deltoid is supplied by the axillary nerve (C5 and C6) branched from the posterior cord of the brachial plexus. The axillary nerve passes through the quadrilateral space from anterior to posterior direction and splits into two branches (the anterior and the posterior branches) within the quadrilateral space. Anterior branch travels around the surgical neck of the humerus and supplies the middle and anterior portions of the deltoid (Fig. 26.3). Moreover, sub-branches from the anterior branch supplies the posterior portion [18, 35]. The posterior branch runs posteriorly and supplies the posterior portion of the deltoid. The posterior branch also sends a motor branch to the teres minor muscle and a sensory branch (superior lateral brachial cutaneous nerve) to the superolateral area of the shoulder called "regimental badge." The posterior portion of the deltoid has double-supply from the anterior and posterior branches of the axillary nerve in 89.1 % of the cadaveric specimens [18]. In their series, the posterior portion is supplied by the anterior branch alone in 2.3 % and by the posterior branch alone in 8.5 % [18]. The mean diameter of the anterior and posterior branches are 4.0 and 3.3 mm, respectively [29]. The mean distance between the acromial edge and the axillary nerve varies among the reporters [20] reported that the distance from the humeral head and the axillary nerve ranged from 4.0 to 6.7 cm, however, the axillary nerve moves superiorly during abduction. The distance between the acromion and the axillary nerve ranged from 66.6 to 72.6 mm in the hanging arm position and from 53.9 to 61.6 mm in 60° of abduction [7].

Electromyographic assessment suggests that the deltoid consists of at least seven segments

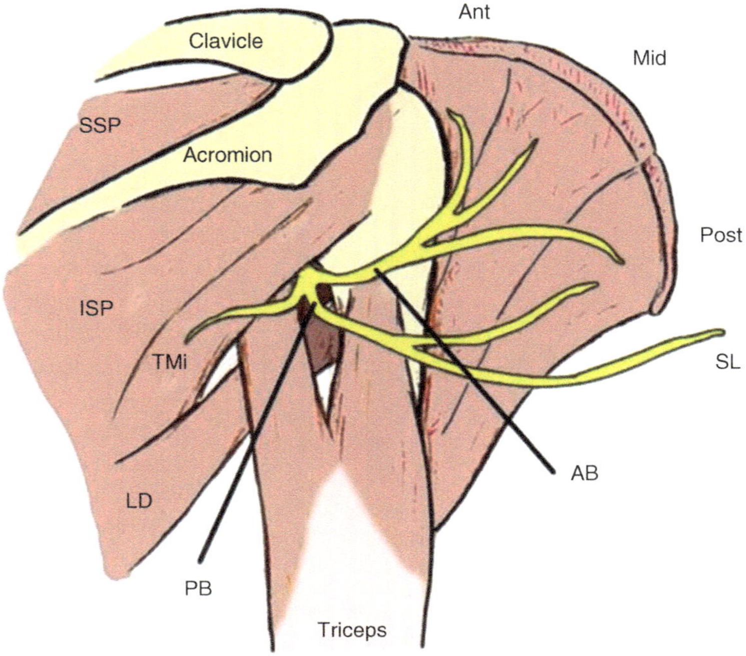

Fig. 26.3 Axillary nerve. The axillary nerve is divided into the anterior and the posterior branches in the quadrilateral space. The anterior branch innervates the whole deltoid, and the posterior branch provides a motor branch to the teres minor muscle and a sensory branch to the superolateral area of the shoulder. *AA* anterior branch of the axillary nerve, *PA* posterior branch of the axillary nerve, *SL* superior lateral brachial cutaneous nerve, *ant* anterior portion, *mid* middle portion, *post* posterior portion

coordinated by the central nervous system [3, 4, 34]. However, these functional seven segments may differ from the anatomical seven segments divided by the intramuscular tendons. In other words, the relation between the innervated segments and the anatomical segments is still unclear and needs further investigation.

26.3 Vascularity

The vascular supply of the deltoid is largely derived from the posterior circumflex humeral artery, which travels with the axillary nerve through the quadrilateral space (Fig. 26.4). The posterior circumflex humeral artery supplies the middle and the posterior portions of the deltoid. The thoracoacromial artery also provides a supply to the deltoid muscle. The thoracoacromial artery is branched from the axillary artery and separates into two branches: the deltoid artery and the acromial artery. The deltoid artery runs near the deltopectoral groove and supplies the anterior portion of the deltoid [5]. The acromial artery travels in a deep layer of the deltoid near the clavicle and acromion and gives some branches to the middle portion. The anterior circumflex humeral artery sends a branch to the anterior portion in 63 % of cadaveric specimens [11].

The venous branches accompany the arterial branches, except the cephalic vein, which runs in the deltopectoral groove.

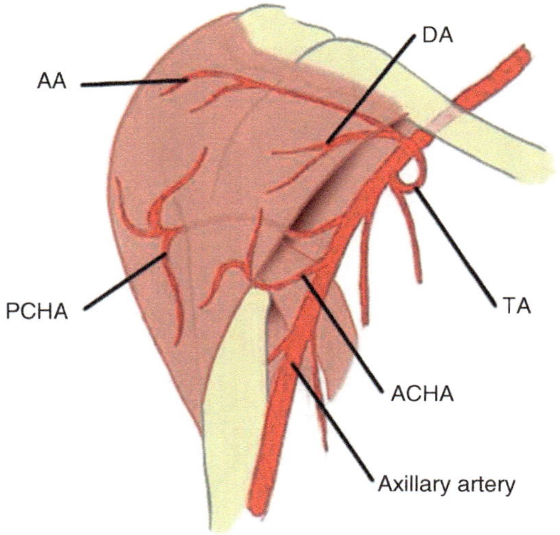

Fig. 26.4 The blood supply of the deltoid. The acromial and deltoid arteries, which branched from the thoracoacromial artery, and the anterior and posterior circumflex humeral arteries supply the deltoid muscle. *TA* thoracoacromial artery, *AA* acromial artery, *DA* deltoid artery, *ACHA* anterior circumflex humeral artery, *PCHA* posterior circumflex humeral artery

26.4 Function

The motion of the shoulder is a complex of actions in many directions such as flexion/extension, abduction/adduction, internal rotation/external rotation. In addition, many muscles co-work in any single motion. Therefore, it is quite difficult to assess the participation of each muscle to the shoulder movement. The muscle function is evaluated by various methods, such as the physiological cross-sectional area (PCSA), the moment arm, the potential moment, and the electromyographic (EMG) activity [8, 15, 25, 33]. The deltoid has a largest PCSA in the shoulder girdle and is a key muscle in abduction of the shoulder. Moreover, the deltoid co-works with the rotator cuff muscles and constructs a force couple to move the shoulder joint smoothly.

The anterior portion of the deltoid elevates the arm forward with some contribution by the clavicular portion of the pectorals major and the biceps brachii. The middle portion elevates the arm laterally. The posterior portion works with the teres minor and the latissimus dorsi and elevates the arm backward [22]. In a moment arm study, the posterior portion was estimated to provide 14 % of the shoulder extension torque in addition to 20 % of the shoulder extension torque [8].

The deltoid used to be thought to push the humeral head superiorly during arm elevation. However, recent biomechanical studies have demonstrated that the deltoid also stabilizes the humeral head against the glenoid fossa during arm elevation [2, 23, 31] (Fig. 26.5). This function is very important in a rotator cuff deficient shoulder.

Biomechanical studies during the throwing motion have demonstrated that the deltoid acts as an anterior stabilizer along with the rotator cuff muscles at 90° of abduction and 90° of external rotation. Although this anterior stabilizing effect is

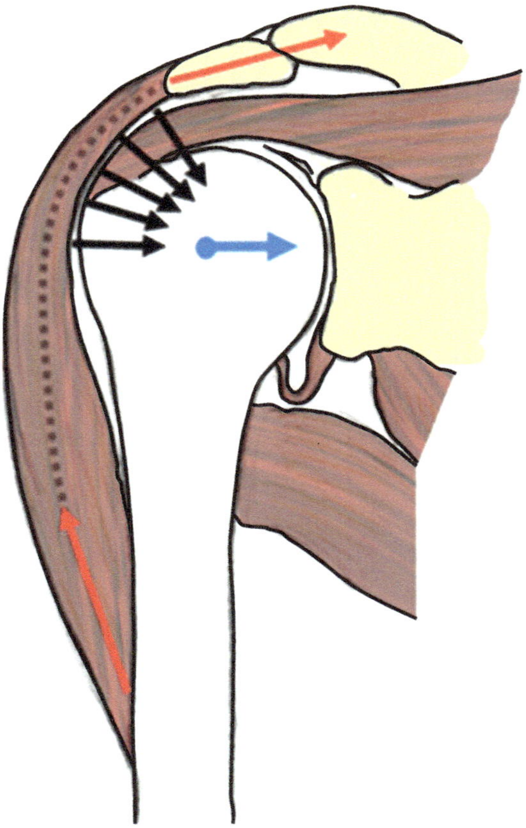

Fig. 26.5 The deltoid muscle as a stabilizer of the humeral head. The contraction force of the deltoid (*red arrows*) is converted into the stabilizing force (*black arrows*), which compresses the humeral head against the glenoid fossa (*blue arrow*). (Modified from Nyffeler et al. [23])

stronger in 90° of elevation in the scapular plane than in the coronal plane, it is still weaker than the stability provided by the rotator cuff [13, 17].

26.5 Pathology of the Deltoid

Axillary nerve palsy is not uncommon, but the other conditions of the deltoid are rare.

26.5.1 Rupture of Deltoid Muscle

A spontaneous rupture of the deltoid muscle is uncommon and usually considered as a complication of the shoulder surgery. In case after acromioplasty, the deltoid origin is partially detached from the acromion. The open acromioplasty sacrifices 40–70 % of the deltoid origin [30], whereas the arthroscopic acromioplasty preserves most of the footprint of the deltoid.

An isolated tear of the deltoid is reported in sports injuries and motor vehicle accidents. A tear occurs most frequently near the musculotendinous junction or at the origin from the acromion. A rupture at the middle of the muscle belly is rare [12]. A massive rotator cuff tear causes a deltoid tear, especially the middle portion. In a shoulder with a massive rotator cuff tear, the humeral head migrates superiorly and impinges against the acromial edge. At the same time, the undersurface of the deltoid muscle suffers a friction with the greater tuberosity, which may lead to a tear of the deep fascia of the deltoid. In a series of massive rotator cuff tears operated using a lateral para-acromial incision, the deltoid detachment occurred in 8 % of the patients during the first three months after surgery [9].

26.5.2 Axillary Nerve Palsy

The axillary nerve palsy is well-recognized as a posttraumatic complication, especially after anterior dislocation of the shoulder [24]. In the dislocated shoulder, the axillary nerve is stretched by the displaced humeral head until it is reduced [32]. The length of time during which the dislocation goes unreduced is related to the incidence of the transient or permanent injury of the axillary nerve. The incidence of the axillary nerve injury after the anterior shoulder dislocation ranges 3–21 % based on the clinical findings, while it ranges 19–65 % based on EMG abnormality [27, 36]. The axillary nerve injury is also observed as an iatrogenic complication of the shoulder surgery [1, 20]. During shoulder surgery, surgeons should be aware of the anatomy of the axillary nerve.

26.5.3 Other Pathologies

The contracture of the deltoid muscle is also a well-documented pathological problem. The contracture of the deltoid is a rare condition, however, it is observed as a consequence of the intra-deltoid injections of vaccines, antibiotics, or other pharmaceutics [6, 10] reported the mechanisms of developing the contracture. The injection or puncture into the deltoid may cause the direct muscle disruption, local edema, vascular damage, and focal myositis of the deltoid muscle. These pathological changes induce the muscle ischemia and the abnormality of the collagen synthesis, which causes local fibrosis. These fibrotic bands may induce contracture of the deltoid. The injected drugs also present a myotoxicity and may cause the same condition. The hypertrophic fibrotic cords are detectable on ultrasonography or MR images.

The calcific tendonitis around the shoulder is a common disease, especially it is observed in the rotator cuff tendons. The deltoid is also affected by the calcific tendonitis, but it is usually asymptomatic. The inflammation caused by the calcification at the rotator cuff and the subacromial bursa may extend and affect the deep surface of the deltoid [22].

The enthesopathy of the deltoid is observed in patients with ankylosing spondylitis. In an MRI study, the bone marrow edema was found at the acromial enthesis, clavicular enthesis, as well as the deltoid tuberosity [16].

Various kinds of benign and malignant tumors may affect the deltoid. Examples of benign tumors include lipoma, hemangioma, desmoid tumor. Malignant tumors include liposarcoma and fibrosarcoma (Fig. 26.6).

a b

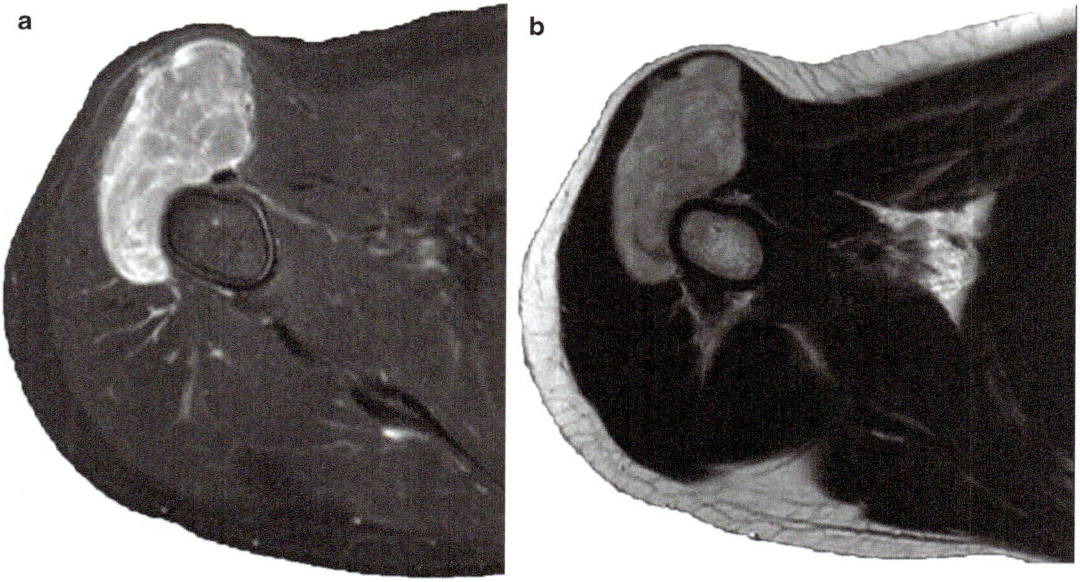

Fig. 26.6 Synovial sarcoma of the deltoid (10-year-old girl). (**a**) T2-weighted axial image shows high signal intensity of oval shaped tumor with irregular surface in the anterior to middle part of the deltoid muscle. (**b**) T1-weighted, fat suppressed, Gad-enhanced axial image shows irregular enhancement inside the tumor with clear margin

References

1. Apaydin N, Tubbs RS, Loukas M, Duparc F. Review of the surgical anatomy of the axillary nerve and the anatomic basis of its iatrogenic and traumatic injury. Surg Radiol Anat. 2010;32(3):193–201. doi:10.1007/s00276-009-0594-8.
2. Billuart F, Devun L, Skalli W, Mitton D, Gagey O. Role of deltoid and passives elements in stabilization during abduction motion (0°–40°): an ex vivo study. Surg Radiol Anat. 2008;30(7):563–8. doi:10.1007/s00276-008-0374-x.
3. Brown JM, Wickham J. Neuromotor coordination of multisegmental muscle during a change in movement direction. J Musculoskelet Res. 2006;10:63–74.
4. Brown JM, Wickham J, McAndrew DJ, Huang XF. Muscles within muscles: coordination of 19 muscle segments within three shoulder muscles during isometric motor tasks. J Electromyograph Kinesiol. 2007;17:57–73.

5. Bunker TD, Cosker TD, Dunkerley S, Kitson J, Smith CD. Anatomical variations of the deltoid artery: relevance to the deltopectoral approach to the shoulder. Bone Joint J. 2013;95-B(5):657–9. doi:10.1302/0301-620X.95B5.31408.
6. Chen SS, Chien CH, Yu HS. Syndrome of deltoid and/or gluteal fibrotic contracture: an injection myopathy. Acta Neurol Scand. 1988;78(3):167–76.
7. Cheung S, Fitzpatrick M, Lee TQ. Effects of shoulder position on axillary nerve positions during the split lateral deltoid approach. J Shoulder Elbow Surg. 2009;18(5):748–55. doi:10.1016/j.jse.2008.12.001.
8. Crouch DL, Plate JF, Li Z, Saul KR. Biomechanical contributions of posterior deltoid and teres minor in the context of axillary nerve injury: a computational study. J Hand Surg Am. 2013;38(2):241–9. doi:10.1016/j.jhsa.2012.11.007.
9. Gumina S, Di Giorgio G, Perugia D, Postacchini F. Deltoid detachment consequent to open surgical repair of massive rotator cuff tears. Int Orthop. 2008;32(1):81–4.
10. Huang CC, Ko SF, Ko JY, Wan YL, Cheng YF, Wang CJ. Imaging factors related to rotator cuff tear in patients with deltoid contracture. J Formos Med Assoc. 2006;105(2):132–8.
11. Hue E, Gagey O, Mestdagh H, Fontaine C, Drizenko A, Maynou C. The blood supply of the deltoid muscle. Application to the deltoid flap technique. Surg Radiol Anat. 1998;20(3):161–5.
12. Ilaslan H, Iannotti JP, Recht MP. Deltoid muscle and tendon tears in patients with chronic rotator cuff tears. Skeletal Radiol. 2007;36(6):503–7.
13. Kido T, Itoi E, Lee SB, Neale PG, An KN. Dynamic stabilizing function of the deltoid muscle in shoulders with anterior instability. Am J Sports Med. 2003; 31:399–403.
14. Klepps S, Auerbach J, Calhon O, Lin J, Cleeman E, Flatow E. A cadaveric study on the anatomy of the deltoid insertion and its relationship to the deltopectoral approach to the proximal humerus. J Shoulder Elbow Surg. 2004;13:322–7.
15. Kronberg M, Nemeth G, Brostrom LA. Muscle activity and coorination in the normal shoulder. An Electromyographic Study. Clin Orthop Relat Res. 1990;257:76–85.
16. Lambert RG, Dhillon SS, Jhangri GS, Sacks J, Sacks H, Wong B, Russell AS, Maksymowych WP. High prevalence of symptomatic enthesopathy of the shoulder in ankylosing spondylitis: deltoid origin involvement constitutes a hallmark of disease. Arthritis Rheum. 2004;51(5):681–90.
17. Lee SB, An KN. Dynamic glenohumeral stability provided by three heads of the deltoid muscle. Clin Orthop Relat Res. 2002;400:40–7.
18. Leechavengvongs S, Teerawutthichaikit T, Witoonchart K, Uerpairojkit C, Malungpaishrope K, Suppauksorn S, Chareonwat B. Surgical anatomy of the axillary nerve branches to the deltoid muscle. Clin Anat. 2015; 28(1):118–22. doi:10.1002/ca.22352.
19. Leijnse JN, Han SH, Kwon YH. Morphology of deltoid origin and end tendons – a generic model. J Anat. 2008;213:733–42.
20. Liu KY, Chen TH, Shyu JF, Wang ST, Liu JY, Chou PH. Anatomic study of the axillary nerve in a Chinese cadaveric population: correlation of the course of the nerve with proximal humeral fixation with intramedullary nail or external skeletal fixation. Arch Orthop Trauma Surg. 2011;131(5):669–74. doi:10.1007/s00402-010-1184-2.
21. Lorne E, Gagey O, Quillard J, Hue E, Gagey N. The fibrous frame of the deltoid muscle. Its functional and surgical relevance. Clin Orthop Relat Res. 2001;386: 222–5.
22. Moser T, Lecours J, Michaud J, Bureau NJ, Guillin R, Cardinal E. The deltoid, a forgotten muscle of the shoulder. Skeletal Radiol. 2013;42(10):1361–75. doi:10.1007/s00256-013-1667-7.
23. Nyffeler RW, Werner CML, Sukthankar A, Schmid MR, Gerber C. Association of a large lateral extension of the acromion with rotator cuff tears. J Bone Joint Surg Am. 2006;88:800–5.
24. Perlmutter GS. Axillary nerve injury. Clin Orthop Relat Res. 1999;368:28–36.
25. Peterson SL, Rayan GM. Shoulder and upper arm muscle architecture. J Hand Surg Am. 2011;36(5):881–9. doi:10.1016/j.jhsa.2011.01.008.
26. Rispoli DM, Athwal GS, Sperling JW, Cofield RH. The anatomy of the deltoid insertion. J Shoulder Elbow Surg. 2009;18:386–90.
27. Robinson CM, Shur N, Sharpe T, Ray A, Murray IR. Injuries associated with traumatic anterior glenohumeral dislocations. J Bone Joint Surg Am. 2012;94(1):18–26. doi:10.2106/JBJS.J.01795.
28. Sakoma Y, Sano H, Shinozaki N, Itoigawa Y, Yamamoto N, Ozaki T, Itoi E. Anatomical and functional segments of the deltoid muscle. J Anat. 2010;218(2):185–90. doi:10.1111/j.1469-7580.2010.01325.x.
29. Stecco C, Gagliano G, Lancerotto L, Tiengo C, Macchi V, Porzionato A, De Caro R, Aldegheri R. Surgical anatomy of the axillary nerve and its implication in the transdeltoid approaches to the shoulder. J Shoulder Elbow Surg. 2010;19(8):1166–74. doi:10.1016/j.jse.2010.05.010.
30. Torpey BM, Ikeda K, Weng M, van der Heeden D, Chao EY, McFarland EG. The deltoid muscle origin. Histologic characteristics and effects of subacromial decompression. Am J Sports Med. 1998;26(3): 379–83.
31. Torrens C, López JM, Puente I, Cáceres E. The influence of the acromial coverage index in rotator cuff tears. J Shoulder Elbow Surg. 2007;16:347–51.
32. Tuckman GA, Devlin TC. Axillary nerve injury after anterior glenohumeral dislocation: MR findings in three patients. AJR Am J Roentgenol. 1996;167(3):695–7.
33. Webb JD, Blemker SS, Delp SL. 3D finite element models of shoulder muscles for computing lines of actions and moment arms. Comput Methods Biomech Biomed Engin. 2014;17(8):829–37. doi:10.1080/10255842.2012.719605.

34. Wickham J, Brown JM. Muscles within muscles: the neuromotor control of intra-muscular segments. Eur J Appl Physiol. 1998;78:219–25.
35. Wysiadecki G, Polguj M, Krasucki K, Zytkowski A, Smigielski J, Topol M, Orkisz S. Morphology and a proposed model of innervation of the human deltoid muscle: a pilot study. Folia Morphol (Warsz). 2014;73(2):216–23. doi:10.5603/FM.2014.0032.
36. Yeap JS, Lee DJ, Fazir M, Kareem BA, Yeap JK. Nerve injuries in anterior shoulder dislocations. Med J Malaysia. 2004;59(4):450–4.
37. Meyer DC, Rahm S, Farshad M, Lajtai G, Wieser K. Deltoid muscle shape analysis with magnetic resonance imaging in patients with chronic rotator cuff tears. BMC Musculoskelet Disord. 2013;14:247. doi: 10.1186/1471-2474-14-247.

Periscapular Muscles

27

William Ben Kibler and Aaron Sciascia

27.1 Osseous Anatomy

A thorough understanding of osteology and the muscular attachments of the scapula is required to grasp the complex movement patterns and function [1, 2]. The anatomy is predicated on the developmental advantages of mobility, such as prehension and overhead use. This is reflected in several primary changes noted through time in the hominid scapula. First, the acromion has broadened and lateralized to allow mechanical advantage for the deltoid muscle. The coracoid enlarged in a manner theorized to assist in the prevention of anterior dislocation at 90° of abduction. Finally, broadening and alteration in the force vector of the infraspinatus and teres minor are postulated to increase both external rotation strength and humeral head depression.

The scapula is a large flat bone which forms from a collection of mesenchymal cells. It shows signs of ossification by the 5th week of embryologic development. The scapula follows a predictable course in descending from the paracervical region to the thorax. Failure of this process leads to Sprengel's deformity. By the 7th week, the scapula has descended to its final position and the glenoid is easily identified.

Grossly, the scapula is a thin sheet of bone which serves as a critical site of muscle attachment. The blood supply is primarily through a network of periosteal vessels which take origin from muscular insertions. Thickening of the bone is notable at the lateral border, and superior and inferior angles. Ventral concavity creates a smooth articulating surface against the ribs. Small oblique ridges exist ventrally for the tendinous insertions of the subscapularis. Similarly, small fibrous septa are present dorsally to attach and separate the infraspinatus, teres minor, and teres major. The dorsal surface is traversed by the scapular spine which divides two concavities; the supraspinatus and infraspinatus fossas. The medial two-thirds of these fossas give rise to the supraspinatus and infraspinatus muscles. The spine contains two important notches. First, the suprascapular notch at the base of the coracoid contains the suprascapular nerve, and compression at this location will affect both the supraspinatus and infraspinatus muscles. Second, the spinoglenoid notch is present at the lateral border of the spine. Various causes can lead to compression of the suprascapular nerve here as well, producing isolated atrophy of the infraspinatus.

Anatomic interest in the scapula is frequently directed at the coracoid, acromion, or glenoid. The name coracoid derives from the Greek word *korakodes* meaning "like a crow's beak." The bent shape resembles a finger pointed toward the glenoid. From the Greek word *akros* for point, the acromion is often referred to as the point of

W.B. Kibler, MD (✉) • A. Sciascia, MS, ATC, PE
Shoulder Center of Kentucky, 1221 South Broadway, Lexington, KY 40504, USA
e-mail: wkibler@aol.com; ascia@lexclin.com

G.I. Bain et al. (eds.), *Normal and Pathological Anatomy of the Shoulder*,
DOI 10.1007/978-3-662-45719-1_27, © ISAKOS 2015

the shoulder. The morphology of the acromion is among the most studied in the body. Considerable cadaveric research has been directed at the relative frequency and postulated causes of the different acromion types. However, the relationship between acromial shape and "impingement syndrome" or rotator cuff tear has not borne out in literature. Similarly, the glenoid has been the subject of intensive study in an effort to define bony anatomy in shoulder instability. Average values for size include a height of 35 mm and width of 25 mm, but considerable variability exists. Comparison to the contralateral side may be required to precisely define bone loss. Glenoid version may also range widely. Retroversion, up to 6° is most common, as seen in 75 % of the population, but anteversion up to 2° is reported.

27.2 Muscle Anatomy and Function

The function of the scapula is dependent on the complex recruitment patterns of the numerous muscular attachments [1, 3, 5, 6]. These muscles can generally be categorized as: axioscapular and scapulohumeral muscles of the upper arm (coracobrachialis, biceps brachii, and triceps brachii).

The axioscapular muscles serve to anchor the scapula for its role as the foundation of the shoulder. In addition, they guide the scapula through the requisite degrees of freedom. These muscles include the serratus anterior, levator scapula, pectoralis minor, rhomboids, and trapezius. The trapezius is the largest and most superficial axioscapular muscle. The expansive muscle originates from the occiput, nuchal ligament, and spinous processes of C7 through T12. The upper trapezius inserts across the distal third of the clavicle and acromion. The middle trapezius inserts across the scapular spine and the lower portion at the base of the spine. The broad muscle allows complex function in scapular retraction, elevation, and posterior tilting based upon the recruitment pattern [2, 5]. Frequently, the upper and lower trapezius muscles are associated separately. Motor innervation is through cranial nerve 11, the spinal accessory nerve.

The rhomboids are divided into major and minor portions. The rhomboid minor originates from the spinous processes of C7 and T1 and inserts at the medial scapular border at the base of the spine. The rhomboid major begins from T2 through T5 and inserts along the posterior aspect of the medial border from the base of the spine caudally to the inferior angle. This orientation allows an important role in scapular retraction. The dorsal scapular nerve (C5) provides innervation.

The serratus anterior is comprised of three divisions taking origin from the anterolateral aspect of the first through ninth ribs. Innervation of the serratus is provided by the long thoracic nerve. The serratus produces protraction which is a composite of scapular translation and multidirectional rotation. It is not uncommon for upward rotation, posterior tilting, and external rotation of the scapula to occur simultaneously during protraction [9]. The role of the serratus during arm elevation is to provide a critical stabilization function against excessive internal rotation throughout nearly all positions of arm forward flexion and elevation.

The levator scapula is intimately associated with the serratus as a stabilizer but also serves a role to elevate and upwardly/downwardly rotate the scapula. The levator originates from the transverse processes of C1 through C3, and at times C4. Insertion is found upon the superior angle. Innervation comes from the deep branches of C3 and 4.

The pectoralis minor is often overlooked in its role in scapular position. The muscle originates from the second through fifth ribs and courses superolateraly to insert upon the coracoid. It is ideally located to assist the serratus anterior in scapular protraction and anterior tilt. Chronic tightness can contribute to protracted, anteriorly tilted scapular positioning.

The scapulohumeral muscles produce glenohumeral motion and are composed of the deltoid, supraspinatus, infraspinatus, subscapularis, teres minor, and teres major. The deltoid originates broadly across the acromion and scapular spine while inserting on the deltoid tubercle of the humerus. This structure allows it to power elevation in multiple planes. As previously noted, the supraspinatus and infraspinatus originate from the medial two-thirds of their respective fossas while inserting in a complex arrangement on the greater tuberosity. The subscapularis originates from the anterior aspect of the scapula and attaches on the lesser tuberosity. The pennation pattern of the rotator cuff fibers, inserting obliquely to the tendon, allow them to exert a strong compressive force on the humerus, increasing concavity/compression. The teres minor takes origin from the middle section of the lateral scapula and is innervated by the posterior branch of the axillary nerve. The teres major emerges from a more inferior position on the lateral scapula and shares a common tendinous insertion with the latissimus dorsi on the medial aspect of the bicipital groove. It shares innervation from the subscapular nerve and functions in internal rotation, adduction, and extension of the humerus.

Scapular stabilization on the thorax involves coupling of the upper and lower fibers of the trapezius muscle with the serratus anterior and rhomboid muscles [10]. Elevation of the scapula with arm elevation is accomplished through activation and coupling of the serratus anterior and lower trapezius with the upper trapezius and rhomboids [10, 11]. During this motion, the lower trapezius helps maintain the instant center of rotation of the scapula through its attachment to the medial scapular spine, which is mechanically advantageous. Its attachment to the scapular spine allows for a straight line of pull as the arm elevates and the scapula upwardly rotates [12]. The lower trapezius has been often identified as an upward rotator of the scapula because it maintains its long moment arm during the full range of upward rotation [12]. However, it also has a role as a scapular stabilizer when the arm is lowered from an elevated position. During the descent or return from upward elevation, the well-positioned lower trapezius, when operating efficiently, helps maintain the scapula against the thorax.

The serratus anterior also has a role as a stabilizer of the scapula. This muscle has been historically identified as a protractor of the scapula due to high EMG activity elicited during various push-up maneuvers [13, 14]. Other evidence suggests the serratus muscle helps upwardly rotate the scapula [11, 12]. The serratus anterior is actually multi-faceted in that it contributes to all components of three-dimensional motion of the scapula during arm elevation [15]. The serratus muscle helps produce scapular upward rotation, posterior tilt, and external rotation while stabilizing the medial border and inferior angle which prevents scapular winging [4, 15]. This is most likely due to the variable fiber orientation of the serratus anterior on the scapula and thorax. The highest level of serratus anterior activation occurs in the cocking phase of the throwing motion [16–19], and serratus anterior activation occurs in the earliest stages of arm elevation [20]. It would appear that a prime role of the serratus in these activities is as an external rotator/stabilizer of the scapula with arm motion.

The scapular position that allows optimal muscle activation to occur is that of retraction and external rotation. Scapular retraction is an obligatory and integral part of normal scapulohumeral rhythm in coupled shoulder motions and functions [12, 21]. It results from synergistic muscle activations in patterns from the hip and trunk through the scapula to the arm, which then facilitates maximal muscle activation of the muscles attached to the scapula [22–26]. The retracted scapula then can act as a stable base for the origin of all the rotator cuff muscles [27, 28]. Positions of protraction have been shown to be limiting to both muscle strength and motion [7, 8].

27.3 Summary

The anatomy of the scapula allows the integrated complex patterns of arm movement to occur. The bony anatomy serves as a platform for the numerous muscular attachments and contributes to the many degrees of freedom at the glenohumeral joint. Simultaneous stabilization and movement occurs as a result of the multiple periscapular muscles which contribute to the complexity of scapulohumeral motion. Understanding the scapular roles is best achieved through an understanding of the anatomical design and function of the scapula which in turn would assist clinicians in identifying deviations from normal function. Evaluation of any patient with shoulder injury should include assessment of scapular resting position and dynamic motion as well as strength and function of the periscapular muscles.

References

1. Kibler WB, Sciascia AD. Current concepts: scapular dyskinesis. Br J Sports Med. 2010;44(5):300–5.
2. Kibler WB, Sciascia A, Wilkes T. Scapular dyskinesis and its relation to shoulder injury. J Am Acad Orthop Surg. 2012;20(6):364–72.
3. Lukasiewicz AC, McClure P, Michener L, Pratt N, Sennett B. Comparison of 3-dimensional scapular position and orientation between subjects with and without shoulder impingement. J Orthop Sports Phys Ther. 1999;29(10):574–86.
4. McClure PW, Michener LA, Sennett BJ, Karduna AR. Direct 3-dimensional measurement of scapular kinematics during dynamic movements in vivo. J Shoulder Elbow Surg. 2001;10:269–77.
5. Ludewig PM, Cook TM. Alterations in shoulder kinematics and associated muscle activity in people with symptoms of shoulder impingement. Phys Ther. 2000;80(3):276–91.
6. Kibler WB, McMullen J. Scapular dyskinesis and its relation to shoulder pain. J Am Acad Orthop Surg. 2003;11:142–51.
7. Kebaetse M, McClure PW, Pratt N. Thoracic position effect on shoulder range of motion, strength, and three-dimensional scapular kinematics. Arch Phys Med Rehabil. 1999;80:945–50.
8. Smith J, Kotajarvi BR, Padgett DJ, Eischen JJ. Effect of scapular protraction and retraction on isometric shoulder elevation strength. Arch Phys Med Rehabil. 2002;83:367–70.
9. Baugh DD, McClure PW, Karduna AR. Three-dimensional scapulothoracic motion during active and passive arm elevation. Clin Biomech (Bristol, Avon). 2005;20(7):700–9.
10. Speer KP, Garrett WE. Muscular control of motion and stability about the pectoral girdle. In: Matsen Iii FA, Fu F, Hawkins RJ, editors. The shoulder: a balance of mobility and stability. Rosemont: American Academy of Orthopaedic Surgeons; 1994. p. 159–73.
11. Bagg SD, Forrest WJ. Electromyographic study of the scapular rotators during arm abduction in the scapular plane. Am J Phys Med. 1986;65(3):111–24.
12. Bagg SD, Forrest WJ. A biomechanical analysis of scapular rotation during arm abduction in the scapular plane. Am J Phys Med Rehabil. 1988;67:238–45.
13. Moseley JB, Jobe FW, Pink MM, Perry J, Tibone JE. EMG analysis of the scapular muscles during a shoulder rehabilitation program. Am J Sports Med. 1992;20(2):128–34.
14. Decker MJ, Hintermeister RA, Faber KJ, Hawkins RJ. Serratus anterior muscle activity during selected rehabilitation exercises. Am J Sports Med. 1999;27(6):784–91.
15. Ludewig PM, Cook TM, Nawoczenski DA. Three-dimensional scapular orientation and muscle activity at selected positions of humeral elevation. J Orthop Sports Phys Ther. 1996;24(2):57–65.
16. Glousman R, Jobe FW, Tibone JE, Moynes D, Antonelli D, Perry J. Dynamic electromyographic analysis of the throwing shoulder with glenohumeral instability. J Bone Joint Surg. 1988;70:220–6.
17. Glousman R. Electromyographic analysis and its role in the athletic shoulder. Clin Orthop Relat Res. 1993;288:27–34.
18. Bast SC, Mohr KJ, Glousman RE. Electromyographic analysis in overhead activities in athletes. Sports Med Arthrosc Rev. 2000;8(2):135–40.
19. DiGiovine NM, Jobe FW, Pink M, Perry J. An electromyographic analysis of the upper extremity in pitching. J Shoulder Elbow Surg. 1992;1(1):15–25.
20. Kibler WB, Chandler TJ, Shapiro R, Conuel M. Muscle activation in coupled scapulohumeral motions in the high performance tennis serve. Br J Sports Med. 2007;41:745–9.
21. Happee R, Van der Helm FC. The control of shoulder muscles during goal directed movements, an inverse dynamic analysis. J Biomech. 1995;28(10):1179–91.
22. Prilutsky BI, Isaka T, Albrecht AM, Gregor RJ. Is coordination of two-joint leg muscles during load lifting consistent with the strategy of minimum fatigue. J Biomech. 1998;31:1025–34.
23. Hirashima M, Kadota H, Sakurai S, Kudo K, Ohtsuki T. Sequential muscle activity and its functional role in the upper extremity and trunk during overarm throwing. J Sports Sci. 2002;20:301–10.
24. Hirashima M, Kudo K, Watarai K, Ohtsuki T. Control of 3D limb dynamics in unconstrained overarm throws of different speeds performed by skilled baseball players. J Neurophysiol. 2007;97(1):680–91.
25. Hirashima M, Yamane K, Nakamura Y, Ohtsuki T. Kinetic chain of overarm throwing in terms of joint rotations revealed by induced acceleration analysis. J Biomech. 2008;41:2874–83.
26. Cordo PJ, Nashner LM. Properties of postural adjustments associated with rapid arm movements. J Neurophysiol. 1982;47(2):287–308.
27. Kibler WB, Sciascia AD, Dome DC. Evaluation of apparent and absolute supraspinatus strength in patients with shoulder injury using the scapular retraction test. Am J Sports Med. 2006;34(10):1643–7.
28. Tate AR, McClure P, Kareha S, Irwin D. Effect of the scapula reposition test on shoulder impingement symptoms and elevation strength in overhead athletes. J Orthop Sports Phys Ther. 2008;38(1):4–11.

Kinematics of Scapular Motion

28

William Ben Kibler and Aaron Sciascia

28.1 Biomechanics of Normal Scapular Motions

Scapular motion has traditionally been described in a single planar two-dimensional model, with scapular upward rotation and acromial elevation as the end point [1, 2]. Upper trapezius activation to pull the acromion up and serratus anterior activation to move the inferior border laterally were described as the key muscle force couples [1, 3, 4]. The magnitude of the upward rotation varied between studies but averaged 60°, establishing the 1:2 scapula/humerus motion ratio for total scapulohumeral rhythm (SHR).

Recent studies show that scapular motion is actually multiplanar and three-dimensional. This work, using motion tracking systems and indwelling bone pins [5–8] demonstrates that total scapular movement is a composite of motions (rotations around axes) and translations (sliding along a surface).

The three observable rotary motions are upward/downward rotation around an axis perpendicular to the scapular body, internal/external rotation around a vertical axis along the medial border, and anterior/posterior tilt around a horizontal axis along the scapular spine. The absolute amount of each rotation varies within the studies. The indwelling bone pin study by Ludewig is probably the most accurate [7]. It shows that the resting position, in relation to the thorax, is 5.4° of upward rotation, 41.1° of internal rotation, and 13.5° of anterior tilt. As the arm moves to maximum elevation, the scapula moves in all three motions. It upwardly rotates 45°, posteriorly tilts 21°, and moves into internal then external rotation with a net change towards external rotation of 2° [9]. The largest part of these motions occurs in arm elevations above 80°.

Two translations can occur in the presence of an intact clavicular strut and acromioclavicular (AC) joint. They are upward/downward sliding on the thorax due to clavicular upward/downward motion at the sternoclavicular (SC) joint and anterior/posterior sliding around the curvature of the thorax due to clavicular anterior/posterior motion at the SC joint.

The clavicle and the SC and AC joints are major factors in creating the scapular positions, motions, and translations. The clavicle is the only bony connection of the scapula to the axial skeleton. This creates a stable strut, allowing controlled motion in several planes. To maximize scapular movement and scapulohumeral motion during maximal arm elevation, the clavicle retracts 16°, elevates 6°, and posteriorly rotates on its long axis 31° [9]. All these motions are

---

Electronic supplementary material The online version of this chapter 10.1007/978-3-662-45719-1_28 contains supplementary material, which is available to authorized users.

W.B. Kibler, MD (✉) • A. Sciascia, MS, ATC, PES
Shoulder Center of Kentucky,
1221 South Broadway, Lexington, KY 40504, USA
e-mail: wkibler@aol.com; ascia@lexclin.com

G.I. Bain et al. (eds.), *Normal and Pathological Anatomy of the Shoulder*,
DOI 10.1007/978-3-662-45719-1_28, © ISAKOS 2015

based on the SC joint. AC joint motions resulting from acromial motion on the clavicle demonstrated 8° of internal rotation, 11° of upward rotation, and 19° of posterior tilting [9]. These constrained motions create a reproducible efficient screw axis of motion between the clavicle and scapula through the AC joint and allows the three-dimensional motions [10].

Because of the relatively limited bony attachment, the scapula is mainly dependent upon muscle activation for mobility and stability. This activation allows controlled dynamic motion around the clavicular strut. The only connections of the entire medial scapular border to the axial skeleton are muscular. This allows a great amount of mobility and accommodation to many demands for different arm positions, but also creates large demands for eccentric muscle activation to withstand high distraction loads in activities involving forward arm motions or withstanding loading in the arm. Multiple periscapular muscles are activated to maintain this stability and activate the mobility.

Most research reveals that the upper and lower trapezius muscles, which usually are activated independently, and the serratus anterior muscle, contribute the most to scapular stability and mobility [3, 4, 11]. The upper trapezius acts on the acromion and upper lateral border of the scapular spine, and the serratus anterior acts on the inferior medial border. Coupling of activation of these two muscles initiates upward rotation and posterior tilt [4]. This force couple is especially active at the beginning of arm elevation [1] and with arm elevation below 90°. As the arm elevation exceeds 90° the lower trapezius is precisely positioned to increase and maintain upward rotation through a direct line of pull [1, 9]. In this arm position, the serratus anterior works to stabilize the medial border against the thorax, acting as a scapular external rotator. Lower trapezius activation is also important in the descent from maximum elevation, being activated eccentrically to control excessive anterior tilt. Other intrinsic muscles, the rhomboids and pectoralis minor, play important but not primary roles. The rhomboids are important muscles to initiate and maintain external scapular rotation in arm elevation and to control against excessive internal scapular rotation in arm descent. The pectoralis minor assists the serratus anterior as a scapular protractor in pushing activities but frequently becomes tight and shortened. Extrinsic muscles, mainly the latissimus dorsi and pectoralis major, create scapular motion through their effect as prime movers of the arm. Humeral motion also can create scapular motion by tension on the glenohumeral capsule and muscles, especially when glenohumeral internal rotation deficit (GIRD) is present. Finally, the scalenes and sternocleidomastoid muscles may affect scapular position due to their effect on clavicle motion.

Since these stability and mobility muscles all attach to the axial skeleton, control of posture and stability in the core is as important for maximal activation of these muscles as control of the scapula is important for maximal rotator cuff activation. Maximal activation of these muscles and force couples only occurs through patterns of activation that start from the core and proceed to the extremities [12–14]. These patterns coordinate co-contractions and force couples and synergize activation to maximize the developed strength [15–17]. Recent research shows that lower trapezius and serratus anterior activation is maximized when the recruitment is in a diagonal direction, from the contralateral hip through the lumbodorsal fascia to the lower trapezius [18].

28.2 Roles of the Scapula in Normal Shoulder Function

The complex motions and translations are necessary to allow the scapula to function as part of SHR, the integrated coupled motion of the moving arm and scapula that is the basis for effective upper extremity use. This creates the biomechanically advantageous position of stabilized retraction, and facilitates efficiency in SHR and glenohumeral function. Dr Carter Rowe has characterized this coupling as "a ball on a sea lion's nose." The sea lion moves its body (the scapula) to keep its nose (the glenoid) in line with the moving ball (the humerus) so there is no instability. This dynamic analogy is more physiologic than the static "golf ball on a tee." The scapula plays several roles to achieve efficient SHR.

First, it allows a congruent ball and socket arrangement through the full ranges of arm motion by keeping the alignment of the humerus and glenoid within physiologic limits. These limits have been calculated to be ±29.3° of glenohumeral angulation [19]. This arrangement maximizes the concavity/compression capability of the joint [20]. The relatively straight alignment of the bones also allows the rotator cuff muscles to maximize their efficiency in co-contraction and compression, pulling in straight lines on both sides of the joint. The relatively equivalent amounts of joint compression around the glenoid allow the labrum to work most effectively as a washer for the joint, decreasing peak joint loads and spreading compression effects [21].

Second, it creates a stable base for optimal activation of the scapular based muscles. Studies in asymptomatic subjects have documented that maximal demonstrated rotator cuff strength can be developed when the scapula is stabilized in a position of neutral retraction [22]. Excessive protraction or retraction decreased the developed strength by 11 %. In symptomatic subjects, the change was even greater. Stabilization of the scapula in retraction increased the developed strength by 24 % [23, 24]. These changes result from improved stability of the scapula and from the facilitation of rotator cuff activation by increased muscle activation. The use of the stabilized retraction position in the clinical evaluation of rotator cuff strength will increase the test/retest reliability and lead to accurate assessment of changes in strength with treatment.

A third role is one of clearance of the acromion as the arm elevates. Most kinematic studies show that posterior tilt is necessary, in addition to upward rotation, to allow maximum arm flexion [6, 9, 25]. This position allows optimal function in overhead activities, and reduces the occurrence of external impingement symptoms.

The final role is of optimal force transfer from the site of largest force development – the core – to the most common force delivery site, the hand, as part of the kinetic chain of all integrated dynamic body activities. The shoulder acts as a funnel, transmitting and concentrating the developed forces. This function requires dynamic stability for the efficient transfer of energy. The dynamic stability is created by the actions of the scapular stabilizers which are maximized when hip and trunk strength is maximized [18].

28.3 Alterations of Scapular Motions and Roles – Scapular Dyskinesis

Most discussions regarding alterations in scapular motion center on the terms "winged scapula" [26] and "snapping scapula" [27]. The "winged" scapula is a descriptive term usually used to identify the patient with an asymmetrically prominent medial scapular border, either at rest or upon arm motion [26]. There is commonly a deficit in shoulder function due to the scapular instability. In past literature, it was assumed that the large majority of cases were due to injury to one of the nerves supplying the scapular stabilizing musculature, either the long thoracic nerve (serratus anterior), accessory nerve (trapezius), or dorsal scapular nerve (rhomboids), or an underlying neuromuscular problem such as muscular dystrophy [28–31]. More recent research has shown that this biomechanical position or motion is more frequently associated with alterations in the supporting bony structure, in the joints of the thoracoscapulohumeral complex, and/or in the strength, flexibility, activation sequencing, and attachment of the stabilizing musculature [32–34]. Therefore, the evaluation of a patient with a winged scapula must be comprehensive enough to identify which of these factors may be causing the altered position and motion.

Similarly, the "snapping" scapula is a descriptive term identifying a patient with painful crepitus along the medial scapular border with arm motion [27]. These symptoms have been traditionally ascribed to osteochondromas or other bony pathology, or thickened bursitis in the thoraco-scapular space. More recent research has shown that alterations in normal SHR underlie most of the cases of snapping scapula [27]. These alterations create increased compressive pressure along the medial border and contribute to the symptoms. Once again, a comprehensive evaluation of the flexibility and strength of all the surrounding musculature must be done to identify the causative factors.

Since most of the clinical problems associated with the scapula involve some type of alteration of scapular resting position and dynamic motion, it appears that a more general framework should be developed to provide a more effective understanding of the roles of the scapula in shoulder pathology. The most basic concept to unify these thoughts would be the biomechanical term "scapular dyskinesis."

"Dys" (alteration of) "kinesis" (motion) is a more general term that reflects loss of control of normal scapular physiology, mechanics, and motion. It has the advantage of unifying the thoughts on the scapula and grouping all of the resulting biomechanical deficits and the possible causative factors into a framework for evaluation and treatment, and indicates the possible connections with shoulder pathology.

Dyskinesis by itself is not an injury, not a guarantee of an injury, or directly related to a specific injury. It is characterized by medial or inferior medial border prominence, early scapular elevation or shrugging upon arm elevation, and/or rapid downward rotation upon arm lowering [35]. These all contribute to the posture of scapular protraction, the biomechanically inferior position that limits optimal SHR and glenohumeral function. The alteration of motion reduces the efficiency of shoulder function in several ways, including changes in three-dimensional glenohumeral angulation, AC joint strain, subacromial space dimensions, maximal muscle activation, and optimal arm position and motion. If these suboptimal capabilities are combined with functional demands such as joint stability, maximal muscle activation, repetitive joint motions, high joint loads, or specific overhead or forward flexion activities, there may be problems with decreased performance or increased injury risk. In this way it should be considered like a capsular laxity or a sulcus sign, in that it can be a factor producing symptoms, and it should be ruled in or out as part of the comprehensive evaluation.

28.4 Causative Factors for Scapular Dyskinesis

Normal scapular motion and the alterations that create dyskinesis have been demonstrated by many biomechanical studies utilizing Moire topographic analysis [36], skin electrode monitors [5, 6, 34, 37, 38], and indwelling bone pins [7], and it is well established that scapular motion and position are truly altered in many shoulder and arm conditions.

The scapula is stabilized on the axial skeleton and the curved ellipsoid thoracic wall by the bony strut of the clavicle, the intact AC and SC joints, and coordinated integrated muscle activation patterns [9]. Alterations in all the areas can create dyskinesis.

Excessive thoracic kyphosis or scoliosis can change scapular motion. Acquired thoracic kyphosis creates increased scapular internal rotation and elevation and decreased scapular upward rotation and posterior tilt. These kinematic alterations lead to decreased maximal arm elevation in forward flexion and abduction [25, 39].

The loss of clavicular strut function can be seen as a result of clavicular fracture, nonunions, or shortened/rotated malunions. In the presence of an intact AC joint, scapular dyskinesis will result as the scapula follows the position of the distal clavicle fragment. Most frequently, there will be increased scapular internal rotation, increased anterior tilt, and varying degrees of interference with upward rotation. Maximal rotator cuff strength and decreased functional outcomes scores related to decreased strength have been reported in these patients [40].

High grade (types III-VI) AC separations also impair clavicular strut function by allowing the scapula to move inferiorly and medially to the clavicle creating the "third translation." This motion results in increased internal rotation and inhibits posterior tilt, creating all of the unfavorable scapular kinematics. It has been documented in greater than 70 % of high grade AC separations [32].

Lesions of the nerves that supply the scapular supporting, stabilizing, or moving musculature will create dyskinesis. These include the long thoracic nerve, the accessory nerve, and the dorsal scapular nerve. In addition, cervical disc disease with involvement of the C5–C6 nerve roots can affect scapular muscle function. A rare but frequently missed neurogenic cause is the acute brachial plexitis, or Parsonnage–Turner Syndrome, most commonly associated with viral illness. Dyskinesis can be associated with both acute nerve deficit and the chronic muscle weakness seen as the nerve injury is healing.

The most common causative mechanisms for scapular dyskinesis involve alterations in the soft tissues. Inflexibility and stiffness of the pectoralis minor and biceps short head create anterior tilt and protraction due to their pull on the coracoid. The most common soft tissue inflexibility is glenohumeral internal rotation deficit (GIRD), which creates a "wind-up" of the scapula on the thorax with arm internal rotation or horizontal abduction. GIRD has been frequently associated with impingement and other rotator cuff disease [41–43].

Several studies have documented alterations in periscapular muscle activation in patients with scapular dyskinesis and rotator cuff disease. Serratus anterior activation and strength is decreased in patients with impingement and shoulder pain, contributing to loss of posterior tilt and upward rotation causing dyskinesis [34, 44, 45]. In addition, the upper trapezius/lower trapezius force couple may be altered, with delayed onset of activation in the lower trapezius, which alters upward rotation and posterior tilt [46, 47]. Finally, fatigue in the periscapular muscles will alter kinematics. The lower portion of the stabilizing force couple, the lower trapezius and serratus anterior, is most susceptible to the effects of fatigue [48, 49].

An uncommon muscle problem creating dyskinesis is scapular muscle detachment. In this problem, the lower trapezius and rhomboids are anatomically or functionally detached from their scapular attachment sites, almost always secondary to acute tensile trauma. The resulting dyskinesis is associated with pain and impairment of shoulder function because of the deficit in muscle control.

The end result of almost all of these causative factors is the protracted scapula at rest, or the excessively protracting scapula with arm motion. This position which usually results from increased internal rotation and anterior tilt is unfavorable for every shoulder function except the "plus" position in weight lifting. It creates decreased subacromial space and increases impingement symptoms, decreases demonstrated rotator cuff strength, increases strain on the anterior glenohumeral ligaments, increases the risk of internal impingement, and increases strain on the scapular stabilizing muscles. Most of the major goals of treatment of scapular dyskinesis relate to regaining functional retraction capability.

The relationship between dyskinesis and shoulder symptoms is not clear in all cases. In cases of nerve injury, fracture, AC separation, or muscle detachment, the injury creates the dyskinesis, which affects shoulder function. In other cases, such as rotator cuff disease, labral injury, and multidirectional instability, it may be that the dyskinesis is causative; creating pathomechanics that predispose the arm to injury, or it may be response to the injury, creating pathomechanics that increase the dysfunction. In either case, dyskinesis is present and must be addressed.

These multiple possible causative mechanisms are frequently not isolated, and several may be present in the same patient. Careful examination for the presence or absence of scapular dyskinesis and each of the causative mechanisms should be done as part of the comprehensive evaluation of patients with shoulder injury.

28.5 Associated Pathologies

28.5.1 Impingement and Rotator Cuff Disease

Studies have almost uniformly identified dyskinesis in patients with rotator cuff impingement or rotator cuff tendinopathy [34, 50, 51]. The exact nature of the alterations is not consistent, with varying combinations of changes in upward rotation (most showing a decrease), posterior tilt (most showing a decrease), and internal/external rotation (no change or increased internal rotation).

All studies in patients with demonstrated rotator cuff tears showed increased upward rotation of some magnitude; with most also showing decreased posterior tilt [51, 52]. Also, in a large prospective study of patients with MRI-proven full thickness rotator cuff tears, scapular dyskinesis was identified as a major factor associated with lower functional scores.

It is not clear whether the observed dyskinesis is a cause, an effect, or a compensation of the shoulder pathology. If it is a cause, it could be due to decreased upward rotation and posterior tilt altering rotator cuff clearance under the coraco-acromial arch, producing mechanical abrasion and wear; decreased external rotation creating glenoid ante-tilting during arm motion, leading to internal impingement; or increasing strain within the rotator cuff tendon [51, 53, 54]. If dyskinesis is an effect, it is probably due to the inhibitory effect of pain on individual muscle activation and the disruption of normal activation patterns, and on the effect of pain avoidance upon kinematic patterns. It appears that increased upward rotation in patients with rotator cuff tears may be a compensation in an attempt to increase or maximize arm elevation or positioning in the face of weakened or absent rotator cuff activation. In any of these cases, dyskinesis is associated with lower functional scores.

28.5.2 Labral Injuries

Dyskinesis has been associated frequently with labral injury [55]. The increased internal rotation and anterior tilt changes glenohumeral alignment, placing increased tensile strain on the anterior ligaments, increasing "peel-back" of the biceps/labral complex on the glenoid [42], and creating pathological internal impingement [56]. Evaluation of dyskinesis in patients with suspected labral injury can focus rehabilitation protocols. The presence of dyskinesis as part of the pathophysiology suggests the need for scapular rehabilitation to improve scapular retraction, including mobilization of tight anterior muscles and institution of the scapular stability series of exercises [57].

28.5.3 Clavicle Fracture

Fracture fragment alignment may disrupt the relationship of the scapula to the axial skeleton and in turn affect scapulohumeral kinematics. Alterations in clavicle anatomy include true shortening due to fragment overlap or butterfly fragments, anterior/posterior or inferior/superior angulation, or external rotation of the distal fragment. Altered scapular mechanics can result as a result of a protracted and tilted position of the scapula. Malunited fractures with 15 mm of shortening demonstrated significant scapular protraction and anterior tilting along with lower subjective scores and significant decreases in strength [58, 59].

Plating versus nonoperative management of displaced mid shaft clavicle fractures revealed higher satisfaction and significantly better Constant and DASH scores for the operative group [60]. The components of the outcomes scores that were most correlated with poor scores were abduction strength and endurance, and flexion range of motion, which can be related to dyskinetic position and motion.

28.5.4 AC Joint Injuries

Dyskinesis has been demonstrated in 73 % of patients with high grade AC symptoms [32]. High grade (Type III and V) AC separations alter the strut function of the clavicle on the scapula and change the biomechanical screw axis of SHR, allowing excessive scapular internal rotation and protraction as the acromion slides inferior and medial under the clavicle and decreased dynamic acromial elevation when the arm is elevated. This motion is termed the "third translation" of the scapula on the thorax. The protracted scapular position creates many of the dysfunctional problems associated with chronic AC separations, including impingement and decreased demonstrated rotator cuff strength. In addition, scapular and shoulder dysfunction can also occur in type II injuries if the AC ligaments are torn. This creates an anterior/posterior AC joint laxity around the axis of the intact coracoclavicular

(CC) ligaments and can be associated with symptoms of pain, clicking, decreased arm elevation, and decreased shoulder function.

If dyskinesis is not demonstrated on clinical exam, then the joint can be regarded as functionally stable, and the patient can progress as rapidly as tolerated through physical therapy. If dyskinesis is demonstrated on the clinical exam, attention should be directed towards correcting the scapuloclavicular biomechanical abnormality. Bracing should include clavicle/scapula retraction with a figure of eight brace. Physical therapy should be directed towards achieving scapular retraction and external rotation first, followed by posterior tilt. Those that fail a supervised 3–6 week program will frequently continue to demonstrate dyskinesis and functional symptoms, and should be counseled regarding surgical options. Operative treatment should include not only CC ligament reconstruction but also AC ligament reconstruction to completely restore the screw axis mechanism and stabilize both inferior/superior and anterior/posterior motions.

28.5.5 Multidirectional Instability

The inherent capsular and ligamentous laxity is only one component of the unstable shoulder in multidirectional instability (MDI). Many patients have increased protraction of the scapula and simultaneous humeral head migration away from the center of the joint as the arm moves [37]. When patients with MDI elevate the arm, the scapula deviates from the normal kinematic pattern of upward rotation, posterior tilting, and minimal internal rotation, and instead follows a pattern of upward rotation, anterior tilting, and excessive internal rotation [53]. This position allows the glenoid to face inferiorly and diminishes the bony constraint to inferior translation, allowing the humeral head to translate inferiorly out of the glenoid socket, creating the instability. Altered scapular muscle activation patterns produce scapular protraction and increased humeral head motion. Inhibition of the subscapularis, lower trapezius, and serratus anterior activation, coupled with increased activation of pectoralis minor and latissimus dorsi, have been demonstrated to place the scapula in a protracted position [61]. In addition, the hyperactive latissimus is the main dynamic deforming force to pull the humeral head inferiorly. Increased rotator cuff activation and biceps activation occurs as a compensation for this altered scapula-humeral rhythm which tends to allow the humeral head to migrate away from the joint center, translate inferiorly, and then move anteriorly or posteriorly [37].

The seeming paradox of a protracting scapula in the face of a posterior directed instability is explained by the same mechanical alterations. As the scapula protracts and the posterior cuff muscles are weakened and/or inhibited, the lax capsular structures cannot constrain the latissimus dorsi's action to pull the humeral head into internal rotation and horizontal adduction, and then pulls the humeral head posteriorly. Patients can frequently reduce their subluxations by externally rotating their arms and placing their scapulae in retraction, which allows the dynamic stabilization.

28.6 Clinical Evaluation of the Scapula

The clinical evaluation of the scapula should be inclusive for all possible local and distant contributors to dyskinesis, and also dynamic, since the motion of the scapula is the key component of dyskinesis.

28.6.1 Important Questions

History: Important questions to ask in the history include past or present trauma to the scapula, clavicle or AC joint, chronic or acute spinal symptoms, recent or remote hip or leg injuries, any surgical procedures on the shoulder, neck symptoms, or viral illness. All of these may have effects on the bony or muscular stabilizing mechanisms. Many patients will have had "therapy," but it is important to find out the specific content and results of the therapy programs. Physical therapy protocols that emphasize modalities, early open chain rotator cuff exercises with resistance, shoulder shrugs, and shoulder protraction exercises have not been found to be effective for scapular dyskinesis.

28.6.2 "Non-Scapula Examination"

The "non-scapula" part of the scapular exam can be done mainly by a screening exam. Areas proximal to the scapula to be emphasized include knee, hip, and trunk. Leg and trunk muscle activity and flexibility are important in scapula/shoulder function, as the stable proximal base for distal mobility, as the core for developing force for the arm and hand, and as facilitation for scapula/shoulder muscle activity.

An effective "non-scapula" screening exam includes a one leg stability series and tests for hip rotation, lumbar flexion/extension, lumbar lordosis/thoracic kyphosis, and cervical lordosis. More detailed analysis can be focused if deficits are found on the screening exam.

The one leg stability series assesses the patient's capability of controlling the trunk and body over the planted leg. It has two components, the single leg standing balance test, which evaluates dynamic control of position, and the single leg squat test, which evaluates dynamic control of motion. In the standing balance test, the patient is asked to place their hands over their chest and stand on one leg with no other verbal cue. Deviations such as a Trendelenburg posture or internally or externally rotating the weight-bearing limb indicates inability to control the posture and has been found to correlate with proximal core weakness especially in the gluteus medius [62]. The single-leg squat is the next progressive evaluation. Assuming the same starting point as the standing balance test, the patient is asked to do repetitive partial half-squats going down and returning to the standing position with no other verbal cues. Similar deviations in the quality of the movement are assessed as in the standing balance test. A Trendelenburg posture which may not be noted on standing balance may be brought out with a single-leg squat. The patient may also use their arms for balance or may go into an exaggerated flexed or rotated posture – "corkscrewing" – in order to put the gluteal or short rotator muscles on greater tension to compensate for muscular weakness [63].

28.6.3 The Scapula Examination

The scapular examination should concentrate on the evaluation for the presence or absence of scapular dyskinesis and determining the possible effects on the symptoms and signs of the dysfunction. Six main components should comprise the examination of the scapula [35]: (1) localization of periscapular symptoms; (2) observational scapular assessment; (3) manual muscle testing; (4) posture; (5) muscle tightness; and (6) symptom/sign alteration maneuvers. The results of the exam will aid in establishing the involvement of the scapula and some of the causative factors of the dyskinesis, and will help guide treatment and rehabilitation.

28.6.3.1 Localization of Symptoms

Localization of pain is helpful in the clinical exam. Pain to palpation is commonly found along the medial scapular border, especially close to the scapular spine. Other common areas are the upper trapezius/levator scapulae area along the superior edge, the serratus anterior, the latissimus dorsi along the lateral scapular border, and the anterior coracoid muscles, pectoralis minor, and short head of the biceps. These point-tender areas are thought to represent tight, shortened, or spastic muscles and are managed by mobilization techniques. In patients with scapular muscle detachment, the pain is localized along the medial border, and is quite intense, averaging 8/10 numeric pain rating [64].

28.6.3.2 Observational Scapular Assessment

Observational scapular assessment can be difficult due to the overlying tissue and lack of reliable imaging techniques. Much effort has been directed towards developing clinically useful methods for this. A panel of experienced researchers and clinicians reviewed the literature

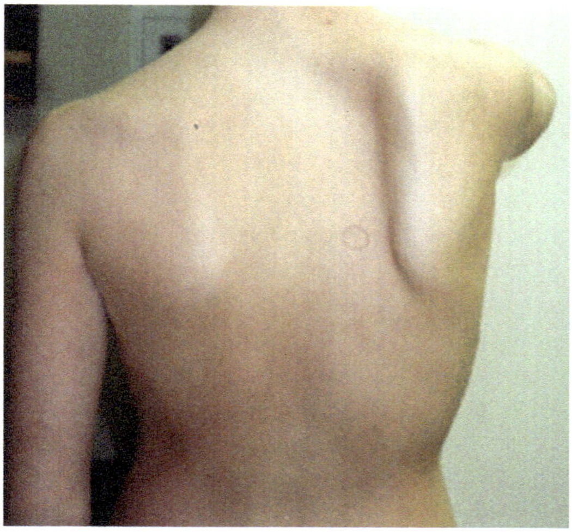

Fig. 28.1 Dynamic assessment to identify presence of scapular dyskinesis

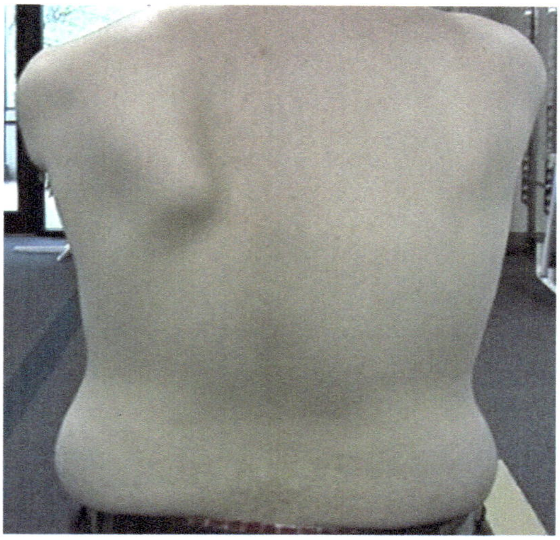

Fig. 28.2 Low row maneuver to isometrically assess lower trapezius strength

and developed a consensus document reflecting best practices for the scapular exam. This evaluation method involves using the medial border of the scapula as the landmark for scapular orientation, and uses both static position at rest and dynamic motion with arm elevation as observation points. Medial border prominence patterns may be predominantly inferior medial border (Type I), entire medial border (Type II), or superior medial border pattern (Type III), or may be a combination of these patterns.

The scapular exam should largely be accomplished from the posterior aspect. The scapula should be exposed for complete visualization. This can be down by gowning, a tank top, or by removing the shirt. The resting posture should be checked for side to side asymmetry but especially for evidence of inferior medial or medial border prominence. If there is difficulty with determining scapular positions, marking the superior and inferior medial borders may help ascertain the position.

Dynamic scapular motions may be evaluated by having the patient move the arms in ascent and descent three to five times. This will usually bring out any weakness in the muscles and display the dyskinetic patterns. Motion in forward flexion is most likely to demonstrate medial border prominence (Fig. 28.1). Prominence of any part of the medial border is recorded in a "yes" (present) or "no" (absent) fashion which has strong clinical utility [38]. If necessary, more repetitions, up to ten, or addition of 3–5 lb weights will highlight the weakness even more [65, 66]. Once this has been demonstrated, tests for strength and flexibility can help determine some of the causative factors.

28.6.3.3 Manual Muscle Testing

One test advocated to assess the integrity of the lower trapezius and serratus anterior muscles is that of the low row [57]. To perform this maneuver, the patient is standing with the involved arm resting at the side with the palm facing posteriorly (Fig. 28.2). The patient is instructed to extend their trunk and push their hand maximally against an examiner's resistance in the direction of shoulder extension and instructed to retract and depress the scapula. This maneuver assesses both muscles' ability to actively stabilize the scapula while providing the examiner a visual depiction of lower trapezius muscle contraction. Other tests such as active scapular squeezing or pinching (rhomboids and middle trapezius) and the wall push-up (serratus anterior) have also been advocated as maneuvers to employ to assess scapular muscle function.

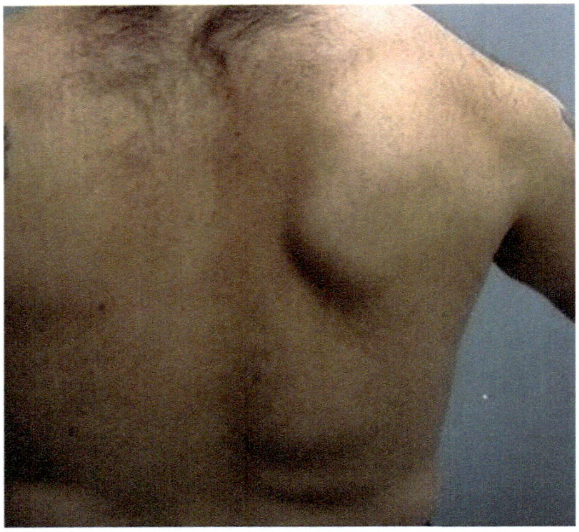

Fig. 28.3 Preferred patient for humeral internal rotation goniometric measurement

28.6.3.4 Posture and Flexibility

Coracoid based inflexibility can be assessed by palpation of the pectoralis minor and the short head of the biceps brachii at their insertion on the coracoid tip. They will usually be tender to palpation, even if they are not symptomatic in use, can be traced to their insertions as taut bands, and will create symptoms of soreness and stiffness when the scapulae are manually maximally retracted and the arm is slightly abducted to approximately 40–50. A rough measurement of pectoralis minor tightness may be obtained by standing the patient against the wall and measuring the distance from the wall to the anterior acromial tip. This can be done using a "double square" device with a patient standing with his or her back against a wall [67]. A bilateral measurement is taken (in inches or centimeters) to determine if there is a notable difference between the involved and noninvolved shoulder, with a side to side asymmetry >3 cm considered abnormal.

To obtain accurate glenohumeral internal rotation measurements, the patient should be positioned supine on a flat level surface. A second examiner should be positioned behind the patient in order to properly stabilize the scapula by applying a posteriorly directed force to the coracoid and humeral head to ensure that scapular movement does not occur (Fig. 28.3). The humerus is supported on the surface with the elbow placed at 90° and the arm on a bolster in the plane of the scapula. A measurement is obtained using a standard bubble goniometer where the fulcrum is set at the olecranon process of the elbow, the stationary arm perpendicular to the table as documented by the bubble on the goniometer, and the moving arm in line with the styloid process of the ulna. The clinician passively moves the arm into internal and external rotation. Rotation is taken to "tightness," a point where no more glenohumeral motion would occur unless the scapula would move or the examiner applies rotational pressure. This measurement should be taken bilaterally, and side to side differences are calculated. Side-to-side differences in internal rotation greater than 20° are considered a pathologic glenohumeral internal rotation deficit (GIRD) [68].

28.6.3.5 Symptom Alteration (Corrective Maneuvers)

If scapular dyskinesis is demonstrated on the clinical exam of patients with shoulder injury, different types of corrective maneuvers may be employed to determine the effect of the altered motion on symptoms or signs of shoulder injury. The goal of the maneuvers would be to alter or reduce some of the signs or symptoms.

The scapular assistance test (SAT) and scapular retraction test (SRT) are corrective maneuvers that may alter the injury symptoms and provide information about the role of scapular dyskinesis in the total picture of dysfunction that accompanies shoulder injury and needs to be restored [35]. The SAT helps evaluate scapular contributions to impingement and rotator cuff strength, and the SRT evaluates contributions to rotator cuff strength and labral symptoms. In the SAT, the examiner applies gentle pressure to assist scapular upward rotation and posterior tilt as the patient elevates the arm (Fig. 28.4) [69]. A positive result occurs when the painful arc of impingement symptoms is relieved and the arc of motion is increased. This test has good test/retest reliability [70]. The test has been found to alter scapular motion by increasing scapular posterior tilt [71], so a positive test would point to the need for

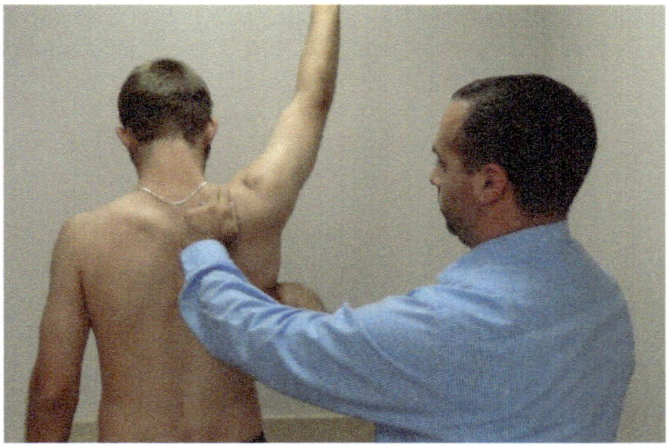

Fig. 28.4 Scapular assistance test to demonstrate scapular involvement in shoulder dysfunction

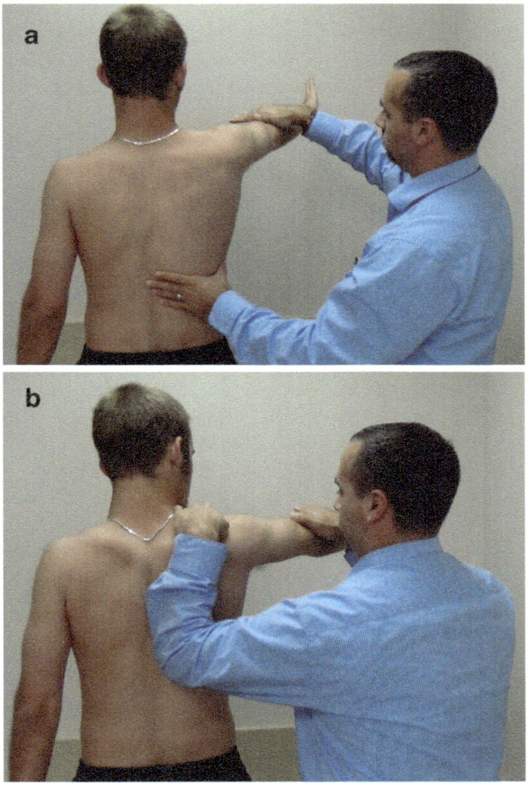

Fig. 28.5 (**a**) Scapular retraction test begins by manually muscle testing forward elevation. (**b**) The scapula is manually stabilized in retraction and the arm is retested to determine if scapular positioning is affecting the manual muscle test result

improvement in pectoralis flexibility and lower trapezius strength. In the SRT, the examiner grades the supraspinatus muscle strength following standard manual muscle testing procedures or by a hand held dynamometer (Fig. 28.5a). The clinician then manually places and stabilizes the scapula in a retracted position (Fig. 28.5b). A positive test occurs when the demonstrated supraspinatus strength is increased or the symptoms of internal impingement in the labral injury are relieved in the retracted position [23]. The major kinematic result of this test is to increase scapular external rotation and posterior tilt, so a positive test would indicate that rotator cuff strengthening is not necessary, and focus should be on rhomboid strengthening and serratus function in retraction. Although these tests are not capable of diagnosing a specific form of shoulder pathology, a positive SAT or SRT shows that scapular dyskinesis is directly involved in producing the symptoms and indicates the need for inclusion of early scapular rehabilitation exercises to improve scapular control.

28.7 Summary

Normal scapular mobility and stability are key and basic components of normal shoulder function. Scapular dyskinesis, alteration in mobility and stability, has been associated with clinical dysfunction in a variety of shoulder injuries. It may be a cause or an effect. Evaluation for and restoration of scapular dyskinesis can help provide information regarding treatment guidelines, demonstrate areas of emphasis in rehabilitation, and indicate progress towards functional restoration.

References

1. Bagg SD, Forrest WJ. A biomechanical analysis of scapular rotation during arm abduction in the scapular plane. Am J Phys Med Rehabil. 1988;67:238–45.
2. Inman VT, Saunders JB, Abbott LC. Observations of the function of the shoulder. Clin Orthop Relat Res. 1996;330:3–13.
3. Bagg SD, Forrest WJ. Electromyographic study of the scapular rotators during arm abduction in the scapular plane. Am J Phys Med. 1986;65(3):111–24.
4. Speer KP, Garrett WE. Muscular control of motion and stability about the pectoral girdle. In: Matsen Iii FA, Fu F, Hawkins RJ, editors. The shoulder: a

balance of mobility and stability. Rosemont: American Academy of Orthopaedic Surgeons; 1994. p. 159–73.
5. Lukasiewicz AC, McClure P, Michener L, Pratt N, Sennett B. Comparison of 3-dimensional scapular position and orientation between subjects with and without shoulder impingement. J Orthop Sports Phys Ther. 1999;29(10):574–86.
6. McClure PW, Michener LA, Sennett BJ, Karduna AR. Direct 3-dimensional measurement of scapular kinematics during dynamic movements in vivo. J Shoulder Elbow Surg. 2001;10:269–77.
7. Ludewig PM, Cook TM, Nawoczenski DA. Three-dimensional scapular orientation and muscle activity at selected positions of humeral elevation. J Orthop Sports Phys Ther. 1996;24(2):57–65.
8. Ludewig PM, Behrens SA, Meyer SM, Spoden SM, Wilson LA. Three-dimensional clavicular motion during arm elevation: reliability and descriptive data. J Orthop Sports Phys Ther. 2004;34(3):140–9.
9. Ludewig PM, Phadke V, Braman JP, Hassett DR, Cieminski CJ, LaPrade RF. Motion of the shoulder complex during multiplanar humeral elevation. J Bone Joint Surg Am. 2009;91A(2):378–89.
10. Sahara W, Sugamoto K, Murai M, Yoshikawa H. Three-dimensional clavicular and acromioclavicular rotations during arm abduction using vertically open MRI. J Orthop Res. 2007;25:1243–9.
11. Kibler WB, Chandler TJ, Shapiro R, Conuel M. Muscle activation in coupled scapulohumeral motions in the high performance tennis serve. Br J Sports Med. 2007;41:745–9.
12. Nichols TR. A biomechanical perspective on spinal mechanics of coordinated muscular action. Acta Anat. 1994;15:1–13.
13. Hirashima M, Kadota H, Sakurai S, Kudo K, Ohtsuki T. Sequential muscle activity and its functional role in the upper extremity and trunk during overarm throwing. J Sports Sci. 2002;20:301–10.
14. Zattara M, Bouisset S. Posturo-kinetic organisation during the early phase of voluntary upper limb movement. 1 Normal subjects. J Neurol Neurosurg Psychiatry. 1988;51:956–65.
15. Young JL, Herring SA, Press JM, Casazza BA. The influence of the spine on the shoulder in the throwing athlete. J Back Musculoskelet Rehabil. 1996;7:5–17.
16. Kibler WB, Press J, Sciascia AD. The role of core stability in athletic function. Sports Med. 2006;36(3):189–98.
17. Kibler WB, McMullen J, Uhl TL. Shoulder rehabilitation strategies, guidelines, and practice. Orthop Clin North Am. 2000;8(4):258–67.
18. De May K, Danneels L, Cagnie B, Cools A. Are kinetic chain rowing exercises relevant in shoulder and trunk injury prevention training? Br J Sports Med. 2011;45(4):320.
19. Nieminen H, Niemi J, Takala EP, Viikari-Juntura E. Load-sharing patterns in the shoulder during isometric flexion tasks. J Biomech. 1995;28(5):555–66.
20. Lippitt S, Vanderhooft JE, Harris SL, Sidles JA, Harryman Ii DT, Matsen Iii FA. Glenohumeral stability from concavity-compression: a quantitative analysis. J Shoulder Elbow Surg. 1993;2(1):27–35.
21. Veeger HEJ, van der Helm FCT. Shoulder function: the perfect compromise between mobility and stability. J Biomech. 2007;40:2119–29.
22. Smith J, Kotajarvi BR, Padgett DJ, Eischen JJ. Effect of scapular protraction and retraction on isometric shoulder elevation strength. Arch Phys Med Rehabil. 2002;83:367–70.
23. Kibler WB, Sciascia AD, Dome DC. Evaluation of apparent and absolute supraspinatus strength in patients with shoulder injury using the scapular retraction test. Am J Sports Med. 2006;34(10):1643–7.
24. Tate AR, McClure P, Kareha S, Irwin D. Effect of the scapula reposition test on shoulder impingement symptoms and elevation strength in overhead athletes. J Orthop Sports Phys Ther. 2008;38(1):4–11.
25. Kebaetse M, McClure PW, Pratt N. Thoracic position effect on shoulder range of motion, strength, and three-dimensional scapular kinematics. Arch Phys Med Rehabil. 1999;80:945–50.
26. Kuhn J, Plancher K, Hawkins R. Scapular winging. J Am Acad Orthop Surg. 1995;3:319–25.
27. Kuhne M, Boniquit N, Ghodadra N, Romeo AA, Provencher MT. The snapping scapula: diagnosis and treatment. Arthroscopy. 2009;25(11):1298–311.
28. Herzmark MH. Traumatic paralysis of the serratus anterior relieved by transplantation of the rhomboidei. J Bone Joint Surg Am. 1951;33:235–8.
29. Steinman S, Wood M. Pectoralis major transfer for serratus anterior paralysis. J Shoulder Elbow Surg. 2003;12:555–60.
30. Wright TA. Accessory spinal nerve injury. Clin Orthop Relat Res. 1975;108:15–8.
31. Bigliani LU, Perez-Sanz IR, Wolfe IN. Treatment of trapezius paralysis. J Bone Joint Surg Am. 1985;67:871–7.
32. Gumina S, Carbone S, Postacchini F. Scapular dyskinesis and SICK scapula syndrome in patients with chronic type III acromioclavicular dislocation. Arthroscopy. 2009;25(1):40–5.
33. Borstad JD, Ludewig PM. The effect of long versus short pectoralis minor resting length on scapular kinematics in healthy individuals. J Orthop Sports Phys Ther. 2005;35(4):227–38.
34. Ludewig PM, Cook TM. Alterations in shoulder kinematics and associated muscle activity in people with symptoms of shoulder impingement. Phys Ther. 2000;80(3):276–91.
35. Kibler WB, Ludewig PM, McClure PW, Uhl TL, Sciascia AD. Scapula summit 2009. J Orthop Sports Phys Ther. 2009;39(11):A1–13.
36. Warner JJP, Micheli LJ, Arslanian LE, Kennedy J, Kennedy R. Patterns of flexibility, laxity, and strength in normal shoulders and shoulders with instability and impingement. Am J Sports Med. 1990;18(4):366–75.

37. Ogston JB, Ludewig PM. Differences in 3-dimensional shoulder kinematics between persons with multidirectional instability and asymptomatic controls. Am J Sports Med. 2007;35:1361–70.
38. Uhl TL, Kibler WB, Gecewich B, Tripp BL. Evaluation of clinical assessment methods for scapular dyskinesis. Arthroscopy. 2009;25(11):1240–8.
39. Endo K, Yukata K, Yasui N. Influence of age on scapulo-thoracic orientation. Clin Biomech (Bristol, Avon). 2004;19:1009–13.
40. McKee MD, Pedersen EM, Jones C, Stephen DJG, Kreder HJ, Schemitsch EH, et al. Deficits following nonoperative treatment of displaced midshaft clavicular fractures. J Bone Joint Surg Am. 2006;88:35–40.
41. Borich MR, Bright JM, Lorello DJ, Cieminski CJ, Buisman T, Ludewig PM. Scapular angular positioning at end range internal rotation in cases of glenohumeral internal rotation deficit. J Orthop Sports Phys Ther. 2006;36:926–34.
42. Burkhart SS, Morgan CD, Kibler WB. The disabled throwing shoulder: spectrum of pathology part I: pathoanatomy and biomechanics. Arthroscopy. 2003; 19(4):404–20.
43. Harryman II DT, Sidles JA, Clark JM, McQuade KJ, Gibb TD, Matsen Iii FA. Translation of the humeral head on the glenoid with passive glenohumeral motion. J Bone Joint Surg Am. 1990;72(9):1334–43.
44. Lin JJ, Hanten WP, Olson SL, Roddey TS, Soto-Quijano DA, Lim HK, et al. Functional activities characteristics of shoulder complex movements: exploration with a 3-D electromagnetic measurement system. J Rehabil Res Dev. 2005;42(2):199–210.
45. Cools AM, Witvrouw EE, DeClercq GA, Vanderstraeten GG, Cambier DC. Evaluation of isokinetic force production and associated muscle activity in the scapular rotators during a protraction-retraction movement in overhead athletes with impingement symptoms. Br J Sports Med. 2004;38:64–8.
46. Cools AM, Witvrouw EE, DeClercq GA, Danneels LA, Cambier DC. Scapular muscle recruitment pattern: trapezius muscle latency with and without impingement symptoms. Am J Sports Med. 2003;31:542–9.
47. Cools AM, Dewitte V, Lanszweert F, Notebaert D, Roets A, Soetens B, et al. Rehabilitation of scapular muscle balance. Am J Sports Med. 2007;35(10):1744–51.
48. Ebaugh DD, McClure PW, Karduna AR. Effects of shoulder muscle fatigue caused by repetitive overhead activities on scapulothoracic and glenohumeral kinematics. J Electromyogr Kinesiol. 2006;16:224–35.
49. Tsai NT, McClure P, Karduna AR. Effects of muscle fatigue on 3-dimensional scapular kinematics. Arch Phys Med Rehabil. 2003;84:1000–5.
50. McClure P, Michener LA, Karduna AR. Shoulder function and 3-dimensional scapular kinematics in people with and without shoulder impingement syndrome. Phys Ther. 2006;86:1075–90.
51. Graichen H, Stammberger T, Bonel H, Wiedemann E, Englmeier KH, Reiser M, et al. Three-dimensional analysis of shoulder girdle and supraspinatus motion patterns in patients with impingement syndrome. J Orthop Res. 2001;19(6):1192–8.
52. Mell AG, LaScalza S, Guffey P, Carpenter JE, Hughes RE. Effect of rotator cuff pathology on shoulder rhythm. J Shoulder Elbow Surg. 2005;14:S58–64.
53. Ludewig PM, Reynolds JF. The association of scapular kinematics and glenohumeral joint pathologies. J Orthop Sports Phys Ther. 2009;39(2):90–104.
54. Karduna AR, Kerner PJ, Lazarus MD. Contact forces in the subacromial space: effects of scapular orientation. J Shoulder Elbow Surg. 2005;14:393–9.
55. Kibler WB. What is a clinically important superior labrum anterior to posterior tear? In: Pagano MW, Hart RA, editors. Instructional course lectures, vol. 62. Rosemont: American Academy of Orthopaedic Surgeons; 2013. p. 483–9.
56. Kibler WB, Dome DC. Internal impingement: concurrent superior labral and rotator cuff injuries. Sports Med Arthrosc. 2012;20(1):30–3.
57. Kibler WB, Sciascia AD, Uhl TL, Tambay N, Cunningham T. Electromyographic analysis of specific exercises for scapular control in early phases of shoulder rehabilitation. Am J Sports Med. 2008;36(9):1789–98.
58. Ledger M, Leeks N, Ackland T, Wang A. Short malunions of the clavicle: an anatomic and functional study. J Shoulder Elbow Surg. 2005;14(4):349–54.
59. Matsumura N, Ikegami H, Nakamichi N, Nakamura T, Nagura T, Imanishi N, et al. Effect of shortening deformity of the clavicle on scapular kinematics: a cadaveric study. Am J Sports Med. 2010;38(5):1000–6.
60. Canadian Orthopaedic Trauma Society. Nonoperative treatment compared with plate fixation of displaced midshaft clavicular fractures. A multicenter, randomized clinical trial. J Bone Joint Surg Am. 2007;89:1–10.
61. Morris AD, Kemp GJ, Frostick SP. Shoulder electromyography in multidirectional instability. J Shoulder Elbow Surg. 2004;13:24–9.
62. Hardcastle P, Nade S. The significance of the Trendelenburg test. J Bone Joint Surg Br. 1985;67(5): 741–6.
63. Crossley KM, Zhang WJ, Schache AG, Bryant A, Cowan SM. Performance on the single-leg squat task indicates hip abductor muscle function. Am J Sports Med. 2011;39(4):866–73.
64. Kibler WB, Sciascia A, Uhl T. Medial scapular muscle detachment: clinical presentation and surgical treatment. J Shoulder Elbow Surg. 2014;23(1):58–67.
65. McClure PW, Tate AR, Kareha S, Irwin D, Zlupko E. A clinical method for identifying scapular dyskinesis: part 1: reliability. J Athl Train. 2009;44(2):160–4.
66. Tate AR, McClure PW, Kareha S, Irwin D, Barbe MF. A clinical method for identifying scapular dyskinesis: part 2: validity. J Athl Train. 2009;44(2):165–73.
67. Kluemper M, Uhl TL, Hazelrigg H. Effects of stretching and strengthening shoulder muscles on forward shoulder posture in competitive swimmers. J Sport Rehabil. 2006;15(1):58–70.

68. Wilk KE, Macrina LC, Fleisig GS, Porterfield R, Simpson Ii CD, Harker P, et al. Loss of internal rotation and the correlation to shoulder injuries in professional baseball pitchers. Am J Sports Med. 2011; 39(2):329–35.
69. Kibler WB. The role of the scapula in athletic function. Am J Sports Med. 1998;26:325–37.
70. Rabin A, Irrgang JJ, Fitzgerald GK, Eubanks A. The intertester reliability of the scapular assistance test. J Orthop Sports Phys Ther. 2006;36(9):653–60.
71. Seitz AL, McClure P, Lynch SS, Ketchum JM, Michener LA. Effects of scapular dyskinesis and scapular assistance test on subacromial space during static arm elevation. J Shoulder Elbow Surg. 2012;21(5):631–40.

29 Anatomy of Scapula Winging

William Ben Kibler and Aaron Sciascia

29.1 Introduction

Injury to the scapula and its musculature is not uncommon. Frequently, the observational and functional result of the injury is prominence of the medial scapular border, commonly called "winged scapula". Clinical presentations of scapular winging include neurologically based scapular winging, scapular muscle detachment, snapping scapula, and kinetic chain or muscle inhibition based scapular dyskinesis. These can all be organized as manifestations of scapular dyskinesis (abnormal scapular motion during arm movement) which can be seen as causes or results of various types of shoulder injury.

29.2 Neurologically Based Scapular Winging

The most common perception of the "winged" scapula, with prominence of the medial scapular border, is that it occurs as a result of damage to the nerves supplying the scapular stabilizing muscles. The exact frequency of neurologically based winging is poorly defined. Injury may be traumatic, iatrogenic, or idiopathic and the clinical interpretation requires careful history and examination. The three nerves most often involved include the long thoracic, spinal accessory, and dorsal scapular nerves.

The *long thoracic nerve* innervates the serratus anterior muscle and originates from the ventral rami of C5, C6, and C7. It is a purely motor nerve. The course of the nerve is clinically relevant as it predisposes the nerve to frequent injury. After passing through the middle scalene, the nerve crosses under the clavicle and remains superficial along the lateral chest wall. Here, it can be subjected to blunt trauma, diverse athletic injuries, or traction [1, 2]. Numerous reports of compressive neuropathy exist from lateral positioning or prolonged convalescence. Iatrogenic injury should be considered in patients' status recovering from invasive lateral thoracic surgery such as radical mastectomy or axillary lymph node dissection. Finally, Parsonage–Turner type neuritis may be the underlying cause in patients experiencing significant pain or with recent viral infection. Recovery from idiopathic causes is usually apparent by 1 year but has been reported up to 2.

Long thoracic nerve injury creates loss of the serratus anterior muscle function which results in translation of the scapula superiorly and medially and disruption of normal scapulohumeral kinematics. Clinically, this rotation causes the inferior angle to become notably prominent in both static and dynamic examination (Fig. 29.1). The diagnosis can be confirmed on EMG around 6 weeks following initial trauma. Initial treatment

W.B. Kibler, MD (✉) • A. Sciascia, MS, ATC, PES
Shoulder Center of Kentucky, Lexington Clinic,
1221 South Broadway, KY 40504, USA
e-mail: ascia@lexclin.com

G.I. Bain et al. (eds.), *Normal and Pathological Anatomy of the Shoulder*,
DOI 10.1007/978-3-662-45719-1_29, © ISAKOS 2015

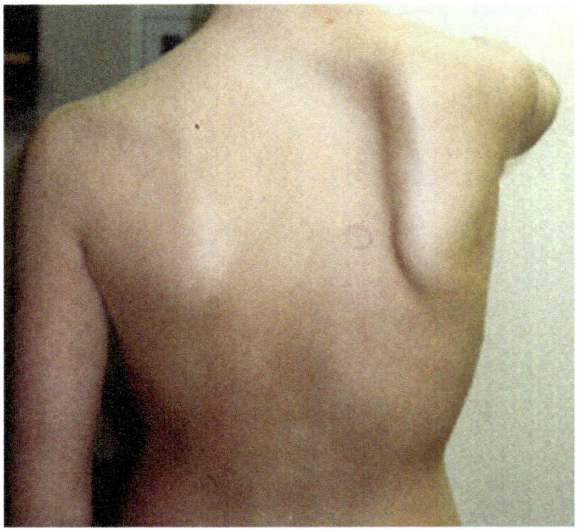

Fig. 29.1 Example of long thoracic nerve palsy creating scapular winging

measures include observation with supportive care, rehabilitation, and possibly follow-up EMG at 3-month intervals. Rehabilitation should focus upon preservation of glenohumeral motion and scapular stabilization by activation of the rhomboids and lower trapezius. Strengthening of the surrounding scapular stabilizers can be challenging; however, clinicians should avoid the use of long-lever exercises which can be over taxing and exacerbate symptoms. Rhomboid and lower trapezius strengthening is best achieved through the use of short-lever maneuvers which encourage scapular retraction and depression.

Surgery for long thoracic nerve palsy may be indicated for patients with 1 year of symptoms, functional deficits, and no signs of recovery. Scapulothoracic fusion has been considered and may result in satisfactory results. However, due to the morbidity of the procedure it is generally reserved for high demand laborers and salvage situations. Fasciodesis procedures (sling) have also been utilized but can result in high rates of attritional loosening and poor function. Therefore, muscle transfers can be performed in an effort to restore a semblance of dynamic kinematics. Transfer of the sternocostal head of the pectoralis major has found the greatest measure of success [3, 4]. The selected portion of the tendon is reflected from its insertion on the humerus, tunneled ventral to the scapula and attached to the inferior angle of the scapula by various techniques. The tendon length generally requires augmentation with fascia latae or other graft for length. Published results have generally been favorable with an increase in variety of postsurgical outcomes scores [3, 5, 6].

Scapular winging can also be the product of injury to the *spinal accessory nerve* (cranial nerve 12) [4]. Upon loss of the trapezius activation, the scapula assumes a more inferior and lateral, or "drooping" posture. Winging is often less prominent than with serratus anterior palsy; however, the atrophy in the upper trapezius, loss of muscle tone and inability to shrug are easily discernible. Inability of the lower trapezius to achieve and maintain the retracted "functional" position of the scapula leads patients to report pain and weakness with forward elevation and abduction. Compensatory muscle spasm in the rhomboids and levator scapula are common. The diagnosis can be corroborated by EMG. Interpretation of EMG findings must be carefully performed as a false-negative report can occur due to placement of the recording electrode in the normal underlying rhomboids instead of the thin, atrophic overlying trapezius. For idiopathic causes or neuritis, supportive management and observation are recommended up to 1 year. In cases of iatrogenic transection or penetrating trauma, surgery is often advocated.

Scapulothoracic fusion may provide static stability and relieve pain but the loss of motion and complications may be unacceptable. The Eden-Lange Transfer was developed to provide a dynamic medial and superior restraint [3, 4]. In this procedure, the levator scapula and rhomboids are transferred approximately 5 cm laterally and secured via drill holes into the scapular body to improve mechanical advantage and substitute for trapezius function. Both short-term and long-term positive outcomes have been previously reported following this procedure [5, 7].

Less commonly, weakness of the major and minor rhomboid muscles is encountered in deficits of the *dorsal scapular nerve*. The nerve is a branch of the C5 nerve root and may be involved in radiculopathy. With loss of the rhomboids, unopposed pull of the serratus anterior results in lateral rotation of the inferior angle of the scapula. Atrophy of the rhomboids may be observed. EMG and supportive observation as outlined earlier are recommended. Fasciodesis procedures have met with limited success.

29.3 Scapular Muscle Detachment

This problem is not well known or well categorized with limited results reported [8]. As a result, these patients have experienced symptoms for months and years. The pathoanatomy appears to be an anatomic or physiologic detachment of the lower trapezius and rhomboids from the spine and medial border of the scapula. The large majority of cases present after an acute traumatic tensile load, half involving seat belt restrained motor vehicle accidents but there are multiple other causes such as throwing, catching, or lifting a heavy object with the arm at full extension, pulling against a heavy object, hanging on the rim after dunking a basketball, and electrical shock such as electrocution or cardioversion. The presenting symptom cluster is very uniform with early posttraumatic onset of localized pain along the medial scapular border. The pain increases in intensity as the condition evolves and averages 8/10 numeric pain rating. There are major limitations of arm use away from the body in forward flexion or overhead positions. Increased upper trapezius activity and spasm, resulting from lack of lower trapezius activity, creates migraine-like headaches. Neck and shoulder joint symptoms may be present due to dyskinesis and will often become the focus of treatment, including surgery with infrequent positive results.

The physical exam also exhibits a consistent cluster of findings including the localized tenderness, often a noticeable and palpable soft tissue defect, either due to the detachment or the muscle atrophy, altered scapular resting position as well as dynamic dyskinesis including snapping scapula, shoulder impingement and weakness in forward flexion, and clinical decrease or relief of symptoms with scapular corrective maneuvers. MRI and CT imaging are of minimal benefit. The cuts may not be in the correct plane, the detachment scar is not easily demonstrated, the lower trapezius is detached from the spine but then lays back over in the supine retracted imaging position, or the chronic changes are too subtle for the static imaging processes. Consistency of the history and physical exam findings allows for a reliable clinical diagnosis.

Most of these patients have had workups to rule out neurologic or bony causation, and have had varying types of treatment, including local or distant surgery and various rehabilitation protocols. If they have failed an appropriate scapular rehabilitation program and do not demonstrate other anatomic defects, surgical reattachment is indicated. This is accomplished by direct reattachment through pairs of drill holes in the medial scapular border and scapular spine (Fig. 29.2) [3, 8]. The detached and scarred rhomboids are mobilized and reattached onto the dorsal aspect of the scapula about 1 cm from the medial edge (Fig. 29.3). The lower trapezius is mobilized and reattached along the proximal scapular spine.

Postoperatively, the arm is protected in neutral rotation for 4 weeks but gentle scapular retraction is encouraged immediately. At 4 weeks, closed chain activation up to 90° abduction with the hand stabilized is started. By 6–8 weeks, as the repair has healed and early strength is gained, motion over 90° is allowed and the patient is started on the standard scapular strengthening program. Maximum strength is not regained for about 6–9 months, probably reflecting the chronic muscle disuse and atrophy. Results from a 2-year follow-up of a small cohort show that pain scores decreased from 8/10 to 2/10, and ASES scores improved from 38/100 to 68/100 [8]. These results are durable at 2-year follow-up.

29.4 Snapping Scapula

The diagnosis and management of snapping scapula can be clinically challenging. The condition has been estimated to be present in up to 30 % of asymptomatic individuals yet can cause crippling pain. Snapping scapula most frequently represents a disruption of the smooth gliding of the scapula

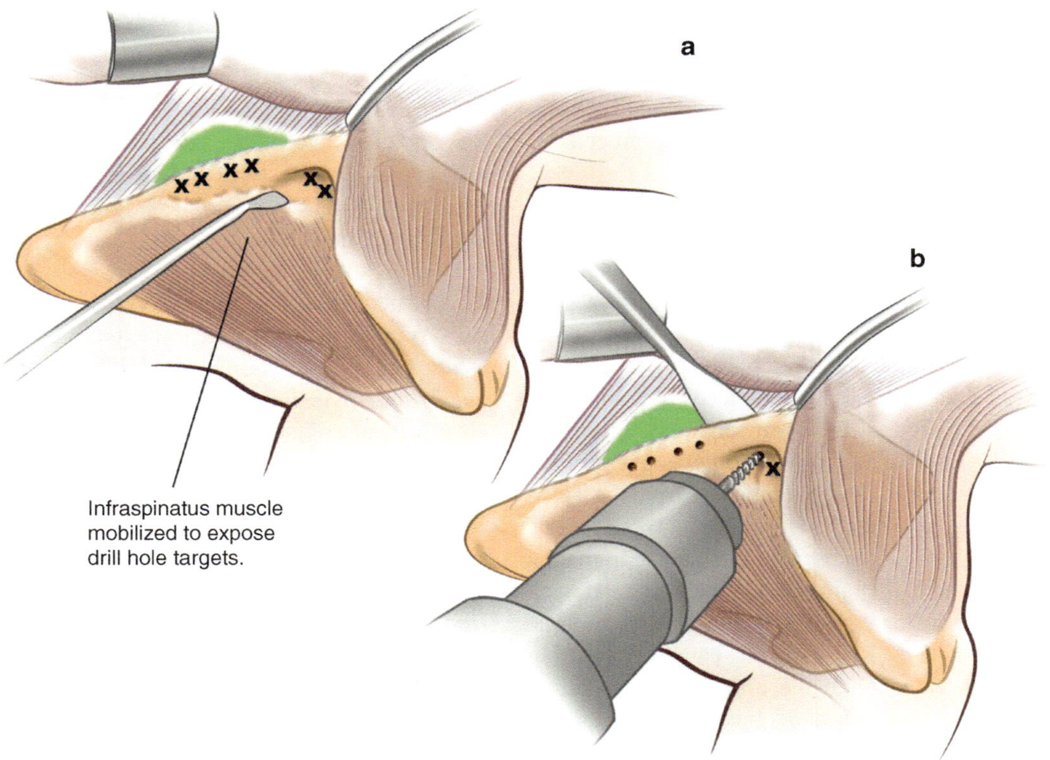

Fig. 29.2 Illustration of drill hole placement when performing the scapular muscle reattachment procedure. Mobilization of the infraspinatus away from the medial border of the scapula (**a**) in order for placement of drill holes for muscle reattachment (**b**)

over the thoracic cage and periscapular muscles [9]. A thorough understanding of the three-dimensional kinematics and anatomy is required to appreciate the varied etiologies. The normal coupled scapular motions of posterior tilt and external rotation are decreased as the arm elevates often due to tissue tightness, muscle weakness, and in some instances compensatory movement patterns following injury. Consequently, the normal movement of the instant center of rotation of the scapula from the superior medial border to the AC joint is disrupted, causing the scapula to rotate around the medial border, creating excessive pressure and leading to symptoms [10].

Patients generally complain of periscapular pain with overhead activities. History may reveal recent overuse or a single traumatic event. Symptoms often limit sports and activities of daily living. The throwing motion, with its large scapular excursion, is particularly affected. Patients often will note an audible grinding or snapping, amplified by the thoracic cavity that may be precipitated by variable active range of motion or even shrugging. However, patients usually do not note symptoms with isometric contraction. The superomedial border is most commonly cited as the location of pain, yet this may be variable.

The cause of crepitus is thought to be the chronically inflamed bursa or anatomic abnormalities. The infraserratus bursa resides between the serratus anterior and the chest wall. The supraserratus bursa occupies space between the subscapularis and serratus anterior. In addition, several minor bursae may be present at the superomedial border, the inferior angle, or the medial base of the spine. Bursa may become persistently inflamed through overuse and/or the abnormalities in mechanics.

In a minority of patients, anatomic abnormalities may exist which predispose the patient to

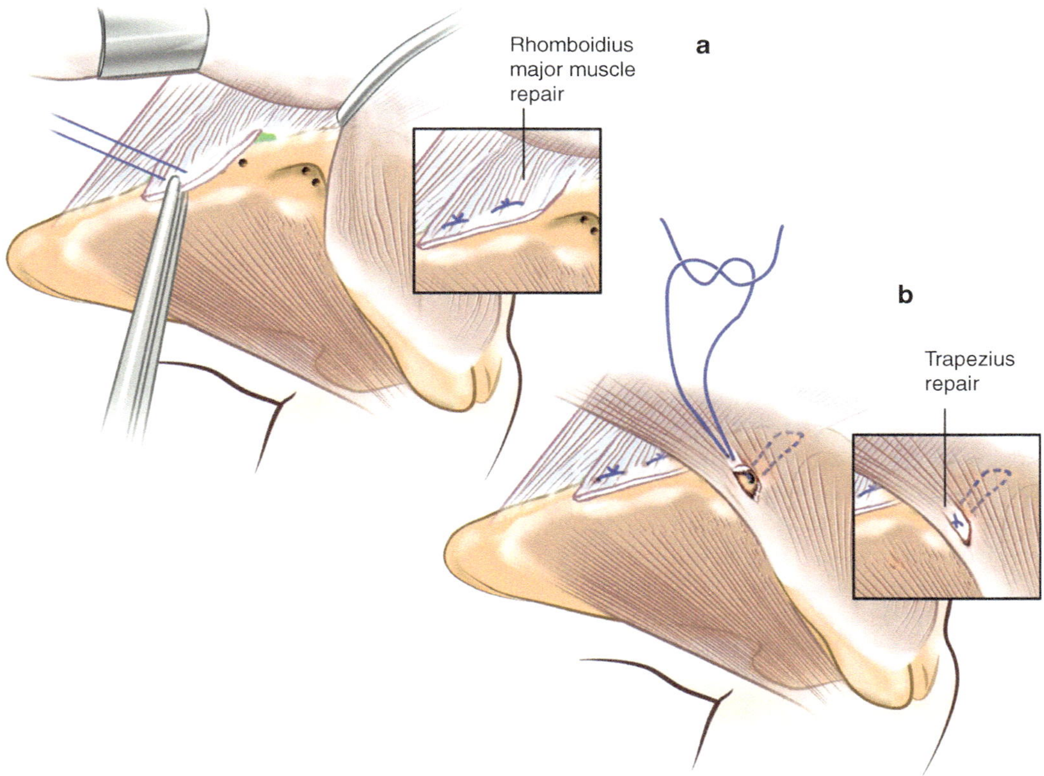

Fig. 29.3 Illustration of the reattachment of the rhomboid and lower trapezius muscles. Reattachment of the rhomboid muscle is performed initially (**a**) followed by reattachment of the lower trapezius muscle (**b**)

snapping scapula by disrupting the scapulothoracic articulation. Examples of such may vary widely from osteochondroma of the scapula to malunited rib fractures. Several authors have postulated excessive curvature of the ventral surface of the scapula as a cause of incongruity [4, 11, 12]. However, good normative data is limited and values to define pathologically increased curvature are undetermined. Luschka's tubercle has been described as a prominence or hook at the superomedial corner. Examinations of normal scapular anatomy have differed in reporting of the relative frequencies of such anatomic variations. One infrequent cause of snapping scapula is post-traumatic scapular muscle detachment [8]. The major physical finding is a palpable soft tissue defect along the medial scapular border at the scapular spine.

The examination of the snapping scapula must be comprehensive to evaluate all possible causes of change in scapular position and motion and to investigate possible bony causes of the crepitus. Resting posture should be evaluated and medial border prominence, indicating increased anterior tilt and internal rotation should be noted [13]. Palpation of the pectoralis minor, scalenes, and sternocleidomastoid muscles will often reveal contracted and tight muscles that will need to be mobilized. Dynamic scapular motion as the arms are elevated will frequently demonstrate the lack of smooth motion and point out muscle weakness in the lower trapezius. Comparison to the contralateral side is critical. Tenderness, swelling, or palpable defects in the periscapular muscles should also be noted. The scapular assistance test can be very helpful in evaluating relief of symptoms by alteration of scapular position and motion. Plain radiographs should include a scapular Y to appraise the dorsal and ventral surfaces. The use of three-dimensional CT has been studied to gauge anatomic abnormalities. MRI may be useful in elucidating inflamed bursa and muscle.

The treatment of snapping scapula should begin with comprehensive nonoperative management. Each program should be individually directed at the demonstrated etiologic factors. However, the cornerstone is generally physical therapy to address proper postural and periscapular mechanics. Focus is placed on strengthening the lower trapezius and serratus anterior through both isometric and dynamic endurance training. All of the contracted anterior muscles should be mobilized by massage and stretching. Activity modification and modalities may be included. Scapular bracing may have a place in certain situations. Bursal inflammation can be addressed through precise injections placed with appropriate technique and caution.

Surgery may be indicated for patients who fail a thorough conservative program, are sufficiently disabled, and appear willing to comply and carry out postoperative care. Patients with neurological injury or severe muscle atrophy are contraindicated for isolated snapping scapula surgery. Good success rates have been published with surgery, despite the fact that techniques vary widely [9]. Both open and arthroscopic techniques have shown success although concern has been raised for the morbidity and cosmesis of open procedures [9]. Arthroscopic techniques have demonstrated faster recovery but are demanding and carry an increased neurovascular risk. Through either technique, surgeons may choose simple bursectomy or partial scapulectomy. No clearly defined indications for bony removal exist and the variability in reported amount of resection is remarkable.

29.5 Kinetic Chain and Muscle Inhibition Based Scapular Dyskinesis

Optimal shoulder function is the result of physiological motor activations, acting on intact anatomical structures, creating specific biomechanical motions and positions to generate forces and actions and task-specific skills. Task-specific function occurs when the activations, motions, and resultant forces are specific and efficient for the needs of that task. For example, intricate biomechanical tasks such as overhead throwing, serving, or lifting requires sequential muscle activation from both the upper and lower extremities. These segments are collectively known as the links in the kinetic chain. The kinetic chain is a coordinated sequencing of activation, mobilization, and stabilization of body segments to produce an athletic activity [14, 15].

An effective kinetic chain is characterized by three components: (1) optimized anatomy (intact and noninjured) and physiology (strength, flexibility, endurance, and power generation); (2) well-developed, efficient task-specific motor patterns for muscle activation; and (3) sequential generation of forces appropriately distributed across motions that result in the desired athletic function [15]. Dysfunction within a particular segment in the chain can result in either altered performance or injury to a more distal segment. For example, scapular stability is essential to proper kinetic chain function since the scapula is the link within the kinetic chain which connects the energy-producing core with the energy-transferring arm. Deficits at or around the scapula such as muscle weakness and/or tightness can negatively impact the desired biomechanical output during arm-specific tasks.

Weakness or tightness within other kinetic chain segments can create a dysfunctional scapula. Deficits at the hip or knee can create scapular dyskinesis by altering force generation or decreasing proximal kinetic chain motion. The lower trapezius attaches to the lumbodorsal fascia and hip extensors, and is maximally activated through gluteal activity, especially in diagonal patterns [16]. When the hip and trunk do not adequately rotate, the scapula must compensate, either by increased retraction/protraction or increased stabilization. Both compensations are inefficient and difficult due to the requirement for increased muscle activity. Strength imbalances within different segments have been demonstrated by many studies in many sports and activities, which suggest that these deficits may play a role in the dysfunction of the kinetic chain [17–22].

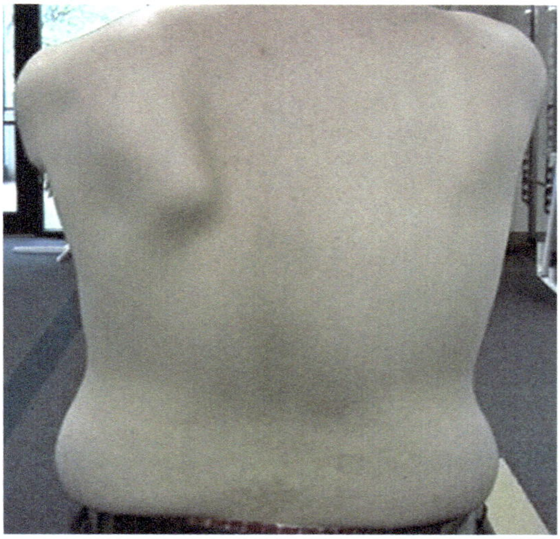

Fig. 29.4 Example of nonpathologic scapular dyskinesis caused by kinetic chain dysfunction

Scapular dyskinesis is a nonspecific response to this loss of proximal activation rather than a specific response to specific pathology [13]. Scapular dyskinesis has multiple kinetic chain causative factors, including previous leg or hip injury, muscle weakness/imbalance, nerve injury, and poor mechanics. The medial border prominence seen on observation appears to be the result of abnormal muscle activations, either directly due to muscle involvement such as inflexibility, weakness, fatigue, or nerve injury, and is usually treated by rehabilitation (Fig. 29.4).

Muscle tightness of the upper extremity portion of the kinetic chain including tightness of the inert and contractile tissue at and around the shoulder is another factor which can result in scapular dyskinesis. Inflexibility of the muscles that insert on the coracoid process, the pectoralis minor, and the short head of the biceps is common as well as tightness of the posterior rotator cuff muscles. Tightness in these muscles leads to scapular protraction, decrease in scapular posterior tilt, and a decrease in the subacromial space height, which is associated with external impingement [23]. Internal rotation deficits are common in overhead athletes with the throwing arm having approximately 10° less of internal rotation than the nonthrowing arm [24–26]. This rotation loss has been proposed to occur as a result of bony, capsular, and/or muscular changes over time [27–33]. The discovery of dyskinesis with physical impairments but without obvious pathology should prompt the evaluation of kinetic chain causative factors. Treatment is based on correcting all of the kinetic chain deficits [34, 35].

Loss of coordinated muscle activation, producing scapular dyskinesis, is also seen due to inhibition from glenohumeral joint injury. External impingement, labral injury, biceps pathology, rotator cuff disease, glenohumeral instability, and adhesive capsulitis have been associated with alterations in lower trapezius, upper trapezius, and serratus anterior activation that result in scapular protraction and dyskinesis [13, 36]. The dyskinesis can increase the dysfunction from the injury, and can delay functional recovery if not included in the comprehensive rehabilitation plan.

29.6 Summary

Multiple pathologies can result in the alterations in position and motion that have been called scapular winging. The neurologically based factors traditionally thought to be the causation are less common than other causative factors. The clinical observation of medial border prominence should be the start of a diagnostic process to investigate all the possible causative factors.

References

1. Aval SM, Durand P, Shankwiler JA. Neurovascular Injuries to the athlete's shoulder: part 1. J Am Acad Orthop Surg. 2007;15(4):249–56.
2. Aval SM, Durand P, Shankwiler JA. Neurovascular injuries to the athlete's shoulder: part 2. J Am Acad Orthop Surg. 2007;15(5):281–9.
3. Kibler WB. Scapular surgery I–IV. In: Reider B, Terry MA, Provencher MT, editors. Sports medicine surgery. Philadelphia: Elsevier Saunders; 2010. p. 237–67.
4. Kuhn J, Plancher K, Hawkins R. Scapular winging. J Am Acad Orthop Surg. 1995;3:319–25.
5. Galano GJ, Bigliani LU, Ahmad CS, Levine WN. Surgical treatment of winged scapula. Clin Orthop Relat Res. 2008;466(3):652–60.

6. Tauber M, Moursy M, Koller H, Schwartz M, Resch H. Direct pectoralis major muscle transfer for dynamic stablization of scapular winging. J Shoulder Elbow Surg. 2008;17(1):S29–34.
7. Romero J, Gerber C. Levator scapulae and rhomboid transfer for paralysis of trapezius: The Eden-Lange procedure. J Bone Joint Surg Br. 2003;85(8):1141–5.
8. Kibler WB, Sciascia A, Uhl T. Medial scapular muscle detachment: clinical presentation and surgical treatment. J Shoulder Elbow Surg. 2014;23(1):58–67.
9. Kuhne M, Boniquit N, Ghodadra N, Romeo AA, Provencher MT. The snapping scapula: diagnosis and treatment. Arthroscopy. 2009;25(11):1298–311.
10. Bagg SD, Forrest WJ. A biomechanical analysis of scapular rotation during arm abduction in the scapular plane. Am J Phys Med Rehabil. 1988;67:238–45.
11. Lehtinen JT, Tingart MJ, Apreleva M, Warner JJP. Quantitative morphology of the scapula: normal variation of the superomedial scapular angle, and superior and inferior pole thickness. Orthopedics. 2005;28(5):481–6.
12. Ebraheim NA, Rongming X, Haman SP, Miedler JD, Yeasting RA. Quantitative anatomy of the scapula. Am J Orthop. 2000;29(4):287–92.
13. Kibler WB, Sciascia AD. Current concepts: scapular dyskinesis. Br J Sports Med. 2010;44(5):300–5.
14. Putnam CA. Sequential motions of body segments in striking and throwing skills: description and explanations. J Biomech. 1993;26:125–35.
15. Sciascia AD, Thigpen CA, Namdari S, Baldwin K. Kinetic chain abnormalities in the athletic shoulder. Sports Med Arthrosc Rev. 2012;20(1):16–21.
16. De May K, Danneels L, Cagnie B, Cools A. Are kinetic chain rowing exercises relevant in shoulder and trunk injury prevention training? Br J Sports Med. 2011;45(4):320.
17. Reeser JC, Joy EA, Porucznik CA, Berg RL, Colliver EB, Willick SE. Risk factors for volleyball-related shoulder pain and dysfunction. Phys Med Rehabil. 2010;2(1):27–35.
18. Radwan A, Francis J, Green A, Kahl E, Maciurzynski D, Quartulli A, et al. Is there a relation between shoulder dysfunction and core instability? Int J Sports Phys Ther. 2014;9(1):8–13.
19. Vad VB, Bhat AL, Basrai D, Gebeh A, Aspergren DD, Andrews JR. Low back pain in professional golfers: the role of associated hip and low back range-of-motion deficits. Am J Sports Med. 2004;32(2):494–7.
20. Ellenbecker TS, Roetert EP. Age specific isokinetic glenohumeral internal and external rotation strength in elite junior tennis players. J Sci Med Sport. 2003;6(1):63–70.
21. Nadler SF, Malanga GA, Feinberg JH, Prybicien M, Stitik TP, DePrince M. Relationship between hip muscle imbalance and occurrence of low back pain in collegiate athletes: a prospective study. Am J Phys Med Rehabil. 2001;80(8):572–7.
22. Nadler SF, Malanga GA, Bartoli LA, Feinberg JH, Prybicien M, DePrince M. Hip muscle imbalance and low back pain in athletes: influence of core strengthening. Med Sci Sports Exerc. 2002;34(1):9–16.
23. Kibler WB, Ludewig PM, McClure PW, Michener LA, Bak K, Sciascia AD. Clinical implications of scapular dyskinesis in shoulder injury: the 2013 consensus statement from the "scapula summit". Br J Sports Med. 2013;47:877–85.
24. Wilk KE, Meister K, Andrews JR. Current concepts in the rehabilitation of the overhead throwing athlete. Am J Sports Med. 2002;30(1):136–51.
25. Reagan KM, Meister K, Horodyski M, Werner DW, Carruthers C, Wilk KE. Humeral retroversion and its relationship to glenohumeral rotation in the shoulder of college baseball players. Am J Sports Med. 2002;30(3):354–60.
26. Myers JB, Laudner KG, Pasquale MR, Bradley JP, Lephart SM. Glenohumeral range of motion deficits and posterior shoulder tightness in throwers with pathologic internal impingement. Am J Sports Med. 2006;34:385–91.
27. Burkhart SS, Morgan CD, Kibler WB. The disabled throwing shoulder: spectrum of pathology part I: pathoanatomy and biomechanics. Arthroscopy. 2003; 19(4):404–20.
28. Crockett HC, Gross LB, Wilk KE, Schwartz ML, Reed J, O'Mara J, et al. Osseous adaptation and range of motion at the glenohumeral joint in professional baseball pitchers. Am J Sports Med. 2002;30(1):20–6.
29. Kibler WB, Sciascia AD, Moore SD. An acute throwing episode decreases shoulder internal rotation. Clin Orthop Relat Res. 2012;470:1545–51.
30. Reinold MM, Wilk KE, Macrina LC, Sheheane C, Dun S, Fleisig GS, et al. Changes in shoulder and elbow passive range of motion after pitching in professional baseball players. Am J Sports Med. 2008; 36(3):523–7.
31. Wilk KE, Macrina LC, Fleisig GS, Porterfield R, Simpson Ii CD, Harker P, et al. Loss of internal rotation and the correlation to shoulder injuries in professional baseball pitchers. Am J Sports Med. 2011;39(2):329–35.
32. Tyler TF, Nicholas SJ, Roy T, Gleim GW. Quantification of posterior capsule tightness and motion loss in patients with shoulder impingement. Am J Sports Med. 2000;28:668–73.
33. Shanley E, Rauh MJ, Michener L, Ellenbecker TS, Garrison JC, Thigpen CA. Shoulder range of motion measures as risk factors for shoulder and elbow injuries in high school softball and baseball players. Am J Sports Med. 2006;39:1997–2006.
34. Kibler WB, McMullen J, Uhl TL. Shoulder rehabilitation strategies, guidelines, and practice. Oper Techn Sports Med. 2000;8(4):258–67.
35. Sciascia A, Cromwell R. Kinetic chain rehabilitation: a theoretical framework. Rehabil Res Pract. 2012; 2012:1–9.
36. Kibler WB, Sciascia A, Wilkes T. Scapular dyskinesis and its relation to shoulder injury. J Am Acad Orthop Surg. 2012;20(6):364–72.

Pectoralis Major and Minor Muscles 30

Alberto de Castro Pochini,
Eduardo Antonio Figueiredo,
Bernardo Barcellos Terra, Carina Cohen,
Paulo Santoro Belangero,
Carlos Vicente Andreoli,
Benno Ejnisman, and Levi Morse

30.1 Gross Anatomy

The clavicular head of the pectoralis major arises from the medial half of the anterior surface of the clavicle. The sternocostal head arises from the body of the sternum and the lateral half of the anterior surface of the manubrium, the upper six costal cartilages and the aponeurosis of the external oblique muscle Figs. 30.1, 30.2, 30.3, 30.4, 30.5, 30.6, and 30.7).

From this extensive origin the fibers converge toward their insertion; those arising from the clavicle pass obliquely downward and outwards (laterally), and are usually separated from the rest by a slight interval. Those from the sternocostal origin run upward and laterally, while the middle fibers pass horizontally. The pectoralis major tendon is bilaminar and approximately 5 cm in breadth. It is inserted into the lateral lip of the bicipital groove of the humerus (Figs 30.1, 30.2, 30.3, 30.4, 30.5, 30.6, and 30.7).

The pectoralis minor is a small triangular muscle which arises from the third, fourth, and fifth ribs under the cover of pectoralis major. It inserts via a short thick tendon onto the upper surface and medial border of the coracoid process.

30.2 Laminae

Many articles describe the pectoralis major tendon as consisting of two laminae, placed one in front of the other. They are usually blended together inferiorly.

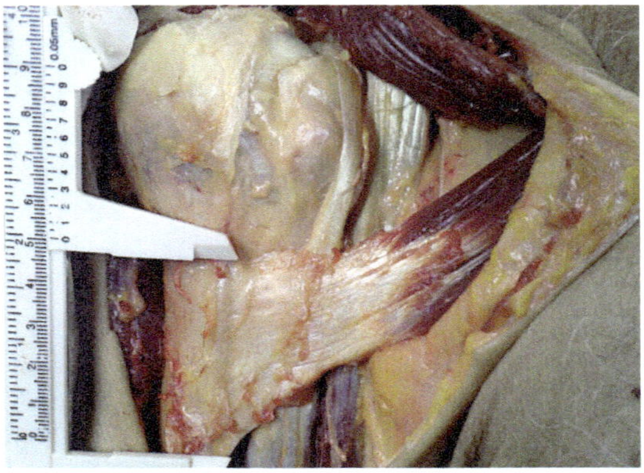

Fig. 30.1 Tendon of pectoralis major at its insertion onto the humerus

A. de Castro Pochini (✉) • E.A. Figueiredo • B.B. Terra
C. Cohen • P.S. Belangero • C.V. Andreoli • B. Ejnisman
Department of Orthopaedic Surgery,
Federal University of São Paulo, São Paulo, Brazil
e-mail: apochini@uol.com.br

L. Morse
Department of Orthopaedic Surgery,
Flinders University of South Australia,
Bedford Park, Australia

G.I. Bain et al. (eds.), *Normal and Pathological Anatomy of the Shoulder*,
DOI 10.1007/978-3-662-45719-1_30, © ISAKOS 2015

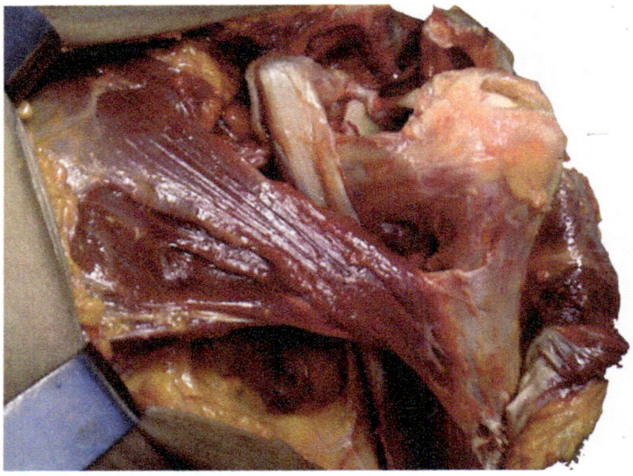

Fig. 30.2 Gross anatomy of pectoralis major muscle. Note the anterior lamina formed by the clavicular head, and the sternocostal head curling underneath and inserting progressively higher as the posterior lamina

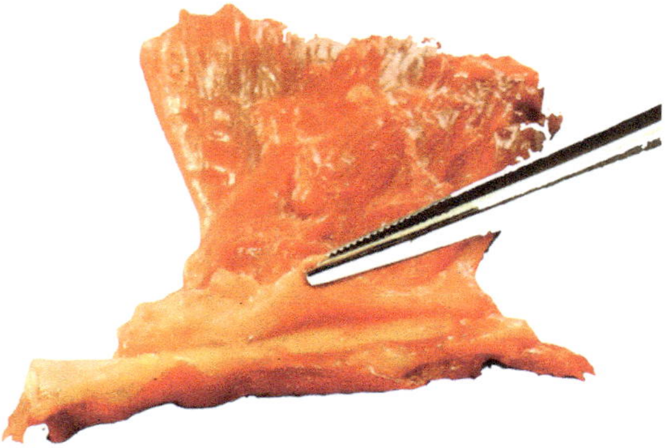

Fig. 30.3 Anatomical relationship between the pectoralis major tendon and biceps tendon

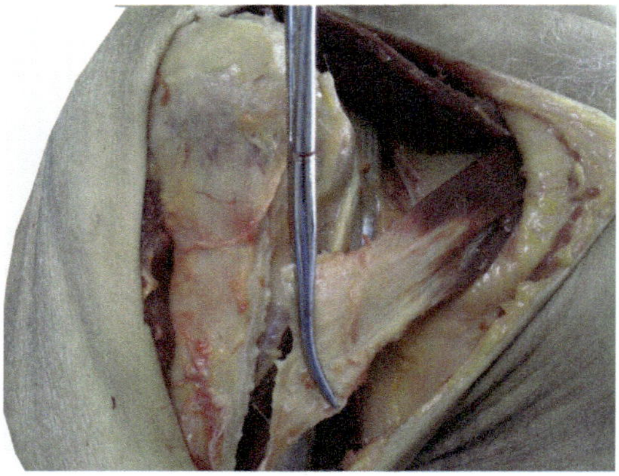

Fig. 30.4 The pectoralis major detached from its insertion demonstrating the underlying proximal biceps tendon

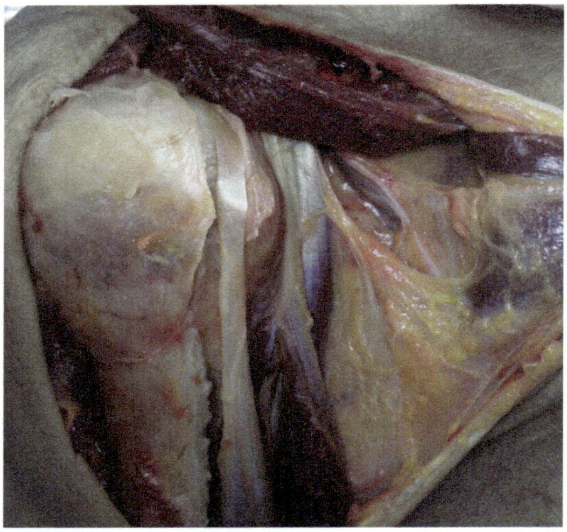

Fig. 30.5 The pectoralis major detached and reflected from its insertion demonstrating the underlying proximal biceps tendon

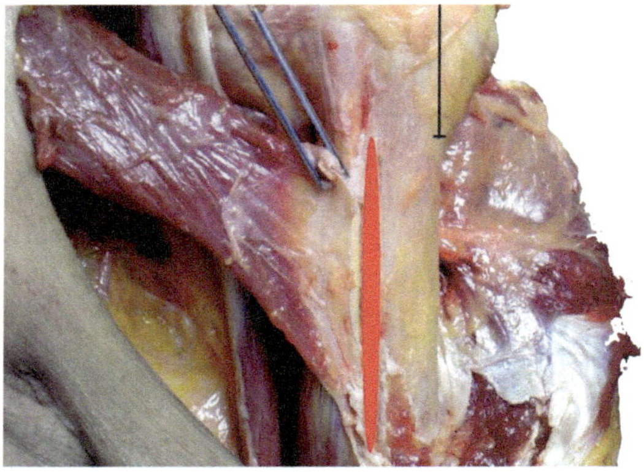

Fig. 30.6 The tendon insertion of the pectoralis major

- The anterior lamina is thicker, and consists of the clavicular and the uppermost sternal fibers. They are inserted in the same order as that in which they arise. That is, the most lateral of the clavicular fibers are inserted at the upper part of the anterior lamina, while the uppermost sternal fibers pass to the lower part of the lamina. This lamina extends as low as the tendon of the deltoid muscle and joins with it (Fig. 30.2).

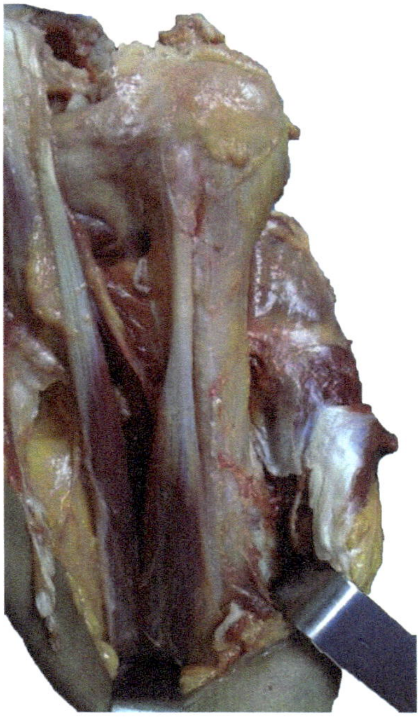

Fig. 30.7 Complete removal of pectoralis major demonstrating the underlying biceps brachii muscle

- The posterior lamina of the tendon receives the attachment of the greater part of the sternal portion and the deep fibers from the costal cartilages. As they course upwards and laterally, they insert progressively higher into the posterior lamina of the tendon, producing the rounded appearance of the anterior axillary fold and the twisted appearance of the tendon.

The posterior lamina reaches higher on the humerus than the anterior, and from it an expansion extends over the intertubercular groove of the humerus and blends with the capsule of the shoulder joint. There is a close anatomical relationship between the pectoralis major and biceps tendons. This is consistent with our observations that in some instances following total rupture of the pectoralis major tendon, the patient also demonstrates subluxation of the long head of the biceps. It may be that the posterior lamina of the pectoralis major tendon is an important stabilizer of the biceps tendon (Fig. 30.3).

From the deepest fibers of this lamina at its insertion, an expansion is given off which lines the intertubercular groove, while from the lower border of the tendon a third expansion passes downward to the fascia of the arm.

30.3 Innervation

The pectoralis major muscle receives motor innervation by the medial and lateral pectoral nerves, so named because of their origins from the medial and lateral cords of the plexus. The medial pectoral nerve originates from the C7, C8, and T1 nerve roots from the lower trunk of the brachial plexus. It then communicates the action potential across the neuromuscular junction by releasing acetylcholine into the neuromuscular junction, inciting a proportional muscle contraction of the sternal head of the pectoralis major muscle. The second source of innervation of the pectoralis major originates from the C5 and C6 nerve roots which merge to form the upper trunk, splits off into the anterior division of the upper trunk which joins with the middle trunk to form the lateral cord. The lateral pectoral nerve branches off of the lateral cord of the brachial plexus and is distributed over the deep surface of the pectoralis major. At the neuromuscular junction, the lateral pectoral nerve provides motor input to the clavicular head of the pectoralis major.

The pectoralis minor muscle also receives innervation from both the medial and lateral pectoral nerves, from the C6 to C8 nerve roots.

The sensory feedback from the pectoralis major follows the reverse path, returning via first-order neurons to the spinal nerves at C5, C6, C8, and T1 through the anterior rami. After the synapse in the posterior horn of the spinal cord, sensory information concerning movement of the muscle, proprioception, and pressure then travels through a second-order neuron in the dorsal column medial lemniscus tract to the medulla. There, the fibers decussate to form the medial lemniscus which carries the sensory information the rest of the way to the thalamus, the "gateway to the cortex." The thalamus diverts some sensory information to the cerebellum and the basal nuclei to complete the motor feedback loop while some sensory information ascends directly to the postcentral gyrus of the parietal lobe of the brain via third-order neurons. Sensory information for the pectoralis major is processed in the superior portion of the sensory homunculus, adjacent to the longitudinal fissure that divides the two hemispheres of the brain.

Electromyography suggests that it consists of at least six groups of muscle fibers that can be independently coordinated by the central nervous system.

30.4 Function

The pectoralis major muscle is a powerful adductor and medial rotator of the arm, and also assists in flexion of the shoulder joint. It is also responsible for keeping the arm attached to the trunk of the body, and with the upper limb fixed in abduction the muscle is a useful accessory muscle of inspiration. The two different parts are responsible for different actions. The clavicular part is in close proximity to the deltoid muscle and contributes to flexion, horizontal adduction, and inward rotation of the humerus. When at an approximate 110° angle, it contributes to adduction of the humerus. The sternocostal part is antagonistic to the clavicular part, contributing to downward and forward movement of the arm and inward rotation when accompanied by adduction. The sternal fibers can also contribute to extension, but not beyond anatomical position.

The pectoralis minor muscle is of less functional significance, but has a protective role by providing a tight band across the front of the axillary neurovascular and lymphatic contents. It assists the serratus anterior in scapular protraction and can also assist gravity in restoring the scapula to its resting position following full abduction of the arm. However, pectoralis minor has been implicated as a factor in the development of frozen shoulder.

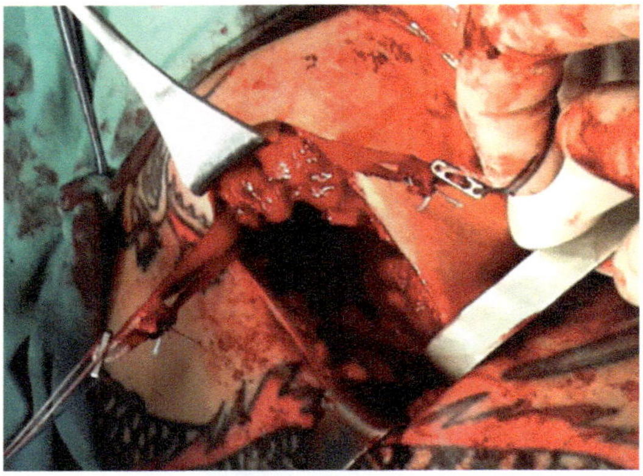

Fig. 30.8 Operative photo with the pectoralis muscle and a hamstring tendon graft

30.5 Injuries and Imaging

Tears of the pectoralis major are uncommon, and congenital absence is rare (Figs. 30.8 and 30.9) [1–7, 10]. The most common mechanism of injury is weight lifting, in particular performing bench-press maneuvers. Most lesions are located at the musculotendinous junction and result from violent, eccentric contraction of the muscle. We have, however, also described the rupture as occurring at the transition between eccentric and concentric contractions during a bench press in athletes with stronger muscles as a result of anabolic steroids [29, 8]. A less frequent rupture site is the muscle belly, usually as a result of a direct blow [31]. Most lesions occur in male athletes, especially those practicing contact sports and weight lifting [11–27]. Women are less susceptible to these tears because of larger tendon-to-muscle diameter, greater muscular elasticity, and less energetic injuries [9]. The injury is characterized by pain in the chest wall, bruising and loss of strength of the muscle, and loss of definition of the anterior axillary fold. High-grade partial or full thickness tears often require surgery, particularly in the athletic population. Most patients are able to return to activity following surgery with high patient satisfaction and only slight reduction in strength compared to pre-injury function [28–46]. Both ultrasound and magnetic resonance imaging can confirm the diagnosis [29, 31].

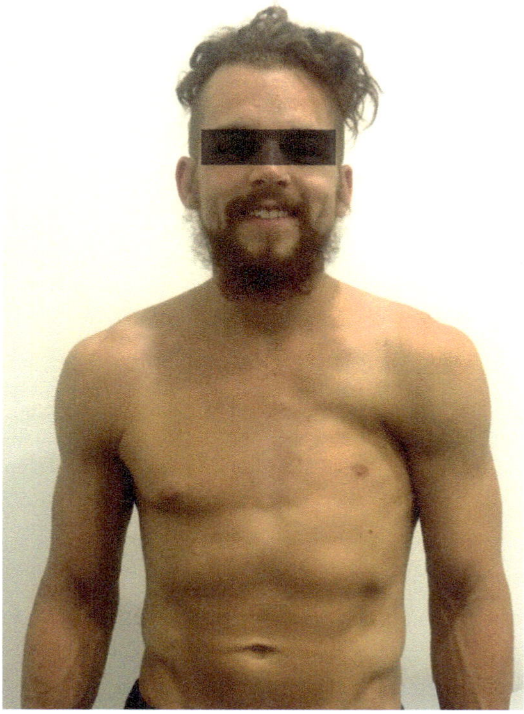

Fig. 30.9 Congenital absence of the costal head of the pectoralis major muscle (Copyright Dr Gregory Bain)

References

1. Aarimaa V, Rantanen J, Heikkila J, Helttula I, Orava S. Rupture of the pectoralis major muscle. Am J Sports Med. 2004;32(5):1256–62.
2. Alho A. Ruptured pectoralis major tendon. A case report on delayed repair with muscle advancement. Acta Orthop Scand. 1994;65(6):652–3.
3. Anbari A, Kelly 4th JD, Moyer RA. Delayed repair of a ruptured pectoralis major muscle. A case report. Am J Sports Med. 2000;28(2):254–6.
4. Bak K, Cameron EA, Henderson IJ. Rupture of the pectoralis major: a meta-analysis of 112 cases. Knee Surg Sports Traumatol Arthrosc. 2000;8(2):113–9.
5. Beloosesky Y, Grinblat J, Katz M, Hendel D, Sommer R. Pectoralis major rupture in the elderly: clinical and sonographic findings. Clin Imaging. 2003;27(4):261–4.
6. Berson BL. Surgical repair of pectoralis major rupture in an athlete. Case report of an unusual injury in a wrestler. Am J Sports Med. 1979;7(6):348–51.
7. Carek PJ, Hawkins A. Rupture of pectoralis major during parallel bar dips: case report and review. Med Sci Sports Exerc. 1998;30(3):335–8.
8. Connell DA, Potter HG, Sherman MF, Wickiewicz TL. Injuries of the pectoralis major muscle: evaluation with MR imaging. Radiology. 1999;210(3):785–91.
9. Chiavaras MM, Jacobson JA, Smith J, Dahm DL. Pectoralis major tears: anatomy, classification, and diagnosis with ultrasound and MR imaging. Skeletal Radiol. 2015;44:157–64.
10. Dunkelman NR, Collier F, Rook JL, Nagler W, Brennan MJ. Pectoralis major muscle rupture in windsurfing. Arch Phys Med Rehabil. 1994;75(7):819–21.
11. Dvir Z. Isokinetics—muscle testing, interpretation and clinical applications. Edinburgh: Churchill Livingstone; 1995.
12. Egan TM, Hall H. Avulsion of the pectoralis major tendon in a weight lifter: repair using a barbed staple. Can J Surg. 1987;30(6):434–5.
13. ElMaraghy AW, Devereaux MW. A systematic review and comprehensive classification of pectoralis major tears. J Shoulder Elbow Surg. 2012;21(3):412–22.
14. Fleury AM, Silva AC, Pochini A, Ejnisman B, Lira CA, Andrade Mdos S. Isokinetic muscle assessment after treatment of pectoralis major muscle rupture using surgical or non-surgical procedures. Clinics (Sao Paulo). 2011;66(2):313–20.
15. Garrigues GE, Kraeutler MJ, Gillespie RJ, O'Brien DF, Lazarus MD. Repair of pectoralis major ruptures: single-surgeon case series. Orthopedics. 2012;35(8):e1184–90.
16. Hanna CM, Glenny AB, Stanley SN, Caughey MA. Pectoralis major tears: comparison of surgical and conservative treatment. Br J Sports Med. 2001;35(3):202–6.
17. Hart ND, Lindsey DP, McAdams TR. Pectoralis major tendon rupture: a biomechanical analysis of repair techniques. J Orthop Res. 2011;29(11):1783–7.
18. Jones MW, Matthews JP. Rupture of pectoralis major in weight lifters: a case report and review of the literature. Injury. 1988;19(3):219.
19. Joseph TA, Defranco MJ, Weiker GG. Delayed repair of a pectoralis major tendon rupture with allograft: a case report. J Shoulder Elbow Surg. 2003;12(1):101–4.
20. Law WB. Closed incomplete rupture of pectoralis major. Br Med J. 1954;4886:499.
21. Lee J, Brookenthal KR, Ramsey ML, Kneeland JB, Herzog R. MR imaging assessment of the pectoralis major myotendinous unit: an MR imaging-anatomic correlative study with surgical correlation. AJR Am J Roentgenol. 2000;174(5):1371–5.
22. Lindenbaum BL. Delayed repair of a ruptured pectoralis major muscle. Clin Orthop. 1975;109:120–1.
23. Marmor L, Becthol CO, Hall CB. Pectoralis major muscle function of esternal portion and mechanism of rupture of normal muscle: case reports. J Bone Joint Surg Am. 1961;43:81–7.
24. Miller MD, Johnson DL, Fu FH, Thaete FL, Blanc RO. Rupture of the pectoralis major muscle in a collegiate football player. Use of magnetic resonance imaging in early diagnosis. Am J Sports Med. 1993;21(3):475–7.
25. Motzkin N, Cahalan TD, Morrey BF, Chao EYS. Isometric and isokinetic endurance testing of the forearm complex. Am J Sports Med. 1991;19:107–11.
26. Ohashi K, El-Khoury GY, Albright JP, Tearse DS. MRI of complete rupture of the pectoralis major muscle. Skeletal Radiol. 1996;25(7):625–8.

27. Park JY, Espiniella JL. Rupture of pectoralis major muscle. A case report and review of literature. J Bone Joint Surg Am. 1970;52(3):577–81.
28. Pochini A, Ejnisman B, Andreoli CV, Cohen M. Reconstruction of the pectoralis major tendon using autologous grafting and cortical button attachment: description of the technique. Tech Should Elbow Surg. 2012;13(5):77–80.
29. Pochini AC, Ejnisman B, Andreoli CV, et al. Exact moment of tendon of pectoralis major muscle rupture captured on video. Br J Sports Med. 2007;41(9):618–9.
30. Pochini AC, Ejnisman B, Andreoli CV, et al. Pectoralis major muscle rupture in athletes: a prospective study. Am J Sports Med. 2010;38(1):92–8.
31. Pochini AC, Andreoli CV, Ejnisman B. Clinical considerations for the surgical treatment of pectoralis major muscle ruptures based on 60 cases: a prospective study and literature review. Am J Sports Med. 2014;42(1):95–102.
32. Quinlan JF, Molloy M, Hurson BJ. Pectoralis major tendon ruptures: when to operate. Br J Sports Med. 2002;36(3):226–8.
33. Rabuck SJ, Lynch JL, Guo X, et al. Biomechanical comparison of 3 methods to repair pectoralis major ruptures. Am J Sports Med. 2012;40(7):1635–40.
34. Schachter AK, White BJ, Namkoong S, Sherman O. Revision reconstruction of a pectoralis major tendon rupture using hamstring autograft: a case report. Am J Sports Med. 2006;34(2):295–8.
35. Schepsis AA, Grafe MW, Jones HP, Lemos MJ. Rupture of the pectoralis major muscle. Outcome after repair of acute and chronic injuries. Am J Sports Med. 2000;28(1):9–15.
36. Scott BW, Wallace WA, Barton MA. Diagnosis and assessment of pectoralis major rupture by dynamometry. J Bone Joint Surg Br. 1992;74(1):111–3.
37. Sherman SL, Lin EC, Verma NN, Mather RC. Biomechanical analysis of the pectoralis major tendon and comparison of techniques for tendo-osseous repair. Am J Sports Med. 2012;40(8):1887–94.
38. Silverstein JA, Goldberg B, Wolin P. Proximal humerus shaft fracture after pectoralis major tendon rupture repair. Orthopedics. 2011;34(6):222.
39. Tietjen R. Closed injuries of the pectoralis major muscle. J Trauma. 1980;20(3):262–4.
40. Uchiyama Y, Miyazaki S, Tamaki T, et al. Clinical results of a surgical technique using endobuttons for complete tendon tear of pectoralis major muscle: report of five cases. Sports Med Arthrosc Rehabil Ther Technol. 2011;3:20.
41. Wheat H, Bugg B, Lemay K, Reed J. Tears of pectoralis major in steer wrestlers: a novel repair technique using the EndoButton. Clin J Sport Med. 2013;23(1):80–2.
42. Wolfe SW, Wickiewicz TL, Cavanaugh JT. Ruptures of the pectoralis major muscle. An anatomic and clinical analysis. Am J Sports Med. 1992;20(5):587–93.
43. Wong B, Fung L, Ravichandiran K, Agur A, Rindlisbacher T, Elmaraghy A. Three-dimensional study of pectoralis major muscle and tendon architecture. Clin Anat. 2009;22(4):500–8.
44. De Zordo T, Chhem R, Smekal V, Feuchtner G, Reindl M, Fink C, Faschingbauer R, Jaschke W, Klauser AS. Real-time sonoelastography: findings in patients with symptomatic achilles tendons and comparison to healthy volunteers. Ultraschall Med. 2010;31(4):394–400.
45. Turan A, Tufan A, Mercan R, Teber MA, Tezcan ME, Bitik B, Goker B, Haznedaroğlu S. Real-time sonoelastography of Achilles tendon in patients with ankylosing spondylitis. Skeletal Radiol. 2013;42(8):1113–8.
46. Drakonaki EE, Allen GM, Wilson DJ. Ultrasound elastography for musculoskeletal applications. Br J Radiol. 2012;85(1019):1435–45.

Part VI

Nervovascular Structures

Brachial Plexus

31

Akimoto Nimura, Keiichi Akita, and Hiroyuki Sugaya

31.1 Development of the Brachial Plexus

31.1.1 Basic Structure of the Muscles of the Shoulder Girdle

The upper limbs of humans correspond to the anterior limbs of four-legged animals and are described to originate from the pectoral fin of the fish. Thus, to understand the principal of the structure of the shoulder girdle, let us consider the developmental process of the fin. The pectoral fin originates from the continuous ventrolateral skin fold that bilaterally appeared at the ventral side of the body trunk of the agnathonae. The fin fold is thought to be separated into two parts to form the set of paired appendages in the primitive gnathostomes [6]. The muscles of fins are constructed by the trunk muscles entering into fins. The muscles of fins are innervated by ventral rami of the spinal nerves, because the muscles of fins originate from the muscles of the ventral trunk (V). The dorsal rami of the spinal nerves and the muscles of the dorsal trunk (D) do not contribute to form fins.

The main functions of the fin are elevation and depression. The intermuscular septa appear at the middle line of fins and correspond to interosseous membranes and bones of the upper limb. These structures divided fin muscles into elevators (dorsal side; d) and depressors (ventral side; v). Furthermore, additional septa appear between the muscles of the fin and the body trunk. They correspond to the limb girdle of the four-legged animals. The shoulder girdle of a human has a glenoid fossa at the center of the scapula and it is connected with the proximal end of the humerus to form a joint. The shoulder girdle itself is divided into d and v by the glenoid fossa, in other words, the body of the scapula is included in d, and the coracoid process and the clavicle is in v (Fig. 31.1) [4]. In addition, muscles are distributed into categories from proximal to distal according to the connection of bones. Based on the above rules, muscles of the shoulder girdle can be listed as Table 31.1.

31.1.2 Morphological Consideration of the Brachial Plexus

During the development of the location of muscles, the relationship between each muscle and the innervating nerve has been described to be phylogenetically preserved. Therefore, morphological

A. Nimura, MD, PhD (✉) • K. Akita, MD, PhD
Department of Clinical Anatomy,
Graduate School of Medical and Dental Sciences,
Tokyo Medical and Dental University, Tokyo, Japan
e-mail: nimura.orj@tmd.ac.jpl; akita.fana@tmd.ac.jp

H. Sugaya, MD, PhD
Shoulder and Elbow Service,
Funabashi Orthopaedic Sports Medicine Center,
Funabashi, Japan
e-mail: Hsugaya@nifty.com

G.I. Bain et al. (eds.), *Normal and Pathological Anatomy of the Shoulder*,
DOI 10.1007/978-3-662-45719-1_31, © ISAKOS 2015

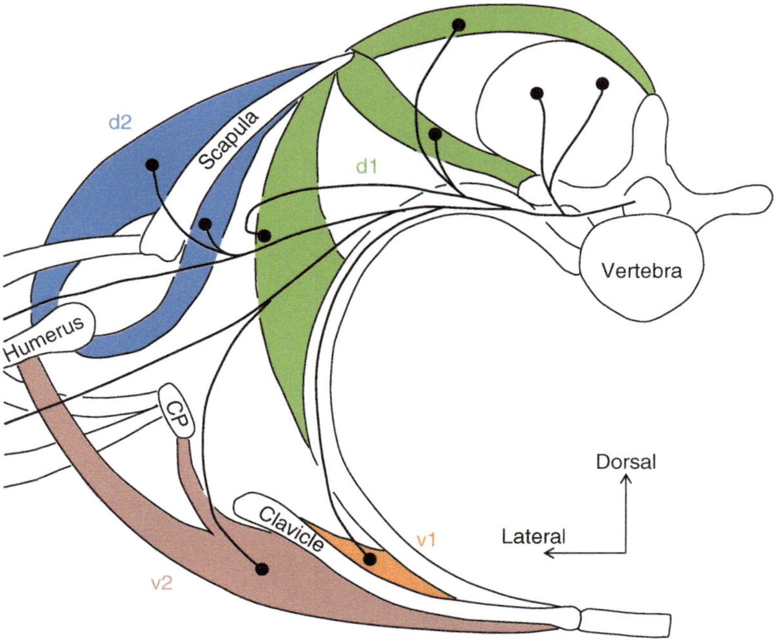

Fig. 31.1 Schematic transverse section through the thorax and arm demonstrating the classification of muscles included in the shoulder girdle, depending on their locations and the courses of the nerves. Muscles in d1 of Table 31.1 are demonstrated as green. Muscles in d2, v1, and v2 are demonstrated as *blue*, *orange*, and *pink*, respectively. Muscles in each group (*d1*, *d2*, *v1*, and *v2*) are shown in Table 31.1. *CP* coracoid process

Table 31.1 Location of shoulder girdle muscles

Location		Muscle	Nerve	Roots
Dorsal (d)	d1	Levator scapulae	Dorsal scapular	C5
		Rhomboids		C5
		Anterior serratus	Long thoracic	C567
	d2	Subscapularis	Subscapular	C56
		Teres major		C56
		Latissimus dorsi	Thoracodorsal	C678
Ventral (v)	v1	Subclavius	Subclavian	C56
	v2	Pectoralis major	Med and lat pectoral	C567
		Pectoralis minor	Medial pectoral	C8T1

consideration of muscles of the shoulder girdle may give a clue to the understanding of the complex structure of the brachial plexus.

A spinal nerve firstly separates into the anterior and posterior branch to respectively innervate the ventral and dorsal trunk muscles of the same segment. The brachial plexus should be composed of the anterior branch, because the upper extremity emerges in the ventral side of the body trunk. The consideration of the development of the shark's fin makes it easier to understand why anterior branches at the several segmental levels complicatedly form a plexus (Fig. 31.2). If the development of the fin slowly progressed while the base of the fin kept opening to the body trunk, the segmental order of the innervation should be preserved and the concentration should not take place. Yet, while the base of the fin is actually closed during the rapid development, myotomes are mixed and innervating nerves make anastomosis to form a plexus. During these processes, the segmental orders of nerves are significantly destroyed [1].

However, following two rules of innervating nerves have been approximately preserved during the development of the brachial plexus. Firstly, branches of the brachial plexus are separated into a ventral and dorsal layer based on locations of

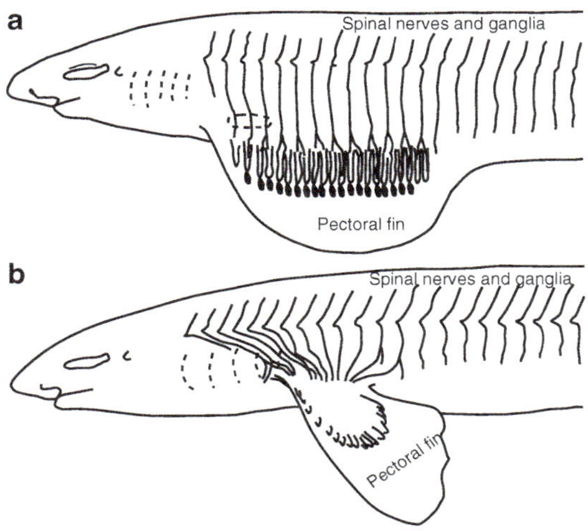

Fig. 31.2 Schemes of the upper part of adult sharks showing the nerve-supply of the fin. (**a**) The pectoral fin is expanded, and its nervous, muscular, and skeletal elements are distributed. (**b**) The base of the fin is closed, and concentration of myotomes innervating nerves take place

innervated muscles. In other words, the dorsal muscles (d) are innervated by nerves of the dorsal layer, and the ventral muscles (v) are by nerves of the ventral layer. Second, the closer the location of the muscle is to the body trunk, the more proximal the origin of the innervating nerve from the brachial plexus is. To be specific, the dorsal scapular and long thoracic nerve (d1 in Table 31.1), and subclavian nerve (v1) arise near the intervertebral foramina, while the subscapular and thoracodorsal nerve (d2), and pectoral nerve (V2) arise at even the infraclavicular area (Figs. 31.1 and 31.3).

31.2 Description of Structure

31.2.1 Overview of the Brachial Plexus

The brachial plexus is formed by the union of the ventral rami of the fourth to eighth cervical nerves and the most part of the first thoracic ventral ramus (Fig. 31.3). The most common arrangement of the brachial plexus is as follows: the fifth and sixth rami mix at the lateral border of middle scalene muscle as the superior trunk; the eighth cervical and first thoracic rami join behind anterior scalene muscle as the inferior trunk; the seventh cervical ramus becomes the middle trunk. These trunks pass laterally, and bifurcate into anterior and posterior divisions. The anterior divisions of the superior and middle trunks form a lateral cord that lies lateral to the axillary artery. The anterior division of the inferior trunk descends at first behind and then medial to the axillary artery and forms the medial cord; it often receives a branch from the seventh cervical ramus. Posterior divisions of all three trunks form the posterior cord, which is at first above, and then behind, the axillary artery [3].

31.2.2 The Dorsal Scapular Nerve and the Long Thoracic Nerve (d1)

The dorsal scapular nerve comes from the fifth cervical ramus. The long thoracic nerve is usually formed by roots from the fifth to the seventh cervical rami, although the last one may be absent [7]. The levator scapulae, rhomboid, and serratus anterior muscles broadly attach to the medial border of the scapula from the superior to inferior angle. Thus, these three muscles are thought to be homologous, and innervation patterns of these muscles were connected with each other.

The levator scapulae muscle is innervated directly by branches of the third and fourth cervical rami, and the branch of the dorsal scapular nerve (Fig. 31.4). The branch of the fourth cervical ramus pierces the levator scapulae muscle and communicates with the branch from the fifth cervical ramus which passes behind the levator scapulae muscle, then these innervate the rhomboid muscles as the dorsal scapular nerve.

The serratus anterior muscle is composed of three parts: the superior, middle, and inferior ones. The superior part is innervated by the independent branches which formed the common trunk with the nerve innervating the rhomboid muscles, and short branches from the long thoracic nerve [2, 5]. The middle and inferior parts are mainly innervated by the long thoracic nerve descending the dorsal side of the serratus anterior muscle. Therefore, the superior part is thought to have an

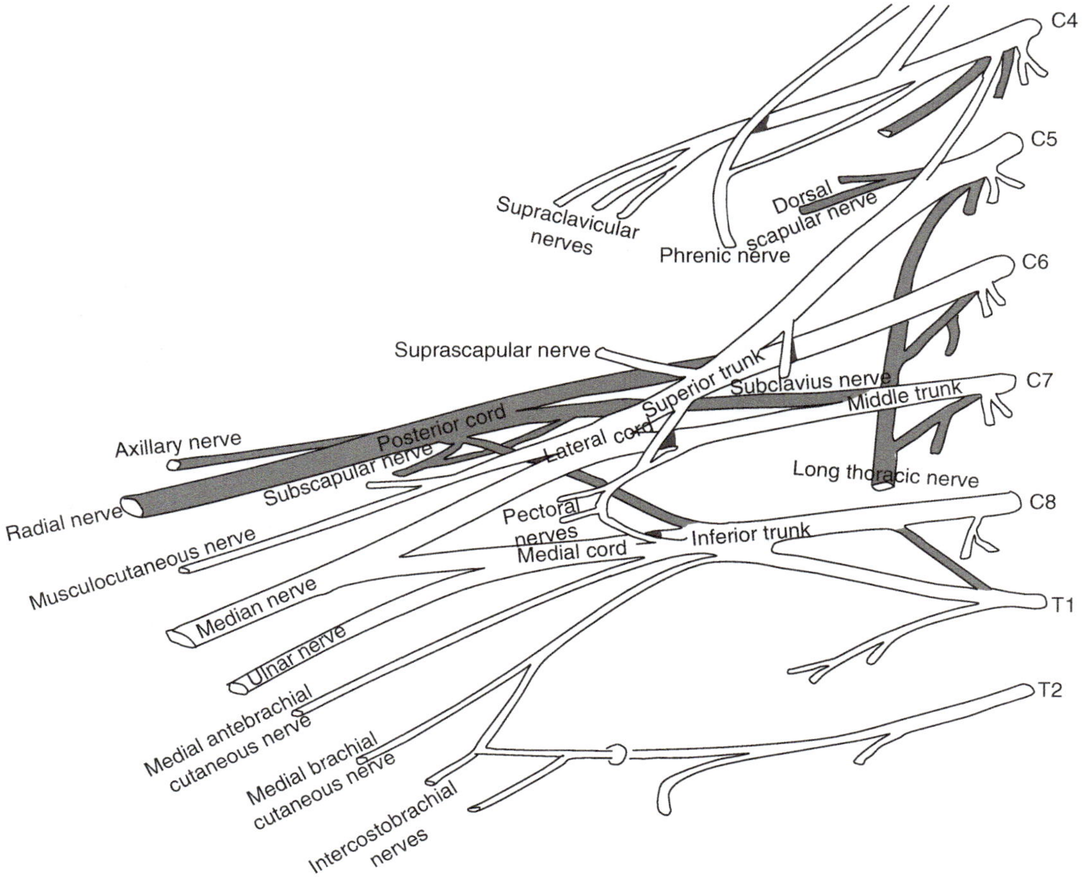

Fig. 31.3 Overview of the brachial plexus. Nerves passing dorsal are indicated as *gray color*. *C4–8* 4–8th cervical nerves, *T1, 2* first and second thoracic nerves

intimate relation with the levator scapulae and the rhomboid muscles, while the middle and inferior parts could be the actual serratus anterior muscle.

31.2.3 The Subscapular Nerve and the Thoracodorsal Nerve (d2)

The latissimus dorsi muscle is innervated from the brachial plexus, while it is one of the superficial back muscles. The scapular origin of the latissimus dorsi muscle preserves the primitive fact that it is in the group of the shoulder girdle muscles. Origins from spinous processes, the iliac crest, and ribs developed secondary to the scapular origin. Taking into consideration of the innervation, the muscle of the shoulder girdle could be thought to expand to the back.

The superior and inferior subscapular nerves and the thoracodorsal nerve arise from the posterior cord and could be categorized into a same group. The superior and inferior subscapular nerves arise from C5 and C6, and supply the subscapularis muscle and the teres major muscle, respectively. The

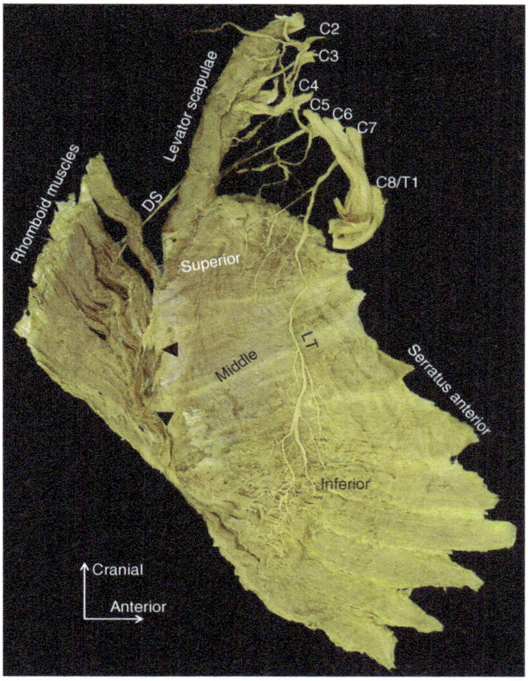

Fig. 31.4 The levator scapulae muscle, rhomboid muscles, and serratus anterior muscles innervated by the brachial plexus. The lateral view of the right muscles with preserving the insertion to the medial border of the scapula (*arrow heads*) is shown. The serratus anterior muscle is divided into the superior, middle, and inferior parts. *DS* the dorsal scapular nerve, *LT* the long thoracic nerve. *C2–8* second – 8th cervical nerves, *T1* first thoracic nerve

thoracodorsal nerve arises from mainly C7 (C6–8) between the subscapular nerves. It supplies the latissimus dorsi muscle, reaching its distal border.

31.2.4 The Subclavius Nerve (v1)

The subclavius nerve arises near the junction of the fifth and sixth cervical ventral rami and descends anterior to the plexus to supply the subclavius muscle. The infrahyoid muscles, diaphragm, and subclavius muscles are thought to be similarly included in the group of the abdominal rectus muscle. Thus, the subclavius muscle strongly has a developmental connection with the diaphragm. Therefore, the subclavius nerve is sometimes connected to the phrenic nerve.

31.2.5 The Pectoral Nerves (v2)

The pectoral major and minor muscles are innervated by pectoral nerves. Pectoral nerves arise from the ventral side of the plexus and descend anterior to it. Pectoral nerves are separated into the lateral and medial pectoral nerves. The lateral pectoral nerve arises from the lateral cord which is a unit from the fifth to seventh cervical rami. The medial pectoral nerve arises from the medial cord which is derived from the eighth cervical and first thoracic rami. These are why they are called as the "lateral" and "medial" pectoral nerves. The lateral pectoral nerve passes superior to the pectoral minor muscle, pierces the clavipectoral fascia and supplies the deep surface of the pectoral major muscle. The medial pectoral nerve pierces or passes inferior to the pectoral minor muscle to end in the pectoral major muscle. Both pectoral nerves anastomose to form a loop anterior to the axillary artery.

References

1. Goodrich ES. Studies on the structure & development of vertebrates. Chicago: The University of Chicago Press; 1930.
2. Hamada J, Igarashi E, Akita K, Mochizuki T. A cadaveric study of the serratus anterior muscle and the long thoracic nerve. J Shoulder Elbow Surg. 2008;17(5):790–4. doi:10.1016/j.jse.2008.02.009. S1058-2746(08)00344-3.
3. Johnson D. Pectoral girdle and upper limb: overview and suface anatomy. In: Standring S, editor. Gray's anatomy. 14th ed. London: Churchill Livingstone; 2008. p. 777–90.
4. Kato K, Sato T. Innervation of the levator scapulae, the serratus anterior, and the rhomboideus in crab-eating macaques and its morphological significance. Anat Anz. 1984;157(1):43–55.
5. Nasu H, Yamaguchi K, Nimura A, Akita K. An anatomic study of structure and innervation of the serratus anterior muscle. Surg Radiol Anat. 2012;34(10):921–8. doi:10.1007/s00276-012-0984-1.
6. Tabin CJ. Why we have (only) five fingers per hand: hox genes and the evolution of paired limbs. Development. 1992;116(2):289–96.
7. Tytherleigh-Strong G. Pectoral girdle, shoulder region and axilla. In: Standring S, editor. Gray's anatomy. 14th ed. London: Churchill Livingstone; 2008. p. 791–822.

Axillary Nerve

32

Ian J. Galley

32.1 Anatomy

The axillary nerve is one of the two large terminal branches of the posterior cord (C5 and C6 ventral rami) [57]. It lies behind the axillary artery and vein, superior to the radial nerve, and lateral to the median and ulnar nerves. It then descends inferolaterally anterior to the subscapularis muscle. It divides into anterior and posterior branches, passing through the quadrilateral space at the inferior border of subscapularis. It then gives branches to the deltoid muscle, teres minor muscle, superior lateral brachial cutaneous nerve and to the glenohumeral joint.

The axillary nerve follows Hilton's law where the nerve supplying the muscles extending directly across and acting at a given joint also innervate the joint [32].

Duparc et al. [22] suggested that the nerve has five segments.

1. From its origin to the inferior border of the subscapularis
2. From subscapularis to the long head of triceps
3. From the long head of triceps to the surgical neck of the humerus
4. From the humerus to the entry into the deltoid
5. Its intramuscular course through the deltoid

32.1.1 Axillary Nerve

The relationship of the axillary nerve within the brachial plexus is relatively constant. The axillary nerve is the most superior nerve in most cases (musculocutaneous nerve superior in 20 %) [4, 62]. Usually there are no branches within the first segment.

As the nerve passes anterior to the subscapularis muscle, medial and lateral fasciculi bundles are contained by loose perineurium. The glenohumeral articular branch passes from the lateral fasciculi bundles.

The lateral fasciculi bundle becomes the anterior branch while the medial bundle becomes the posterior branch of the axillary nerve [70]. The axillary nerve usually divides into the anterior and posterior branches in the quadrilateral space, but this can occur within the deltoid muscle (35 %) [45].

The posterior branch is more superficial and gives rise to branches directed posteriorly, whereas the anterior branch winds around the

Electronic supplementary material The online version of this chapter 10.1007/978-3-662-45719-1_32 contains supplementary material, which is available to authorized users.

I.J. Galley
Department of Orthopaedic, Tauranga Hospital
Tauranga, Tauranga, New Zealand

Grace Orthopaedic Centre Tauranga,
Tauranga, New Zealand
e-mail: Ian.galley@orthocentre.co.nz

G.I. Bain et al. (eds.), *Normal and Pathological Anatomy of the Shoulder*,
DOI 10.1007/978-3-662-45719-1_32, © ISAKOS 2015

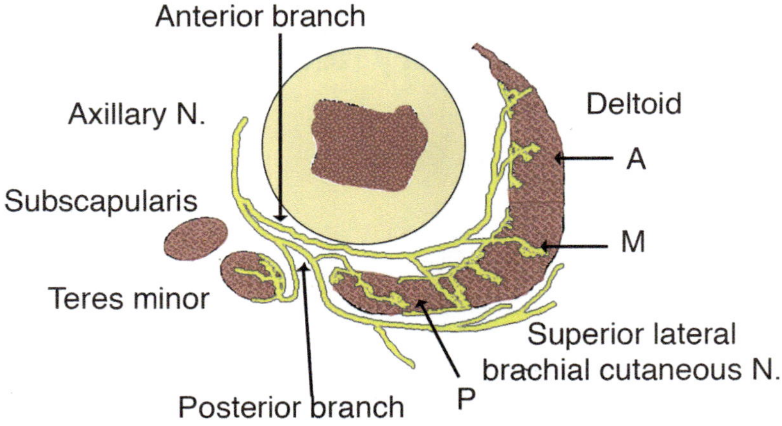

Fig. 32.1 Schematic drawing of the course of the axillary nerve after the quadrilateral space. Branches *A* anterior deltoid, *M* middle deltoid, *P* posterior deltoid (Modified with permission from Zhao et al. [70])

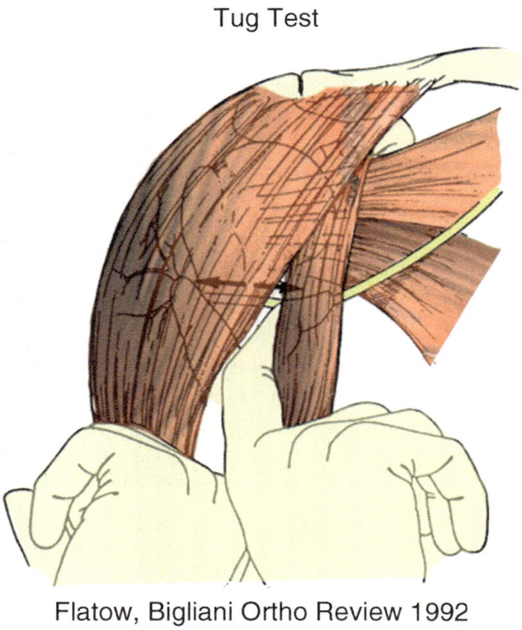

Fig. 32.2 'Tug Test'. One finger is placed on the axillary nerve at the front of subscapularis, and the other finger under the deltoid on the anterior branch. A gentle tug applied by either finger will be transmitted to the other finger and is usually easily palpable (Modified with permission from Flatow and Bigliani [25])

neck of the humerus and gives rise to branches directed anteriorly [70] (Fig. 32.1).

The posterior branch of the axillary nerve always supplies teres minor and the superior lateral brachial cutaneous nerve. The anterior branch passes on the deep surface of the deltoid to supply the anterior and middle deltoid. The posterior deltoid can be supplied by either or both branches. The middle deltoid can also be supplied by the posterior branch.

Flatow and Bigliani [25] described the tug test to identify the axillary nerve during a deltopectoral approach to the shoulder. One index finger is passed under the deltoid laterally to palpate the anterior branch. The other index finger is placed along the superficial surface of the subscapularis medially under the coracoid muscles (coracobrachialis and the short head of biceps) and then rotated inferiorly. The gentle tug applied by either finger will be transmitted to the other finger and is usually palpable (Fig. 32.2) (Video 32.1).

32.1.2 Quadrilateral Space

The quadrilateral space is bound by the medial border of the humerus, teres major, teres minor and the long head of triceps. The axillary nerve and posterior circumflex vessels traverse the space [18] (Fig. 32.3).

32.1.3 Posterior Branch Axillary Nerve

All branches that supply the teres minor and the superior lateral brachial cutaneous nerve arise from the posterior branch. There are variable branches supplying the posterior and middle deltoid.

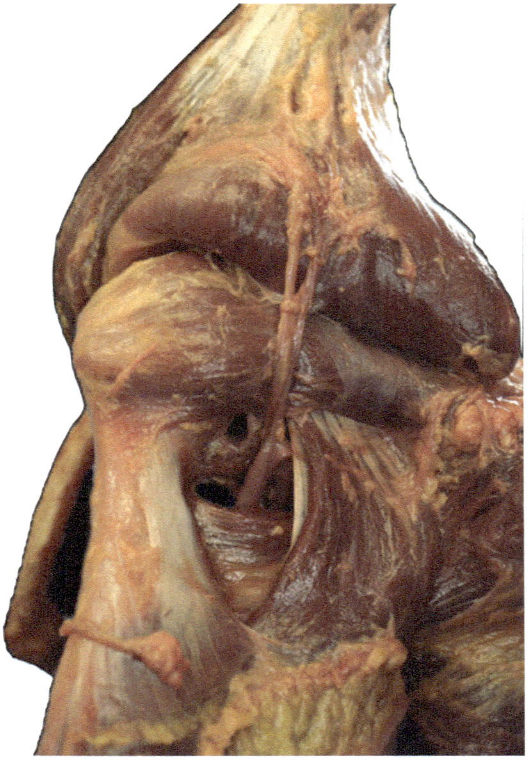

Fig. 32.3 Posterior view of the Quadrilateral space, bounded by the humerus laterally, teres major inferiorly, teres minor superiorly and the long head of triceps medially. It is accompanied by the posterior circumflex humeral artery (Image courtesy of Dr Mark Ross, Brisbane, Australia)

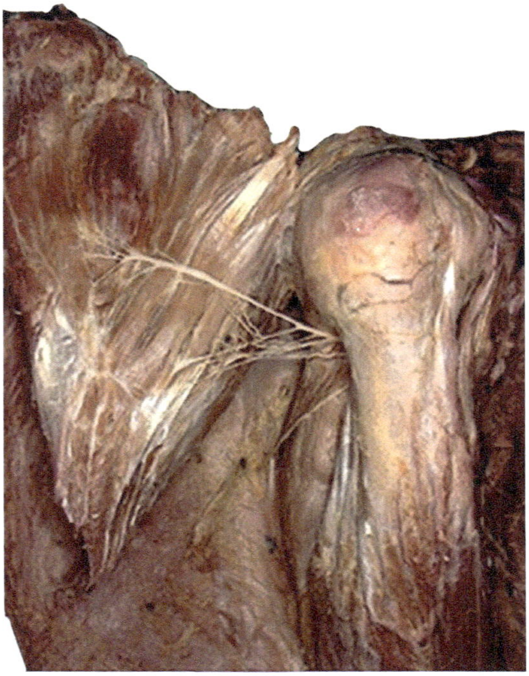

Fig. 32.4 Anterior branch of the axillary nerve. The deep surface of the deltoid muscle was exposed and the subdeltoid fascia removed to expose the axillary nerve (Reproduced with permission from Cetik et al. [14])

Loukas et al. [45] found the posterior branch gave a branch to the teres minor and the superior lateral brachial cutaneous nerve in 100 % of cases. A branch to the posterior part of the deltoid muscle was present in 90 % of cases and a branch to the middle part of the deltoid in 38 %.

Ball et al. [5] reported that the axillary nerve divided just anterior to the origin of the long head of the triceps at the 6 o'clock position on the glenoid face. Within the quadrilateral space, the posterior branch is medial to the anterior branch, at an average of only 1 mm lateral to the glenoid rim. The posterior branch courses medially and posteriorly along the lateral edge of the long head of triceps origin, inferior to the glenoid rim for 2–17 mm. The nerve to teres minor originated at the lateral edge of the long head of triceps origin. At this level the nerves lay directly on the joint capsule at the level of the glenoid rim. The nerve to teres minor coursed medially for 11–25 mm before entering the muscle in a medial direction on its inferior surface. A branch to the posterior deltoid was seen in 79 % of cases.

32.1.4 Anterior Branch Axillary Nerve

The anterior branch of the axillary nerve lies lateral to the posterior branch in the quadrilateral space and courses laterally from the glenoid rim toward the neck of the humerus. It travels around the neck of the humerus penetrating the subdeltoid fascia and runs anteriorly distal to the sub acromial bursa to supply the deltoid (Figs. 32.4 and 32.5).

Zhao et al. [70] described three innervation patterns of the posterior deltoid. Type 1 (47.5 %)

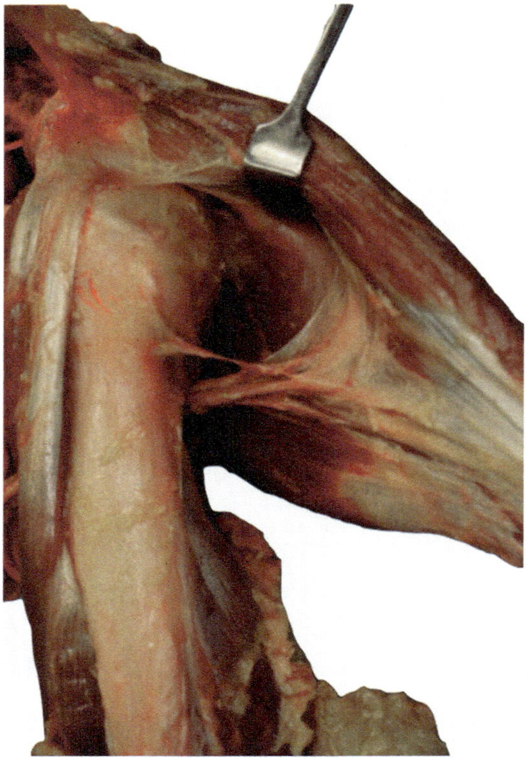

Fig. 32.5 Anterior view of the anterior branch of the axillary nerve traversing laterally from the glenoid rim around the neck of the humerus to supply the anterior and middle deltoid +/− posterior deltoid. (Image courtesy of Dr Mark Ross, Brisbane, Australia)

innervation by the anterior and posterior branches; Type 2 (42.5 %) innervation entirely by the posterior branch and Type 3 (10 %) innervation entirely by the anterior branch.

Zhao et al. [70] found that the whole of the middle and anterior deltoid was supplied by the anterior branch. The second articular branch to the glenohumeral joint, if present, usually arises from the anterior branch at the inferior border of the teres minor.

Loukas et al. [45] found the anterior branch provided a branch to the joint capsule, a branch to the anterior part of the deltoid muscle and the middle part of the deltoid muscle in 100 % of cases. In 18 % of cases the anterior branch of the axillary nerve provided a branch to the posterior part of the deltoid muscle. The middle part of the deltoid muscle received dual innervation in 38 % of cases and the posterior part of the deltoid muscle in 8 %.

Uz et al. [64] showed innervation of the posterior deltoid was from the posterior branch of the axillary nerve only in 70 %, both anterior and posterior branches in 26.7 % and from the anterior branch only in 3.3 %

Kulkarni et al. [41] studied the position of the axillary nerve in the deltoid muscle. The nerve was found to lie 2.2–2.6 cm superior to a midpoint on the vertical plane of the muscle (Videos 32.2 and 32.3).

32.1.5 Muscles Supplied

The teres minor is supplied by the posterior branch the axillary nerve.

The deltoid is supplied by the anterior and posterior branches of the axillary nerve. The deltoid arises from the anterior border and upper surface of the lateral third of the clavicle, from the whole of the lateral border of the acromion and from the inferior lip of the crest of the scapula spine. On the lateral border of the scapula four ridges may be seen; from them four fibrous septa pass down into the muscle. The deltoid tuberosity on the lateral aspect of the humerus is V-shaped, with a central vertical ridge. From the ridge and limbs of the V three fibrous septa pass upwards between the four septa from the acromion. The spaces between the septa are filled with a fleshy mass of muscle fibres which are attached to contiguous septa. The multipennate middle third of the deltoid so formed has a diminished range of contraction, but a correspondingly increased force of pull. The anterior and posterior fibres, arising from the clavicle and the scapula spine, are not multipennate. They converge on the anterior and posterior margins of the deltoid tuberosity, and their range of movement is greater but the force of their pull is less [57].

32.1.6 Superior Lateral Brachial Cutaneous Nerve

Zhao et al. [70] described two branching patterns of the superior lateral brachial cutaneous nerve. The more common pattern is that all branches of the nerve pierce the fascia on the posterior border of the deltoid. The second pattern has the superior lateral brachial cutaneous nerve dividing into two parts, with one part piercing the fascia on the posterior border of the deltoid, and another piercing the intermuscular septum between the posterior and middle thirds of the deltoid muscle [70].

Ball et al. [5] found the nerve with the arm in adduction pierced the deep fascia at the medial border of the posterior aspect of the deltoid 6.3–10.9 cm inferior to the posterolateral corner of the acromion. The nerve becomes subcutaneous at this level before crossing in an anterolateral direction into the overlying subcutaneous tissue. In no specimen did the nerve pass directly through the posterior deltoid to reach the subcutaneous tissue.

The nerve supplies an area of skin over the inferior deltoid also known as the 'regimental badge area' due to its location on the lateral upper arm.

32.1.7 Articular Branch and Proprioception

The innervation of the glenohumeral joint capsule has been widely described and is variable [22, 44, 64] Gelber et al. [27] pointed out the importance of the inferior glenohumeral ligament as a static stabilizer of the glenohumeral joint. The innervation of the inferior glenohumeral ligament is important as being one of the determinants of proprioceptive properties of the glenohumeral joint. Whether it is injured in shoulder joint lesions or not, it may determine changes in processing the mechanical stimuli signals created by natural movements [27, 29].

Uz et al. [64] reported that the articular branch arose from the main nerve trunk in 30 % of cases, from the posterior branch in 33 % and from the anterior branch in 16.6 %. Similarly Aszman et al. [4] described an articular branch from the anterior division of the axillary nerve. Subsequently, two more articular branches arose from the anterior branch as it crossed the inferior border of the subscapularis muscle. However Gelber et al. [27] demonstrated in the majority of cases (65.7 %) the posterior branch of the axillary nerve gave rise to the main articular branch. In 41 % it was from the branch to teres minor and the remaining cases from the initial portion of the posterior branch. Duparc et al. [22] stated that the articular branch may arise from different segments of the axillary nerve and Zhao et al. [70] showed the articular branch arising from the anterior trunk. Loukas et al. [45] reported the same branch to be arising from the anterior trunk in only 18 % of cases [2].

32.1.8 Relationship with Subdeltoid Bursa

The lower edge of the subdeltoid bursa is also considered a safe limit above the axillary nerve [6, 31], but is not a reliable landmark, due to its mobility and variant position related to the level of pressure of the arthroscopy fluid [2]. Beals et al. showed the mean minimum distance from the subdeltoid bursal reflection to the axillary nerve was 0.8 cm (0–1.4). The bursal reflection was always cephalad to the axillary nerve [6].

32.1.9 Relationship to Acromion

Ball et al. [5] looked at the relationship of the posterior branch of the axillary nerve to the posterolateral acromion. Posterior superficial dissection around the medial border of the posterior aspect of the deltoid will only endanger the superior lateral brachial cutaneous nerve when continued for an average of >80 mm. Hence the sensory branch should be at very little risk during routine exposure. Excessive lateral retraction of the deltoid during exposure of the posterior aspect of the shoulder can specifically injure the motor supply from the posterior branch of the axillary nerve to the deltoid. This branch enters the muscle at an average of only 5 cm directly inferior to the posterolateral corner of the acromion. This branch is present in 79–90 % of cases [5, 45]. Uz et al. [64] found the mean distance from the posterolateral corner of the acromion to the axillary nerve and its branches was 7.8 cm.

Traditional teaching suggests the safe zone of the axillary nerve is 5 cm from the middle of the acromiun. This is relevant in protecting the anterior branch of the axillary nerve from procedures involving a split deltoid approach, arthroscopy, intramedullary nailing and intra-muscular injections.

Several reference landmarks have been used to clarify the location of the axillary nerve. These include the acromion, greater tuberosity, bicipital groove, subdeltoid bursa and deltoid tuberosity [5, 14, 31, 38, 40, 46]. Of these landmarks the acromion is most frequently used because its border, located just below the skin and easily palpable, can be clarified most easily [12, 14, 46, 60, 65].

The acromion has some limitations as a reference landmark. First the morphological discrepancy between its angulated anterolateral and posterolateral corners and the deltoid's round surface make the determination of a constant perpendicular point on the deltoid muscle difficult. Second, often the shape of the anterolateral and posterolateral acromial corners are bluntly, rather than acutely, angulated. Third, because the axillary nerve at the posterolateral corner is not imposed in the fascia of the deltoid, it is not stationary. Fourth, because the anterior branch of the axillary nerve runs superiorly within the muscle, the vertical distance from that branch to the anterolateral corner of the acromion can change abruptly. Fifth, the slope of the acromion is not constantly parallel to the axillary nerve's course [10]. It is mostly divergent, sometimes convergent and rarely parallel [60, 65].

Burkhead et al. [12] and Duparc et al. [22] reported relatively short mean distances from the acromion to the anterior branch of the axillary nerve of 5.8 cm (4.3–7.3) and 3.4 cm (3.0–4.8). Prince and Hoppenfeld et al. [33, 55] reported mean distances of 5.87 and 7.0 cm respectively. Duparc et al. [22] measured the distance of the nerve to the upper insertion of the deltoid so the height of the acromion is not included.

Sung et al. [60] showed a mean distance between the anterior axillary nerve and the acromion to be 6.5 cm. There was a strong correlation between acromial axillary nerve distance and cadaver height ($r = 0.767$) and humeral length ($r = 0.797$).

Cetik et al. [14] found the average distance of the axillary nerve from the anterior edge of the lateral acromion was 6.08 cm (5.2–6.9). The average distance of the axillary nerve from the posterior edge of the lateral acromion was 4.87 cm (4.3–5.5). The average arm length was 30.4 cm (28.1–32.9). There was a significant correlation between arm length and anterior distance ($r = 0.79$, $p < 0.001$) and posterior distance ($r = 0.61$, $p = 0.001$). The ratio between arm length and distance to the nerve was expressed as an anterior and posterior index. The anterior index averaged 0.20 (0.19–0.22). The posterior index averaged 0.16 (0.15–0.17). Using these measurements a quadrangular safe zone was calculated (Figs. 32.6 and 32.7).

Kontakis [39] in a letter to the editor raised concerns over the safe zones described by Cetik et al. [14]. Kontakis et al. [40] found in 25 % of cadavers the vertical distance from the upper border of the deltoid was <4 cm. Also Burkhead et al. [12] found, in nearly one-fifth of cadavers

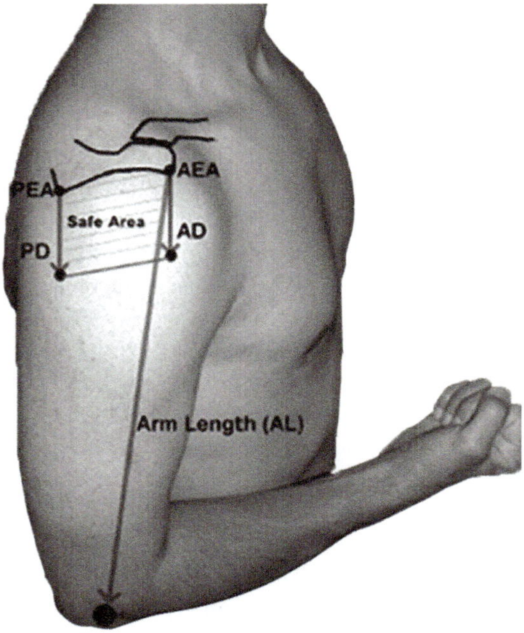

Fig. 32.6 The safe area is quadrangular in shape, with the length of the lateral edges being dependant on arm length. *AD* anterior distance, *PD* posterior distance, *AEA* anterior edge of acromion, *PEA* posterior edge of acromion (Reproduced from Cetik et al. [14])

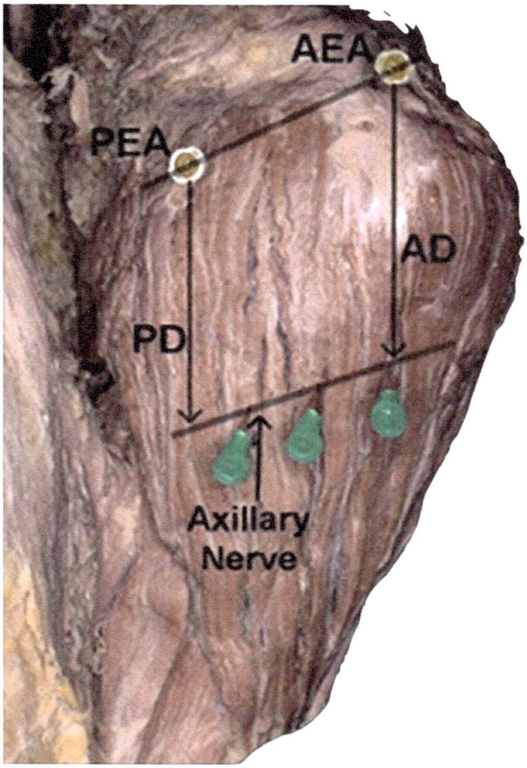

Fig. 32.7 The deltoid muscle was replaced in its original anatomic position. The course of the nerve is shown with the needles, and the anterior distance (*AD*) and posterior distance (*PD*) were measured. *AEA* anterior edge of acromion, *PEA* posterior edge of acromion (Reproduced from Cetik et al. [14])

the nerve was, 5 cm from the palpable edge of the acromion and at a minimal distance of 3.1 cm.

Stecco et al. [59] found a mean distance from axillary nerve as it enters the deltoid and the humeral head of 5 cm and to the acromion of 6.8 cm. They found when a 5 hole 114 mm PHILOS proximal humeral plate (Synthes, Stratec Medical Ltd, Mezzovico, Switzerland) is inserted with a lateral approach, in 100 % of cases the two distal holes in the plate dedicated to the humeral head coincided with the passage of the axillary nerve. Smith et al. [58] showed that the individuated three holes of the same plate, between those dedicated at the head of the humerus, could damage the integrity of the axillary nerve.

Nijs et al. [51] showed that bent nails with oblique head interlocking bolts appeared to be most dangerous in relation to the axillary nerve. The two designs featuring such a bend and oblique bolt showed a mean distance of the locking screw to the axillary nerve of 1 and 2.7 mm respectively. Sirus (Zimmer®) and (Stryker®) T2PHN (Proximal Humeral Nail).

Injury to the anterior branch of the axillary nerve has been reported with IM injections into the deltoid. Safe zones for prevention of such events have been suggested [17, 20].

There is variability in reported measurements. The measurements regarding safe zones should be used as a guide. The nerve is likely to be closer in subjects of short stature and short arm length.

32.1.10 Relationship to Capsule

Ball et al. [5] found the posterior branch of the axillary nerve to lay inferior to the capsule in the 6 o'clock position only 1 mm lateral to the glenoid rim. The nerve to teres minor and superior lateral brachial cutaneous nerve lay on the capsule at the lateral edge of the triceps origin at the level of the glenoid margin (Fig. 32.8). These nerves are at risk during arthroscopic and open surgery on the posteroinferior capsule. This may affect only the teres minor, as the deltoid is largely supplied by the anterior branch of the axillary nerve which lies lateral to the posterior branch in the quadrilateral space and courses laterally from the glenoid rim toward the neck of the humerus. The teres minor is an important dynamic stabilizer and external rotator of the humerus in an adducted position

Sensory changes in the distribution of the superior lateral brachial cutaneous nerve have been noted following arthroscopic thermal shrinkage procedures on the posterior and inferior aspects of the shoulder. Similar findings have been noted following anterior dislocation of the shoulder and in association with Quadrilateral space syndrome [26, 38, 42]. Sensory changes in the distribution of the superior lateral brachial cutaneous nerve may indicate dysfunction in teres minor even if deltoid function is normal [5].

Apaydin et al. [3] found the distance from the most inferior part of the glenohumeral ligament

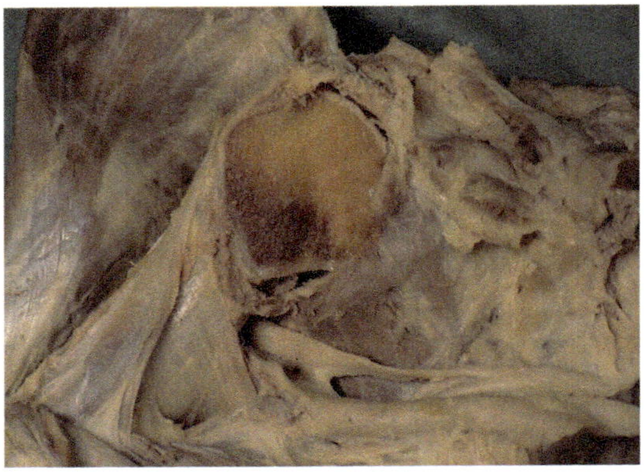

Fig. 32.8 Anterior view of cadaveric specimens showing the axillary nerve crossing over the subscapularis to enter the quadrilateral space. The relationship to the capsule is seen (Copyright Dr Gregory Bain)

to the axillary nerve averaged 1.1 cm (0.3–2.5 cm).

Price et al. [54] found the distance between the glenoid labrum and the axillary nerve was closest at the 6 o'clock position 12.4 mm (11.6–13.2). The nerve was 1.78 mm closer at the 6 o'clock position than 10 mm anterior or posterior to that position. The distance between the inferior glenohumeral ligament and the axillary nerve was also closest at the 6 o'clock position, 2.3 mm (1.7–2.9).

Yoo et al. [69] examined the axillary nerve from an arthroscopic perspective in a position of 30° abduction and 20° forward elevation. The axillary nerve entered the arthroscopic visual field around the 4 o'clock position (right) or 8 o'clock position (left) below the inferior margin of the subscapularis muscle. The boundaries of the nerve were the subscapularis muscle anteriorly, long head of triceps inferiorly, teres minor and major muscle posteriorly, inferior glenoid rim medially and humeral head laterally. Viewed arthroscopically the nerve was closest at the 5:30–6 o'clock position (right) or 6–6:30 o'clock position (left). The nerve had the closest distance range from 10 to 25 mm. Eleven specimens showed only the main trunk in the operative field, the remaining 12 showed the anterior and posterior branches.

Eakin et al. [24] examined the proximity of the axillary nerve to arthroscopically placed circumferential capsulolabral sutures. All sutures entered the capsule approximately 1 cm away from the glenoid and exited beneath the labrum. The shoulders were frozen in a position of 45° abduction and 20° flexion. The average distance from the axillary nerve was found to be 16.7 mm at the anterior position, 12.5 mm at the anteroinferior position, 14.4 mm at the inferior position, 24.1 mm at the posteroinferior position and 32.3 mm at the posterior position. In no specimen was a suture closer to the axillary nerve than 7 mm.

The axillary nerve can be vulnerable inferior to the glenohumeral capsule during open and arthroscopic procedures, such as capsulolabral repair, capsular plication and shift, capsular release, shoulder arthroplasty, internal fixation and historically thermal capsular shrinkage. Burkhart et al. described releasing the axillary pouch at least 5 mm lateral to the labrum during capsular release to increase the safe zone [11] (Fig. 32.9) (Video 32.4).

32.1.11 Effect of Arm Position on Nerve Position

Inferior humeral translation, as in the patient with multidirectional instability reduces the distance between the axillary nerve and the anteroinferior capsule. It also draws the nerve more taut across subscapularis making it more difficult to palpate [25].

Yoo et al. [69] examined the axillary nerve from an arthroscopic perspective in a position of 30° abduction and 20° forward elevation. In 45° abduction with neutral rotation the nerve moved further away from the capsule.

Uno et al. [63] showed that arm position affects the position of the axillary nerve from an arthroscopic perspective. Abduction of the shoulder causes the axillary nerve to become taut and displace laterally and superiorly. External rotation of the shoulder stretched the subscapularis and pulled the axillary nerve taut anteriorly. Internal rotation resulted in the subscapularis and axillary nerve becoming lax (Video 32.5). Flexion of the shoulder caused the

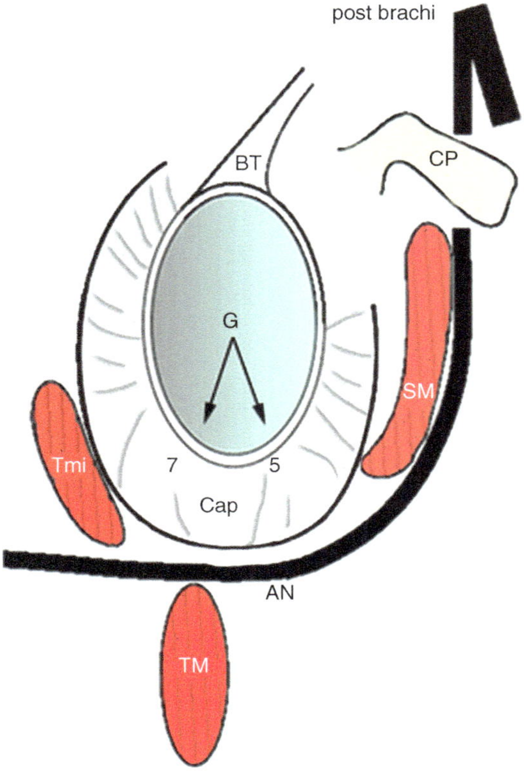

Fig. 32.9 The axillary nerve passes approximately 1 cm below the inferior inferior capsule. *AN* axillary nerve, *BT* biceps tendon, *CP* coracoid process, *SM* subscapularis muscle, *TM* teres major, *Tmi* teres minor (Modified with permission from Uno et al. [63])

nerve to become lax and extension made it taut. Longitudinal traction made the axillary nerve taut but was not enough to displace it. Perpendicular traction made the axillary nerve taut and displaced it laterally.

With shoulder abduction, external rotation and perpendicular traction, the capsule becomes taut and the axillary nerve moves away from the glenoid [63].

Cheung et al. [16] showed the distance from the mid acromion to the superior border of the axillary nerve was 66.6 mm and inferior axillary nerve was 75.7 mm. Vertical abduction to 60° significantly moved the superior and inferior borders of the axillary nerve to a distance of 53.9 and 61.6 mm respectively. Forward flexion had no significant effect and rotation had a variable effect.

32.1.12 Relationship to Coracoid Process

Apaydin et al. [3] found the closest distance between the anteromedial aspect of the coracoid tip and the axillary nerve averaged 3.7 cm (3.1–4.8 cm).

32.1.13 Association with Circumflex Humeral Vessels

Duval et al. [23] found the anterior humeral circumflex vessels cross anterior over the axillary nerve an average of 2.6 cm (2.1–3.6 cm) from a line perpendicular to the medial border of the bicipital groove. This can be a useful guide for finding the axillary nerve in a deltopectoral approach to the shoulder.

The anterior humeral circumflex artery runs deep to coracobrachialis and both heads of biceps, giving here an ascending branch which runs up the intertubercular sulcus and is an important source of blood supply to the head of the humerus. It then passes around the surgical neck of the humerus to anastomose with the posterior humeral circumflex artery [57].

The posterior humeral circumflex artery like the anterior humeral circumflex artery arises from the third part of the axillary artery. It is larger and passes through the quadrilateral space accompanied above by the axillary nerve. It also supplies the deltoid, giving branches to the long and lateral head of triceps and the shoulder joint. It anastomoses with the profunda brachii artery [57].

32.1.14 Relationship to Arthroscopy Portals

Meyer et al. [47] found the 5 o'clock portal was the closest portal to the axillary nerve with a mean distance of 15 mm (6–22) in the beach chair position (Video 32.6).

Lo et al. [43] examined the proximity of posterior, posterolateral, anterior, 5 o'clock, anterosuperiolateral and port of Wilmington portals to the axillary nerve in the lateral decubitus position. All nerves were more than 20 mm from the portals. The 5 o'clock portal was a mean distance of 33.3 mm (23–40) from the axillary nerve.

Bhatia et al. [8] examined the proximity of posteroinferior portals to the axillary nerve. Inferior portals such as the axillary pouch portal were safe in respect to axillary nerve as it entered the posterior deltoid however were within 5 mm of the intradeltoid continuation of the axillary nerve.

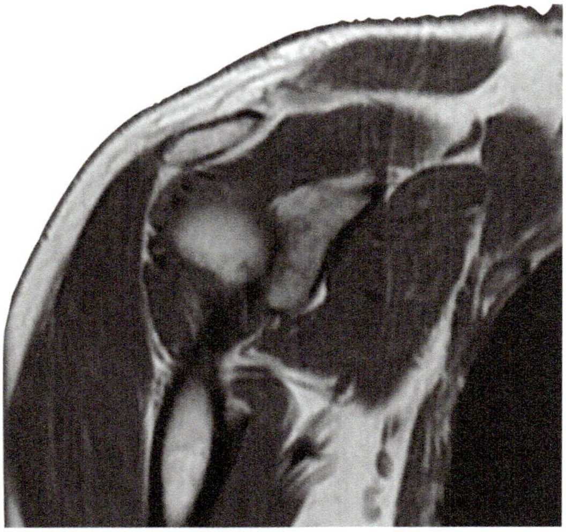

Fig. 32.11 Coronal T1 FSE MRI image showing axillary nerve and posterior circumflex humeral artery traversing through the quadrilateral space (I J Galley. Used with permission)

32.2 Imaging

32.2.1 MRI Scan

High resolution 3.0 T MRI provides a more detailed visualization of the brachial plexus including terminal nerves [37] (Figs. 32.10, 32.11 and 32.12). It can also visualize the axillary nerve and its branches, and the posterior circumflex humeral artery MRI is also useful for detecting fatty changes and atrophy of the deltoid muscle and teres minor associated with denervation. Quadrilateral space syndrome is manifest on MR imaging by abnormal signal or T1 hyper intense fatty atrophy of the teres minor on oblique sagittal images [8]. The scan also excludes space occupying lesions in the quadrilateral space.

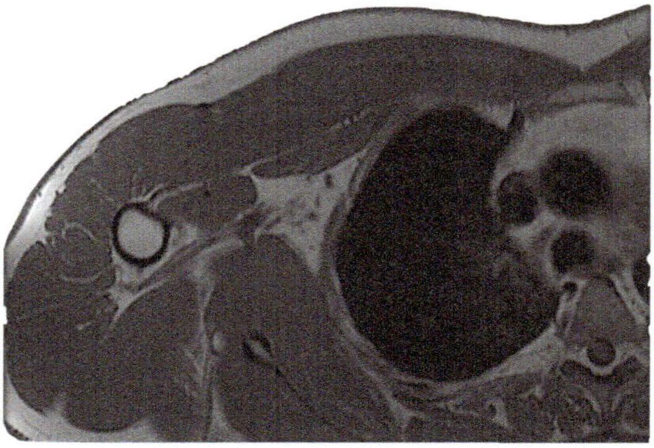

Fig. 32.10 Axial T1 FSE MRI image showing axillary nerve and posterior circumflex humeral artery traversing through the quadrilateral space (I J Galley. Used with permission)

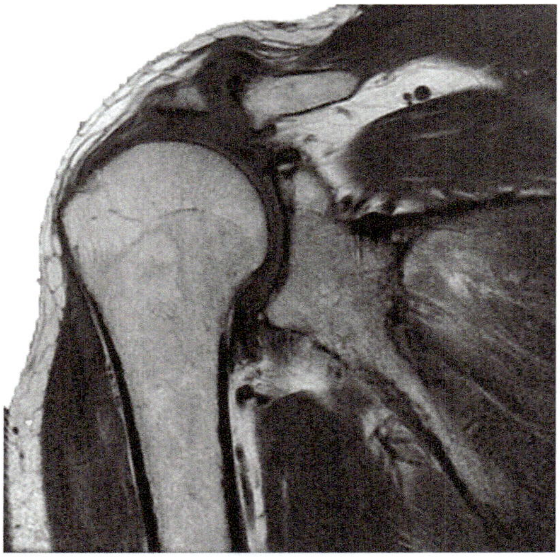

Fig. 32.12 Coronal T1 FSE MRI image. Patient has a deltoid avulsion from the acromion with associated massive cuff tear. Axillary nerve and posterior circumflex humeral artery have moved inferiorly as they are intimately attached to the undersurface of the deltoid by the subdeltoid fascia (I J Galley. Used with permission)

32.2.2 Clinical Case of Traumatic Axillary Nerve Injury

A 50-year-old male involved in a high speed motor vehicle injury. He suffered an isolated Gustilo grade 3A compound fracture to the right shoulder. His limb had distal pulses. He had complete motor and sensory loss of the upper limb. Radiographs showed a displaced proximal humeral fracture, scapula and acromial fracture and scapulo-thoracic dissociation (Fig. 32.13).

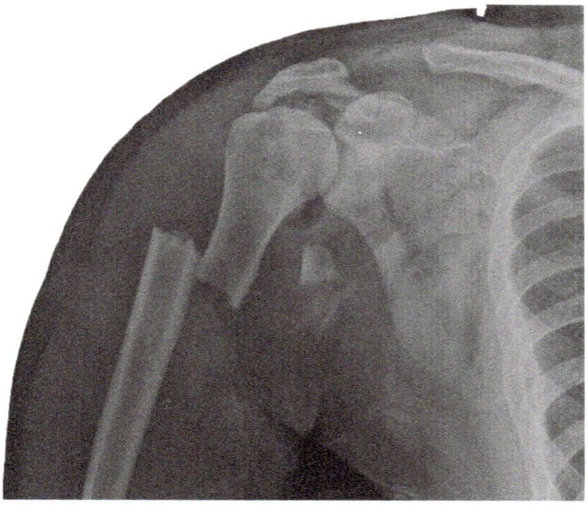

Fig. 32.13 Clinical case of traumatic injury to axillary nerve following motorvehicle accident. AP radiograph showing displaced proximal humeral fracture, scapula and acromial fracture and scapulo-thoracic dissociation (I J Galley. Used with permission)

He was managed with debridement, open reduction and internal fixation of the proximal humerus and acromion. Bridge plating of the acromioclavicular joint was performed to stabilize the scapulothoracic dissociation. Limited exploration of the brachial plexus through the compound wound was performed (Fig. 32.14a, b).

The AC joint bridging plate was removed at 3 months. He made an excellent neurological recovery, with the exception of complete loss of axillary nerve function. Plain radiographs confirmed inferior humeral subluxation associated with axillary nerve palsy (Fig. 32.15).

At 6 months he showed no signs of recovery of the axillary nerve clinically or on NCS and EMG studies (Fig. 32.16).

MRI scan images showed deltoid atrophy and fatty changes with relative preservation of teres minor muscle indicating predominantly an injury to the anterior branch of the axillary nerve (Fig. 32.17a, b). He was treated with selective neurotisation using a radial nerve branch to triceps. Recovery is incomplete.

32.3 Pathoanatomy

32.3.1 Quadrilateral Space Syndrome

Cahill and Palmer described quadrilateral space syndrome (QSS), which is a chronic compression of the axillary nerve +/− artery within the

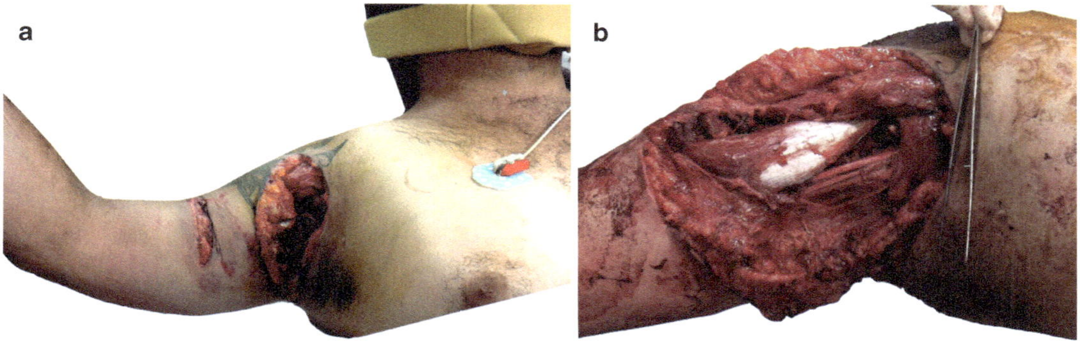

Fig. 32.14 (a, b) Clinical pictures of initial injury and limited exploration of brachial plexus (I J Galley. Used with permission)

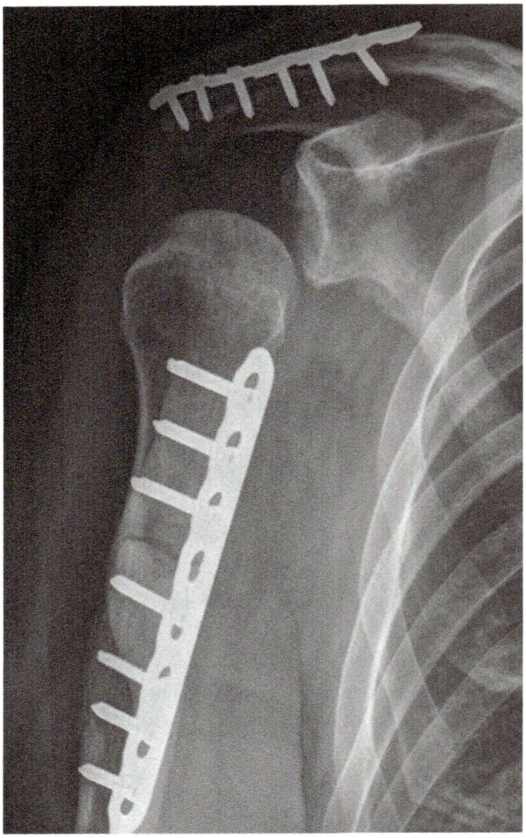

Fig. 32.15 AP Radiograph showing inferior subluxation of the glenohumeral joint associated with axillary nerve injury (I J Galley. Used with permission)

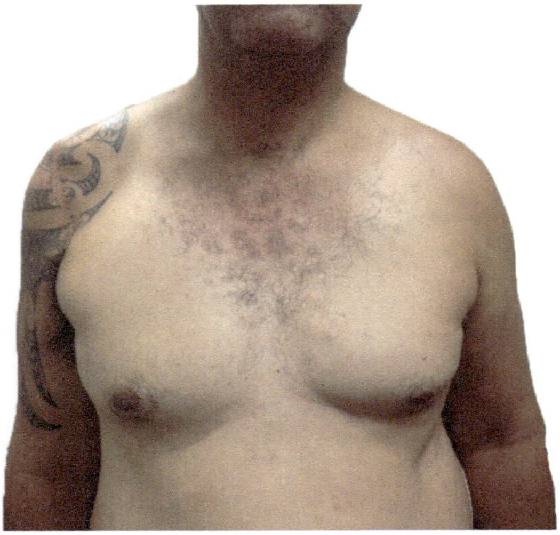

Fig. 32.16 Clinical picture showing deltoid wasting after traumatic axillary nerve injury (I J Galley. Used with permission)

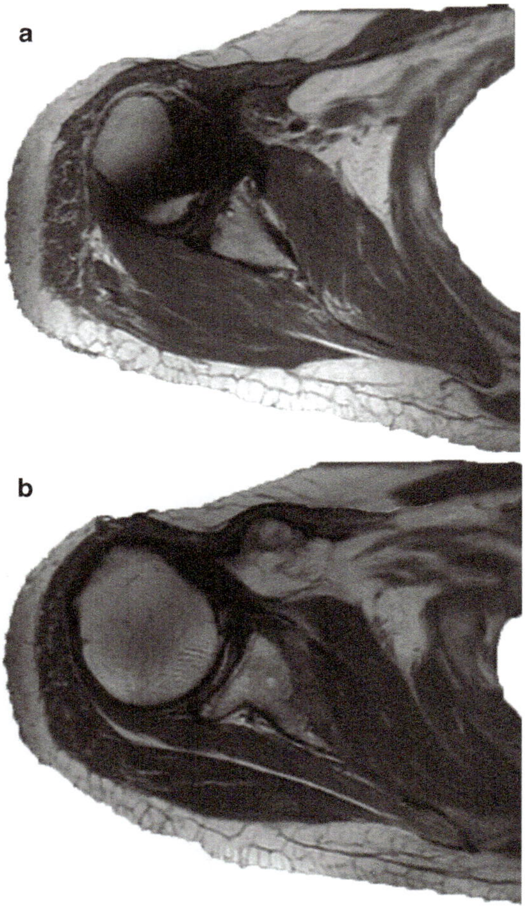

Fig. 32.17 (**a**, **b**) Axial PD FS MRI scan images showing deltoid atrophy and fatty changes, with relative preservation of teres minor muscle indicating predominantly an injury to the anterior branch of the axillary nerve (I J Galley. Used with permission)

quadrilateral space [13]. It presents with a poorly localized pain and paraesthesia affecting the shoulder and upper limb with discrete point tenderness posteriorly over the quadrilateral space. The symptoms are aggravated with abduction and abduction or forward flexion [13]. The true incidence of QSS is unknown. Of patients undergoing a shoulder MRI scan, 0.8 % had evidence of findings consistent with a diagnosis of quadrilateral space syndrome [19]. Most however were clinically asymptomatic.

The size of the quadrilateral space decreases during shoulder abduction and the axillary artery has been shown to occlude during abduction in some patients [15, 48, 52]. Whether this syndrome is neurological in origin or due to vascular compression of the posterior circumflex humeral artery is debated.

Cahill and Palmer [13] were the first to attribute compression of quadrilateral space contents to fibrous bands at surgery, but they stated that they were unable to identify any fibrous bands in cadaveric dissections. McKowen and Voorrhies [49] and Francel et al. [26] have reported the presence of fibrous bands in vivo during dissection of the quadrilateral space. McKowen and Voorrhies [49] noted that these bands of connective tissue were entrapping the neurovascular structures. The location of these fibrous bands is poorly described in the literature [13, 26, 49].

McClelland and Paxinos [48] identified fibrous bands both visually and by palpation in 14/16 random cadaveric specimens. The fibrous bands were multiple in most cases and in different directions. In 11 shoulders, the most significant fibrous band consisted of a fascial thickening overlying the long head of triceps, which ran from the proximal end of the long head of triceps, as it approaches the infraglenoid tubercle, to the teres major and onto the humerus, forming a sling adjacent to the axillary nerve. If present the fibrous bands were always bilateral. Rotation of the shoulder was tested with the shoulder abducted to 90°. The quadrilateral space reduced in volume with rotation in 11 of 16 shoulders. The fibrous sling between the teres major and long head of the triceps tightened in rotation in 11 of the 14 shoulders where a sling was present. The fibrous sling was tightest in external rotation in 7 of 16 shoulders and in internal rotation in 4 shoulders. No vascular abnormalities or other space occupying lesions were noted in any cadaver.

Reported anatomic causes of QSS also include glenoid labral cysts, [56] a ganglion, [35] muscle hypertrophy [26] and a spike of bone after a scapula fracture [1].

QSS presents with nonspecific symptoms, so the diagnosis can be elusive. MRI may demonstrate abnormal signal within the teres minor, or T1 hyper intense fatty atrophy of the teres minor on oblique sagittal images [18]. EMG studies demonstrate denervation potential within the teres minor +/– deltoid if affected.

32.3.2 Trauma

The axillary nerve may be injured during acute trauma, such as glenohumeral dislocation, proximal humeral fracture, penetrating injury or a direct anterolateral blow to the deltoid muscle [53]. EMG studies have demonstrated that focal axillary neuropathy is common following anterior shoulder dislocation (35–65 %) [9, 21, 36, 61, 67, 68]. Neuropathy is more common in patients older than 50 years [30, 34]. In the older patient the incidence of rotator cuff tears is also higher following anterior dislocation (38–100 %) [61, 7]. Gonzalez and Lopez [28] described the 'unhappy triad', of shoulder dislocation, peripheral nerve injury and rotator cuff tear. Once the shoulder has been reduced, this combination should be considered in the older patient, if the patient cannot lift the arm, there is any paraesthesia of the arm, wasting of the deltoid and/or rotator cuff. EMG and MRI can assist in making the diagnosis (Fig. 32.18).

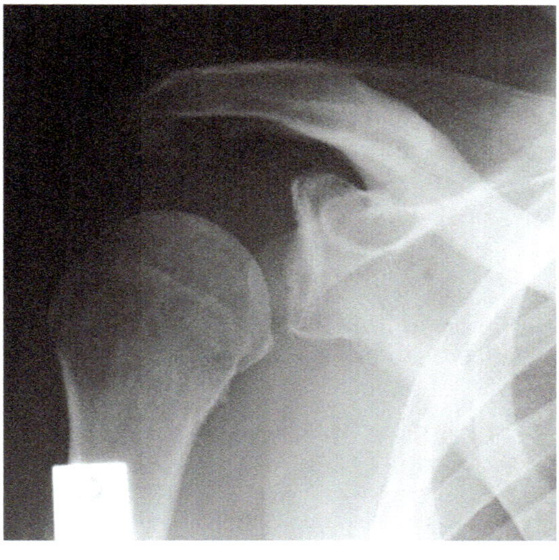

Fig. 32.18 AP Radiograph demonstrating inferior subluxation of the glenohumeral joint due to axillary nerve injury (I J Galley. Used with permission)

The incidence and prognosis of brachial plexus and axillary nerve injury is dramatically worsened if the shoulder remains dislocated for more than 12 h [2].

The mechanism of injury in shoulder dislocations is one of traction and compression as the axillary nerve becomes stretched across the humerus as it dislocates anteriorly and inferiorly [53, 66, 68]. The free course of the axillary nerve is relatively short. It is tethered by the branch to teres minor and the fascia on the deep surface of the deltoid. Overstretching of the axillary nerve over the humeral head may cause elongation of the free portion of the axillary nerve [2].

Neurophysiology studies can be useful. Reduction of the axillary nerve compound motor action potential amplitude can be seen. Needle EMG studies can also show evidence of acute denervation [20].

Nagda et al. [50] performed intraoperative nerve monitoring during shoulder arthroplasty. Of the patients, 17 (56.7 %) had 30 episodes of nerve dysfunction (i.e. nerve alerts) during surgery. In 23 (76.7 %) alerts returned to baseline when the arm was repositioned into a neutral position. Postoperative EMG studies were positive in four of seven patients (57.1 %) who did not return to baseline. The axillary nerve was involved alone in 16.7 and 46.7 % involved a mixed plexopathy. Therefore almost half the events involved portions of the brachial plexus rather than end nerve such as the axillary nerve.

References

1. Amin MF, Berst M, el-Khoury GY. An unusual cause of the quadrilateral space impingement caused by a bone spike. Skeletal Radiol. 2006;35:956–8.
2. Apaydin N, Tubbs RS, Loukas M, Dupare F. Review of the surgical anatomy of the axillary nerve and the anatomic basis of its iatrogenic and traumatic injury. Surg Radiol Anat. 2010;32:193–201.
3. Apaydin N, Uz A, Bozkurt M, Elhan A. The anatomic relationships of the axillary nerve and surgical landmarks for its localization from the anterior aspect of the shoulder. Clin Anat. 2007;20:273–7.
4. Aszman OC, Dellon AL, Birely BT, McFarland EG. Innervation of the human shoulder joint and its implications for surgery. Clin Orthop Relat Res. 1996;330:202–7.
5. Ball C, Steger T, Galatz L, Yamaguchi K. The posterior branch of the axillary nerve: an anatomic study. J Bone Joint Surg Am. 2003;85:1497–501.
6. Beals T, Harryman D, Lazarus M. Useful boundaries of the subacromial bursa. Arthroscopy. 1998;14:465–70.
7. Berbig R, Weishaupt D, Prim J, Shahin O. Primary anterior shoulder dislocation and rotator cuff tears. J Shoulder Elbow Surg. 1999;8:220–5.
8. Bhatia D, de Beer J, Dutoit D. An anatomic study of inferior glenohumeral recess portals: comparative anatomy at risk. Arthroscopy. 2008;24(5):529–36.
9. Brown TD, Newton PM, Steinmann SP, Levine WN, Bigliani LU. Rotator cuff tears and associated nerve injuries. Orthopedics. 2000;23:329–32.
10. Bryan WJ, Schauder K, Tullos HS. The axillary nerve and its relationship to common sports medicine shoulder problems. Am J Sports Med. 1986;14:113–6.
11. Burkhart S, Lo I, Brady P, Denard P. The cowboy's companion. A Trail Guide for the arthroscopic shoulder surgeon. Philadelphia: Lippincott Williams & Wilkins; 2012. p. 404.
12. Burkhead WZ, Scheinberg RR, Box G. Surgical anatomy of the axillary nerve. J Shoulder Elbow Surg. 1992;1:31–6.
13. Cahill BR, Palmer RE. Quadrilateral space syndrome. J Hand Surg. 1983;8:65–9.
14. Cetik O, Uslu M, Acar H, Comert A, Tekdemir I, Cift H. Is there a safe ara for the axillary nerve in the deltoid muscle. A Cadaveric study. J Bone Joint Surg Am. 2006;88:2395–9.
15. Chautems RC, Glauser T, Waeber-Fey MC, Rostan O, Barraud GE. Quadrilateral space syndrome: a case report and review of the literature. Ann Vasc Surg. 2000;14:673–6.
16. Chung S, Fitzpatrick M, Lee T. Effects of shoulder position on axillary nerve positions during the split lateral deltoid approach. J Shoulder Elbow Surg. 2009;18:748–55.
17. Cook I. An evidence based protocol for the prevention of upper arm injury rleated to vaccine administration (UAIRVA). Hum Vaccin. 2011;7(8):845–8.
18. Cook T, Stein J, Simonson S, Kim W. Normal and variant anatomy of the shoulder on MRI. Magn Reson Imaging Clin N Am. 2011;19:581–94.
19. Cothran Jr RL, Helms C. Quadrilateral space syndrome: incidence of imaging findings in a population referred for MRI of the shoulder. AJR Am J Roentgenol. 2005;184:989–92.
20. Davidson LT, Carter GT, Lkilmer DD, Han JJ. Iatrogenic axillary neuropathy after intramuscular injection of the deltoid muscle. Am J Phys Med Rehabil. 2007;86:507–11.
21. De Laat EA, Visser CP, Coene LN, Pahlplatz PV, Tavy DL. Nerve lesions in primary shoulder dislocations and humeral neck fractures. A prospective clinical and EMG study. J Bone Joint Surg Br. 1994;76:381–3.
22. Duparc F, Bocquet G, Simonet J, Freger P. Anatomical basis of the variable aspects of injuries of the axillary nerve (excluding the terminal branches in the deltoid muscle). Surg Radiol Anat. 1997;19:127–32.

23. Duval M, Parker A, Drez Jr D, Hinton M. The anterior humeral circumflex vessels and the axillary nerve. An anatomic study. Orthop Rev. 1993;22(9):1023–6.
24. Eakin C, Dvimak P, Miller C, Hawkins R. The relationship of the axillary nerve to arthroscopically place capsulolabral sutures. An anatomic study. Am J Sports Med. 1998;26(4):505–9.
25. Flatow E, Bigliani L. Locating and protecting the axillary nerve in shoulder surgery: the tug test. Orthop Rev. 1992;21(4):503–5.
26. Francel TJ, Dellon AL, Campbell JN. Quadrilateral space syndrome: diagnosis and operative decompression technique. Plast Reconstr Surg. 1991;87:911–6.
27. Gelber PE, Reina F, Monllau JC, Yema P, Rodriguez A, Caceres E. Innervation patterns of the inferior glenohumeral ligament: anatomical and biomechanical relevance. Clin Anat. 2006;19:304–11.
28. Gonzalez D, Lopez R. Concurrent rotator cuff tear and brachial plexus palsy associated with anterior dislocation of the shoulder. A report of two cases. J Bone Joint Surg Am. 1991;73:620–1.
29. Guanche C, Knatt T, Solomonow M, Lu Y, Baratta R. The synergistic action of the capsule and shoulder muscles. Am J Sports Med. 1995;23:301–6.
30. Gumina S, Postacchini F. Anterior dislocation of the shoulder in elderly patients. J Bone Joint Surg Br. 1997;79:540–3.
31. Haddo O, Rossouw D, Copeland S, Levy O. Subdeltoid bursa is a safe marker for avoiding injury to the axillary nerve. Should Elb. 2009;1:29–30. http://dx.doi.org/10.1111/j.1758-5740.2009.00005.x.
32. Hilton J. On rest and pain: a course of lectures on the influence of mechanical and physiological rest in the treatment of accidents and surgical diseases, and the diagnostic value of pain, delivered at the Royal College of Surgeons of England in the years 1860, 1861, and 1862. 1863.
33. Hoppenfeld S, de Boer P. The shoulder. In: Hoppenfeld S, de Boer P, editors. Surgical exposures in orthopaedics: the anatomic approach. Philadelphia: Lippencott Williams & Wilkins; 2003. p. 1–66. ISBN 978-0-7817-4228-3.
34. Hovelius L, Augustini BG, Fredin H, Johansson O, et al. Primary anterior dislocation of the shoulder in young patients. A ten year prospective study. J Bone Joint Surg Am. 1996;78:1677–84.
35. Ishima T, Usui M, Satoh E, Sakahashi H, Okamura K. Quadrilateral space syndrome caused by a ganglion. J Shoulder Elbow Surg. 1998;7:80–2.
36. Khan LA, Robinson CM, Eill E, Whittaker R. Assessment of axillary nerve function and functional outcome after fixation of complex proximal humeral fractures using the extended deltoid splitting approach. Injury. 2009;40:181–5.
37. Kjelstrup T, Courivaud F, Klaastad O, Breivik H, Hol P. High resolution MRI demonstrates detailed anatomy of the axillary brachial plexus. A pilot study. Acta Anaesthesiol. 2012;56:914–9.
38. Klepps S, Auerbach J, Calhon O, Lin J, Cleeman E, Faltow E. A cadaveric study on the anatomy of the deltoid insertion and its relationship to the deltopectoral approach to the proximal humerus. J Shoulder Elbow Surg. 2004;13:322–7.
39. Kontakis G. Is there a safe area for the axillary nerve in the deltoid muscle? J Bone Joint Surg Am. 2007;89(5):1134–5.
40. Kontakis GM, Steriopoulos K, Damilakis J, Michalodimitrakis E. The position of the axillary nerve in the deltoid muscle. A cadaveric study. Acta Orthop Scand. 1999;70:9–11.
41. Kulkarni R, Nandedkar A, Myosorekar VR. Position of the axillary nerve in the deltoid muscle. Anat Rec. 1992;232:316–7.
42. Lester B, Jeong GK, Welland AJ, Wicklewicz TL. Quadrilateral space syndrome: diagnosis, pathology and treatment. Am J Orthop. 1999;28(718–22):725.
43. Lo I, Lind C, Burkhart S. Glenohumeral arthroscopy portals established using an outside-in technique: neurovascular anatomy at risk. Arthroscopy. 2004;20(6):596–602.
44. Loomer R, Graham B. Anatomy of the axillary nerve and its relation to inferior capsular shift. Clin Orthop Relat Res. 1989;243:100–5.
45. Loukas M, Grabska J, Tubbs RS, Apaydin N, Jordan R. Mapping the axillary nerve within the deltoid muscle. Surg Radiol Anat. 2009;31:43–7.
46. Maman E, Morag G, Safir O, Benifla M, Mozes G, Boynton E. The anterior trunk of the axillary nerve: surgical anatomy and guidelines. A fresh cadavers study. J Orthopaedics. 2008;5:e7.
47. Meyer M, Graveleau N, Hardy P, Landreau P. Anatomic risks of shoulder arthroscopy portals: anatomic cadaveric study of 12 portals. Arthroscopy. 2007;23(5):506–13.
48. McClelland D, Paxinos A. The anatomy of the Quadrilateral space with reference to quadrilateral space syndrome. J Shoulder Elbow Surg. 2008;17:162–4.
49. McKowen HC, Voorrhies RM. Axillary nerve entrapment in the quadrilateral space: a case report. J Neurosurg. 1987;66:932–4.
50. Nagda S, Rogers K, Sestokas A, Getz C, et al. Neer Award 2005: peripheral nerve function during shoulder arthroplasty using intraoperative nerve monitoring. J Shoulder Elbow Surg. 2007;16:2S–8.
51. Nijs S, Sermon A, Broos P. Intramedullary fixation of proximal humeral fractures: do locking bolts endager the axillary nerve or the ascending branch of the anterior circumflex artery? A cadaveric study. Patient Saf Surg. 2008;2(33):1–6.
52. Pecina MM, Krmpotic-Nemanic J, Markiewitz AD. Tunnel syndromes. 2nd ed. Boca Raton: CRC press; 1997.
53. Perlmutter GS. Axillary nerve injury. Clin Orthop Relat Res. 1999;368:28–36.
54. Price M, Tillett E, Acland R, Nettleton G. Determining the relationship of the axillary nerve to the shoulder joint capsule from an arthroscopic perspective. J Bone Joint Surg Am. 2004;86:2135–42.

55. Prince EJ, Breien KM, Fehringer EV, Mormino MA. The relationship of proximal locking screws to the axillary nerve during antegrade humeral nail insertion of four commercially available implants. J Orthop Trauma. 2004;18:585–8.
56. Robinson P, White IM, Lax M, Salonen D, Bell RS. Quadrilateral space syndrome caused by a glenoid labral cyst. AJR Am J Roentgenol. 2000;175:1103–5.
57. Sinnatamby C. Lasts Anatomy Regional and applied. 11th ed. London: Elsevier Ltd; 2006. p. 48–57.
58. Smith J, Berry G, Laflamme Y, Blain-Pare E, et al. Percutaneous insertion of a proximal humeral locking plate: an anatomic study. Injury. 2007;38:206–11.
59. Stecco C, Gagliano G, Lancerotto L, Tiengo C, Macchi V, et al. Surgical anatomy of the axillary nerve and its implication in the transdeltoid approaches to the shoulder. J Shoulder Elbow Surg. 2010;19:1166–74.
60. Sung C, Roh G, Sohn H, Park H. Prediction of the location of the anterior branch of the axillary nerve, using correlations with physical factors: a cadaveric study. J Shoulder Elbow Surg. 2013;22:e9–16.
61. Toolanen G, Hildingsson C, Hedlund T, Knibestol M, Oberg L. Early complications after anterior dislocation of the shoulder in patients over 40 years. An ultrasonographic and electromyographic study. Acta Orthop Scand. 1993;64:549–52.
62. Tubbs RS, Oakes WJ, Blount JP, Elton S, Salter G, Grabb PA. Surgical landmarks for the proximal portion of the axillary nerve. J Neurosurg. 2001;95:998–1000.
63. Uno A, Bain G, Mehta J. Arthroscopic relationship of the axillary nerve to the shoulder capsule: an anatomic study. J Shoulder Elbow Surg. 1999;8:226–30.
64. Uz A, Apaydin N, Bozkurt M, Elhan A. The anatomic branch pattern of the axillary nerve. J Shoulder Elbow Surg. 2007;16:240–4.
65. Vathana P, Chiarapattanakom P, Ratanalaka R, Vorasatit P. The relationship of the axillary nerve and the acromion. J Med Assoc Thai. 1998;81:953–7.
66. Visser CP, Coene LN, Brand R, Tavy DL. The incidence of nerve injury in anterior dislocation of the shoulder and its influence on functional recovery. A prosepctive clinical and EMG study. J Bone Joint Surg Br. 1999;81:679–85.
67. Visser CP, Coene LN, Brand R, Tavy DL. Nerve lesions in proximal humeral fractures. J Shoulder Elbow Surg. 2001;10:421–7.
68. Visser CP, Tavy DL, Coene LN, Brand R. Electromyographic findings in shoulder dislocations and fractures of the proximal humerus: comparison with clinical neurological examination. Clin Neurol Neurosurg. 1999;101:86–91.
69. Yoo J, Kim J, Hwan Ahn J, Hyun Lee S. Arthroscopic perspective of the axillary nerve in relation to the glenoid and arm position: a cadaveric study. Arthroscopy. 2007;23(12):1271–7.
70. Zhao X, Hung L, Zang G, Lao J. Applied anatomy of the axillary nerve for selective neurotization of the deltoid muscle. Clin Orthop Relat Res. 2001;390:244–51.

Suprascapular Nerve

33

Kevin D. Plancher and Stephanie C. Petterson

33.1 Gross Anatomy

The suprascapular nerve arises from the upper trunk of the brachial plexus (C5–C6) at Erb's point. In approximately 25 % of individuals, the suprascapular nerve receives contributions from the C4 nerve root [34, 50]. The suprascapular nerve exits the upper trunk approximately 3 cm above the clavicle to run laterally and parallel to the muscle belly of the omohyoid muscle and deep to the anterior border of the trapezius along the posterior cervical triangle (Fig. 33.1). As it passes through the posterior triangle, it travels with the suprascapular artery and vein. The nerve then travels along the posterior border of the clavicle to reach the superior border of the scapula. The suprascapular nerve then diverges from the artery, taking a posterior approach and diving into the suprascapular notch under the transverse scapular ligament (TSL) (Figs. 33.2 and 33.3a, b). At this point, the nerve is approximately 3 cm away from the supraglenoid tubercle [4]. The suprascapular artery takes an anterior position and enters the suprascapular notch over the TSL. Variations have been reported, although infrequently, with the suprascapular artery staying with the nerve as it passes posterior to the TSL [44].

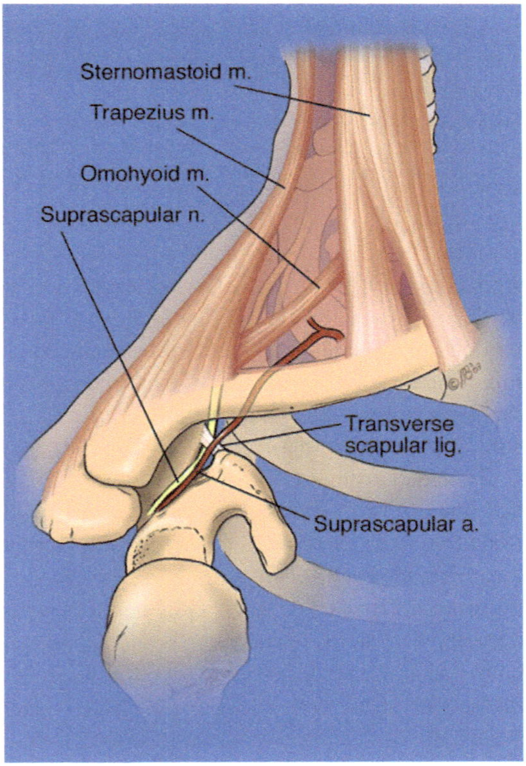

Fig. 33.1 The suprascapular nerve runs parallel to the muscle belly of the omohyoid muscle along the posterior cervical triangle as it exits the upper trunk (Copyright K. Plancher)

S.C. Petterson MPT, PhD
Department of Research, Orthopaedic Foundation, Stamford, CT, USA

K.D. Plancher, MD (✉)
Department of Orthopaedic Surgery, Albert Einstein College of Medicine, Plancher Orthopaedics and Sports Medicine, New York, NY, USA

Orthopaedic Foundation, Stamford, CT, USA
e-mail: kplancher@plancherortho.com

G.I. Bain et al. (eds.), *Normal and Pathological Anatomy of the Shoulder*,
DOI 10.1007/978-3-662-45719-1_33, © ISAKOS 2015

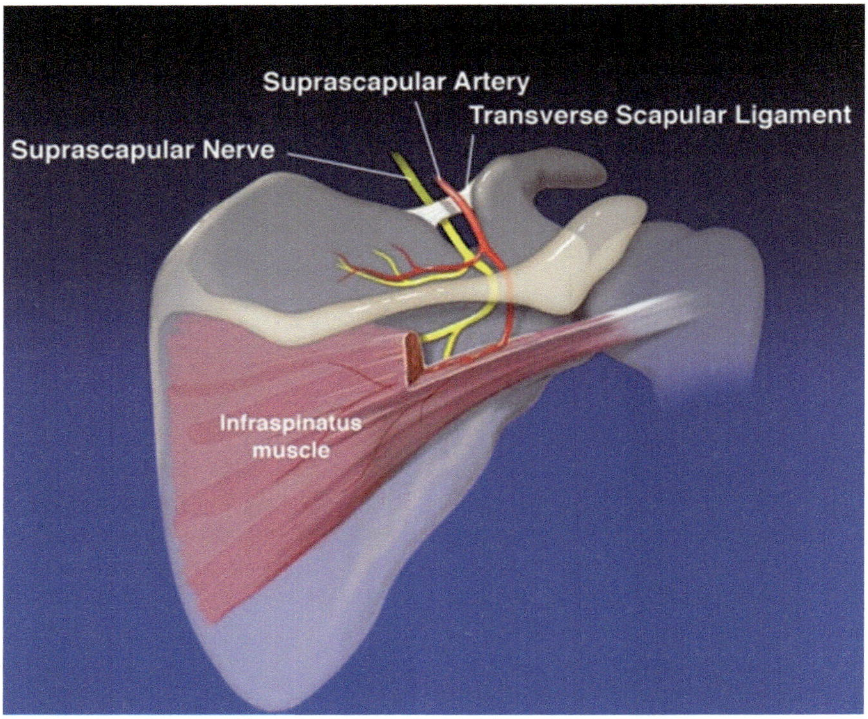

Fig. 33.2 The nerve travels along the posterior border of the clavicle to reach the superior border of the scapula. The nerve diverges from the artery to proceed under the transverse scapular ligament while the artery goes over the transverse scapular ligament diving in to the suprascapular notch (Copyright K. Plancher)

The suprascapular nerve as it enters the supraspinatus fossa gives off 2 motor branches to the supraspinatus muscle belly. The nerve also gives off sensory and sympathetic branches to two-thirds of the glenohumeral joint, the coracoclavicular ligament, the coracohumeral ligament, the subacromial bursa, as well as the posterior capsule of the acromioclavicular (AC) joint [3, 11, 40]. The nerve then travels along the supraspinatus fossa heading laterally and coming within 2 cm of the posterior glenoid rim at the level of the spine of the scapula [47]. The suprascapular nerve travels laterally around the scapular spine to descend into the infraspinatus fossa only to pass under the spinoglenoid ligament (SGL), also known as the inferior transverse scapular ligament (Fig. 33.4a–c). The suprascapular nerve gives off 2–4 branches to the infraspinatus muscle belly. The suprascapular nerve is approximately 2.5 cm away from the glenoid rim and approximately 4 cm from the posterior corner of the spine of the scapula [29].

33.2 Adjacent Structures and Variations

33.2.1 Suprascapular Notch

The morphology of the suprascapular notch, specifically a reduction in the height of the notch, may play a role in the development of suprascapular nerve entrapment [16]. Rengachary first classified 6 variations of the suprascapular notch [34] (Fig. 33.5). Scapula that do not have a notch, but rather a wide depression from the superior angle of the scapula to the base of the scapula, are classified as type I. A type II suprascapular notch is defined as a wide, blunted "V"-shaped notch that occupies approximately one-third of

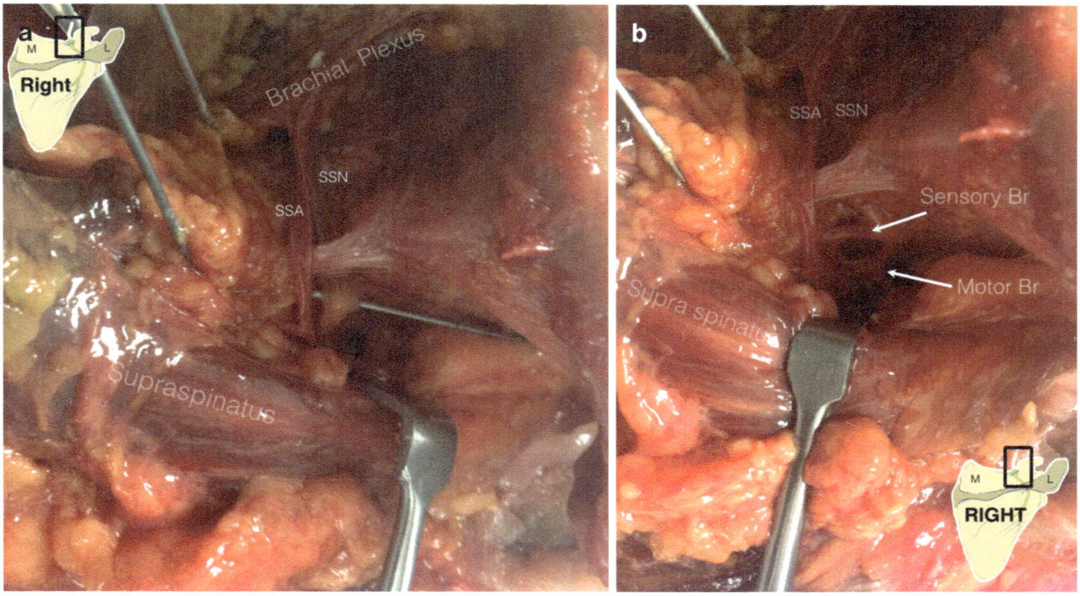

Fig. 33.3 (**a**, **b**) Suprascapular nerve coursing from the brachial plexus under the transverse ligament to enter the supraspinatus fossa and pass deep to the supraspinatus muscle. There, it gives off motor branches to the supraspinatus muscle and sensory fibers to the glenohumeral joint (Cadaveric dissections courtesy of Dr. Felix Savoie, New Orleans, LA, USA) *SSN* = Suprascapular Nerve; *SSA* = Suprascapular Artery; *M* = Medical; *L* = Lateral

the superior border of the scapula. Symmetrical, "U"-shaped suprascapular notches, which have nearly parallel lateral margins, are classified as type III. Very small, "V"-shaped notches are classified as type IV. A type V notch is similar in shape to a type III notch with a partial ossification of the medial part of the TSL resulting in a small diameter along the superior border of the scapula. In a type VI notch, the TSL is completely ossified creating a bony foramen which is variable in size. Other classification systems of the suprascapular notch have also been described; Ticker et al. classified the suprascapular notch as either "U"- or "V"-shaped, evaluating the degree of ossification of the TSL separately, whereas, Iqbal and colleagues reported three types of notches including "U"-, "V"-, and "J"-shaped [17, 42].

33.2.2 Transverse Scapular Ligament

The TSL attaches the base of the coracoid process and the medial end of the suprascapular notch creating a foramen through which the suprascapular nerve traverses in the majority of individuals. The TSL is thin and flat, being narrower at the middle than at its insertions. Ossification of the TSL has been reported to occur in approximately 25 % of clinical cases [42]. Polguj et al. identified 3 variations of the TSL [30]. The majority of specimens exhibited either a fan-shaped ligament (54.6 %) or band-shaped ligament (41.9 %); however, a bifid ligament was also found in 3.5 % of specimens. The anterior coracoscapular ligament was present in only 51 % of specimens and contributed to a smaller area in the suprascapular notch. The presence of the anterior coracoscapular ligament may be an additional etiologic factor to consider in suprascapular nerve compression at the suprascapular notch.

Classically, the suprascapular nerve traverses under the TSL at the suprascapular notch while the suprascapular artery traverses over the TSL. In a 2014 cadaveric study, the

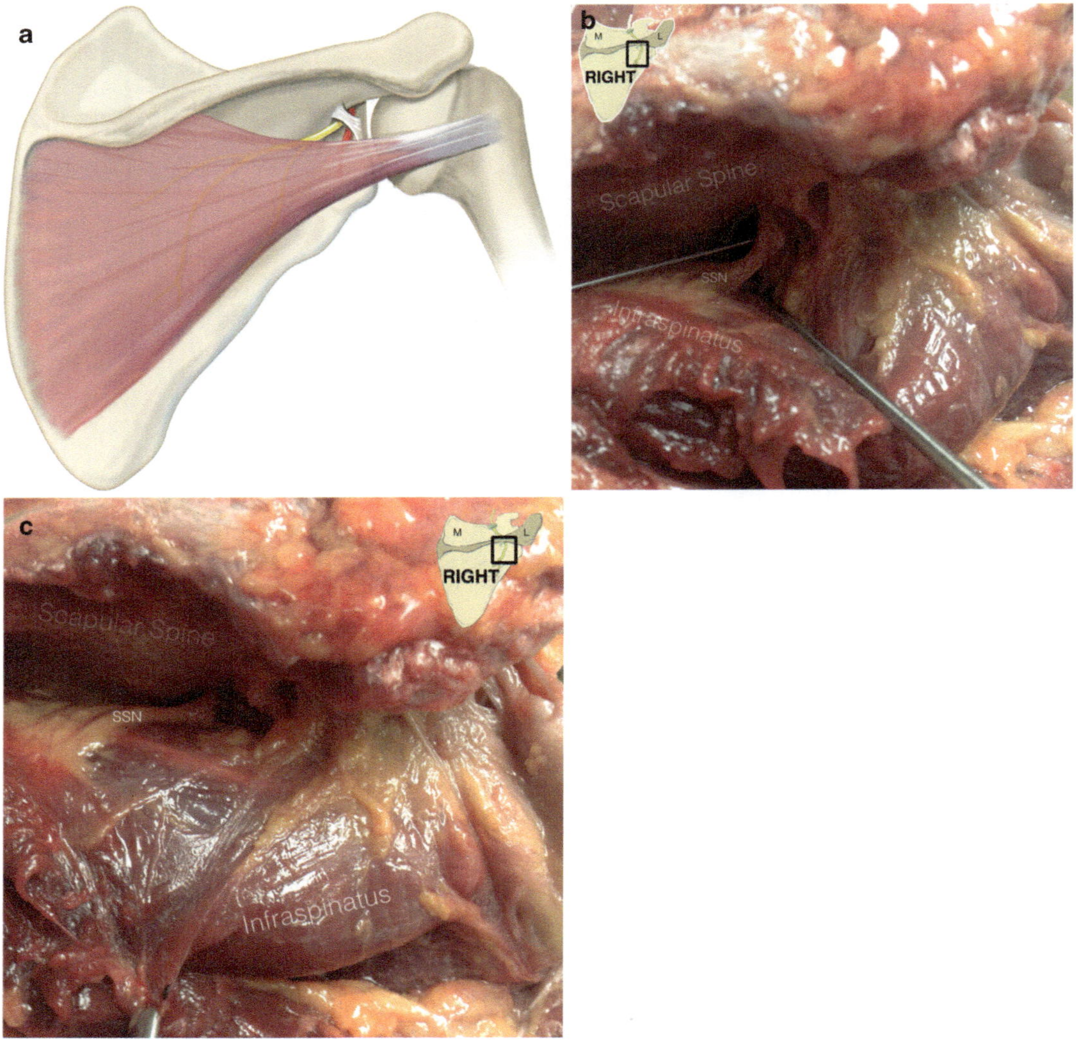

Fig. 33.4 (**a**) The suprascapular nerve descending into the infraspinatus fossa passing under the spinoglenoid ligament also known as the inferior transverse scapular ligament. (Copyright K. Plancher) (**b**, **c**) Cadaveric dissections of the suprascapular nerve at the spinoglenoid notch. (Cadaveric dissections courtesy of Dr. Felix Savoie, New Orleans, LA, USA) *SSN* = Suprascapular nerve; *M* = Medial; *L* = Lateral

suprascapular nerve and vein traveled below the ligament in 61.3 % of specimens [31]. Other arrangements found included (1) the suprascapular artery and vein traveling above the ligament and the nerve coursing below the ligament (17 %), (2) the suprascapular vessels and nerve all traveling below the ligament (12.3 %), and (3) other variations of the suprascapular neurovascular structures occurred in 9.4 % of specimens.

33.2.3 Spinoglenoid Ligament

The SGL is quadrangular in shape and extends from the posterior glenoid neck and posterior glenohumeral joint capsule to insert a bilaminar ligament into the scapular spine [29]. Two types of the SGL have been described: (1) type I, a thin indistinct band of tissue, and (2) type II, a well-formed ligament. The geometric shape has been described as either band-like, triangular, or irregular.

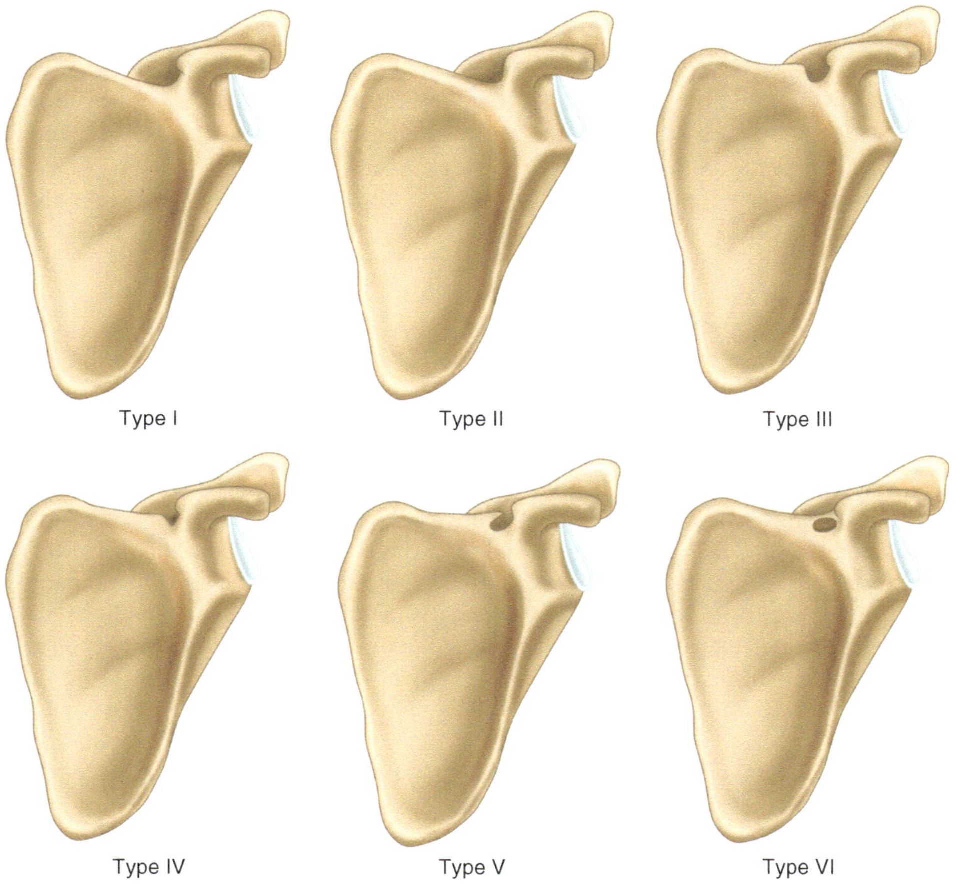

Fig. 33.5 Classification of abnormalities of the suprascapular notch by Rengachary (Adapted from Rengachary et al. [34])

Plancher et al. previously described this ligament to be present in 100 % of fresh-frozen specimens.

33.3 Pathoanatomy

Compression of the suprascapular nerve at the suprascapular notch under the TSL is the most common site of compression of the suprascapular nerve. While hypertrophy of the TSL can lead to stenosis of the suprascapular notch, the variation in the geometry of the suprascapular notch itself (see 33.2.1 Adjacent Structures and Variations – Suprascapular Notch) may also cause compression of the nerve, leading to a neuropraxia. A "V"-shaped notch with a smaller suprascapular foramen has been associated with suprascapular neuropathy [2].

33.4 Biomechanics

The SGL is a dynamic ligament. As previously mentioned, the ligament inserts onto the posterior glenohumeral joint capsule, and therefore, motion at the glenohumeral joint impacts the suprascapular nerve [29]. This insertion has larger effects upon internal rotation of the shoulder [29]. As such, when the arm moves into positions of cross-body adduction and internal rotation, the ligament tightens which can cause compression of the nerve at this distal site of the SGL [10].

On the contrary, when the arm is in positions of excessive shoulder abduction and external rotation, the medial tendinous margin of the infraspinatus and supraspinatus muscles can impinge against the lateral edge of the scapular spine, compressing the infraspinatus branch of the suprascapular nerve [34, 37]. More proximally, excessive shoulder abduction and external rotation can create an angulation against the TSL with resultant irritation to the suprascapular nerve [7, 34]. This tractioning of the nerve has been referred to as the "sling effect" because of the sharp turn the nerve takes.

Whatever the mechanism, when the nerve is subject to excessive stretch, altered nerve conduction velocity and subsequently possible clinical symptomatology can ensue. The threshold for detection of altered nerve conduction velocity due to stretch of a nerve has been shown to be at 6 % of the resting length of the nerve. Nerve stretch greater than 15 % of the resting length of the nerve leads to irreversible nerve damage [5, 41].

33.5 Diagnostic Modalities

Diagnostic modalities include imaging, diagnostic injections, and neurophysiology.

33.5.1 Imaging

While many authors have suggested that the diagnosis of suprascapular neuropathy is difficult, as it is a diagnosis of exclusion, an accurate history, detailed physical examination, and appropriate diagnostic imaging can accurately recognize this disease entity and detect any overt neoplastic disease.

Plain radiographs should always be obtained including true (Grashey) anteroposterior (AP), Y or supraspinatus outlet, axillary lateral, and Stryker notch views as well as a Zanca view to inspect the AC joint. An AP scapular view with the beam aimed 15–30° cephalad obliquely at the TSL can aid in identifying any calcifications, exostosis, or previous trauma in the form of callous formation at the notch of osseous notch variants [33, 51]. The goal of this plain film series is to detect any fractures or minute trauma to the scapula, clavicle, coracoid, or glenoid neck.

Magnetic resonance imaging (MRI) is the best imaging modality in suspected suprascapular nerve pathology because of its soft tissue resolution. Visualization of the course of the suprascapular nerve is possible with T2-weighted sagittal oblique images. Identification of soft tissue masses such as ganglion cysts has also become increasingly important when making the diagnosis of suprascapular neuropathy. The MRI will help to identify their presence, location, and size (Fig. 33.6a, b). Fritz has described the characteristic findings in asymptomatic patients with a ganglion cyst, as a homogenous signal, low T1 signal intensity with high T2 signal intensity, and rim enhancement if contrast is placed [14]. Concomitant pathologies such as labral tears which may produce secondary impingement on the suprascapular nerve, rotator cuff tendinopathy, neoplastic processes whether nerve in origin or not, and glenohumeral joint osteoarthritis can also be detected with this imaging modality. Muscle atrophy and fatty infiltration of both supraspinatus and infraspinatus, more common in chronic cases, should also be evaluated as well as the presence of muscle edema which some have suggested to be one of the earliest signs of suprascapular nerve entrapment [20] (Fig. 33.7a, b).

Lastly, computed tomography (CT) and ultrasound can be valuable tools in making the diagnosis of suprascapular neuropathy. CT can detect or confirm notch variants as described by Rengachary (see 33.2.1 Adjacent Structures and Variation – Suprascapular Notch), fractures of the clavicle or scapula, and evidence of an ossified TSL [35]. Diagnostic ultrasound may also be helpful in identifying ganglion cysts in the office. In addition, ultrasound can be used to perform ultrasound-guided aspirations of a ganglion cyst.

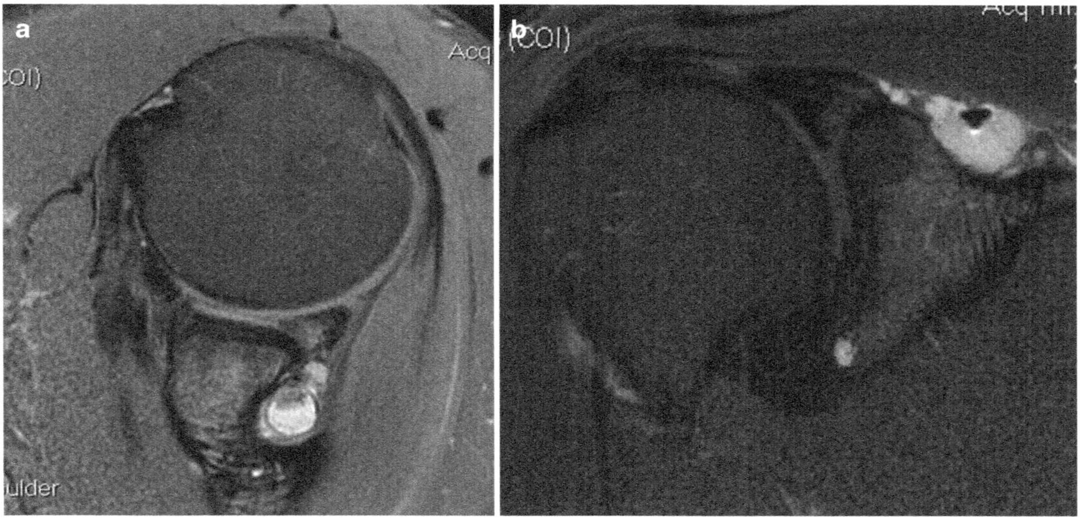

Fig. 33.6 MRI demonstrating a ganglion cyst displacing the suprascapular nerve at the spinoglenoid notch. (**a**) coronal, (**b**) axial images (Copyright K. Plancher)

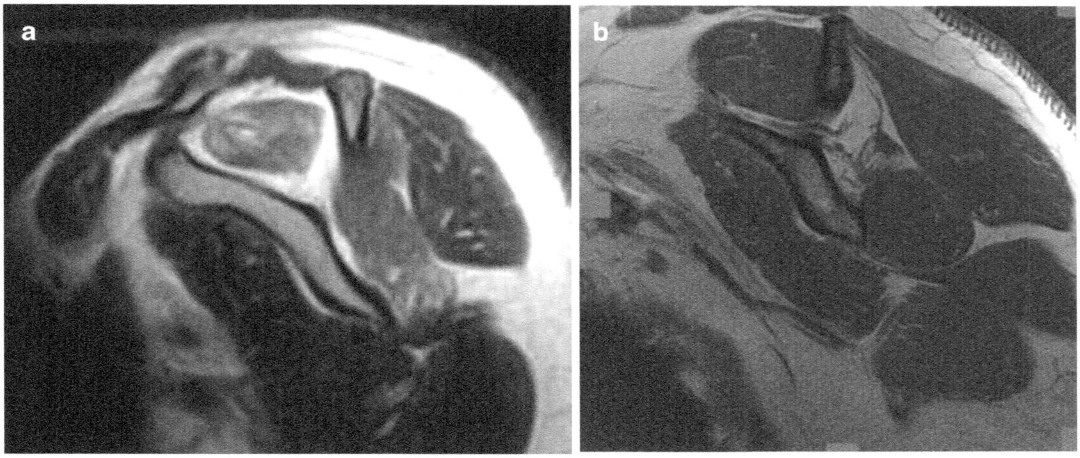

Fig. 33.7 Sagittal oblique MRI demonstrating (**a**) supraspinatus atrophy in a young male and (**b**) isolated infraspinatus atrophy in a volleyball player. Note the course of the nerve in this T2-weighted image (Copyright K. Plancher)

33.5.2 Lidocaine Injection

A 1 % lidocaine anesthetic injection can be immensely helpful to accurately make the diagnosis of suprascapular nerve entrapment at either the TSL or the SGL. The authors routinely use a 25-guage, 1½-inch needle with great success. Ultrasound can be used as an adjunct to guide the needle to ensure accuracy.

When performing the diagnostic injection for compression at the TSL, it is important to understand the relationship of the artery to the nerve as previously described. The needle should be placed into the suprascapular notch from a posterosuperior approach, 3 cm medial to Nevaiser's portal, and aiming anteriorly and aspirating first (Fig. 33.8a, b).

Injection at the spinoglenoid notch is simpler than at the TSL. When injecting into the

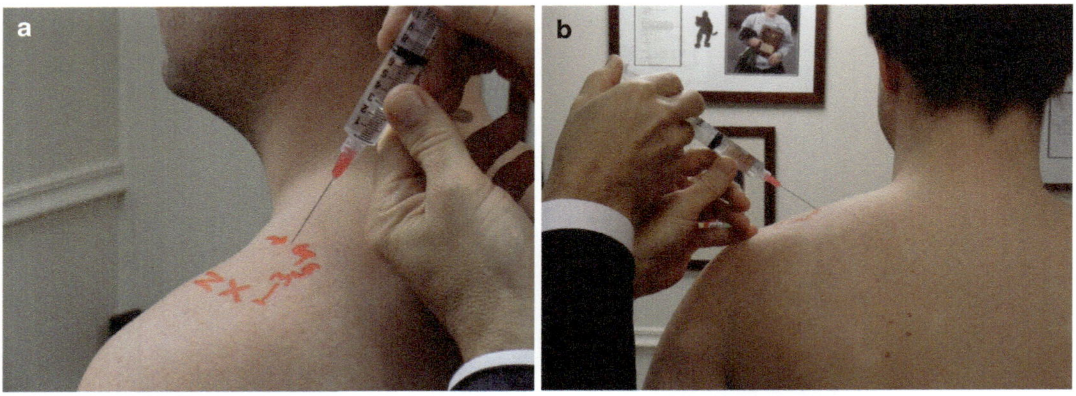

Fig. 33.8 (a) Clinical photo of a lidocaine injection to be placed at the transverse scapular ligament, 3 cm medial to Nevaiser's portal. (b) Posterior view. Please note the angle of the needle (Copyright K. Plancher)

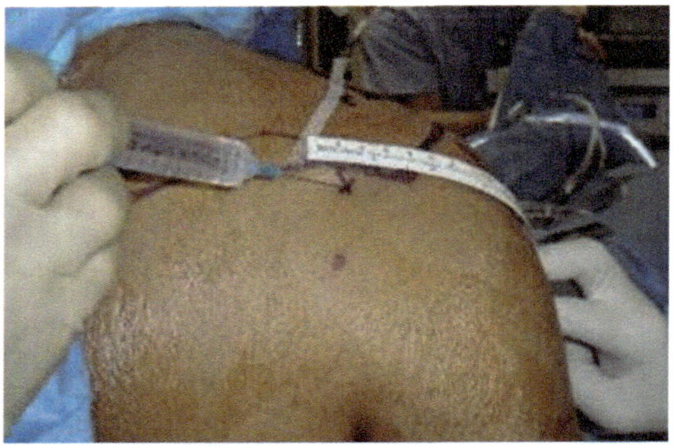

Fig. 33.9 Clinical photo of a lidocaine injection to be placed at the spinoglenoid ligament, 4 cm medial to the posterolateral corner of the acromion (Copyright K. Plancher)

spinoglenoid notch, the needle should be placed 4 cm medial to the posterolateral corner of the acromion (Fig. 33.9). When the injector feels the spine of the scapula, they should drop inferior to it by 1–2 cm and aspirate, and then they should easily fall into the spinoglenoid notch.

Patients should be questioned regarding their pain profile immediately following the injection; pain relief can be dramatic and almost immediate. No different than when using a diagnostic injection for confirmation of impingement syndrome, the cross arm adduction test should be performed both before and after the injection to confirm the diagnosis. If previously positive, the test should now be a non-provocative maneuver. The patient may describe the absence of pain at the AC joint after this intervention, once again helping the physician in ascertaining a definite diagnosis of a suprascapular nerve compression.

A negative test, however, does not rule out the disease in those patients who have a type 4–6 notch as the ability to deliver the lidocaine is quite difficult in those situations. Patients with a negative response when there is no atrophy, a negative EMG, and no evidence of a labral tear or ganglion cyst yet present with weakness and pain require a 3-month course of nonoperative treatment before considering any type of operative intervention.

33.5.3 Electromyogram and Nerve Conduction Velocity Testing

Electrodiagnostic testing with myography and nerve conduction studies can be helpful, if positive, when there is a suspicion of the diagnosis by physical exam, imaging studies are negative (i.e., no soft tissue mass is seen), and atrophy is not present. The suprascapular nerve, as mentioned previously, is a mixed motor and sensory nerve which makes detection of a partial compression extremely difficult. Though evaluation of the sensory velocities is less useful as the sensory innervation of the suprascapular nerve is not as well defined, EMG and nerve conduction velocity testing have been shown to have 91 % accuracy in detecting nerve injury associated with muscle weakness [26, 32]. However, suprascapular nerve dysfunction can still be present with a normal EMG and nerve conduction study, making the diagnosis of suprascapular neuropathy a challenging disease entity.

Electrodiagnostic studies should be performed bilaterally for comparison to the contralateral side. Stimulation is typically performed at Erb's point [18]. Increased latency time often indicates impaired conductivity. The usual latency, or nerve conduction velocity, varies in a range of 1.7–3.7 ms for the supraspinatus. A value beyond 2.7 ms often indicates an abnormality. An increased latency beyond 3.3 ms (range 2.4–4.2 ms) signifies a positive result for compression to the infraspinatus. A classic positive electrodiagnostic study that detects compression at the SGL will demonstrate a motor loss to the infraspinatus without changes in the supraspinatus muscle. A delayed terminal latency to the inferior branch of the suprascapular nerve is expected [27]. EMG testing of the infraspinatus is more difficult as only one branch can be affected and the rest of the muscle may remain unaffected. The clinician should test multiple locations to minimize the risk of a false-negative EMG result. Stimulation of other periscapular muscles leads to volume interference; therefore, needle recording may be the best way of monitoring this disease in lieu of surface recordings.

Other EMG findings have also been reported as confirmatory findings of suprascapular nerve compression. Decreases in the amplitude as well as spontaneous or marked polyphasicity of evoked potentials have been reported to be significant findings in confirming the presence of suprascapular entrapment [33]. Patients who have a long-standing neuropathy often have a reduction in the interference pattern in denervation to the supraspinatus and infraspinatus. The presence of positive sharp waves, fibrillation potentials, and the absence of or decreased number of motor unit action potentials in the infraspinatus and supraspinatus muscles can also result, once again confirming the presence of SSN entrapment.

33.6 Sports Significance of the Anatomy

Suprascapular nerve compression is more commonly found in overhead athletes and overhead workers such as electricians. Overhead athletes may be more susceptible to injury due to the repetitive mechanical stresses placed on the shoulder, often at extreme ranges of motion [6]. The sport-specific motions of baseball, tennis, and volleyball place the arm in positions that increase tension on the suprascapular nerve. For example, the follow-through phase of throwing and an overhead serve in volleyball or tennis place the arm in a cross-body adducted and internally rotated position which, as previously mentioned, increases tension on the suprascapular nerve at the SGL [9, 10]. Cadaveric studies have shown that this position of follow-through (e.g., shoulder internal rotation and adduction in an extended position) increases the tension and pressure on the distal branch suprascapular nerve within the spinoglenoid notch [28] (Fig. 33.10). On the other hand, positions of extreme abduction and external rotation, such as those seen during the cocking phase of throwing, can cause nerve compression more commonly at the TSL compared to the SGL [9, 37]. Martin et al. confirmed that stretch of the suprascapular nerve is dependent on rotation of the scapula at extremes of shoulder motion such as those required in baseball, volleyball, and swimming. These extreme end range positions may potentially lead to increased strain on the nerve, resultant ischemia, irritation, and swelling of the nerve [24, 37].

The reason why some athletes develop this disease entity and others do not remains elusive to the clinician. Some authors have suggested it is increased shoulder mobility in combination with repetitive motions that places the suprascapular nerve at risk, predisposing an athlete to develop a suprascapular neuropathy [37, 48]. Others have suggested that it is the rapid, eccentric contraction of the infraspinatus muscle that increases stress on the suprascapular nerve at its terminal portion [12, 13]. While still others suggest that an indirect ischemic injury to the suprascapular nerve may result from repetitive

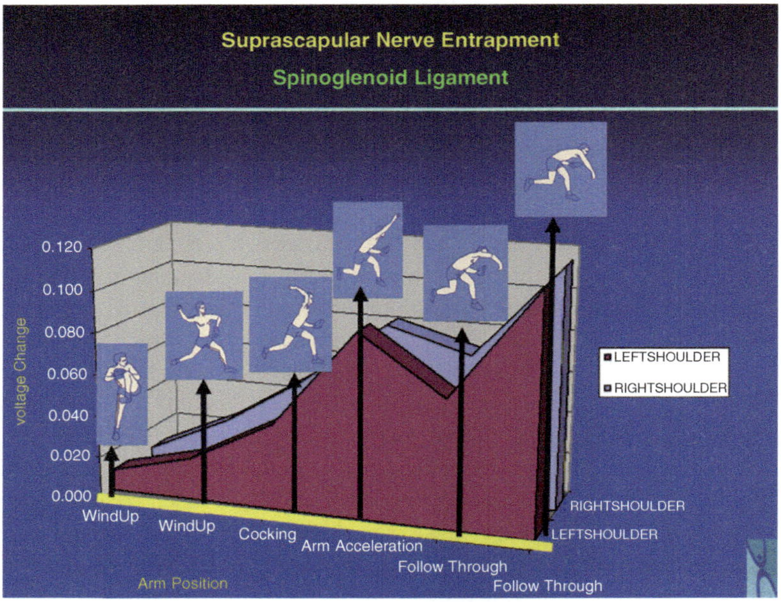

Fig. 33.10 The voltage change with throwing motion with an intact spinoglenoid ligament. Note that follow-through or crossed-arm adduction yields the highest pressure change at the spinoglenoid ligament (Copyright K. Plancher. Previously published in Plancher et al. [28]; Figure 4)

microtrauma and resultant microemboli. We believe though that injury does occur in many of these athletes but often goes unrecognized as this is a disease of young people who have strong serratus anterior, latissimus dorsi, rhomboids, and levator muscles. As the individual ages, these muscles lose power and soon atrophy and weakness becomes noticeable. We believe many, if not all of these athletes, are afflicted with some form of this disease.

33.7 Effect of Trauma

A nerve compression lesion of the suprascapular nerve is often localized to a discrete portion of the length of the nerve, which because of its anatomical position makes it susceptible to entrapment. Early literature noted the nerve was affected with trauma such as a fracture through the scapular notch and even with a proximal humerus fracture caused by a direct blow to the shoulder. The suprascapular nerve has also been reported as the second most common isolated nerve injury seen with shoulder dislocations, second to axillary nerve injuries [46].

33.8 Effect of Disease

Tumors whether benign or malignant can encroach on the suprascapular nerve at the suprascapular notch or SGL. The ganglion cyst represents one of the most common of these space-occupying lesions (Fig. 33.11). Ganglion cysts have also been found in the presence of labral tears which may produce secondary impingement on the suprascapular nerve.

Suprascapular nerve involvement has also been associated with large and massive rotator cuff tears. One study revealed a 40 % incidence of electrodiagnostic evidence of isolated suprascapular nerve dysfunction in patients with massive rotator cuff tears [8]. A significant amount of retraction of the supraspinatus tendon can lead to a reduction in the acute angle that exists between the suprascapular nerve and its first motor branch, increasing tension on the suprascapular nerve at the spinoglenoid notch. Albritton et al. demonstrated this concept in all specimens with 2–3 cm of supraspinatus retraction [1]. Clinical investigations with EMG findings in patients exhibiting suprascapular neuropathy have been varied. Mallon et al. showed positive EMG findings in all patients with massive

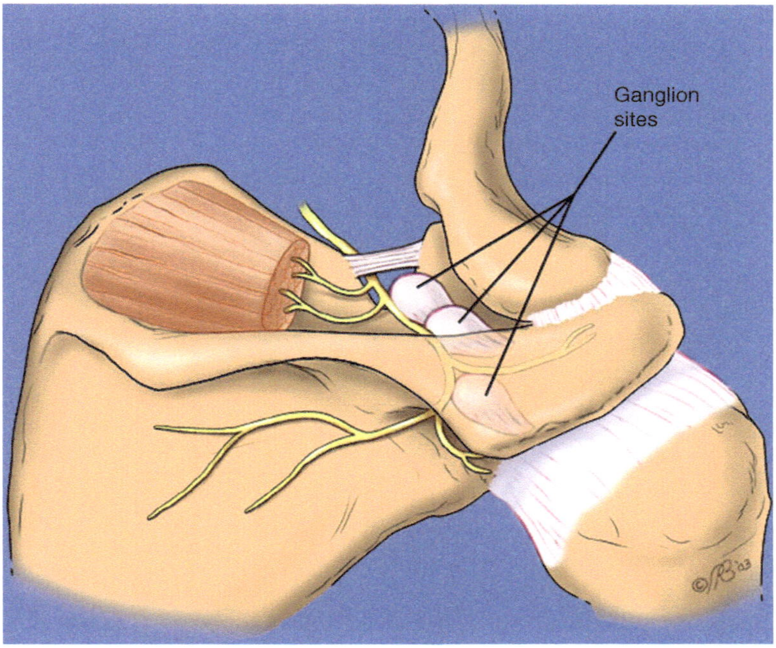

Fig. 33.11 Ganglion cysts causing encroachment and compression of the suprascapular notch at the transverse scapular ligament. A rarer finding than compression at the spinoglenoid ligament (Copyright K. Plancher)

rotator cuff tears, whereas, Shi et al. found that only 37 % of patients with a rotator cuff tear and suspected suprascapular nerve involvement demonstrated suprascapular neuropathy on EMG and nerve conduction velocity [22, 39]. Vad et al. found 28 % of patients with a massive full-thickness rotator cuff tear had abnormal EMG findings with axillary nerve involvement being more common than suprascapular nerve involvement when fat atrophy of the supraspinatus, infraspinatus and/or deltoid exists [45].

33.9 Surgical Significance

Anatomic guidelines are important for the surgeon to have a better appreciation of the close proximity of neurovascular structures for any arthroscopic or open procedure to avoid any undue complications. Iatrogenic injury to the suprascapular nerve has been reported in the literature upon distal clavicle excision, both arthroscopic and open Latarjet, or any posterior approach to the shoulder [21]. In the Latarjet procedure, it is the proximity of the nerve to the exit sites of the AP screws that put the suprascapular nerve at risk [23, 38]. In a study of 23 cadaveric shoulders, Knudsen et al. found that blunt dissection 2.5 cm medial to the AC joint and 5 cm medial to the palpable lateral acromion is safe before encountering the suprascapular nerve or artery [19]. Others have suggested that a minimum distance of 3.86 cm from the suprascapular notch in males and 3.71 cm in females should be maintained during portal placement and soft tissue release during rotator cuff repair [43].

Recent clinical studies together with previous anatomic dissections have convinced many of the larger amounts of sensory innervation of the shoulder by the suprascapular nerve. This sensory innervation may explain pain upon traction or compression of suprascapular nerve as well as upon repair of a massive rotator cuff tear with advancement of the tissue to the footprint [25]. As presented by Albritton et al., retraction of the suprascapular nerve when a full-thickness rotator cuff tear exists can cause increased tension with a more acute angle takeoff at the spinoglenoid notch, but the repair of the full-thickness tear causes increased tension on the nerve at the TSL [1]. Several cadaveric studies have shown that lateral advancement of a retracted rotator cuff tear may be between 1 and 3 cm, and with more advancement the neurovascular pedicle is placed under tension within the substance of the muscle belly [15, 47]. The ensuant stretch on the suprascapular nerve following repair of a massive tear may at times cause the repair to fail. Therefore, we recommend for that reason release of the nerve upon repair of all massive rotator cuff tears.

33.10 Summary

Injury to the suprascapular nerve when recognized leads to a multitude of symptoms including pain and weakness of the shoulder. The pain and disability associated with this condition can be unduly prolonged, but it can be easily treated now with advanced arthroscopic techniques. While this entity may seem to represent a small percentage of the average shoulder surgeon's practice, an understanding of the anatomy, biomechanics, diagnostic testing, and recent advancements in treatment options has brought this diagnosis of exclusion to the forefront and minds of many surgeons.

References

1. Albritton MJ, Graham RD, Richards 2nd RS, Basamania CJ. An anatomic study of the effects on the suprascapular nerve due to retraction of the supraspinatus muscle after a rotator cuff tear. J Shoulder Elbow Surg. 2003;12(5):497–500.
2. Antoniadis G, Richter HP, Rath S, Braun V, Moese G. Suprascapular nerve entrapment: experience with 28 cases. J Neurosurg. 1996;85(6):1020–5.
3. Aszmann OC, Dellon AL, Birely BT, McFarland EG. Innervation of the human shoulder joint and its implications for surgery. Clin Orthop Relat Res. 1996;330:202–7.
4. Bigliani LU, Dalsey RM, McCann PD, April EW. An anatomical study of the suprascapular nerve. Arthroscopy. 1990;6(4):301–5.
5. Bora Jr FW, Pleasure DE, Didizian NA. A study of nerve regeneration and neuroma formation after nerve suture by various techniques. J Hand Surg Am. 1976; 1(2):138–43.
6. Chan KL, Liu S, Maffulli N, Nobuhara K, et al. Controversies in orthopaedic sports medicine. Hong Kong: Williams & Wilkins; 1998.
7. Clein LJ. Suprascapular entrapment neuropathy. J Neurosurg. 1975;43(3):337–42.
8. Costouros JG, Porramatikul M, Lie DT, Warner JJ. Reversal of suprascapular neuropathy following arthroscopic repair of massive supraspinatus and infraspinatus rotator cuff tears. Arthroscopy. 2007; 23(11):1152–61.
9. Cummins CA, Messer TM, Schafer MF. Infraspinatus muscle atrophy in professional baseball players. Am J Sports Med. 2004;32(1):116–20.
10. Demirhan M, Imhoff AB, Debski RE, Patel PR, Fu FH, Woo SL. The spinoglenoid ligament and its relationship to the suprascapular nerve. J Shoulder Elbow Surg. 1998;7(3):238–43.
11. Ebraheim NA, Whitehead JL, Alla SR, et al. The suprascapular nerve and its articular branch to the acromioclavicular joint: an anatomic study. J Shoulder Elbow Surg. 2011;20(2):e13–7.
12. Ferretti A, Cerullo G, Russo G. Suprascapular neuropathy in volleyball players. J Bone Joint Surg Am. 1987;69(2):260–3.
13. Ferretti A, De Carli A, Fontana M. Injury of the suprascapular nerve at the spinoglenoid notch. The natural history of infraspinatus atrophy in volleyball players. Am J Sports Med. 1998;26(6):759–63.
14. Fritz RC, Helms CA, Steinbach LS, Genant HK. Suprascapular nerve entrapment: evaluation with MR imaging. Radiology. 1992;182(2):437–44.
15. Greiner A, Golser K, Wambacher M, Kralinger F, Sperner G. The course of the suprascapular nerve in the supraspinatus fossa and its vulnerability in muscle advancement. J Shoulder Elbow Surg. 2003;12(3): 256–9.
16. Ide J, Maeda S, Takagi K. Does the inferior transverse scapular ligament cause distal suprascapular nerve entrapment? An anatomic and morphologic study. J Shoulder Elbow Surg. 2003;12(3):253–5.
17. Iqbal K, Iqbal R, Khan SG. Anatomical variations in shape of suprascapular notch of scapula. J Morphol Sci. 2010;27:1–2.
18. Khalili AA. Neuromuscular electrodiagnostic studies in entrapment neuropathy of the suprascapular nerve. Orthop Rev. 1974;3:27–8.
19. Knudsen ML, Hibbard JC, Nuckley DJ, Braman JP. Anatomic landmarks for arthroscopic suprascapular nerve decompression. Knee Surg Sports Traumatol Arthrosc. 2014; [Epub ahead of print].
20. Ludig T, Walter F, Chapuis D, Mole D, Roland J, Blum A. MR imaging evaluation of suprascapular nerve entrapment. Eur Radiol. 2001;11(11):2161–9.
21. Mallon WJ, Bronec PR, Spinner RJ, Levin LS. Suprascapular neuropathy after distal clavicle excision. Clin Orthop Relat Res. 1996;329:207–11.
22. Mallon WJ, Wilson RJ, Basamania CJ. The association of suprascapular neuropathy with massive rotator cuff tears: a preliminary report. J Shoulder Elbow Surg. 2006;15(4):395–8.

23. Maquieira GJ, Gerber C, Schneeberger AG. Suprascapular nerve palsy after the Latarjet procedure. J Shoulder Elbow Surg. 2007;16(2):e13–5.
24. Martin SD, Simmons C, Koris M. Suprascapular nerve strain: a biomechanical study. Paper presented at AOSSM 23rd annual meeting. Sun Valley; 1997.
25. Matsumoto D, Suenaga N, Oizumi N, Hisada Y, Minami A. A new nerve block procedure for the suprascapular nerve based on a cadaveric study. J Shoulder Elbow Surg. 2009;18(4):607–11.
26. Nardin RA, Rutkove SB, Raynor EM. Diagnostic accuracy of electrodiagnostic testing in the evaluation of weakness. Muscle Nerve. 2002;26(2):201–5.
27. Ogino T, Minami A, Kato H, Hara R, Suzuki K. Entrapment neuropathy of the suprascapular nerve by a ganglion. A report of three cases. J Bone Joint Surg Am. 1991;73(1):141–7.
28. Plancher KD, Luke TA, Peterson RK, Yacoubian SV. Posterior shoulder pain: a dynamic study of the spinoglenoid ligament and treatment with arthroscopic release of the scapular tunnel. Arthroscopy. 2007; 23(9):991–8.
29. Plancher KD, Peterson RK, Johnston JC, Luke TA. The spinoglenoid ligament. Anatomy, morphology, and histological findings. J Bone Joint Surg Am. 2005;87(2):361–5.
30. Polguj M, Jedrzejewski K, Podgorski M, Majos A, Topol M. A proposal for classification of the superior transverse scapular ligament: variable morphology and its potential influence on suprascapular nerve entrapment. J Shoulder Elbow Surg. 2013;22(9):1265–73.
31. Polguj M, Rozniecki J, Sibinski M, Grzegorzewski A, Majos A, Topol M. The variable morphology of suprascapular nerve and vessels at suprascapular notch: a proposal for classification and its potential clinical implications. Knee Surg Sports Traumatol Arthrosc. 2014; Epub ahead of print].
32. Post M, Grinblat E. Suprascapular nerve entrapment: diagnosis and results of treatment. J Shoulder Elbow Surg. 1993;2(4):190–7.
33. Post M, Mayer J. Suprascapular nerve entrapment. Diagnosis and treatment. Clin Orthop Relat Res. 1987;223:126–36.
34. Rengachary SS, Burr D, Lucas S, Hassanein KM, Mohn MP, Matzke H. Suprascapular entrapment neuropathy: a clinical, anatomical, and comparative study. Part 2: anatomical study. Neurosurgery. 1979;5(4):447–51.
35. Rengachary SS, Neff JP, Singer PA, Brackett CE. Suprascapular entrapment neuropathy: a clinical, anatomical, and comparative study. Part 1: clinical study. Neurosurgery. 1979;5(4):441–6.
36. Rose DL, Kelly CR. Shoulder pain. Suprascapular nerve block in shoulder pain. J Kans Med Soc. 1969;70(3):135–6.
37. Sandow MJ, Ilic J. Suprascapular nerve rotator cuff compression syndrome in volleyball players. J Shoulder Elbow Surg. 1998;7(5):516–21.
38. Sastre S, Peidro L, Mendez A, Calvo E. Suprascapular nerve palsy after arthroscopic Latarjet procedure: a case report and review of literature. Knee Surg Sports Traumatol Arthrosc. 2014; [Epub ahead of print].
39. Shi LL, Boykin RE, Lin A, Warner JJ. Association of suprascapular neuropathy with rotator cuff tendon tears and fatty degeneration. J Shoulder Elbow Surg. 2014;23(3):339–46.
40. Spinner M, Spencer PS. Nerve compression lesions of the upper extremity. A clinical and experimental review. Clin Orthop Relat Res. 1974; 104:46–67.
41. Sunderland S. The anatomy and physiology of nerve injury. Muscle Nerve. 1990;13(9):771–84.
42. Ticker JB, Djurasovic M, Strauch RJ, et al. The incidence of ganglion cysts and other variations in anatomy along the course of the suprascapular nerve. J Shoulder Elbow Surg. 1998;7(5):472–8.
43. Tom JA, Mesfin A, Shah MP, et al. Anatomical considerations of the suprascapular nerve in rotator cuff repairs. Anat Res Int. 2014;2014:674179.
44. Tubbs RS, Smyth MD, Salter G, Oakes WJ. Anomalous traversement of the suprascapular artery through the suprascapular notch: a possible mechanism for undiagnosed shoulder pain? Med Sci Monit. 2003;9(3):BR116–9.
45. Vad VB, Southern D, Warren RF, Altchek DW, Dines D. Prevalence of peripheral neurologic injuries in rotator cuff tears with atrophy. J Shoulder Elbow Surg. 2003;12(4):333–6.
46. Visser CP, Coene LN, Brand R et al. The incidence of nerve injury in anterior dislocation of the shoulder and its influence on functional recovery. A Prospective clinical and EMG Study. J Bone Joint Surg Br. 1999;81(4):679–85.
47. Warner JP, Krushell RJ, Masquelet A, Gerber C. Anatomy and relationships of the suprascapular nerve: anatomical constraints to mobilization of the supraspinatus and infraspinatus muscles in the management of massive rotator-cuff tears. J Bone Joint Surg Am. 1992;74(1):36–45.
48. Witvrouw E, Cools A, Lysens R, et al. Suprascapular neuropathy in volleyball players. Br J Sports Med. 2000;34(3):174–80.
49. Won HJ, Won HS, Oh CS, Han SH, Chung IH, Yoon YC. Morphological study of the inferior transverse scapular ligament. Clin Anat. 2014;27(5): 707–11.
50. Yan J, Horiguchi M. The communicating branch of the 4th cervical nerve to the brachial plexus: the double constitution, anterior and posterior, of its fibers. Surg Radiol Anat. 2000;22(3–4): 175–9.
51. Yoon TN, Grabois M, Guillen M. Suprascapular nerve injury following trauma to the shoulder. J Trauma. 1981;21(8):652–5.

Vascularity of the Shoulder

34

Maritsa Konstantinos Papakonstantinou, Giovanni Di Giacomo, and Gregory I. Bain

34.1 Introduction

The shoulder region is an area that is highly vascularised and contains a multitude of anastomoses between the main supplying arteries. For this reason, occlusion of the main arterial supply to the shoulder and arm either by surgical ligation or atheroma does not usually lead to arterial insufficiency of the upper limb. The same cannot be said of the pelvic girdle and lower limb which in clinical practice are commonly plagued with limb-threatening atheromas. Retrograde filling of arteries is also possible in the shoulder region when a main artery is occluded and vascularity is dependent on anastomoses [1, 2].

The arterial blood supply of the shoulder is a very important topic in orthopaedic surgery especially in regard to proximal humeral fractures. While avascular necrosis can occur spontaneously, it is also frequently observed following proximal humeral fractures both with and without dislocation. The necrosis can be focal or affect the entire humeral head [3]. This can lead to joint surface incongruence, limitation of movement and pain from the articular surfaces. A thorough knowledge of proximal humeral blood supply is key to understanding how humeral head avascular necrosis develops following fractures and fracture-dislocations.

34.2 Arterial Anatomy

The arterial supply to the pectoral girdle and upper limb starts from the subclavian arteries. The subclavian artery is divided into three parts based on its position in relation to the scalenus anterior muscle. Medial to the scalenus anterior, the first part of the artery has three main branches, of which the thyrocervical trunk is involved in the arterial supply of the shoulder. The second part of the artery travelling behind the scalenus anterior gives off one branch, and the third part travelling lateral to the scalenus anterior supplies the dorsal scapular artery which also supplies the shoulder region [1, 4].

M.K. Papakonstantinou (✉)
Taylor Lab, Department of Anatomy and Neurosciences, The University of Melbourne, Parkville, VIC, Australia
e-mail: m.papakonstantinou@ymail.com

G. Di Giacomo
Concordia Hospital for Special Surgery, Via delle Sette Chiese, 90, 00145 Rome, Italy
e-mail: concordia@iol.it

G.I. Bain, MBBS, PhD, FA(Orth)A, FRACS
Department of Orthopaedic Surgery, Flinders University of South Australia, Adelaide, SA, Australia

Department of Orthopaedic Surgery, Flinders Medical Centre, Adelaide, SA, Australia

Professor of Upper Limb and Research, University of Adelaide, Adelaide, SA, Australia
e-mail: greg@gregbain.com.au

G.I. Bain et al. (eds.), *Normal and Pathological Anatomy of the Shoulder*,
DOI 10.1007/978-3-662-45719-1_34, © ISAKOS 2015

The thyrocervical trunk has three branches, with two, the suprascapular and transverse cervical arteries, participating in the arterial supply of the shoulder. The suprascapular artery travels along the posterior border of the clavicle and enters the supraspinous fossa to supply the supraspinatus before curving around the lateral border of the spine of the scapula to supply the infraspinatus and teres minor. The transverse cervical artery supplies the trapezius and the rhomboid muscles [1, 4].

Once the subclavian artery passes behind the clavicle, the axillary artery begins. Again it is divided into three parts depending on its relationship to the pectoralis minor muscle. Medial to the pectoralis minor, the first part of the axillary artery gives off the superior thoracic artery. Deep to the pectoralis minor, the axillary artery gives two branches, the thoracoacromial and lateral thoracic arteries. The thoracoacromial artery, which pierces the clavipectoral fascia, is short and quickly divides into four branches: acromial, deltoid, pectoral and clavicular branches. Their names denote what structures they supply and in which direction they travel. Finally, lateral to the pectoralis minor, the third part of the axillary artery gives rise to three branches. The subscapular artery is the largest of these and travels inferiorly along the lateral border of the scapula supplying the subscapularis and serratus anterior. It divides into two branches, the thoracodorsal and circumflex scapular arteries, which supply the latissimus dorsi, teres major and teres minor [1, 4].

The last two branches of the third part of the axillary artery are the anterior (ACHA) and posterior (PCHA) circumflex humeral arteries. These two arteries provide the main blood supply to the proximal humerus. They share a common origin from the axillary artery but often the PCHA may arise from the subscapular artery or the profunda brachii artery (Fig. 34.1) [5]. Both arteries send branches or groups of branches to supply the four main segments of the proximal humerus: the surgical neck, lesser tuberosity, great tuberosity and humeral head. While the ACHA ramifies along the anterior aspect of the surgical neck of the humerus, the PCHA plunges posteriorly, travelling along the medial surface of the surgical neck to emerge through the quadrangular space and

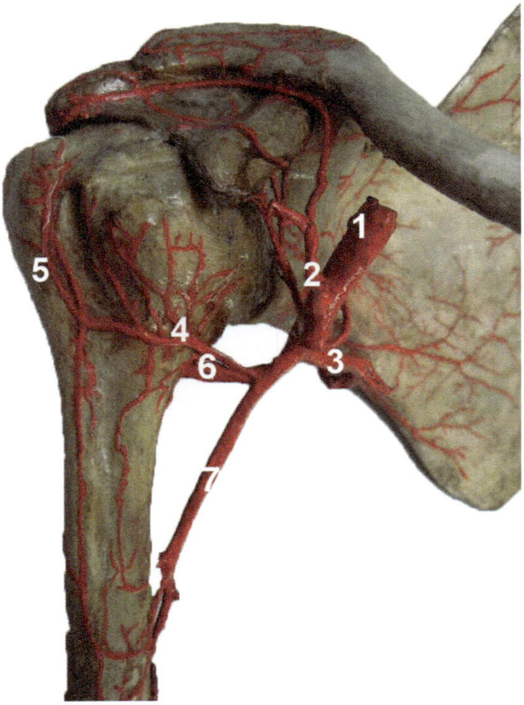

Fig. 34.1 Anterior view of shoulder showing the arterial supply; (*1*) axillary artery, (*2*) thoracoacromial artery, (*3*) subscapular artery, (*4*) ACHA, (*5*) anterolateral artery travelling superiorly on the lateral lip of the intertubercular groove, (*6*) PCHA, (*7*) brachial artery

supply the deltoid, the triceps brachii as well as teres major and minor [5–7].

At the inferior border of the teres major, the axillary artery becomes the brachial artery and continues its course distally in the arm. One of its branches, the profunda brachii, which travels through the triangular interval, partakes in anastomoses between the vasculature of the shoulder region and arm [2, 4].

34.2.1 Anterior Circumflex Humeral Artery and Its Branches

This artery has a mean diameter of 1.2 mm [3] and originates from the axillary artery at the superior edge of the pectoralis major muscle. It travels laterally under the coracobrachialis and both heads of biceps brachii, supplying these and the subscapularis as it continues towards the surgical neck. Just

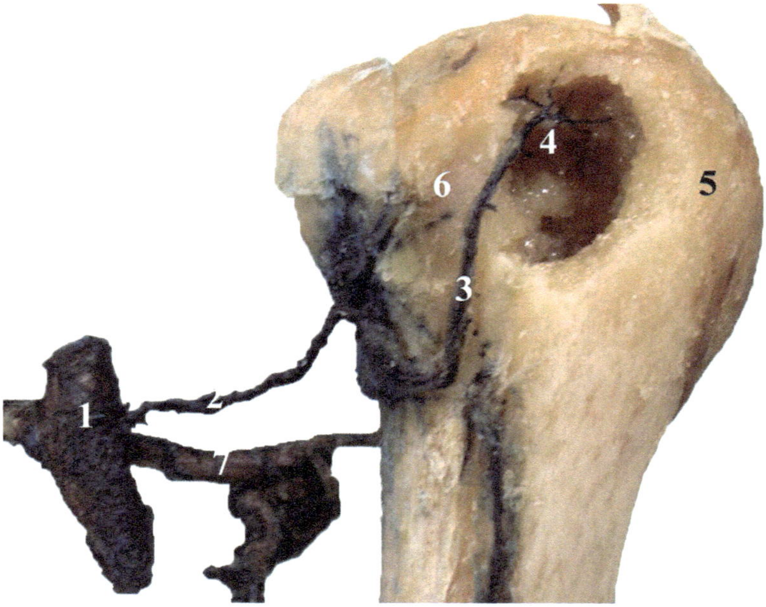

Fig. 34.2 Left proximal humerus of an ink injected specimen, lateral view, showing a continuous stretch of arteries from the axillary artery, to the anterolateral artery and finally the arcuate artery and its branches. The cortex of the greater tuberosity has been breached to follow the intraosseous vasculature: (*1*) axillary artery, (*2*) ACHA, (*3*) anterolateral artery, (*4*) arcuate artery, (*5*) greater tuberosity, (*6*) inter-tubercular groove, (*7*) PCHA

before reaching the surgical neck, it sends a branch to the anterior portion of the axillary recess of the joint capsule [8]. At the surgical neck it divides into four main branches. These include a descending branch to the pectoralis major insertion, a transverse branch to the periosteum of the humerus and greater tuberosity, a muscular branch to the overlying deltoid and an ascending branch known as the anterolateral artery [8]. The anterolateral artery ascends along the lateral lip of the bicipital groove, sending some penetrating branches into the lesser tuberosity along the way, and enters bone at the top of the groove [3, 7–14]. At the point where the anterolateral artery becomes intraosseous, it supplies some branches to the superior facet of the greater tuberosity and to the anterior rotator cuff interval.

The intraosseous continuation of the anterolateral artery and the principal nutrient artery of the proximal humerus is the arcuate artery, named by Laing in 1956 after its "arcuate" posteromedial course in the humeral head (Fig. 34.2) [9]. The arcuate artery supplies the lesser tuberosity, the anterior half of the greater tuberosity and most of the humeral head (Figs. 34.3 and 34.4). Given that it supplies the vast majority of the humeral head, injury or occlusion to this artery is said to increase the risk of avascular necrosis (AVN) of the humeral head [13, 15].

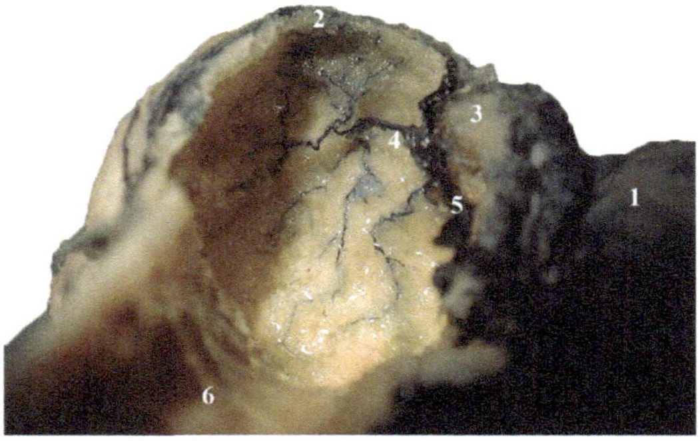

Fig. 34.3 Right shoulder, lateral view facing superiorly into the roof of the proximal humerus. Part of the humeral shaft and most of the cortex of the greater tuberosity have been removed to display the vasculature. The majority of the humeral head is being supplied by the arcuate artery. (*1*) Subscapularis, (*2*) greater tuberosity, (*3*) inter-tubercular groove, (*4*) arcuate artery, (*5*) anterolateral artery, (*6*) humeral shaft

34.2.2 Posterior Circumflex Humeral Artery and Its Branches

The PCHA is three times bigger than its anterior counterpart [3]. After originating from the axillary

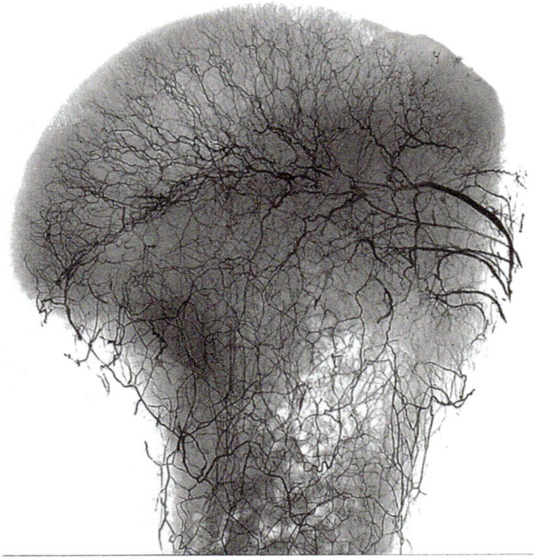

Fig. 34.4 Adult proximal humerus, decalcified and with periosteum removed, showing the intraosseous arterial pattern. The largest artery to the right of the picture likely represents the arcuate artery (Copyright Henry V. Crock 1996 [16])

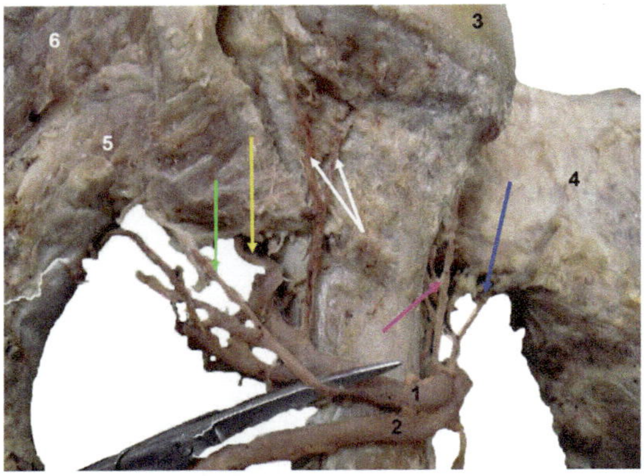

Fig. 34.5 Posteromedial aspect of a left proximal humerus showing the branches of the PCHA. The *white arrows* point to branches of the posteromedial artery, the *pink arrow* shows an anteromedial branch of the PCHA, the *blue arrow* is pointing to a branch supplying the subscapularis, and the branch shown by the *green arrow* supplies the teres minor. The *yellow arrow* points to the posterolateral artery: (*1*) PCHA, (*2*) subscapular artery, (*3*) humeral head, (*4*) subscapularis, (*5*) teres minor, (*6*) infraspinatus

artery, it travels posteriorly along the medial aspect of the surgical neck of the humerus. The PCHA usually provides branches to the inferior border of the subscapularis, the inferior aspect of the joint capsule, the lateral and long heads of the triceps muscle and the teres minor. In addition, it forms the medial and posteromedial vascular groups, a collection of vessels off the PCHA that travel along the joint capsule before ramifying and entering bone just distal to the cartilage of the humeral head (Fig. 34.5) [3, 9, 11]. These intraosseous vessels supply the surgical neck and the inferomedial aspect of the humeral head [13]. They are said to maintain the arterial supply to the humeral head in 4-part valgus impacted fractures and when a medial calcar is present [17].

After travelling along the medial surface of the proximal humerus and emerging through the quadrangular space, the PCHA divides into many branches. Most enter the substance of the deltoid, while one terminal branch, recently named the posterolateral artery, travels along the lateral border of the teres minor, supplying branches to the teres minor and the periosteal network over the greater tuberosity [13]. When it reaches the superior border of the teres minor, the posterolateral artery divides into a number of branches, one of which supplies the connective tissue between the teres minor and infraspinatus, while the others enter bone (Fig. 34.6a, b). Once intraosseous, these branches travel anteromedially across the greater tuberosity and humeral head, supplying the posterior half of the greater tuberosity and a posterolateral section of the humeral head (Fig. 34.7) [13].

34.2.3 Anastomoses

Forming anastomoses with the circumflex humeral arteries and thus contributing to the proximal humeral vasculature are the suprascapular artery, the thoracoacromial artery, the subscapular artery and the profunda brachii artery. The solid network of vasculature formed by these vessels is distributed in four layers via muscular anastomoses, rotator cuff and joint capsule anastomoses, periosteal anastomoses and intraosseous anastomoses.

34.2.3.1 Muscular Anastomoses

The deltoid muscle contains multiple anastomoses between the PCHA, the deltoid branch of the

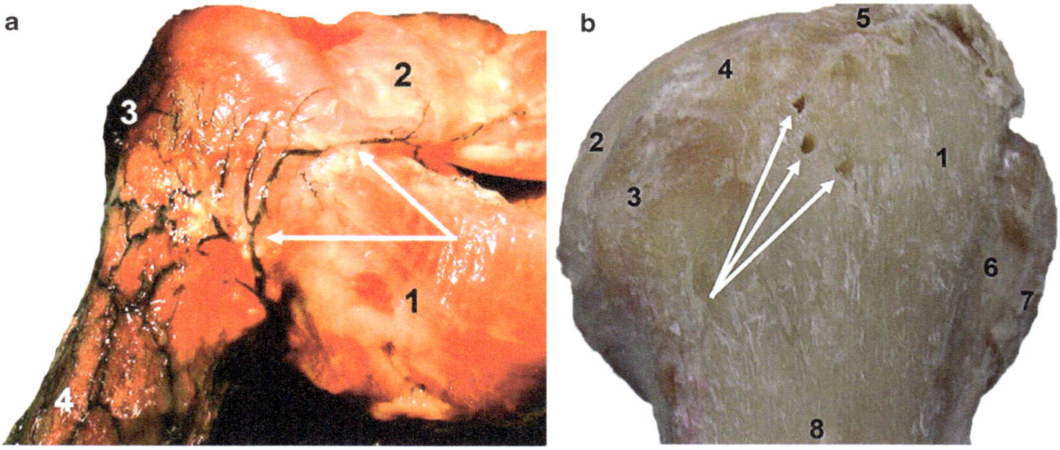

Fig. 34.6 (**a**) Posterolateral view of a left proximal humerus from a fresh specimen. The posterolateral artery (*horizontal white arrow*) highlighting its contributions to the periosteal circulation over the greater tuberosity and the connective tissue it supplies between the teres minor and infraspinatus (*oblique white arrow*). Full thickness rotator cuff tear presents superiorly: (*1*) teres minor, (*2*) infraspinatus, (*3*) greater tuberosity, (*4*) humeral shaft. (**b**) Three vascular foraminae (*white arrows*) on the posterolateral aspect of right proximal humerus for the osseous branches of the posterolateral artery: (*1*) greater tuberosity, (*2*) humeral head, (*3*) inferior facet, (*4*) middle facet, (*5*) superior facet, (*6*) inter-tubercular groove, (*7*) lesser tuberosity, (*8*) humeral shaft

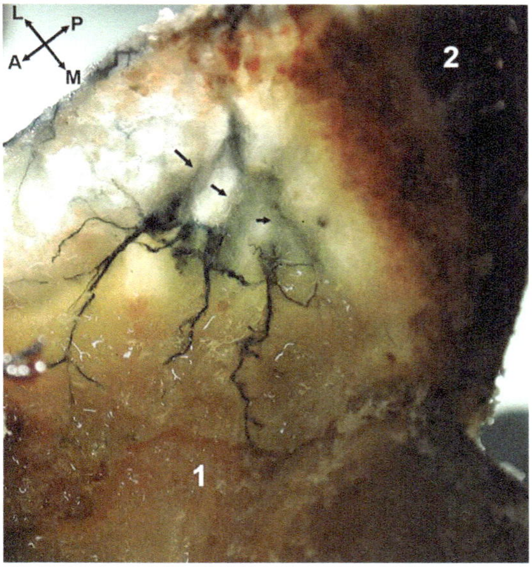

Fig. 34.7 Looking from the inside out towards the cortex of the posterolateral aspect of the greater tuberosity. The specimen has been transilluminated to show the course of the posterolateral artery on the surface of the cortex of the greater tuberosity before it ramifies into three major branches that then become intraosseous: (*1*) infraspinatus, (*2*) teres minor

thoracoacromial artery, the subscapular artery, suprascapular artery and profunda brachii artery with relatively fewer contributions from the ACHA. The supraspinatus muscle contains anastomoses between the suprascapular and dorsal scapular arteries, while the infraspinatus has suprascapular, PCHA and circumflex scapular artery anastomoses. The PCHA anastomoses with branches of the circumflex scapular artery in the teres minor, and in the subscapularis muscle the ACHA and PCHA are involved in a fine network with branches from the subscapular artery [5].

34.2.3.2 Rotator Cuff Tendon and Joint Capsule Anastomoses

The rotator cuff tendons are the site of an anastomosis between four main arteries, the ACHA and PCHA via their terminal branches, the acromial branch of the thoracoacromial artery and the suprascapular artery [18] with minor contributions from the subscapular artery [19]. There are additional anastomoses between the PCHA, subscapular artery and the circumflex scapular artery within the joint capsule.

34.2.3.3 Periosteal Anastomoses

The most significant anastomoses occur via small calibre vessels across the greater tuberosity and lateral aspect of the surgical neck of the humerus

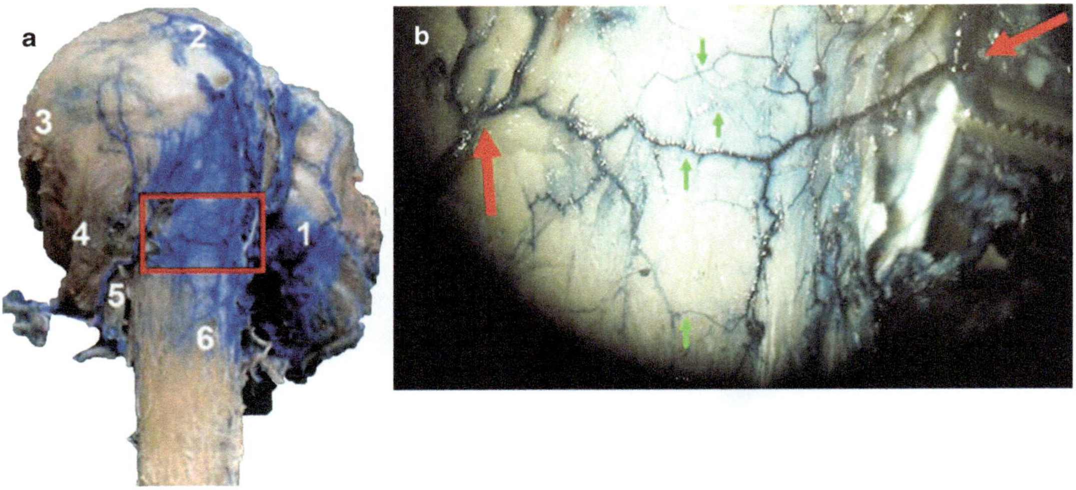

Fig. 34.8 (**a**) Lateral view of a right proximal humerus. A close-up of the area in the *red box* is shown in (**b**). (**b**) Small anastomotic vessels (*green arrows*) are seen across the surgical neck of the humerus laterally between the ACHA held by an instrument (*red arrow on right*) and the posterolateral artery (*red arrow on left*)

between the ACHA and PCHA (Fig. 34.8a, b). While in the majority of anatomic text books the two circumflex humeral arteries are portrayed as participating in one large calibre anastomosis around the surgical neck of the humerus, this finding is not supported by anatomic studies on the subject [7, 8, 11]. Further anastomoses also occur with branches from the profunda brachii artery.

34.2.3.4 Intraosseous Anastomoses

Despite the numerous anastomoses around the shoulder, no anastomoses occur intraosseously in the proximal humerus other than the ones between the terminal branches of the ACHA and PCHA. The vascular border and site of anastomosis between the anterior and posterior circulations occurs through the middle of the greater tuberosity [3]. This is most evident along the undersurface of the superior and middle facets of the greater tuberosity where the vasculature of the arcuate artery anastomoses with the osseous branches of the posterolateral artery directly beneath the border of these two facets (Fig. 34.9a, b). More distally, along the surgical neck of the humerus, the branches of the posteromedial vascular group form anastomoses with both the arcuate artery and posterolateral intraosseous arteries [13].

34.3 Shoulder Dislocations and Fractures

The normally excellent blood supply of the humeral head may be disrupted after a shoulder dislocation. In the case of anterior shoulder dislocations, a number of vascular injuries can occur, including rupture of the axillary artery [26–29], intimal tears, arterial thrombosis [30] and avulsions of the subscapular artery, the ACHA and the PCHA. These vascular injuries are not seen as often in posterior shoulder dislocations. Age may also play a role in AVN development, with older patients having reduced collateral blood supply secondary to tissue degeneration and age-induced arteriosclerotic changes in the vessels [30].

Proximal humeral fractures account for 4–5 % of all fractures in the adult population and less than 1 % in paediatric settings [20, 21]. Up to 85 % are minimally displaced and tend to have good outcomes. Displaced fractures and shoulder dislocations are more likely to be associated with vascular injuries which can eventually result in AVN of the humeral head. The rate of AVN for displaced three-part fractures can be as high as 25 % [20, 22, 23] but is greater for four-part fractures 75 % [20, 24, 25]. This is due to major embarassment of the

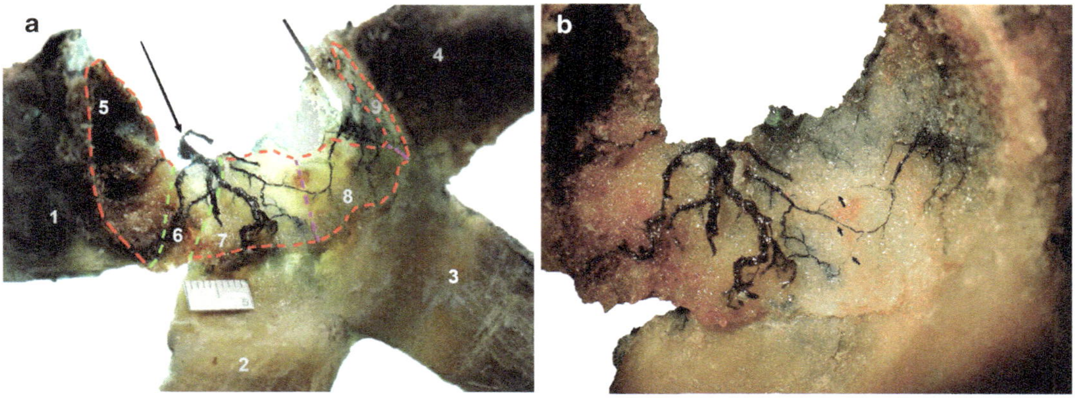

Fig. 34.9 (**a**) Left humerus. The images are of the undersurfaces of the rotator cuff and their insertions on a strip of bone corresponding the lesser tuberosity and facets of the greater tuberosity. The rest of the proximal humerus has been removed. The outlines of the facets and insertions of the rotator cuff are shown in *red* and *pink interrupted lines*. The *green interrupted lines* trace the position of the long head of biceps tendon. The specimen has been trans-illuminated, with opaque areas corresponding to regions of tendinous insertions. The *black arrow* points to the arcuate artery and its branches. The *white arrow* points to the osseous branches of the posterolateral artery: (*1*) subscapularis, (*2*) supraspinatus, (*3*) infraspinatus, (*4*) teres minor, (*5*) deep surface of the lesser tuberosity, (*6*) position of the long head of biceps tendon, (*7*) undersurface of superior facet, (*8*) undersurface of middle facet, (*9*) undersurface of inferior facet. (**b**) A close-up of (**a**), clearly showing three anastomoses between branches of the arcuate artery and intraosseous branches of the posterolateral artery. The point of anastomosis occurs directly beneath the border of the superior and inferior facets

blood supply to the articular fragment which is cut off from the tuberosities and surgical neck through which its blood supply enters.

Hertel reported that the details of the fracture configuration were prognostic for perfusion of the humeral head. Factors associated with compromised perfusion included fractures at the anatomic neck, those with metaphyseal head extension less than 8 mm, disruption of the medial hinge, tuberosity displacement > 10 mm, angular displacement of the head >45° and head splitting fractures [31]. Hertel subsequently published a 5-year follow-up paper, where he reported that the poor perfusion did not invariably lead to avascular necrosis or a poor outcome (2 of the 10 compromised humeral heads collapsed, 4 of the 30 perfused humeral heads collapsed [32].

Reduction manoeuvres and operative treatments for these fractures and dislocations may also lead to iatrogenic vascular injury and contribute to the development of AVN [20]. For this reason, the soft tissues must be respected during surgery and care must be taken to avoid further devitalising the bony fragments.

References

1. Moore KL, Dalley AF. Clinically oriented anatomy. 4th ed. Baltimore: Lippincott Williams & Wilkins; 1999.
2. Taylor GI, Razaboni RM, editors. Arterial anastomotic pathways of the extremities (Translation of M. Salmon, Les Voies Anastomotiques Arte´rielles des Membres, 1939). Quality Medical Publishing: St. Louis; 1994.
3. Gerber C, Schneeberger AG, Vinh TS. The arterial vascularization of the humeral head. An anatomical study. J Bone Joint Surg. 1990;72-A(10):1486–94.
4. McMinn RMH, editor. Last's anatomy: regional and applied. 9th ed. Hong Kong: Churchill Livingstone; 1998.
5. Taylor GI, Razaboni RM, editors. Arteries of the muscles of the extremities and the trunk (translation of M. Salmon and J. Dor, Les Artères des Muscles des Membres et du Tronc, 1939). St Louis: Quality Medical Publishing;1994.
6. Papakonstantinou MK. The arterial supply of the tendinous rotator cuff insertions: an anatomic study. In The Jack Brockhoff Reconstructive Plastic Surgery Research Unit. Melbourne: The University of Melbourne; 2009. p. 220.
7. Sheehan D. Some observation of the course and distribution of the circumflex arteries in Man. Anat Anz. 1932;75:129–45.
8. Kasai T, Suzuki T, Fukushi T, Kodama M, Chiba S. Two circumflex humeral arteries in the arm with

special reference to their communications. Okajimas Folia Anat Jpn. 1984;61(5):347–53.
9. Laing PG. The arterial supply of the adult humerus. J Bone Joint Surg. 1956;38-A(5):1105–16.
10. Duparc F, Muller JM, Fréger P. Arterial blood supply of the proximal humeral epiphysis. Surg Radiol Anat. 2001;23(3):185–90.
11. Meyer C, Alt V, Hassanin H, Heiss C, Stahl J-P, Giebel G, Koebke J, Schnettler R. The arteries of the humeral head and their relevance in fracture treatment. Surg Radiol Anat. 2005;27:232–7.
12. Brooks CH, Revell WJ, Heatley FW. Vascularity of the humeral head after proximal humeral fractures – an anatomical cadaver study. J Bone Joint Surg. 1993;75B(1):132–6.
13. Papakonstantinou MK, Pan W-R, le Roux CM, Richardson MD. Arterial supply of the tendinous rotator cuff insertions: an anatomic study. ANZ J Surg. 2012;82(12):928–34.
14. Moseley HF, Goldie I. The arterial pattern of the rotator cuff of the shoulder. J Bone Joint Surg. 1963;45B(4):780–9.
15. Chansky HA, Iannotti JP. The vascularity of the rotator cuff. Clin Sports Med. 1991;10(4):807–22.
16. Crock HV. An atlas of the vascular anatomy of the skeleton of the spinal cord. London: Martin Dunitz Ltd.; 1996.
17. Jakob RP, Miniaci A, Anson PS, Jaberg H, Osterwalder A, Ganz R. Four-part valgus impacted fractures of the proximal humerus. J Bone Joint Surg. 1991; 73-B(2):295–8.
18. Determe D, Rongieres M, Kany J, Glasson JM, Bellumore Y, Mansat M, Becue J. Anatomic study of the tendinous rotator cuff of the shoulder. Surg Radiol Anat. 1996;18:195–200.
19. Rothman RH, Parke WW. The vascular anatomy of the rotator cuff. Clin Orthop. 1965;41:176–86.
20. Iannotti JP, Williams Jr GR. Disorders of the shoulder-diagnosis and management, vol. 2. 2nd ed. Philadelphia: Lippincott Williams & Wilkins; 2007.
21. Court-Brown CM, Garg A, McQueen MM. The epidemiology of proximal humeral fractures. Acta Orthop Scand. 2001;72(4):365–71.
22. Zyto K. Non-operative treatment of comminuted fractures of the proximal humerus in elderly patients. Injury. 1998;29(5):349–52.
23. Kristiansen B, Christensen SW. Plate fixation of proximal humeral fractures. Acta Orthop. 1986;57(4):320–3.
24. Sturzenegger M, Fornaro E, Jakob RP. Results of surgical treatment of multifragmented fractures of the humeral head. Arch Orthop Trauma Surg. 1982;100:249–59.
25. Leyshon RL. Closed treatment of fractures of the proximal humerus. Acta Orthop Scand. 1984;55(1):48–51.
26. Gibson JMC. Rupture of the axillary artery. J Bone Joint Surg. 1962;44(1):114–5.
27. Max Jardon O, Hood LT, Lynch RD. Complete avulsion of the axillary artery as a complication of shoulder dislocation. J Bone Joint Surg. 1973;55A(1):189.
28. Johnston GW, Lowry JH. Rupture of the axillary artery complicating anterior dislocation of the shoulder. J Bone Joint Surg. 1962;44B(1):116–8.
29. Kirker JR. Dislocation of the shoulder complicated by rupture of the axillary vessels: report of a case. J Bone Joint Surg. 1952;34B(1):72–3.
30. Henson GF. Vascular complications of shoulder injuries. J Bone Joint Surg. 1956;38B(2):528–31.
31. Hertel R, Hempfing A, Stiehler M, Leunig M. Predictors of humeral head ischemia after intracapsular fracture of the proximal humerus. J Shoulder Elbow Surg. 2004;13(4):427–33.
32. Bastian JD, Hertel R. Initial post-fracture humeral head ischemia does not predict development of necrosis. J Shoulder Elbow Surg. 2008;17(1):2–8. doi:10.1016/j.jse.2007.03.026.

35 Neurovascular Injuries with Shoulder Surgery

Harry D.S. Clitherow and Gregory I. Bain

35.1 Introduction

Iatrogenic neurovascular injury following shoulder surgery is uncommon. The reported rates of neurological injury vary between 0.3 and 17 %, with larger series reporting lower incidences [1–7]. Vascular injury is even less common, with multiple large studies reporting none of these complications [2, 4, 7]. However, there are accounts of significant limb and even life-threatening complications in the literature. The authors are aware of occurrences of such complications within our local region and we suspect that they are, in fact, underreported.

To avoid these injuries requires a sound understanding of the surgical anatomy and the pathoanatomical lesions that can increase the risk of neurovascular damage. In addition the surgeon must be aware of the pathoanatomy and presentation of the complications themselves. For example, an axillary artery pseudoaneurysm may not become symptomatic until years after the inciting injury. Failure to appreciate this may lead the surgeon to erroneously discount the significance of a slightly prominent clavicular fixation screw in a patient with dysvascular symptoms years after clavicle fixation.

This chapter will describe the types of complications, where they occur and the pathoanatomical processes behind them.

35.2 Vascular Injury

A summary of the main vessels at risk, the complications sustained and the mechanism of injury is given in Table 35.1.

35.2.1 Mechanism of Injury

Blood vessel compromise may be the result of compression, traction, or perforation of the vessel wall. Perforation may involve a sharp (incision) or blunt force (laceration). The zone of injury may be narrow, such as a stab wound from a blade or wire, or wide, such as a tearing of the

H.D.S. Clitherow, MBChB, FRACS (✉)
Department of Orthopaedics and Trauma, Royal Adelaide Hospital, Adelaide, SA, Australia

Department of Orthopaedics and Trauma, Modbury Public Hospital, Adelaide, SA, Australia
e-mail: harry_sc@hotmail.com

G.I. Bain, PhD, MBBS, FRACS, FA(Ortho)A
Department of Orthopaedic Surgery, Flinders Medical Centre, Adelaide, SA, Australia

Department of Orthopaedic Surgery, Flinders University of South Australia, Adelaide, SA, Australia
e-mail: greg@gregbain.com.au

G.I. Bain et al. (eds.), *Normal and Pathological Anatomy of the Shoulder*,
DOI 10.1007/978-3-662-45719-1_35, © ISAKOS 2015

Table 35.1 Iatrogenic major vessel injuries about the shoulder

Vessel	Vulnerable areas	Injury	Mechanism
Subclavian vein	Medial third of clavicle	Perforation	Drill tearing
		TOS	Callus compression
		DVT	
Subclavian artery	Junction medial and middle third	Pseudoaneurysm	Prominent metalware
Axillary artery and vein	Middle third clavicle	Pseudoaneurysm	Prominent metalware
		AV fistula	Compression
		TOS	
Suprascapular artery	Superomedial glenoid	Perforation	Arthroscopic dissection
Cephalic vein	Anterior open or arthroscopic	Perforation	Instrumentation
		Pseudoaneurysm	

TOS Thoracic outlet syndrome, *DVT* Deep vein thrombosis, *AV* Arteriovenous

vessel over a sharp bone edge. The mechanism is usually acute, although some lesions, such as pseudoaneurysms, have been reported to be the result of attrition injury to the vessel wall over a prolonged period [8].

35.2.2 Pathoanatomy

35.2.2.1 Compression/Thoracic Outlet Syndrome

The thoracic outlet contains the neurovascular structures to the upper limb and can be divided onto three separate anatomical regions. Medially, the scalene triangle is the space occupied by the three scalene muscles (Scalenus anterior, medius and posterior). This space contains the brachial plexus and subclavian artery, which lie in the interval between the anterior and middle scalenes. The subclavian vein is not contained in this space because it lies *anterior* to the scalene muscles (Fig. 35.1). The costoclavicular space is the segment bounded by the clavicle superiorly, first rib inferiorly, scalenus muscles medially and pectoralis minor laterally. The retropectoralis minor (subcoracoid) space is lateral [9, 10].

Thoracic outlet syndrome (TOS) may affect the vascular, as well as neurological, structures to the upper limb. Iatrogenic vascular TOS has been reported following plate fixation of clavicle fractures [11–13].

Iatrogenic TOS most commonly occurs in the costoclavicular space, as distinct from TOS of other aetiologies, which usually occurs more medially in the scalene triangle [9]. Compression of the vascular structures produces variable symptoms of discolouration, swelling and temperature changes, whilst brachial plexus compression produces variable pain, weakness and sensory disturbance. Significant callus formation may encroach upon the costoclavicular space (Fig. 35.2). In this situation, mobilisation and fixation of fragments may further reduce the volume of the costoclavicular space, to the point of a clinically apparent TOS [11, 12, 14].

35.2.2.2 Perforation

The subclavian vessels are in close proximity to the medial two-thirds of the clavicle (Fig. 35.3). Injury to the subclavian vein has been reported following fixation of an acute clavicle fracture [15] and revision fixation for clavicular nonunion [16]. The vein is thin walled, making it easily distendable but difficult to repair (Fig. 35.4). It is located a mean 5 mm from the posterior aspect of the medial clavicle, but can be directly adherent to the periosteum [17]. In contrast, the subclavian artery has comparatively thick walls and is relatively protected by the overlying scalenus anterior muscle.

35.2.2.3 Pseudoaneurysm

A pseudoaneurysm is an extravascular haematoma that freely communicates with the intravascular space through a defect in the vessel wall [18] (Fig. 35.5). The wall of a pseudoaneurysm is

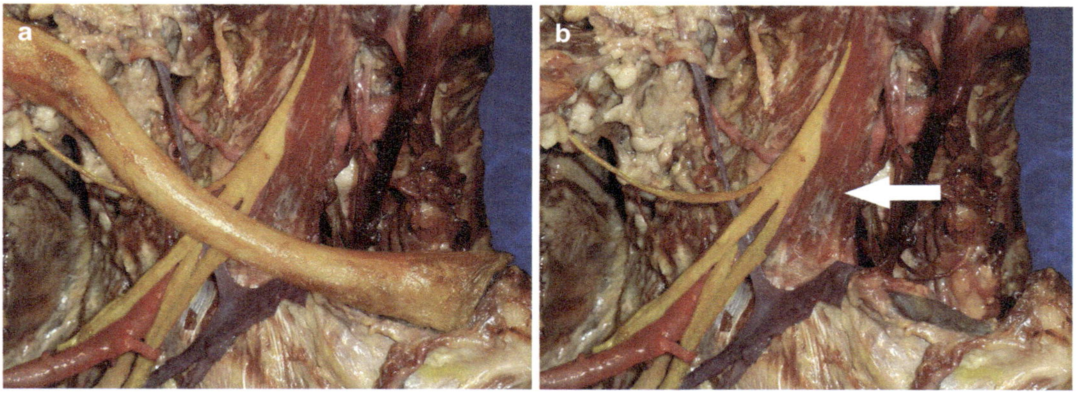

Fig. 35.1 Right anterior neck dissection with clavicle present (**a**) and clavicle removed (**b**). (**a**) Neurovascular structures adjacent to the medial two-thirds of the clavicle. (**b**) Brachial plexus and subclavian artery posterior to scalenus anterior (*white arrow*). Subclavian vein crossing first rib anterior to scalenus anterior and directly posterior to the medial clavicle and sternoclavicular joint (excised). At the medial end of the clavicle, the subclavian artery and brachial plexus are relatively protected by the scalenus anterior. *White arrow*–scalenus anterior muscle, colouring of the neurovascular structures has been enhanced to clearly demonstrate them. Subclavian/axillary artery: *red* Subclavian/axillary vein: *blue* Brachial plexus: *yellow* Sternoclavicular joint (excised): *black* (Copyright Dr Gregory Bain)

Fig. 35.2 Three-dimensional computed tomography reconstruction of left shoulder. Hypertrophic callus around a clavicular fracture site is demonstrated. This mass is decreasing the volume of the costoclavicular space (Copyright Dr Gregory Bain)

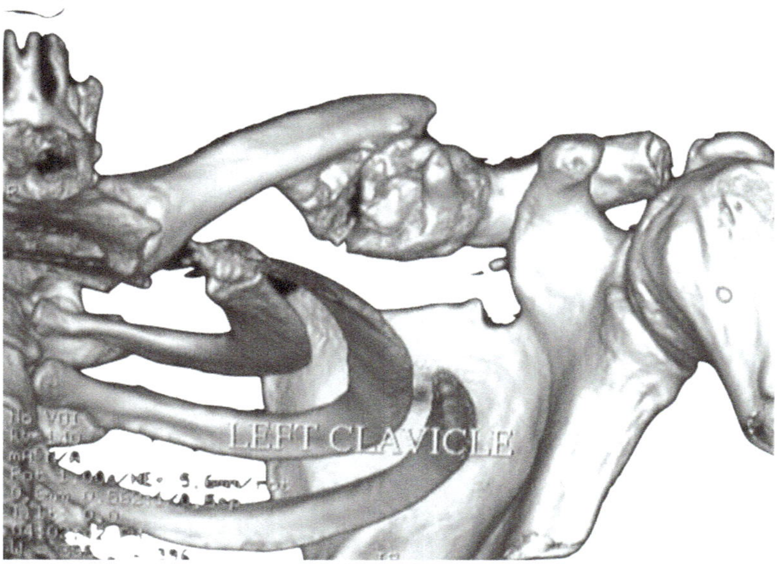

made up of compressed tissues surrounding the haematoma, rather than the three distinct layers of a true vessel wall [19, 20].

They are a result of penetrating trauma, with the most common iatrogenic cause being endovascular cannulation procedures. However, pseudoaneurysms have also been reported in the shoulder in association with prominent clavicular metalware [8, 21–23] (Fig. 35.6) and the following coracoid transfer procedures [2].

Pseudoaneurysms may be clinically silent for many years – a case associated with a prominent screw was reported as presenting 10 years following clavicle fixation [22] (Fig. 35.5). Intermittent, subacute ischaemic symptoms occur distal to the artery due to compression of the

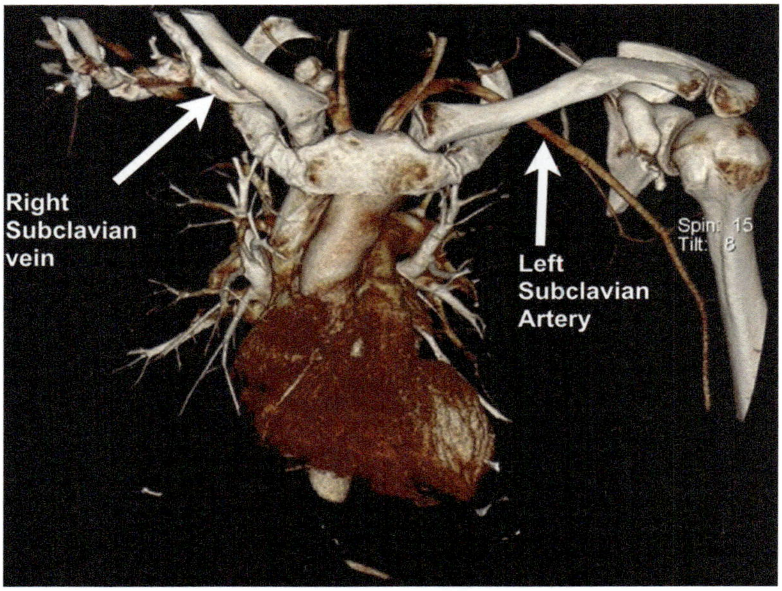

Fig. 35.3 Three-dimensional computed tomography angiogram of the mediastinum and thoracic outlet. Two of the subclavian vessels have been highlighted (right vein and left artery). Note the close proximity of the vessels to the medial end of the clavicle (Copyright Dr Gregory Bain)

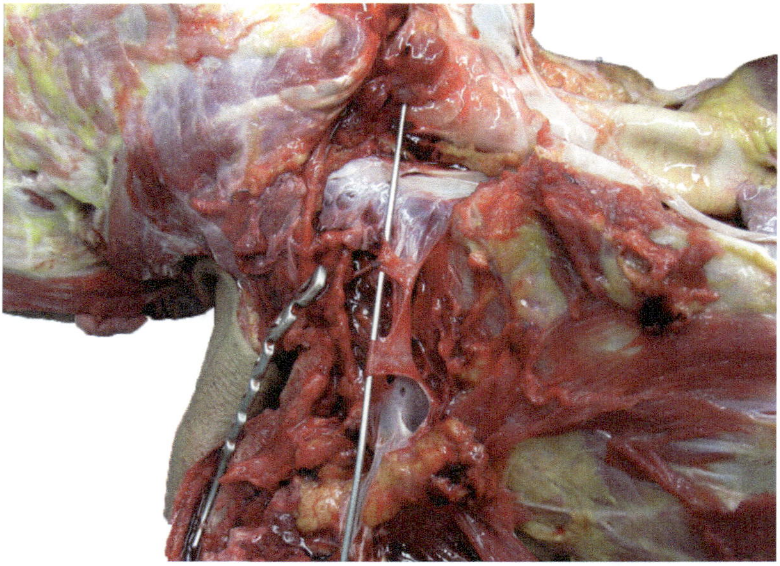

Fig. 35.4 Post-mortem photo of lacerated subclavian vein which occurred during clavicle fixation. A probe is positioned in the lumen of the remaining vein. Note is made of the thin wall of the vessel and its proximity to the medial clavicle. The injury occurred during fixation of a clavicle fracture. Profuse bleeding was noted after one of the holes was drilled. The cause of death was blood loss and air embolism. The injury occurred despite the use of a periosteal elevator positioned along the inferior aspect of the bone, in an attempt to avoid plunging of the drill bit (Image courtesy of the Queensland coroner's court, Australia, Used with permission)

surrounding vessels. Thrombotic events can also occur, which may precipitate acute limb-threatening ischaemia.

35.2.2.4 Arteriovenous Fistula

Arteriovenous (AV) fistulae are a result of injury to both the artery and an adjacent vein [24]. Like

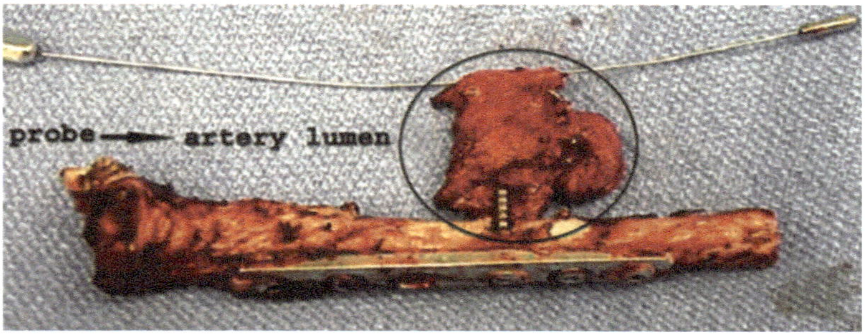

Fig. 35.5 Excised clavicle with pseudoaneurysm at the tip of the prominent screw. A dilator within the true lumen of the excised segment of the subclavian artery. Note risk factors; offending screw is markedly longer than the adjacent screw, screw protrudes from the postero-inferior surface of the middle third of the clavicle. Screw passage is eccentric, such that the surgeon may only have encountered one cortex when drilling (Used with permission from Shackford [22])

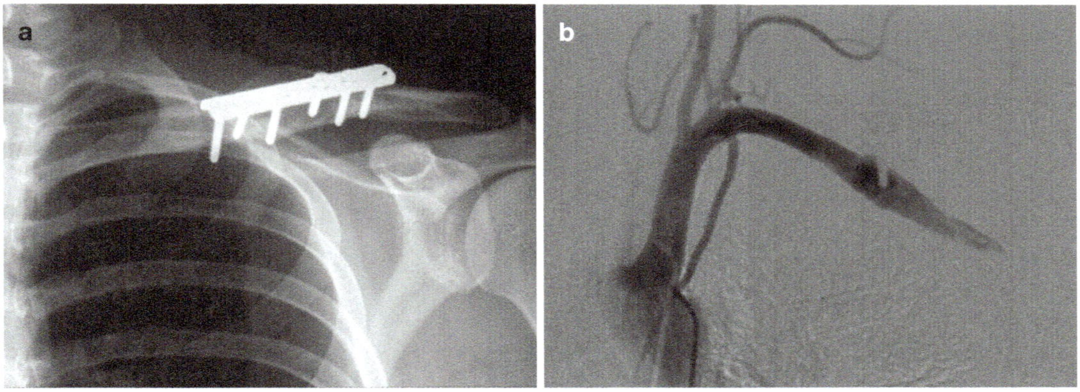

Fig. 35.6 (**a**) Plain radiograph of clavicle and fixation plate 6 years post-surgery for fracture. The patient had intermittent claudication symptoms in the ipsilateral upper limb for the preceding 18 months. (**b**) Angiogram of the same patient. A pseudoaneurysm is demonstrated around the tip of the most medial screw of the fixation plate (Copyright Dr Gregory Bain)

pseudoaneurysms they are commonly associated with penetrating injury.

With time AV fistulae dilate and the pressure in the venous side of the lesion increases, causing venous engorgement and swelling. Persistent, untreated, venous hypertension can cause congestive heart failure or limb-threatening ischaemia.

35.2.2.5 Air Embolism

An air embolism occurs when air or gas is admitted into the vascular system [25] and it most commonly occurs with central venous catheterisation. It is also well recognised in posterior cranial fossa surgery [26], total hip arthroplasty and in spinal surgery in the prone position [27, 28]. Fatal air embolism has been reported following subclavian vein injury sustained during plate fixation of a clavicle fracture [15] (Fig. 35.4).

The surrounding soft tissues adherent to the vein can prevent it from collapsing, and the lumen of the vein has a negative pressure due to the negative intrathoracic pressure. Therefore, any breach of the vessel wall will potentially allow air to be sucked through the defect and into the vessel lumen [29].

Once air enters the subclavian vein, it can have several pathophysiologic consequences. The heart is designed to pump fluid, and struggles to pump the compressible air ("air lock"[30]), which can lead to hypoperfusion and even complete cardiovascular collapse [25, 27, 28, 31]. As in an embolism from any other cause, air in the pulmonary circulation leads to pulmonary vasoconstriction, release of inflammatory mediators, bronchoconstriction and ventilation/perfusion mismatch [27]. If there is a patent foramen ovale, the embolus has the potential to enter the cerebral circulation [27]. The lethal volume of air is estimated to be 200–300 ml (3–5 ml/kg). The closer the vein to the right heart, the smaller this volume [27].

35.2.2.6 Deep Vein Thrombosis

Deep vein thrombosis (DVT) is caused by the classic Virchow triad of vessel wall injury, altered blood flow and hypercoagulability [32]. Surgery and immobility are well known to increase the risk of DVT. There is one reported case of DVT following clavicle fixation [32]. However, it is unclear whether the DVT was due to the initial injury, the surgery or an underlying Paget–Schroetter syndrome. This is a form of thoracic outlet syndrome characterised by venous thrombosis [21, 32]. The risk of pulmonary embolism following upper limb DVT (3–36 %) is similar to that in the lower limb. However, the majority are asymptomatic and it is rarely fatal [33].

35.2.3 Presentation and Prognosis

The clinical effects of a perforation in a vein are usually observed at the time of surgery. Significant bleeding may be obvious; however, if the bleeding is into the chest cavity the only sign may be unexplained hypotension. The effects of air embolism are likewise rapidly apparent. If the cause of hypotension is not identified and addressed, they may prove fatal.

The thoracic outlet syndrome that occurs as a result of vessel compression is usually apparent in the immediate post-operative period.

Arterial wall injuries do not typically present at the time of surgery, but rather after a delay of months to years. Despite this long latent period, once the patient develops symptoms there is potential for them to deteriorate into frank upper limb ischaemia.

35.2.4 Specific Vessels at Risk

35.2.4.1 Subclavian Artery and Vein

The subclavian vessels are at risk during procedures on the medial and middle thirds of the clavicle. The subclavian artery is *posterior* to, and thus relatively protected by, the scalenus anterior muscle. In contrast the subclavian vein is anterior to this muscle and directly *posterior* to the medial clavicle. It lies a mean 5 mm from the bone but may directly appose to the posterior periosteum [17]. The vein wall is thin and can be injured by sharp bone fragments, retractors or drills and screws (Fig. 35.4).

35.2.4.2 Axillary Artery and Vein

The anatomical boundary between the subclavian and axillary vessels is the lateral border of the first rib. The axillary vessels converge and lie *postero-inferior* to the middle third of the clavicle at mean distance of 13–17 mm [17]. Pseudoaneurysms and AV fistulae of these vessels have been reported due to prominent metalware in this region.

The axillary artery descends on the chest wall to the lower border of teres major, where it becomes the brachial artery. Both axillary vessels are at risk in anterior approaches to the shoulder, particularly revision procedures where the anatomical relationships are distorted by scar tissue.

Anterior dislocation of the glenohumeral joint may cause the humeral head to be directly apposed to the axillary artery (Fig. 35.7). This alone may result in either early or late presenting vascular injury [34]. If an anatomic closed reduction cannot be achieved, then a vascular examination is mandatory to ensure that the artery is not interposed between the articular surfaces. Open reduction may be required in chronic cases, and special care must be taken to identify and protect the artery as it will be displaced into the surgical field.

Clavicular surgery warrants particular attention due to the close proximity of these major vessels to the operative field. Care must be taken when drilling to ensure that, wherever possible, the placement of drill holes should be along a trajectory that avoids the vessels. In the medial third of the clavicle, this safe trajectory is in a superior to inferior

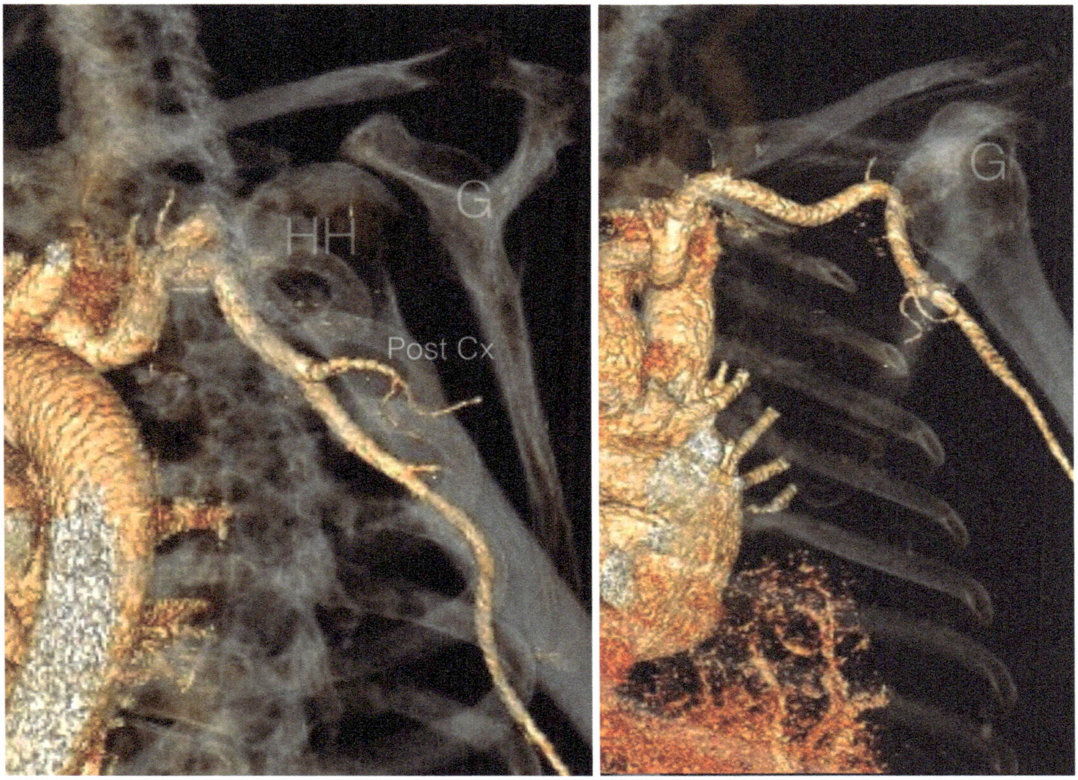

Fig. 35.7 Three-dimensional computed tomography angiogram of a 6-month old chronic anterior glenohumeral dislocation. The axillary artery is directly apposed to the anterior aspect of the humeral head and has been displaced anteriorly as a result. *HH* Humeral head, *G* Glenoid (Copyright Dr Gregory Bain)

direction; and in the middle third, it is anterior to posterior [17]. Dissection around the medial two-thirds of the clavicle must be in the sub-periosteal plane to prevent injury to an adherent subclavian vein. The potential morbidity from compromised blood supply to the clavicle due to this dissection is far less than the morbidity of compromising blood flow to the upper limb or the right atrium due to a vascular injury.

35.3 Neurological Injury

35.3.1 Mechanism of Injury

Nerves may be injured by traction, compression or division. Division may be the result of blunt force (laceration), traction (avulsion) or contact with a sharp surface (incision). The extent of injury depends on the magnitude and duration of the provoking stimulus.

Traction injury occurs when attempts are made to mobilise, distract or retract a nerve, or structures surrounding a nerve, without adequately releasing adjacent tether points. These tethers may be the result of normal anatomy, such as where a nerve passes between two muscle heads, or pathological processes, where the nerve is adherent by scar to adjacent structures. Increasing the resting length of a nerve by 8 % has been reported to cause venous obstruction, with ischaemia occurring at 15 % [35].

Compression injuries may arise from inadvertent placement of retractors, interposition between a fixation plate and the bone or suture entrapment. Compression injury can also occur to prominent nerves that are outside of the shoulder region due to positioning of the patient without adequate padding.

35.3.2 Pathoanatomy

Two commonly used classification systems for nerve injury are based on the pathoanatomical lesion present. Seddon [36] classified nerve injury into three groups, depending on the integrity of the axon and the nerve. Sunderland [37] further delineated these groups according to the integrity of the various anatomical layers of the nerve (Table 35.2).

The common feature of all cases of neurological injury is nerve dysfunction. If all axons are involved the lesion is complete, and if any axons are spared the lesion is partial.

Table 35.2 Sunderland classification and Seddon criteria of nerve injury [36, 37]

Sunderland classification (Seddon criteria)	Pathoanatomy	Prognosis
I (Neuropraxia)	Focal conduction block Axon intact	Recover in hours – weeks
II (Axonotmesis)	Axon ruptured Endo, peri and epineurium all intact	Recovery in months – years
III (Axonotmesis)	Endoneurium ruptured, other layers intact	Recovery variable
IV (Axonotmesis)	Perineurium ruptured but epineurium intact	Recovery variable
V (Neurotmesis)	All layers ruptured – nerve divided	No recovery unless repaired

Sunderland classification (I–V) and the corresponding Seddon criteria (in brackets)

35.3.3 Presentation and Prognosis

Pain is a feature of acute partial nerve injury. In cases where the patient has a persisting noxious stimulus to the nerve – such as an encircling suture or sustained traction over a short segment – the patient may wake from surgery with intense pain in the affected region [38, 39].

The prognosis of the dysfunction is related to the pathoanatomical lesion present in each nerve fibre, that is, the Sunderland class. This in turn directs the management of the injury. The pathoanatomy is influenced by the intensity and the duration of the injury mechanism.

Neurological lesions are usually transient and self-limiting, although permanent injuries have been reported. The brachial plexus or the axillary nerve are the most commonly affected structures. Recovery from a "transient" brachial plexus injury may take 6 months or longer, presenting a significant burden to the patient. Furthermore, it should be noted that the greater the degree of injury to the nerve, the less potential there is for recovery without surgical intervention.

35.3.4 Specific Nerves at Risk

A summary of the major nerves at risk, the areas where they are vulnerable and the reported mechanisms of injury is given in Table 35.3.

35.3.4.1 Brachial Plexus

The brachial plexus is at risk from traction injury during arthroscopic procedures performed in the lateral decubitus position, with a reported 10–30 % incidence of transient paresthesias [40]. Traction is safest when applied with the upper limb positioned at either 0 or 90° of abduction, as the brachial plexus is under the least amount of strain in these positions [41].

Traction plexopathy is an important complication of reverse total shoulder arthroplasty (RTSA). The design of the prosthesis shifts the joint centre inferiorly, thereby increasing the distance between the acromion and the elbow. This has been reported to increase the strain on the plexus by 15–19 % [42].

The plexus is posterior and inferior to the middle third of the clavicle (Fig. 35.1), lying a mean 12 mm from the clavicle at a point 3/5 along the length of the bone (measured from the medial end) [43]. In a clavicle fracture the lateral fragment displaces towards the superior plexus under the weight of the scapula and upper extremity [44]. A pathological tether can develop between the plexus and the fracture callus, which creates a risk of traction injury if the fragments are mobilised [39].

Table 35.3 Iatrogenic major nerve injuries about the shoulder

Nerve	Vulnerable areas	Injury mechanism
Brachial plexus	Lateral position arthroscopy Middle third clavicle (postero-inferior surface) Anterior glenoid	In line traction at 45° abduction Iatrogenic thoracic outlet syndrome Mobilisation of fragments adherent to plexus Prominent metalware
Axillary	Lower border subscapularis Inferior capsule Quadrangular space Proximal humerus	Antero-inferior portals, subscapularis dissection Capsular plication Capsular plication, posterior portals Compression under plate
Suprascapular	Lateral clavicle Suprascapular notch (2.3 cm from articular surface) Base of scapular spine (1.4 cm from articular surface) Postero-inferior glenoid	Distal clavicle excision Arthroscopic dissection, glenoid instrumentation Posterior glenoid exposure Prominent metalware
Musculocutaneous	Between heads of coracobrachialis	Coracoid transfer, biceps tenodesis, reverse TSA
Spinal accessory	Sternocleidomastoid, posterior triangle of neck Superomedial angle of scapula	Lymph node surgery Arthroscopic portal placement
Dorsal scapular	Superomedial angle and medial border of scapula	Scapulothoracic bursoscopy, medial scapular incisions
Upper + lower subscapular	Anterior glenoid (18 mm medial to glenoid rim)	Subscapularis mobilisation during TSA

TSA Total shoulder arthroplasty; portals are arthroscopic portals

Iatrogenic neurological thoracic outlet syndrome has been reported as being due to significant callus [39] or comminuted fragments [45] surrounding the mobilised fracture fragments, or to fragments being fixed in an overlapped position [39]. As in iatrogenic vascular TOS, the pathoanatomical process is due to constriction of the costoclavicular space.

Brachial plexopathy can also occur during scapulothoracic arthrodesis. The correction of the superiorly displaced scapula can tension the plexus over a cervical rib [46] or cause compression between the clavicle and the first rib [47]. The frequency of this complication is highest when the procedure is performed on a scapula deformity that is fixed, rather than mobile [47–50].

The close proximity of the plexus to the clavicle places it at risk from prominent metalware. Injury to brachial plexus elements has been reported following plate and screw fixation of the clavicle, with the plexus entwined, stretched and ultimately ruptured around prominent screws [39].

35.3.4.2 Axillary Nerve

The axillary nerve is at risk during almost all shoulder surgery, both open and arthroscopic.

During anterior surgery, the nerve is at risk as it courses laterally along the inferior margin of the subscapularis muscle, and posteriorly on the inferior glenohumeral capsule. Carofino et al. [38] reported four cases that presented to their peripheral nerve unit with structural injury to the axillary nerve due to suture entrapment during open inferior capsular shift procedures.

The axillary nerve is reported to be the most commonly injured nerve during arthroplasty procedures [5]. The majority of cases are neuropraxias, but nerve laceration has been reported [5, 51]. Traction injury can occur when the procedure is performed through a deltopectoral approach, specifically when the humerus is placed in a position of external rotation and extension to allow preparation of the glenoid [52].

The axillary nerve is potentially at risk during placement of anterior arthroscopic portals. In par-

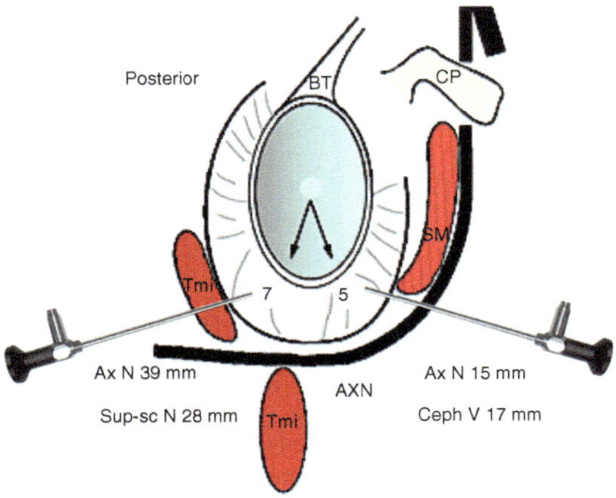

Fig. 35.8 Right glenoid with mean distances of the adjacent neurovascular structures. Arthroscopic antero-inferior (5 o'clock) and the postero-inferior (7 o'clock) portals marked. Ax N, Axillary. The at-risk structures are the cephalic vein (2 mm), axillary nerve (6 mm from the 5 o'clock portal). The portal must be made lateral to the conjoint tendon to avoid injuring the axillary and/or musculocutaneous nerves. *BT* biceps tendon, *CP* coracoid process, *SM* subscapularis, *Tmi* teres minor, *Ax N* axillary nerve, *Sup-sc N* suprascapular nerve, *Ceph V* cepahalic vein (Copyright Dr Gregory Bain)

ticular, it is close to portals inserted at the level of the antero-inferior glenoid (Fig. 35.8). The mean distance from the axillary nerve has been reported to be 15–24 mm, but may be as close as 6 mm. The cephalic vein is a mean 17 mm from the portal, but may be as close as 2 mm.

After exiting the quadrilateral space the nerve courses anteriorly around the humerus on the undersurface of the deltoid. Posterior and lateral deltoid splitting approaches will potentially compromise the nerve if the split is carried further than 4–7 cm distal to the acromion. A quadrilateral "safe zone" distal to the acromion has been described, based on this data [53–55] (see chapter 7.2.18). A deltoid split can be extended beyond this zone if the nerve is carefully undermined off the deltoid prior to splitting the fibres over it.

The subdeltoid bursa provides an aid to locating the nerve during open surgery. The nerve is inferior to the bursa, so if the inferior border of the bursa can be palpated, the surgeon can be confident that the nerve will not be superior to this landmark.

The humeral portion of the axillary nerve can also be damaged during percutaneous humeral plating procedures [56, 57]. This may occur by the nerve being interposed between the plate and the bone, or during placement of the percutaneous screws into the plate [56].

35.3.4.3 Suprascapular Nerve

The suprascapular nerve is at risk of iatrogenic injury in a number of locations along its passage. Mallon et al. [58] described injury to the nerve during distal clavicle excision. The nerve reaches the posterior convexity of the clavicle and then runs parallel with the posterior border of the lateral clavicle for a mean distance of 3.2 cm, before turning posteriorly and inferiorly at a mean 2.2 cm from the lateral end of the bone. Over this course the nerve is always within 1.3 cm of the bone, and may be as close as 6 mm [58]. Excessive dissection posterior to the clavicle introduces the risk of damaging the nerve either during the dissection or the subsequent repair of the soft tissues over the region of the excised distal clavicle.

At the level of the suprascapular notch, the nerve is at risk from arthroscopic dissection further than 2 cm medial to the glenoid [59]. It is also at risk in this region from drills and prominent metalware during repair of superior labrum tears [60] and glenoid baseplate fixation during shoulder arthroplasty (Fig. 35.9).

Where the suprascapular nerve passes inferiorly over the posterior glenoid it is at risk of injury during direct exposure of the posterior glenoid during fracture fixation or osteotomy procedures. Care must be taken when plates are placed on the posterior glenoid to avoid interposing the nerve between the plate and the bone. The nerve is also at risk from drills and screws that perforate the posterior glenoid cortex. This situation may arise during procedures on the inferior glenoid, such as labral repair or coracoid process transfer, but there has also been one report [61] of such a perforation following malposition of a superior labral bone anchor.

A safe zone on the posterior glenoid has been defined that is inferior to the scapular spine and ≤1.4 cm medial to the glenoid articular surface

Fig. 35.9 Photograph of the posterior aspect of the scapula, retrieved post-mortem. A reverse total shoulder arthroplasty had been performed 8 months previously. The superior and posterior glenoid baseplate screws are both protruding from the medial cortex of the scapular neck. The suprascapular nerve passes through the region of both prominent screws and could be injured by them (From Nyffeler et al. [75]. Used with permission, copyright © British Editorial Society of Bone and Joint Surgery)

[62]. To stay in this safe zone, it has been recommended that all anterior drills and screws should be aimed no greater than 10° medial to the plane of the glenoid articular surface [63].

35.3.4.4 Musculocutaneous Nerve

The musculocutaneous nerve perforates the muscles of the conjoined tendon and can be as close as 22 mm to the coracoid [3], placing it at risk of injury during coracoid transfer procedures. The nerve can be tethered as it passes between the heads of coracobrachialis. Mobilisation of the coracoid process and conjoined tendon inferiorly and laterally can either create a kink in the nerve (local compression) or cause a traction injury. Arthroscopic portals must be placed lateral to the conjoined tendon to avoid damaging the nerve as they are advanced in to the joint.

Musculocutaneous nerve palsy has also been reported following open biceps tenodesis [64]. In this case the nerve had been inadvertently wrapped around the long head of biceps tendon during insertion of the tenodesis screw.

35.3.4.5 Spinal Accessory Nerve (Cranial Nerve XI)

The spinal accessory nerve is not part of the brachial plexus, but it is important in shoulder function due to its innervation of trapezius. The nerve passes either through the fires of, or dorsal to, the sternocleidomastoid (SCM). It emerges at the posterior border of SCM, 8 cm cranial to the clavicle [65], then follows an oblique course to the trapezius muscle. The nerve is on the deep surface of trapezius and crosses the superior border of the scapula approximately 2.5 cm lateral to the superomedial angle. At this point it is just lateral to levator scapulae and is directly apposed to the superficial surface of the scapulotrapezial bursa [66, 67].

Spinal accessory nerve injury is most commonly iatrogenic, usually occurring during lymph node biopsy or excision posterior to the SCM [68]. It has been reported that the nerve remains in continuity in up to 50 % of iatrogenic injuries [69]. In order to protect the nerve during open approaches to the superomedial angle of the scapula, the surgeon should avoid splitting the trapezius fibres further than 2.5–3 cm lateral to the superomedial angle [67]. Scapulothoracic arthroscopy portals should be placed below the level of the scapular spine [66, 67, 70].

35.3.4.6 Dorsal Scapular Nerve

The dorsal scapular nerve is medial and parallel to the spinal accessory nerve. It either pierces or runs on the deep surface of the levator scapulae and then runs along the deep surface of the rhom-

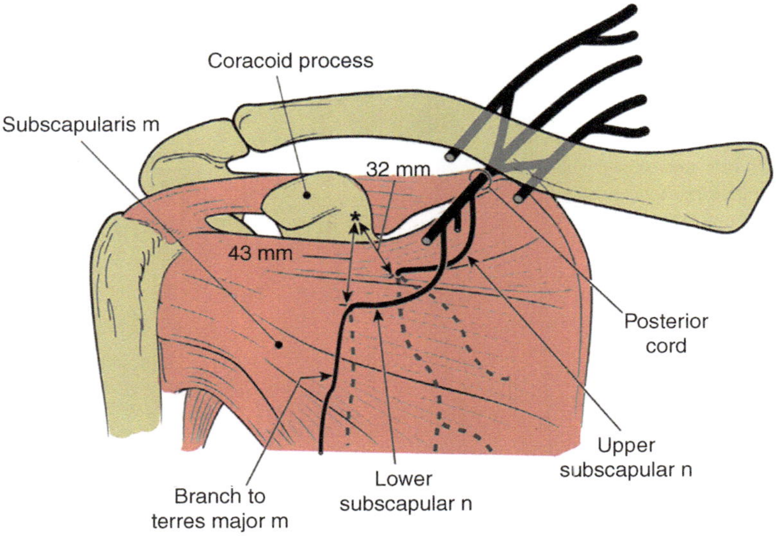

Fig. 35.10 The upper and lower subscapular nerves pierce the subscapularis muscle 32 and 43 mm from the base of the coracoid. This has implications for open and arthroscopic procedures about the coracoid (Used with Permission from Denard et al. [72])

boids, parallel and 1–2 cm medial to the scapular border [66, 67]. It is thus at risk during procedures in this region. The medial viewing portal for scapulothoracic arthroscopy should be placed closer to the midline than the lateral border, both to protect the nerve and to allow better visualisation of the entire scapula [67].

35.3.4.7 Subscapular Nerves

The insertion point of the upper and lower nerves on the anterior aspect of subscapularis has been reported to be a mean 33 and 28 mm medial to the glenoid rim, respectively. The lower nerve may be as close as 18 mm. External rotation of the humerus will translate the nerves laterally by up to 5 mm [71].

The superior and (when present) middle subscapular nerves innervate the superior muscle fibres and the remaining fibres are innervated by the inferior nerve. The boundary between the superior and the inferior fibres is often described as the junction between the superior two-thirds and inferior one-third of the muscle. An alternative is to observe the insertion onto the humerus: the upper and middle nerves supply the portion that forms the tendon and the lower nerve the fibres that insert directly onto the bone [71].

The upper and lower subscapular nerves are at risk around the anterior glenoid during both open procedures where the subscapularis muscle fibres are split, and arthroscopic procedures where dissection is performed medial to the anterior glenoid rim and around the coracoid. The upper and lower subscapular nerves have been reported to be a mean 32 and 43 mm, respectively, from the base of the coracoid (Fig. 35.10) [72].

Injury to the nerves can be avoided by performing muscle-splitting procedures through the internervous plane of the muscle. During arthroscopic procedures, dissection should be carried no more than 2 cm medial to the glenoid rim.

Conclusion

These major neurovascular injuries are fortunately rare. However, a high index of suspicion is required to ensure that these injuries do not occur [73, 74]. We now more commonly perform 3D CT angiograms on complex cases, to ensure that we know the proximity of the vessels (Figs. 35.3 and 35.7).

References

1. Owens BD, Harrast JJ, Hurwitz SR, et al. Surgical trends in Bankart repair: an analysis of data from the american board of orthopaedic surgery certification examination. Am J Sports Med. 2011;39:1865–9. doi:10.1177/0363546511406869.

2. Griesser MJ, Harris JD, McCoy BW, et al. Complications and re-operations after Bristow-Latarjet shoulder stabilization: a systematic review. J Shoulder Elbow Surg. 2013;22:286–92. doi:10.1016/j.jse.2012.09.009.
3. Butt U, Charalambous CP. Complications associated with open coracoid transfer procedures for shoulder instability. J Shoulder Elbow Surg. 2012;21:1110–9. doi:10.1016/j.jse.2012.02.008.
4. Aldinger PR, Raiss P, Rickert M, Loew M. Complications in shoulder arthroplasty: an analysis of 485 cases. Int Orthop (SICOT). 2009;34:517–24. doi:10.1007/s00264-009-0780-7.
5. Bohsali KI, Wirth MA, Rockwood CA. Complications of total shoulder arthroplasty. J Bone Joint Surg Am. 2006;88:2279–92. doi:10.2106/JBJS.F.00125.
6. Cheung E, Willis M, Walker M, et al. Complications in reverse total shoulder arthroplasty. J Am Acad Orthop Surg. 2011;19:439–49.
7. Wijdicks F-JG, Meijden OAJ, Millett PJ, et al. Systematic review of the complications of plate fixation of clavicle fractures. Arch Orthop Trauma Surg. 2012;132:617–25. doi:10.1007/s00402-011-1456-5.
8. Johnson B, Thursby P. Subclavian artery injury caused by a screw in a clavicular compression plate. Cardiovasc Surg. 1996;4:414–5.
9. Perlowski AA, Jaff MR. Vascular disorders in athletes. Vasc Med. 2010;15:469–79. doi:10.1177/1358863X10382944.
10. Atasoy E. Thoracic outlet syndrome: anatomy. Hand Clin. 2004;20:7–14. doi:10.1016/S0749-0712(03)00078-7.
11. Skedros JG, Hill BB, Pitts TC. Iatrogenic thoracic outlet syndrome caused by revision surgery for multiple subacute fixation failures of a clavicle fracture: a case report. J Shoulder Elbow Surg. 2010;19:e18–23. doi:10.1016/j.jse.2009.05.014.
12. Sheikh MA, Bashir R. Iatrogenic thoracic inlet syndrome. Vasc Med. 2005;10:327–8. doi:10.1191/1358863x05vm625xx.
13. Papagelopoulos PJ, Papadopoulos EC, Savvidou OD, et al. Acute thoracic outlet syndrome after internal fixation of a clavicle pseudarthrosis. Orthopedics. 2005;28:606–8.
14. Thavarajah D, Scadden J. Iatrogenic postoperative brachial plexus compression secondary to hypertrophic non-union of a clavicle fracture. Ann R Coll Surg Engl. 2013;95:e55–7. doi:10.1308/003588413X13511609956174.
15. Bain GI, Eng K, Zumstein MA. Fatal air embolus during internal fixation of the clavicle a case report. J Bone Joint Surg Am Case Rep. 2013;3(e24):1–4. doi:10.2106/JBJS.CC.L.00194.
16. Neer CS. Nonunion of the clavicle. JAMA. 1960;172:1006–11.
17. Sinha A, Edwin J, Sreeharsha B, et al. A radiological study to define safe zones for drilling during plating of clavicle fractures. J Bone Joint Surg Br. 2011;93:1247–52. doi:10.1302/0301-620X.93B9.25739.
18. Choo H-J, Kim J-H, Kim D-G. Arterial pseudoaneurysm at the arthroscopic portal site as a complication after arthroscopic rotator cuff surgery: a case report. J Shoulder Elbow Surg. 2013;22:e15–9. doi:10.1016/j.jse.2013.07.056.
19. Corriere MA, Guzman RJ. True and false aneurysms of the femoral artery. Semin Vasc Surg. 2005;18:216–23. doi:10.1053/j.semvascsurg.2005.09.008.
20. Gullo J, Singletary EM, Larese S. Emergency bedside sonographic diagnosis of subclavian artery pseudoaneurysm with brachial plexopathy after clavicle fracture. Ann Emerg Med. 2013;61:204–6. doi:10.1016/j.annemergmed.2012.05.037.
21. Bain GI, Galley IJ, Keogh ARE, Durrant AW. Axillary artery pseudoaneurysm after plate osteosynthesis for a clavicle nonunion: a case report and literature review. Int J Shoulder Surg. 2010;4:79–82. doi:10.4103/0973-6042.76969.
22. Shackford SR. Taming of the screw: a case report and literature review of limb-threatening complications after plate osteosynthesis of a clavicular nonunion. J Trauma. 2003;55:840–3. doi:10.1097/01.TA.0000085862.32648.05.
23. Casselman F, Vanslembroek K, Verougstraete L. An unusual cause of thoracic outlet syndrome. J Trauma. 1997;43:142–3.
24. Ding M, Hu J, Ni J, et al. Iatrogenic subclavian arteriovenous fistula: rare complication of plate osteosynthesis of clavicle fracture. Orthopedics. 2012;35:e287–9. doi:10.3928/01477447-20120123-21.
25. Gordy S, Rowell S. Vascular air embolism. Int J Crit Illn Inj Sci. 2013;3:73–6. doi:10.4103/2229-5151.109428.
26. Jadik S, Wissing H, Friedrich K, et al. A standardized protocol for the prevention of clinically relevant venous air embolism during neurosurgical interventions in the semisitting position. Neurosurgery. 2009;64:533–8; discussion 538–9. doi:10.1227/01.NEU.0000338432.55235.D3.
27. Mirski MA, Lele AV, Fitzsimmons L, Toung TJK. Diagnosis and treatment of vascular air embolism. Anesthesiology. 2007;106:164–77.
28. Austin LS, VanBeek C, Williams GR. Venous air embolism: an under-recognized and potentially catastrophic complication in orthopaedic surgery. J Shoulder Elbow Surg. 2013;22:1449–54. doi:10.1016/j.jse.2013.06.007.
29. Boon JM, Van Schoor AN, Abrahams PH, et al. Central venous catheterization—an anatomical review of a clinical skill—part 1: subclavian vein via the infraclavicular approach. Clin Anat. 2007;20:602–11. doi:10.1002/ca.20486.
30. Albin MS. Venous air embolism: a warning not to be complacent–we should listen to the drumbeat of history. Anesthesiology. 2011;115:626–9. doi:10.1097/ALN.0b013e31822a6408.
31. Muth CM, Shank ES. Gas embolism. N Engl J Med. 2000;342:476–82.
32. Claes T, Debeer P, Bellemans J, Claes T. Deep venous thrombosis of the axillary and subclavian vein after osteosynthesis of a midshaft clavicular fracture: a case report. Am J Sports Med. 2010;38:1255–8. doi:10.1177/0363546509359062.

33. Baarslag H, Koopman MW, Reekers J, Beek ER. Diagnosis and management of deep vein thrombosis of the upper extremity: a review. Eur Radiol. 2004;14:1263–74. doi:10.1007/s00330-004-2252-1.
34. Palcau L, Gouicem D, Dufranc J, et al. Delayed axillary artery pseudoaneurysm as an isolated consequence to anterior dislocation of the shoulder. Ann Vasc Surg. 2012;26:279.e9–12. doi:10.1016/j.avsg.2011.05.039.
35. Lundborg G, Rydevik B. Effects of stretching the tibial nerve of the rabbit. A preliminary study of the intraneural circulation and the barrier function of the perineurium. J Bone Joint Surg Br. 1973;55:390–401.
36. Seddon SH. Surgical disorders of the peripheral nerves. 2nd ed. Edinburgh/New York: Churchill Livingstone; 1975.
37. Sunderland SS. Nerves and nerve injuries. 2nd ed. Edinburgh/New York: Churchill Livingstone; 1978.
38. Carofino BC, Brogan DM, Kircher MF, et al. Iatrogenic nerve injuries during shoulder surgery. J Bone Joint Surg Am. 2013;95:1667–74. doi:10.2106/JBJS.L.00238.
39. Jeyaseelan L, Singh VK, Ghosh S, et al. Iatropathic brachial plexus injury: a complication of delayed fixation of clavicle fractures. Bone Joint J. 2013;95-B:106–10. doi:10.1302/0301-620X.95B1.29625.
40. Peruto CM, Ciccotti MG, Cohen SB. Shoulder arthroscopy positioning: lateral decubitus versus beach chair. Arthroscopy. 2009;25:891–6. doi:10.1016/j.arthro.2008.10.003.
41. Klein AH, France JC, Mutschler TA, Fu FH. Measurement of brachial plexus strain in arthroscopy of the shoulder. Arthroscopy. 1987;3:45–52.
42. Van Hoof T, Gomes GT, Audenaert E, et al. 3D computerized model for measuring strain and displacement of the brachial plexus following placement of reverse shoulder prosthesis. Anat Rec. 2008;291:1173–85. doi:10.1002/ar.20735.
43. Lo EY, Eastman J, Tseng S, et al. Neurovascular risks of anteroinferior clavicular plating. Orthopedics. 2010;33:21. doi:10.3928/01477447-20091124-09.
44. Kitsis CK, Marino AJ, Krikler SJ, Birch R. Late complications following clavicular fractures and their operative management. Injury. 2003;34:69–74.
45. Namdari S, Voleti PB, Huffman GR. Compressive brachial plexopathy after fixation of a clavicular fracture nonunion a case report. J Bone Joint Surg Am Case Rep. 2012;2(e26):1–4. doi:10.2106/JBJS.CC.K.00124.
46. Whitman A. Congenital elevation of scapula and paralysis of serratus magnus muscle: operation. JAMA. 1932;99:1332–4.
47. Mackenzie WG, Riddle EC, Earley JL, Sawatzky BJ. A neurovascular complication after scapulothoracic arthrodesis. Clin Orthop Relat Res. 2003;408:157–61. doi:10.1097/01.blo.0000053342.97749.68.
48. Woodward JW. Congenital elevation of the scapula correction by release and transplantation of muscle origins. J Bone Joint Surg Am. 1961;43:219–28.
49. Grogan DP, Stanley EA, Bobechko WP. The congenital undescended scapula. Surgical correction by the woodward procedure. J Bone Joint Surg Br. 1983;65:598–605.
50. Jeannopoulos CL. Congenital elevation of the scapula. J Bone Joint Surg Am. 1952;34(A):883–92.
51. Wirth MA, Rockwood CA. Complications of total shoulder replacement arthroplasty. J Bone Joint Surg Am. 1996;78:603–16.
52. Nagda SH, Rogers KJ, Sestokas AK, et al. Neer award 2005: peripheral nerve function during shoulder arthroplasty using intraoperative nerve monitoring. J Shoulder Elbow Surg. 2007;16:S2–8. doi:10.1016/j.jse.2006.01.016.
53. Apaydin N, Tubbs RS, Loukas M, Duparc F. Review of the surgical anatomy of the axillary nerve and the anatomic basis of its iatrogenic and traumatic injury. Surg Radiol Anat. 2009;32:193–201. doi:10.1007/s00276-009-0594-8.
54. Uz A, Apaydin N, Bozkurt M, Elhan A. The anatomic branch pattern of the axillary nerve. J Shoulder Elbow Surg. 2007;16:240–4. doi:10.1016/j.jse.2006.05.003.
55. Cetik O, Uslu M, Acar HI, et al. Is there a safe area for the axillary nerve in the deltoid muscle? A cadaveric study. J Bone Joint Surg Am. 2006;88:2395–9.
56. Saran N, Bergeron SG, Benoit B, et al. Risk of axillary nerve injury during percutaneous proximal humerus locking plate insertion using an external aiming guide. Injury. 2010;41:1037–40. doi:10.1016/j.injury.2010.04.014.
57. Park J, Jeong SY. Complications and outcomes of minimally invasive percutaneous plating for proximal humeral fractures. Clin Orthop Surg. 2014;6:146. doi:10.4055/cios.2014.6.2.146.
58. Mallon WJ, Bronec PR, Spinner RJ, Levin LS. Suprascapular neuropathy after distal clavicle excision. Clin Orthop Relat Res. 1996;329:207–11.
59. Warner JP, Krushell RJ, Masquelet A, Gerber C. Anatomy and relationships of the suprascapular nerve: anatomical constraints to mobilization of the supraspinatus and infraspinatus muscles in the management of massive rotator-cuff tears. J Bone Joint Surg Am. 1992;74:36–45.
60. Chan H, Beaupre LA, Bouliane MJ. Injury of the suprascapular nerve during arthroscopic repair of superior labral tears: an anatomic study. J Shoulder Elbow Surg. 2010;19:709–15. doi:10.1016/j.jse.2009.12.007.
61. Yoo JC, Lee YS, Ahn JH, et al. Isolated suprascapular nerve injury below the spinoglenoid notch after SLAP repair. J Shoulder Elbow Surg. 2009;18:e27–9. doi:10.1016/j.jse.2008.10.006.
62. Shishido H, Kikuchi S. Injury of the suprascapular nerve in shoulder surgery: an anatomic study. J Shoulder Elbow Surg. 2001;10:372–6. doi:10.1067/mse.2001.115988.

63. Lädermann A, Denard PJ, Burkhart SS. Injury of the suprascapular nerve during latarjet procedure: an anatomic study. Arthroscopy. 2012;28:316–21. doi:10.1016/j.arthro.2011.08.307.
64. Ma H, Van Heest A, Glisson C, Patel S. Musculocutaneous nerve entrapment: an unusual complication after biceps tenodesis. Am J Sports Med. 2009;37:2467–9. doi:10.1177/0363546509337406.
65. Kierner AC, Zelenka I, Heller S, Burian M. Surgical anatomy of the spinal accessory nerve and the trapezius branches of the cervical plexus. Arch Surg. 2000;135:1428–31.
66. Ruland LJ, Ruland CM, Matthews LS. Scapulothoracic anatomy for the arthroscopist. Arthroscopy. 1995;11:52–6.
67. Williams GR, Shakil M, Klimkiewicz J, Iannotti JP. Anatomy of the scapulothoracic articulation. Clin Orthop Relat Res. 1999;329:237–46.
68. Camp SJ, Birch R. Injuries to the spinal accessory nerve a lesson to surgeons. Bone Joint J. 2011;93:62–7. doi:10.1302/0301-620X.93B1.
69. Kim DH, Cho Y-J, Tiel RL, Kline DG. Surgical outcomes of 111 spinal accessory nerve injuries. Neurosurgery. 2003;53:1106–13. doi:10.1227/01.NEU.0000089058.82201.3D.
70. Bell SN, van Riet RP. Safe zone for arthroscopic resection of the superomedial scapular border in the treatment of snapping scapula syndrome. J Shoulder Elbow Surg. 2008;17:647–9. doi:10.1016/j.jse.2007.12.007.
71. Yung S-W, Lazarus MD, Harryman DT. Practical guidelines to safe surgery about the subscapularis. J Shoulder Elbow Surg. 1996;5:467–70.
72. Denard PJ, Duey RE, Dai X, et al. Relationship of the subscapular nerves to the base of the coracoid. Arthroscopy. 2013;29:986–9. doi:10.1016/j.arthro.2013.02.005.
73. Clitherow HDS, Bain GI. Major neurovascular complications of clavicle fracture surgery. Shoulder Elbow. 2014. doi:10.1177/1758573214546058.
74. Clitherow HDS, Bain GI. Association between screw prominence and vascular complications after clavicle fixation. Int J Shoulder Surg. 2014;8(4):122. doi:10.4103/0973-6042.145261.
75. Nyffeler RW, et al. Analysis of a retrieved Delta III total shoulder prosthesis. J Bone Joint Surg [Br]. 2004;86-B:1187–91.

Part VII

Surgical Anatomy

Surface and Cutaneous Anatomy of the Shoulder

36

Joideep Phadnis and Gregory I. Bain

36.1 Surface Anatomy

Many aspects of the shoulder are amenable to palpation with systematic examination. On inspection, the deltoid contour, muscle wasting, coracoid and acromial shape are visible. A bright tangential light facilitates this by accentuating shadows, which defines the bony outlines and muscle wasting.

36.1.1 Sternoclavicular Joint, Clavicle and Acromioclavicular Joint

The sternoclavicular (SC) joint is usually easy to visualize even in obese patients. It lies adjacent to the sternal notch and the insertion of the sternal head of sternocleidomastoid. It is often swollen and more prominent in SC joint arthritis and less prominent with a posterior SC joint dislocation. This is often best appreciated by palpation and direct comparison with the contralateral SC joint.

The clavicle is a strut connecting the axial skeleton to the shoulder girdle. It is readily visualized and palpated because of its subcutaneous nature. Viewed from superiorly the clavicle is 'S' shaped and widens towards its lateral end which is better appreciated by palpation of its anterior and posterior borders.

The acromioclavicular (AC) joint is visible in slim individuals as the distal clavicle is higher than the acromion and appears as a subtle prominence when viewed from anteriorly. The arthritic AC joint can be even more prominent due to osteophytes. The joint line is usually still palpable but if difficult, the patient can be seated and the humerus pistoned while palpating the AC joint. The orientation and position can be localized by palpating the "V' shape formed by the posterior border of the clavicle and the anterior border of the scapula spine as it forms the acromion. The AC joint lies just anterior to this landmark. It can also be localized by palpation of the lateral coracoid edge. The AC joint lies a fingerbreadth lateral and superior to this point. This landmark is useful when making an anterior arthroscopic portal for AC joint resection. The AC joint is amenable to injection in painful conditions; however, this can be challenging especially in a larger patient or in the presence of osteophytes. In addition, the AC joint has considerable variability in its orientation; therefore, it is advisable to obtain an x-ray prior to injection to better understand this. Unless the examiner is confident he can identify the AC joint, we would recommend injection to be performed under fluoroscopic or ultrasound guidance (Fig. 36.1).

J. Phadnis, FRCS (Tr&Orth) (✉)
Department of Orthopaedic Surgery and Trauma, Flinders University, Adelaide, South Australia
e-mail: joideep@doctors.org.uk

G.I. Bain, PhD, MBBS, FRACS, FA(Ortho)A
Department of Orthopaedics, Flinders University, Adelaide, South Australia
e-mail: greg@gregbain.com.au

G.I. Bain et al. (eds.), *Normal and Pathological Anatomy of the Shoulder*,
DOI 10.1007/978-3-662-45719-1_36, © ISAKOS 2015

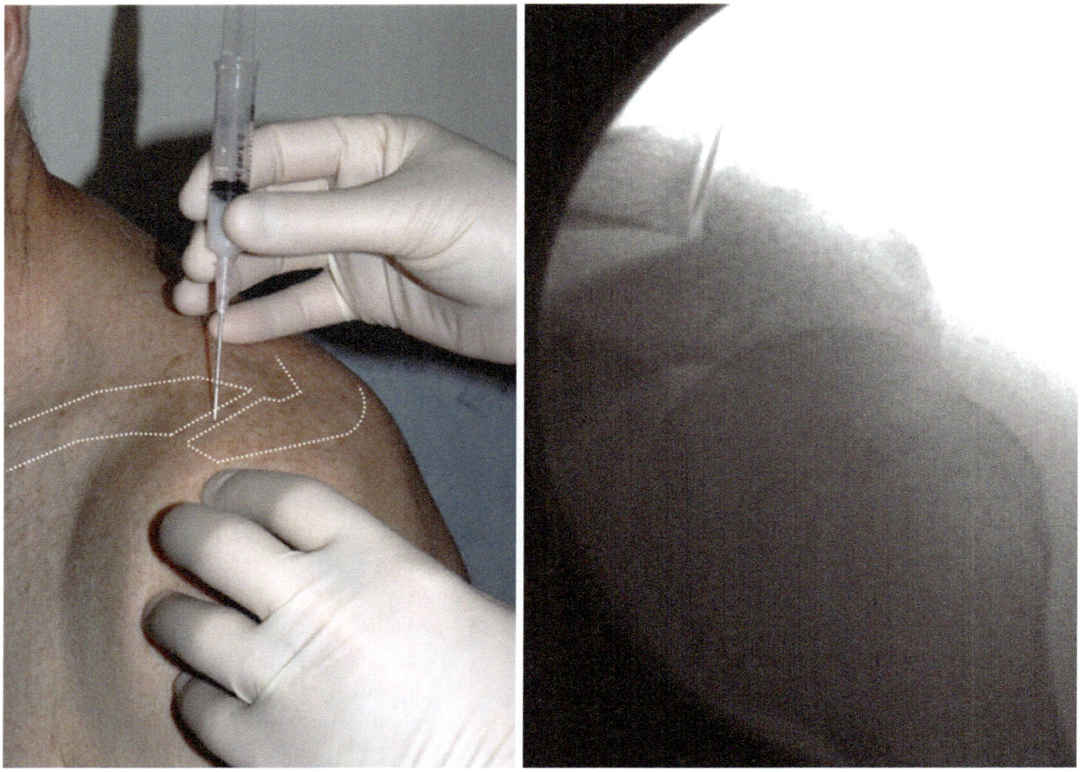

Fig. 36.1 Injection of the AC joint using surface anatomy and fluoroscopic guidance techniques to localize the joint

36.1.2 Acromion

The acromion is a flat trapezoidal process that arises from the scapula spine and then projects anteriorly. Its main function is to provide an origin for the middle deltoid fibres but it also provides attachment for the acromioclavicular ligaments, capsule and the coracoacromial ligament. The posterior border is easily palpable as is the posterolateral corner, which is a key landmark for the standard posterior arthroscopic viewing portal placed just below and medial to the posterolateral corner. The posterior border of the acromion and scapula spine is best palpated by coming from inferior upward with the 'flat of a finger'. The lateral and anterior borders of the acromion are less obvious because of the overlying deltoid but can be identified by using the 'flat of a finger' (Fig. 36.2). A useful surface marking is to draw a line from the 'V' formed by the posterior border of the clavicle and the anterior aspect of the scapula spine extending laterally over the acromion. This bisects the acromion in half and provides a guide for lateral arthroscopic portals (Fig. 36.3). The lateral border can be used to reference the position of the axillary nerve, which lies 4–7 cm inferior to it on the deep surface of the deltoid muscle [1]. The nerve should be marked on the skin in all approaches that split the deltoid and identified during the approach. The anterior lateral corner of the acromion is important for arthroscopy and open surgery. It is the site of most subacromial spurs and is an important landmark when performing an acromioplasty. This corner also marks the raphe between the anterior and middle parts of the deltoid muscle, which is a useful avascular deltoid splitting interval. Subacromial injection is a commonly performed procedure done by physicians, surgeons and allied health professionals. The

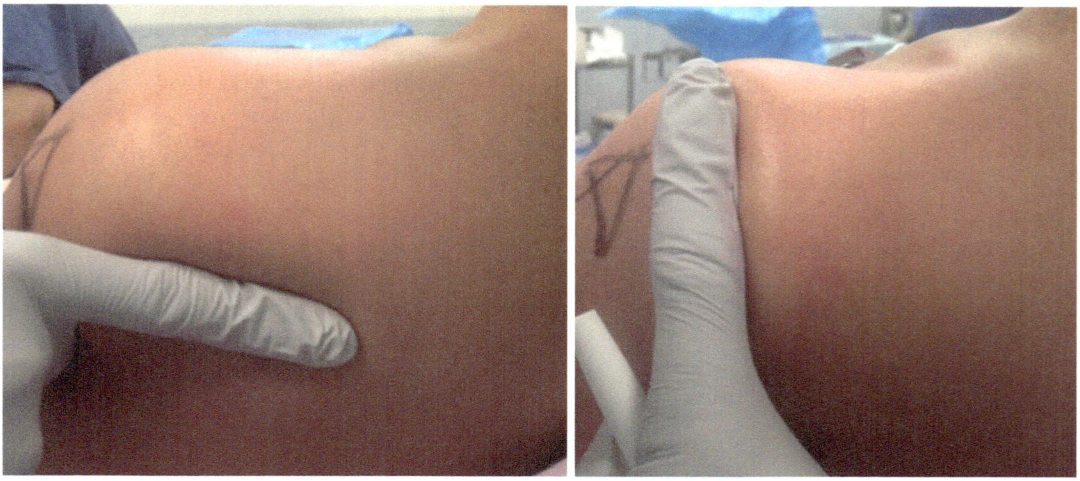

Fig. 36.2 The flat of the finger technique to identify the borders of the acromion

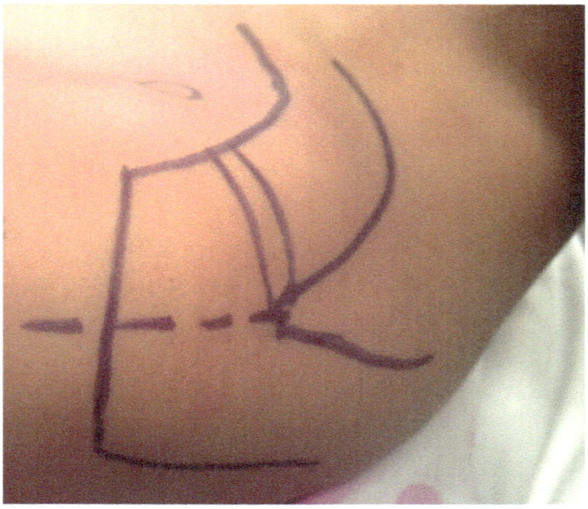

Fig. 36.3 Marking of the skin prior to arthroscopic surgery. Note the dotted line which serves as a guide for lateral portal placement

subacromial bursa is located under the anterior half of the acromion and hence injections are often performed from anterior or lateral. The authors prefer injection from a posterolateral entry point because anteriorly the space between the humeral head and acromion is very small and laterally it is easy to misjudge the slope of the acromion and abut the humeral head or acromion itself. By palpating the posterolateral corner of the acromion, it is easy to guarantee entry into the subacromial space;

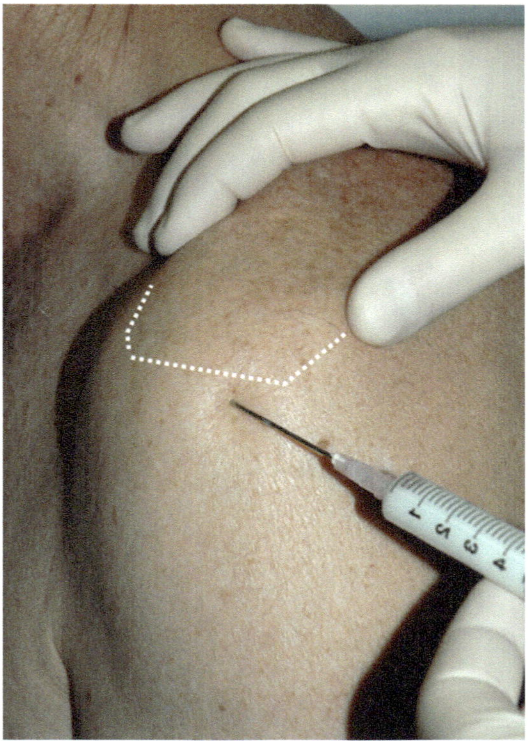

Fig. 36.4 Injection of the subacromial bursa from a posterolateral starting point

and by directing the needle anteriorly, the bursa is very reliably injected with little discomfort or subcutaneous extravasation (Fig. 36.4).

36.1.3 Coracoid Process

The coracoid is a key anatomic landmark in shoulder surgery. It is often visible in the thinner patient and palpable in all. It is often marked as a circle but it is worth remembering that the coracoid is hook shaped with the horizontal component being more oblong and oblique in direction. The brachial plexus, axillary artery and vein are all located on the medial side of the coracoid.

The coracoid thus serves as a safety beacon when performing surgery to the shoulder. The musculocutaneous nerve, pierces the coracobrachialis between 3 and 8 cm distal to the coracoid process [2]. The suprascapular nerve lies 1 cm medial and 2 cm posterior to the coracoid process base as it enters the suprascapular notch (Fig. 36.5).

Arthroscopically the coracoid is important as it provides the trajectory for entry into the glenohumeral joint from the posterior portal. The coracoid is the superficial limit of rotator interval release and can cause impingement to the upper fibres of subscapularis.

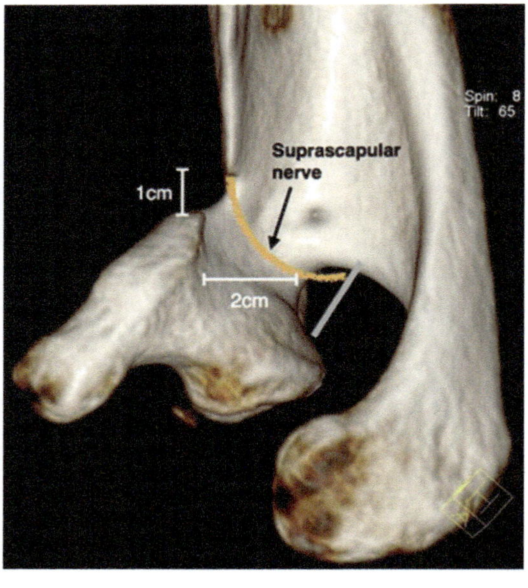

Fig. 36.5 3D CT scan showing superior view of coracoid process and its anatomic relation to the suprascapular nerve as it passes through the suprascapular notch

36.1.4 Glenohumeral Joint

The glenohumeral joint is deep and the joint line is not clearly palpable. However, the humeral head can be readily palpated and balloted to indicate the position of the joint. Aspiration or injection of the joint can be performed from posterior or anterior by using the surface landmarks of the coracoid and acromion. The shortest distance into the joint is via the rotator cuff interval anteriorly, which is located between the coracoid process and bicipital groove. This is located lateral to the coracoid, medial to the biceps tendon and superior to the subscapularis. The subscapularis cannot be palpated therefore staying close to the lower border of the AC joint and just lateral to the coracoid while aiming caudally by 45° will allow safe entry through the rotator interval. Alternatively, a posterior approach starting 2 cm inferior and medial to the posterolateral corner of the acromion and aiming towards the coracoid may be used (Fig. 36.6). A long needle is necessary in larger patients if this method is used.

36.1.5 Scapula and Peri-scapular Muscles

The triangular shape of the scapula body can be identified posteriorly despite circumferential

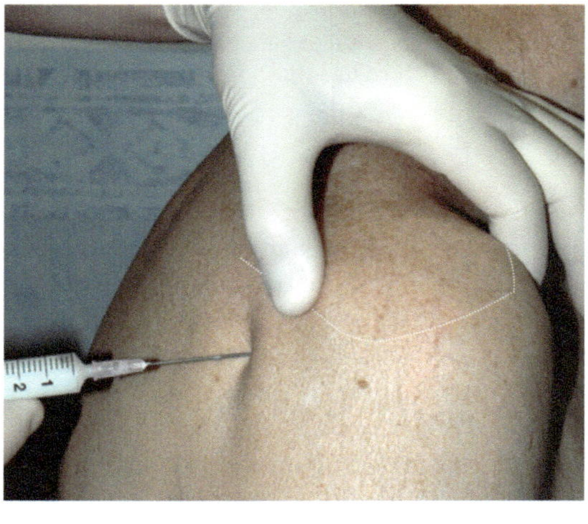

Fig. 36.6 Glenohumeral joint injection using a posterior entry point

attachment of the important peri-scapular muscles. The spine of the scapula separates the supraspinatus and infraspinatus fossae. When examining a patient with shoulder pathology it is essential to identify wasting of the supraspinatus or infraspinatus, which may represent a chronic cuff tear or suprascapular nerve denervation. 'Winging' of the scapula is also evident when viewed from posterior and is important to note as it may be primarily or secondarily related to shoulder pathology. See Chap 29 on the anatomy of scapula winging.

36.1.6 Rotator Cuff

The cuff is difficult to feel, as it lies deep to the deltoid and to bone. The anterior superior rotator cuff insertion (supraspinatus) can be palpated on the humeral head lateral to the anterolateral aspect of the acromion and localized pain in this area can be indicative of cuff pathology. A palpable clicking can often be felt during rotation and elevation of the arm with the opposite hand palpating the cuff insertion on the humeral head. Although not described, this clicking appears to be consistent with the presence of a full thickness rotator cuff tear. The coracoacromial ligament can be felt if thickened just below the anterior edge of the acromion, which may be a source of subacromial impingement.

36.1.7 Muscles

Several large muscles drape the shoulder girdle. An understanding of their surface anatomy is useful in the localization of tenderness, for the diagnosis of pathology and for identification of inter-muscular intervals for surgical approaches to the shoulder. The deltopectoral approach is a workhorse anterior approach to the shoulder utilizing the inter-nervous plane between pectoralis major and deltoid. The coracoid process is palpated and from there the groove can be seen running obliquely across the anterior aspect of the shoulder in most individuals. Its appearance can be enhanced by externally rotating the arm to place the anterior skin under tension. By running the fingers across the anterior deltoid from lateral to medial it can be palpated as the fingers dip into the groove. In some muscular individuals the outline of the cephalic vein, which lies within the groove, can be seen through the skin. Inferior to the deltopectoral groove, the pectoralis major tendon inferior border forms the anterior axillary skin fold. Asymmetry of this fold is indicative of pectoralis major rupture in a patient with an appropriate history. The posterior axillary skin fold is defined by the latismus dorsi. Both these muscles can be made taught for inspection by pushing the hands into the hips with the shoulder abducted. The long head of biceps is a common cause of shoulder pain. It runs in the bicipital groove between the insertions of latissimus dorsi and pectoralis major on the proximal humerus. It can be palpated deeply within the deltopectoral groove especially when the shoulder is externally rotated. On the medial aspect of the upper arm the posterior aspect of the biceps muscle belly forms a groove with the anterior aspect of the triceps. This groove contains the axillary artery, basilic vein and median and ulna nerves as they exit the axilla and course distally. The anterior raphe of the deltoid can also be seen and palpated in certain patients and is an important avascular landmark utilized for dissection during deltoid splitting approaches. The axillary nerve is best evaluated by palpation over the lateral aspect of the upper arm (regimental badge area) and by resisted motor testing of all the parts of the deltoid which are supplied by discrete branches of the nerve.

36.2 Skin Overlying the Shoulder

When planning a skin incision, consideration should be given to the blood supply; the cutaneous nerve supply; the mobility of the skin in allowing access and the cosmesis of the scar that will likely result from the incision.

The skin over the shoulder and its relation to the underlying fascia and muscles varies depending upon the approach chosen.

36.2.1 Structure of the Skin

The two main layers of the skin are the epidermis and dermis (Fig. 36.7). The epidermis is divided into multiple further layers, the deepest of which is the basal cell layer. The most superficial layer of the epidermis is made up of dead squamous cells, which shed from the surface. Squamous shedding from theatre staff and the patient may contribute to post-surgical infection. Several structures traverse the dermis including hair follicles, sweat glands, cutaneous nerves and blood vessels. Hair follicles are intimately related to sebaceous glands, which are distinct from sweat glands. The sebaceous glands secrete sebum, which is the waxy substance that protects the skin. The skin over the face, shoulders and upper trunk is known to have a far more dense concentration of sebaceous glands than other areas of the body. Hormonal imbalances such as during puberty cause increased sebum production. Some bacteria such as Propionibacterium acnes (P acnes) have adapted to live within the sebaceous glands, feeding on sebum. They are linked to the development of adolescent acne because they block the sebaceous glands, causing secondary inflammation. The presence of this elevated density of P acnes in the skin overlying the shoulder is implicated in the pathogenesis of peri-prosthetic infections of the shoulder joint [3]. The fact that P acnes resides in the sebaceous glands could explain why it continues to be a pathogen in shoulder surgery, despite its susceptibility to skin preparations and intravenous antibiotics [4]. The sebaceous glands are intimately related to hair follicles. Hair follicles appear in neat palisading rows rather than haphazardly which suggests the same applies to the sebaceous glands beneath the skin (Fig. 36.8). This raises the possibility that skin incisions could be planned in such a way that they avoid incising through the sebaceous glands and exposing P acnes to the wound.

36.2.2 Skin Incisions

Langer was inspired by the anecdotal findings of Dupytren and Maligne to investigate the behaviour of the skin in response to surgical incisions.

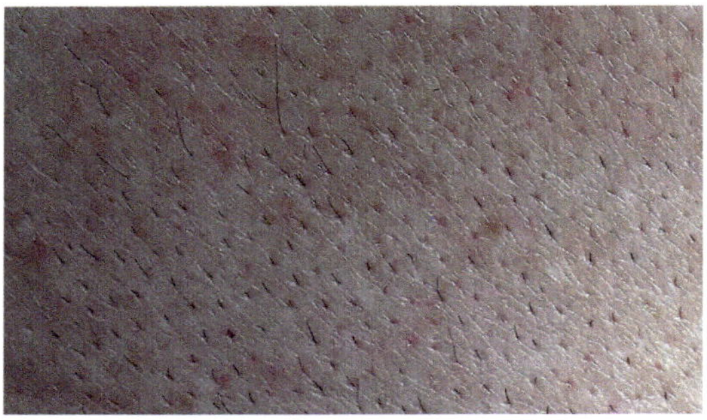

Fig. 36.8 Hair follicles arranged in neat palisading rows

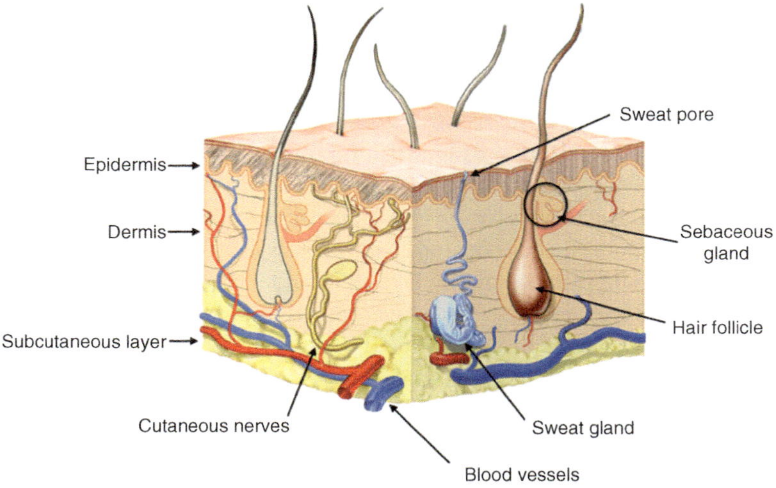

Fig. 36.7 Cross section of the skin

He did this by creating multiple round punctures through the skin all over cadaveric specimens and observing how the skin 'cleaved' [5]. He went on to measure skin tension after circular incisions and observed the behaviour of abdominal skin in postpartum women [6]. This classic research produced the topographic map of skin tension lines known as Langer's lines (Fig. 36.9). Newer techniques have shown that the concept of relaxed tension lines to be more applicable to certain parts of the body including the shoulder [7]. Relaxed tension lines are the linear lines produced when an area of skin is pinched between the thumb and forefinger. The optimal incision from a cosmetic point of view is perpendicular to the most pronounced relaxed tension lines produced with this maneuver. The lines represent the distribution of collagen in the dermis and are important because the collagen realigns itself along the length of the scar regardless of the orientation of the incision. Relaxed tension lines vary between people and are formed over time by the convexities of the underlying muscles and the movement of the joints they cross [7].

36.2.3 Blood Supply to the Skin

The number of cutaneous arteries present in the skin remains the same throughout life [8]. They merely increase in size and length with growth and activity. Figures 36.10 and 36.11 show the vascular territories of the shoulder. There are two horizontal plexuses of blood vessels within the skin. One is the dermal plexus and the other is the fascial plexus. The dermal plexus lies on the deep surface of the dermis and the fascial plexus lies either superficial or deep to the superficial fascia. The path and distribution of the communicating vessels between these plexuses are dependent mainly on the adaptation of the vessels to the relative movement between the epidermis and underlying muscle. This motion occurs through the subcutaneous fatty tissue. The density of vessels coursing towards the skin surface is influenced by the concavity of the skin surface. In concave regions such as the axilla, there tends to be more abundant vascular infiltration, whereas convex regions such as over the deltoid are considered to be relative watershed zones [8, 9].

36.2.4 Application to Surgical Incisions Around the Shoulder

Figure 36.12 shows the different types of skin circulation and their relation to areas of the shoulder. Type A consists of oblique communicating vessels with the facial plexus lying above

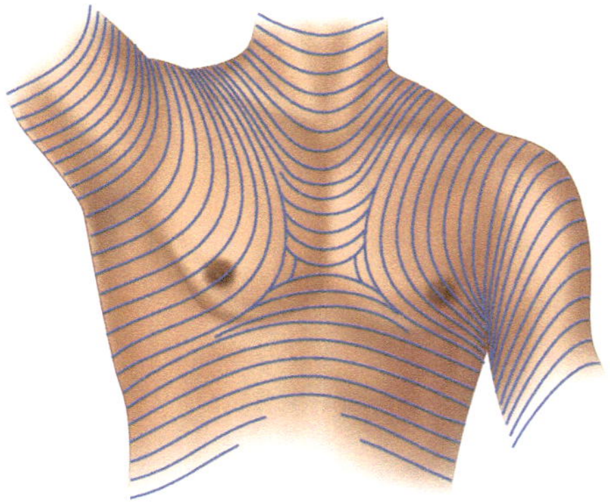

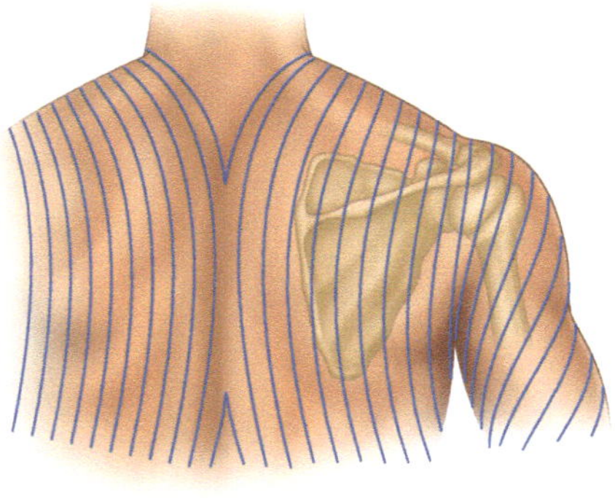

Fig. 36.9 Topographic map of Langer's lines [5]

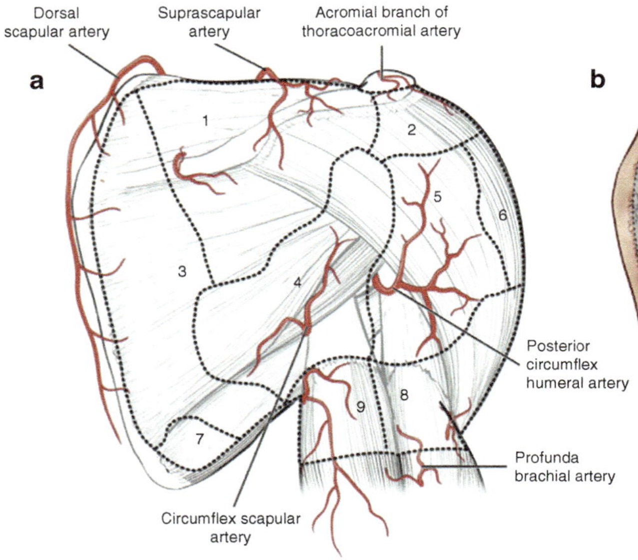

Fig. 36.10 Numbered anatomic territories of the posterior aspect of the shoulder. Posterior shoulder cutaneous vessels and angiosomes. Angiogram of posterior aspect shoulder demonstrating muscular perforators (*red circles*) and septocutaneous perforators (*blue circles*). Deltoid insertion (*green arrow*). Most superior aspect of humerus (*blue arrow*) (Used with permission from Thomas et al. [10], Copyright CCC Republication)

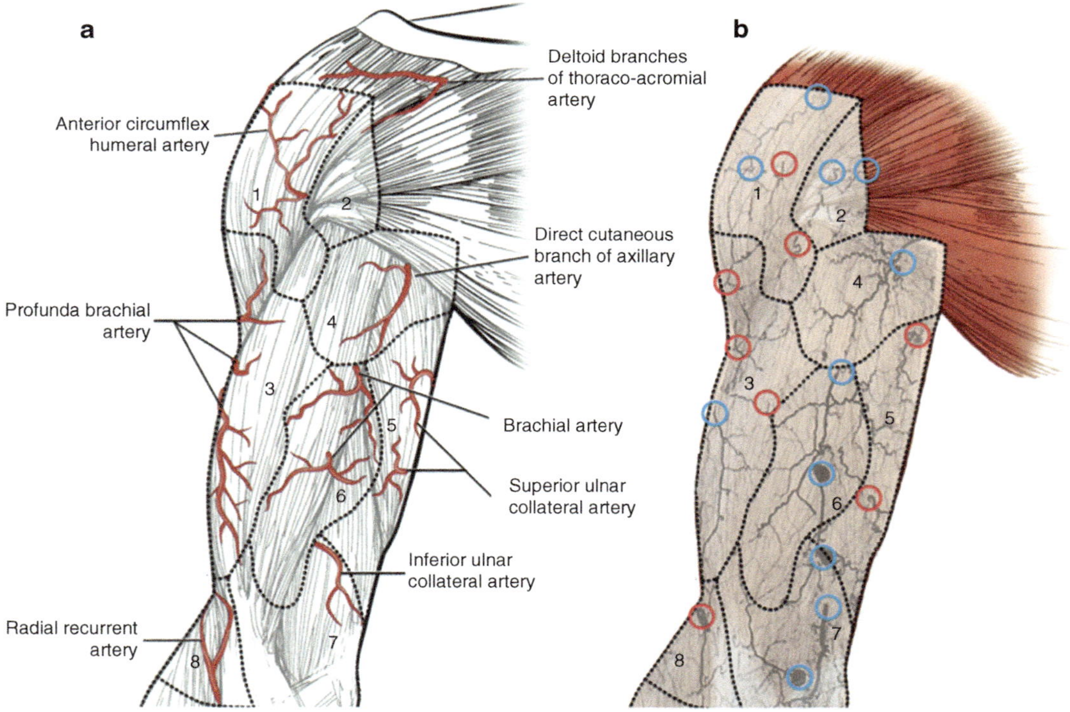

Fig. 36.11 Numbered anatomic territories of the anterior aspect of the shoulder. Anterior cutaneous vessels and angiosomes. (**a**) Angiogram showing the vascular territories. (**b**) *Red circles* are muscular perforators and *blue circles* are septocutaneous perforators. *1* anterior humeral circumflex artery, 2 deltoid branches of thoracoacromial artery, *3* profunda brachii artery, *4* Cutaneous branch of the axillary artery, *5* brachial artery (Used with permission from Thomas et al. [10], Copyright CCC Republication)

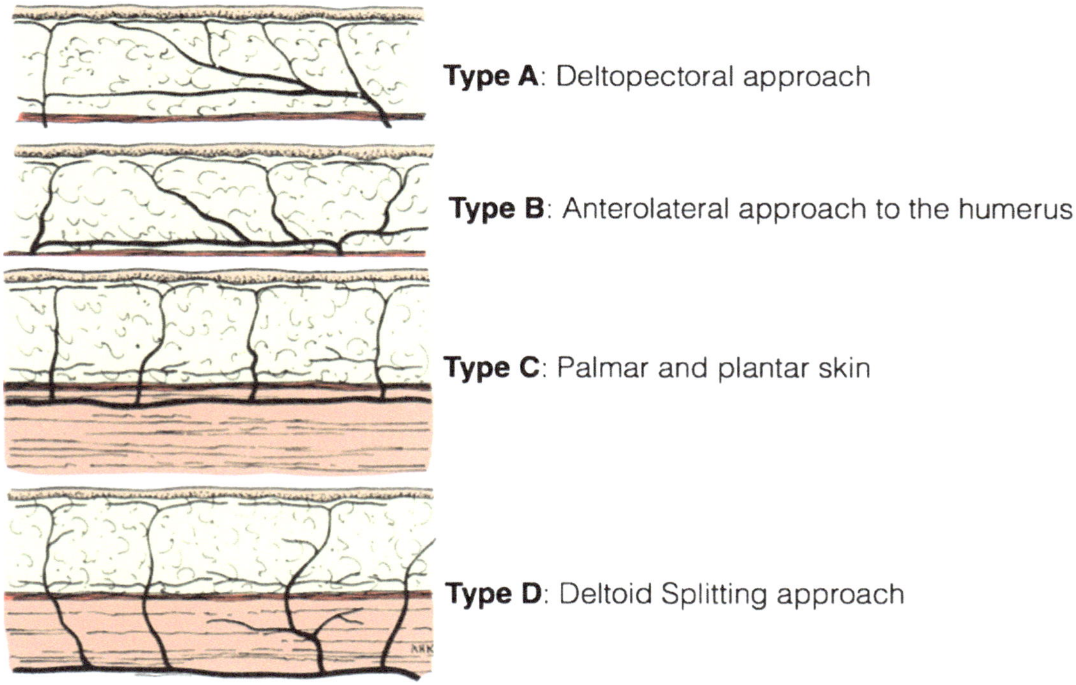

Fig. 36.12 TYPE A: Oblique communicating vessels with the facial plexus lying above the superficial fascia. TYPE B: Branching occurs on the surface of the fascia. TYPE C: Occurs only in the palmar and plantar skin. TYPE D: The fascial plexus lies deep to the fascia [11], (Copyright Elsevier 2009)

the superficial fascia. This represents the mobile skin over the deltopectoral approach. Here it can be seen that raising thick fasciocutaneous flaps offers no advantage in terms of blood supply, as the fascial plexus is superficial to the fascia. In Type B skin, branching occurs on the surface of the fascia because there is more motion between the fascia and the muscle than there is between the fascia and the skin. This means that it is advantageous to reflect full fasciocutaneous flaps as it develops a layer over the underlying muscle. This type of skin is found over the biceps. Type C circulation only occurs in the palmar and plantar skin. Type D circulation is present in the skin over the deltoid. Here the fascial plexus actually lies well deep to the fascia, within the muscle or even beneath the muscle with perforating vessels from the posterior humeral circumflex artery coursing vertically upwards through all the skin layers. This occurs because there is little motion between the epidermis, dermis, subcutaneous tissue and fascia. In this situation, there is no advantage in developing fasciocutaneous flaps, as the fascial vascular plexus lies much deeper [8, 9].

References

1. Burkhead Jr WZ, Scheinberg RR, Box G. Surgical anatomy of the axillary nerve. J Shoulder Elbow Surg. 1992;1(1):31–6. doi:10.1016/S1058-2746(09)80014-1.
2. Hoppenfeld S, deBoer P, Buckley R. Surgical exposures in orthopaedics. Philadelphia: Lippincott Williams & Wilkins; 2012.
3. Aubin GG, Portillo ME, Trampuz A, Corvec S. Propionibacterium acnes, an emerging pathogen: from acne to implant-infections, from phylotype to resistance. Med Mal Infect. 2014;44(6):241–50. doi:10.1016/j.medmal.2014.02.004.
4. Lee MJ, Pottinger PS, Butler-Wu S, Bumgarner RE, Russ SM, Matsen FA. Propionibacterium persists in the skin despite standard surgical preparation. J Bone Joint Surg. 2014;96(17):1447–50. doi:10.2106/JBJS.M.01474.
5. Gibson T. Karl Langer (1819-1887) and his lines. Br J Plast Surg. 1978;31:1–2.

6. Langer K. On the anatomy and physiology of the skin: III. The elasticity of the cutis. Br J Plast Surg. 1978;31(3):185–99.
7. Borges AF. Relaxed skin tension lines (RSTL) versus other skin lines. Plast Reconstr Surg. 1984;73(1): 144–50.
8. Taylor GI, Palmer JH. The vascular territories (angiosomes) of the body: experimental study and clinical applications. Br J Plast Surg. 1987;40(2):113–41.
9. Rockwood Jr CA, Matsen III FA, Wirth MA, Lippitt SB. The shoulder. Philadelphia: Elsevier Health Sciences; 2009.
10. Thomas BP, et al. Chapter 12, vascular supply of the integument of the upper extremity. In: Regional flaps: anatomy and surgical technique/upper extremity.
11. Rockwood CA, Matsen FA. Chapter 2, Bones and joints. In: The shoulder, 4th ed. Lippincott Williams and Wilkins, Philadelphia, PA, USA

Anterior Surgical Approaches to the Shoulder

37

Mark Ross, Kieran Hirpara, Miguel Pinedo, and Vicente Gutierrez

37.1 Introduction

There are several anterior approaches to the shoulder joint, which can be used for shoulder arthroplasty, proximal humeral fracture fixation, rotator cuff repair, soft tissue anterior shoulder stabilisation and bony augmentation of the anterior glenoid.

The *deltopectoral approach* is the workhorse approach and utilises the internervous plane between the deltoid (axillary nerve) and the pectoralis major (medial and lateral pectoral nerves) muscle, and was described in detail by Henry in 1957 [3]. Its principle use is for exposure of the anterior aspect of the glenohumeral joint, and as such can be used for access to the anterior glenoid (superior or inferior), the humeral head and proximal shaft. In addition, the deltopectoral approach allows distal extension for access to the entire humeral shaft. Though all of these procedures utilise the same internervous plane, they vary in the management of the subscapularis tendon and may utilise various releases or extensile measures to improve access to the glenohumeral joint and adjacent structures.

The *anterosuperior approach* splits the deltoid, and gives excellent access to the subacromial space, anterior and superior rotator cuff. With extensile measures it can provide sufficient access for arthroplasty or proximal humeral fracture fixation, but risks denervating the anterior deltoid.

M. Ross, MBBS, FRACS, FAOrthA (✉)
Brisbane Hand and Upper Limb Research Institute, Brisbane, Australia

Orthopaedic Department, Princess Alexandra Hospital, Brisbane, Australia

School of Medicine, The University of Queensland, St Lucia, Australia
e-mail: markross@upperlimb.com; research@upperlimb.com

K. Hirpara, MB, BCh, BAO, MD, FRCS (Orth)
Brisbane Hand and Upper Limb Research Institute, Brisbane, Australia

Orthopaedic Department, Princess Alexandra Hospital, Brisbane, Australia
e-mail: khirpara@gmail.com

M. Pinedo, MD • V. Gutierrez, MD,
Orthopaedic Department, Clínica Las Condes, Santiago, Chile
mpinedo@clinicalascondes.cl

37.2 Patient Positioning

When performing anterior shoulder approaches the patient is usually placed in the beach chair position. This is achieved by placing a patient supine on an operating table, which is then moved into roughly 20–30° of Trendelenburg. The patient's legs are lifted to allow pillows or a positioning wedge to be placed under the thighs and at this stage final caudad positioning is easy to achieve. Finally the upper portion of the table is lifted to the desired angle for surgery. An angle of 30–40° increases venous drainage from the shoulder region to minimise intraoperative

G.I. Bain et al. (eds.), *Normal and Pathological Anatomy of the Shoulder*,
DOI 10.1007/978-3-662-45719-1_37, © ISAKOS 2015

bleeding, and allows blood to drain from the wound to preserve visualisation of the shoulder joint. The head is supported with the cervical spine in relative neutral by a Mayfield support or a gel ring, depending on the operating table used.

Intraoperative bleeding may also be minimised by the use of hypotensive anaesthesia. If there is concern about maintaining cerebral perfusion, cerebral oximeters are available to monitor the blood supply to the brain.

Danger
The desire to minimise bleeding should not take precedence over the preservation of adequate cerebral blood flow.

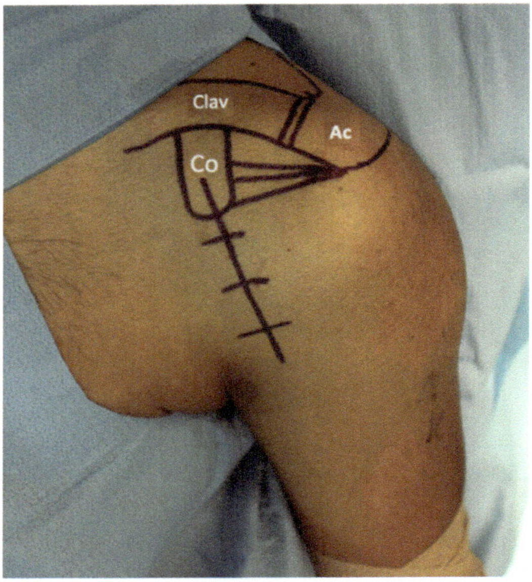

Fig. 37.1 Surface markings for deltopectoral approach. *Ac*, Acromion; *Clav*, clavicle; *Co*, Coracoid process

37.3 Surgical Approaches

37.3.1 The Deltopectoral Approach

37.3.1.1 Surface Anatomy

It is beneficial to mark the bony landmarks of clavicle, acromioclavicular joint, acromion, scapular spine and coracoid prior to planning the skin incision, as this will allow accurate placement of proposed skin incisions. The skin incision used for the deltopectoral approach typically overlies the deltopectoral interval, which may be visualised in the slender patient. If the interval is not easily identifiable then the incision is planned running inferiorly and laterally from the tip of the coracoid process across the corner of the axillary fold and down the upper arm as far as is required. Alternatively, if an extensile exposure is not required, a more cosmetically acceptable vertical incision may be utilised. Typically this incision is 8–10 cm long, 4–5 cm of which lies in the axilla. The skin in this region is mobile enough that with careful mobilisation of wide skin flaps the incision may be retracted to lie over the deltopectoral groove.

37.3.1.2 Superficial Dissection

The key step in the approach to the shoulder is the identification and development of the plane between the medial border of the deltoid muscle and the lateral border pectoralis major muscle, which is usually defined by the presence of the cephalic vein within the deltopectoral groove (Fig. 37.1) [4].

The presence of the vein is usually marked by a longitudinal streak of fat running in the groove. However, this streak of fat may be absent or difficult to define in a particularly slender patient; in the presence of a deep, vestigial or absent vein; or revision surgery. In these situations the interval is easier to define proximally, closer to the clavicle, as the orientation of deltoid and pectoral muscle fibres converge distally. Palpating the underlying coracoid process may also assist in locating the interval.

Tip
Identification of the cephalic vein can be difficult, but it can usually be located in the fat just superficial to the coracoid process in the proximal extent of the deltopectoral interval.

It is easier to dissect between the cephalic vein and the pectoralis major muscle as most of the

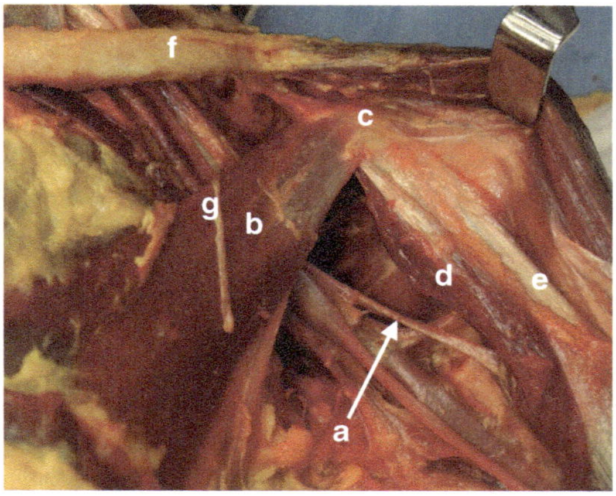

Fig. 37.2 Medial and inferior relations of coracoid process: Anterior view, pectoralis major removed, anterior deltoid retracted laterally. *a* musculocutaneous nerve; *b* pectoralis minor; *c* tip of coracoid process; *d* coracobrachialis; *e* short head of biceps; *f* clavicle; *g* lateral pectoral nerve

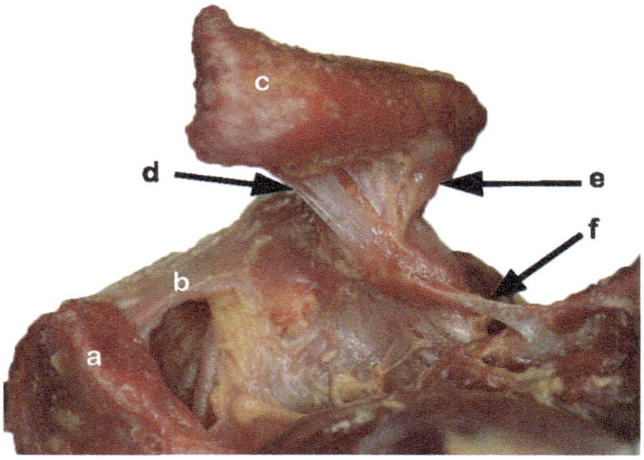

Fig. 37.3 Superior relations to coracoid process: Coracoclavicular ligaments superior view, AC joint capsule divided and clavicle rotated anteriorly. *a* acromion; *b* coracoacromial ligament; *c* lateral clavicle; *d* trapezoid ligament; *e* conoid ligament; *f* transverse scapular ligament across scapular notch

tributaries of the vein run between the lateral aspect of the vein and the deltoid muscle. However, one shortcoming of taking the vein laterally is that proximal extension of the approach may put the cephalic vein at risk where it crosses from lateral to medial in the proximal part of the incision. For that reason if it is anticipated that more extensile exposure is required then it may be preferable to spend a little more time ligating the tributaries on the lateral aspect of the vein and separating the vein from the deltoid and taking it with the pectoralis major.

37.3.1.3 Deep Dissection

Having defined and developed the deltopectoral interval, the principle anatomy of the anterior aspect of the shoulder as it relates to the deltopectoral approach may be defined by two critical bony landmarks. One of these bony landmarks is fixed and the other is mobile and both are of equal importance in developing appropriate surgical exposures in this region.

The fixed bony landmark is the coracoid process. It is encountered in the proximal aspect of the deltopectoral interval as soon as the two muscles are separated. The coracoid process has four key attachments with one from each direction. From the medial aspect is the tendon of pectoralis minor (Fig. 37.2). Superiorly and more toward the root of the coracoid process are the conoid and trapezoid coracoclavicular ligaments (Fig. 37.3). Laterally is the coracoacromial ligament (Fig. 37.4) and inferiorly is the conjoint tendon of the short head of biceps and the coracobrachialis muscle (Fig. 37.2).

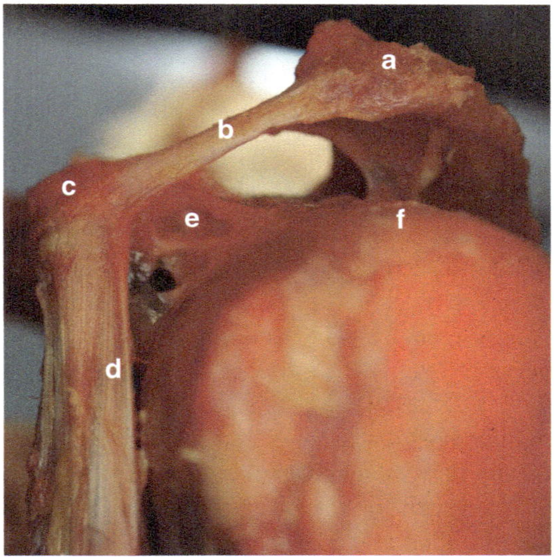

Fig. 37.4 Coracoacromial ligament: Direct lateral view, deltoid removed, clavicle removed. *a* acromion; *b* coracoacromial ligament; *c* coracoid process; *d* conjoint tendon; *e* coracohumeral ligament; *f* supraspinatus

Once the coracoid process has been identified, the clavipectoral fascia is encountered

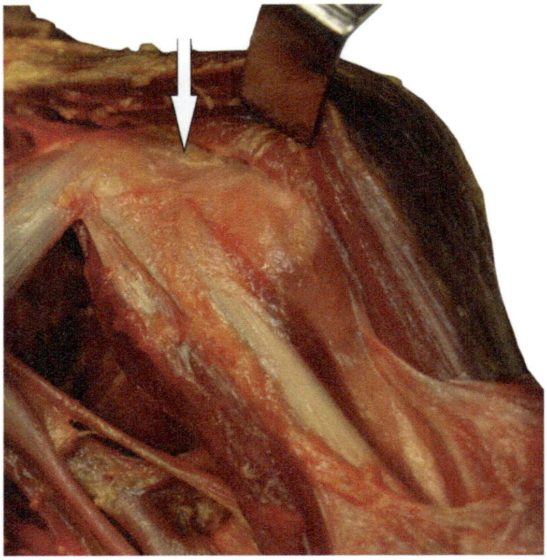

Fig. 37.5 Clavipectoral fascia indicated by *arrow*

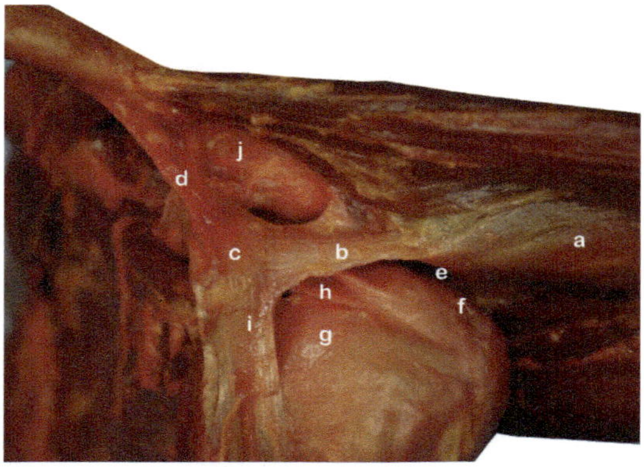

Fig. 37.6 Rotator cuff and coracoacromial ligament: Anterior view, deltoid detached distally and reflected laterally. *a* undersurface of reflected deltoid; *b* coracoacromial ligament; *c* coracoid process; *d* coracoclavicular ligament (conoid); *e* subacromial space; *f* supraspinatus; *g* subscapularis; *h* coracohumeral ligament; *i* conjoint tendon; *j* undersurface lateral clavicle

Danger
The axillary nerve is held to the undersurface of the deltoid by fascia, so is at risk if the deltoid is split.

lateral to the conjoint tendon and inferior to the coracoacromial ligament (Figs. 37.5 and 37.6).

Tip
The safe subdeltoid plane is found by identifying the subacromial space immediately deep to the coracoacromial ligament and sweeping laterally and distally under the deltoid.

Incision of this fascia allows access to the subdeltoid space.

The axillary nerve arises from the posterior cord of the brachial plexus and exits the quadrilateral space posteriorly, primarily to supply the deltoid muscle. The posterior boundaries of the quadrilateral space vary in regard to the anterior boundaries, that is, the superior margin is defined by teres minor instead of subscapularis (Fig. 37.7).

Once the axillary nerve enters the posterior aspect of the shoulder it wraps around the proximal humerus, closely applied to the deep surface of the deltoid muscle. It travels with the posterior circumflex humeral vessels.

There is a constant vessel branching from the circumflex humeral vessels deep to the anterior third of deltoid (Fig. 37.8). The significance of this vessel is twofold. Most surgical procedures via a deltopectoral approach require definition and mobilisation of the subdeltoid space. This vessel is a frequent cause of troublesome bleeding which may be avoided if it is identified and controlled, rather than incidentally disrupted during subdeltoid mobilisation. However, care should be taken not to damage the axillary nerve when controlling this vessel, particularly when using electrocautery. In addition, this vessel can be used as a marker of the level of the axillary nerve on the deep surface of deltoid. The distance from the lateral margin of the acromion to the axillary nerve is variable, but may be as little as 4 cm.

The mobile bony landmark for the deltopectoral approach, which further facilitates identification of critical structures, is the bicipital groove. The location of the groove depends on the rotation of the humerus. However, in general, it is directly anterior when the humerus is in neutral rotation (as

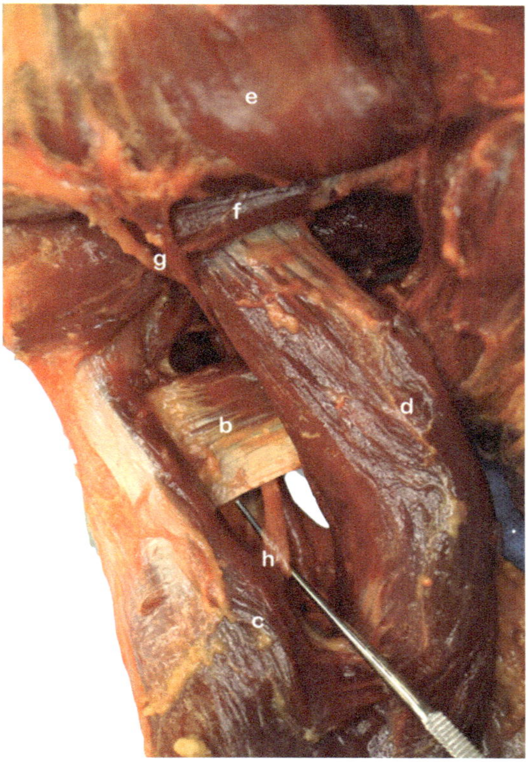

Fig. 37.7 Posterior view of quadrilateral and triangular spaces, teres major and latissimus dorsi: Posterior view, deltoid retracted superiorly. *b* teres major; *c* lateral head of triceps; *d* divided long head of triceps; *e* deltoid reflected superior; *f* teres minor; *g* axillary nerve; *h* radial nerve

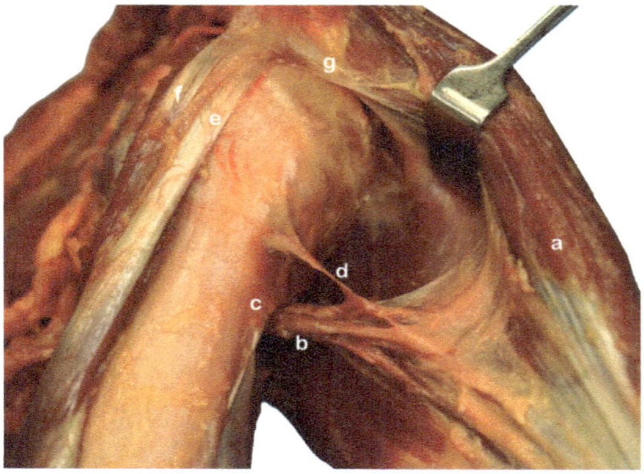

Fig. 37.8 Axillary nerve and deltoid: Lateral view, deltoid detached distally and hinged posteriorly. *a* deltoid; *b* axillary nerve and circumflex humeral vessels; *c* metadiaphyseal junction of humerus; *d* humeral branch from posterior circumflex vessels; *e* long head of biceps; *f* conjoint tendon; *g* coracoacromial ligament

judged by the position of the forearm with the elbow flexed). The bicipital groove and related structures are in a deeper plane than the coracoid process and their identification requires division of the clavipectoral fascia as discussed earlier (Fig. 37.5), particularly lateral to the coracoid and inferior to the coracoacromial ligament.

The groove is occupied by the long head of biceps tendon. The biceps tendon may be traced superiorly to identify the rotator interval between the subscapularis and supraspinatus tendons and is one of the critical intervals for developing exposure to the glenohumeral joint. The medial aspect of the bicipital groove is formed by the lesser tuberosity to which is attached the subscapularis tendon superiorly and subscapularis muscular attachment to the humerus inferiorly (Fig. 37.9a). As the biceps tendon is traced more distally to the shallower portion of the bicipital groove, the tendon is covered by the pectoralis major tendon, which inserts on the lateral lip of the bicipital groove. The pectoralis major tendon insertion is quite complex with crossing over of the fibres such that the costal fibres tend to insert more superiorly and the sternal and clavicular fibres tend to insert more inferiorly. At this level within the floor of the bicipital groove is the insertion of the flat ribbon-like latissimus dorsi tendon, and closely related to the latissimus dorsi insertion and just inferior to the subscapularis muscle insertion is the insertion of the teres major tendon into the medial lip of the bicipital groove. Although closely related and sometimes harvested together as a dual tendon transfer, these tendons are almost completely separate (Fig. 37.9b). There is also a well-defined bursa between the posterior aspect of the latissimus tendon and the medial aspect of the humeral shaft.

If there is a significant internal rotation contracture, such as in osteoarthritis, the pectoralis major insertion may be partially or completely released. This may assist in the dislocation of the joint during arthroplasty. If a complete release is required, then consideration may be given to placing a tagging suture in the pectoralis tendon in order to allow subsequent repair.

37.3.1.4 Clavicular Osteotomy and Deltoid Release

The deltoid origin can run a considerable distance along the clavicle, which may make access

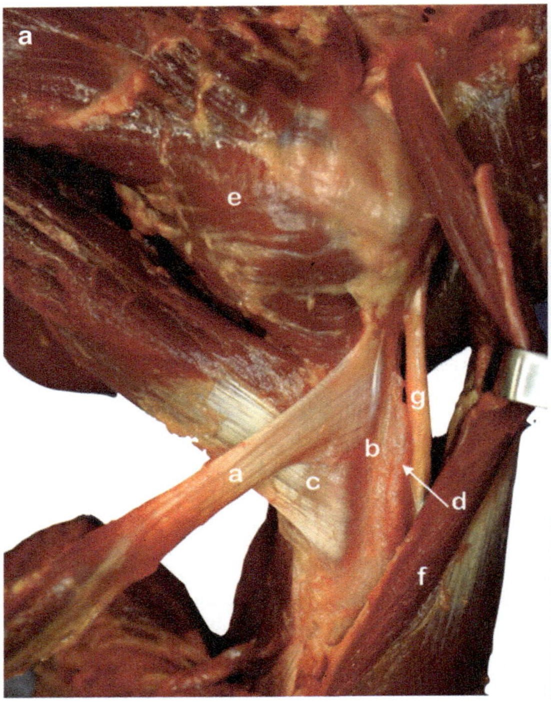

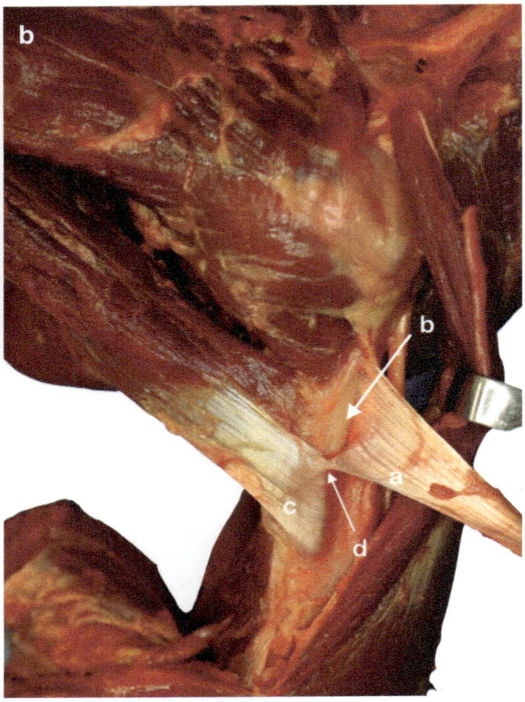

Fig. 37.9 (**a**) Bicipital groove, latissius and teres major: Anterior view, coracobrachialis and short head of biceps retracted laterally. *a* latissimus dorsi tendon; *b* bicipital groove; *c* teres major tendon; *d* divided stump of insertion of pectoralis major; *e* subscapularis; *f* coracobrachialis; *g* long head of biceps displaced laterally out of groove. (**b**) Latissimus and teres major tendons: Anterior view, coracobrachialis and short head of biceps retracted laterally, latissimus dorsi reflected laterally. *a* undersurface of latissimus dorsi tendon; *b* bursa on humeral shaft deep to latissimus tendon; *c* teres major tendon; *d* tenuous connection between latissimus and teres major tendons

to the joint more difficult. The temptation to improve access by aggressive retraction should be avoided, as the retractors can bruise or cut into the anterior deltoid. If access is limited then release of the anterior deltoid can significantly improve access to the joint. This release also allows improved access to the lateral aspect of the proximal humerus, which is of particular value in fracture management, where ideal plate positioning is often compromised by poor access. There are two ways of releasing the anterior deltoid:

1. Clavicular osteotomy
2. Subperiosteal stripping of the deltoid from clavicle

Clavicular Osteotomy

Clavicular osteotomy was described by Redfern in 1989 [9]. The deltopectoral approach is performed, with the skin incision extended proximally over the clavicle to allow the lateral clavicle and the clavicular origin of the deltoid to be exposed. The site of the osteotomy is marked, so that it is the anterior third, extending from the change in curvature of the clavicle to just medial to the acromioclavicular (AC) joint. Therefore all of the deltoid attachment is included, and the AC joint will not be violated. The osteotomy is performed with a narrow blade oscillating saw with irrigation to cool the blade (Fig. 37.10a). The osteotomy is mobilised, with the attached deltoid, allowing the muscle to be reflected laterally. The deltoid raphe can be released, which provides greater exposure of the glenohumeral joint. It is important that the medial corner of the osteotomy is smoothly contoured as opposed to being angular in order to prevent a stress riser that may later lead to a clavicular fracture. It is also important that the lateral extent of the osteotomy does not violate the acromioclavicular

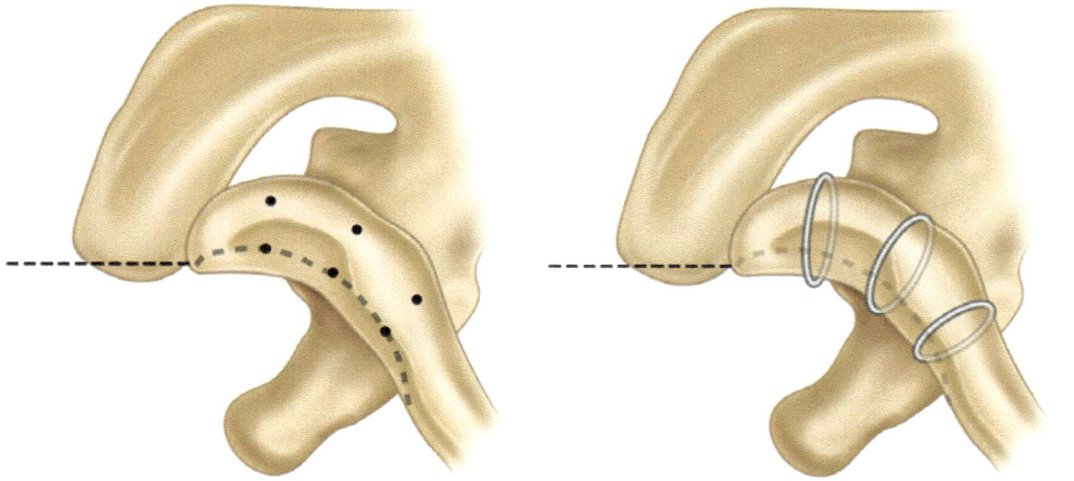

Fig. 37.10 (**a**) Diagramatic representation of transosseous fixation for clavicular osteotomy (Image courtesy of Dr Jeff Hughes, Sydney, Australia). (**b**) Diagramatic representation of cerclage fixation for clavicular osteotomy (Image courtesy of Dr Jeff Hughes)

joint. At the completion of the procedure, the osteotomy is secured with multiple 1-ethibond cerclage (Fig. 37.10a) or transosseous sutures (Fig. 37.10b).

Subperiosteal Deltoid Release

In 1918, Thompson [12] described transverse sectioning of the anterior deltoid from the clavicle and acromion as part of the approach to the shoulder. In 2004, Gill [2] reported excellent results with this technique, with no deltoid detachments and good anterior deltoid function in 81 shoulder arthroplasties. After identifying the deltopectoral interval, the anterior deltoid is dissected directly off the clavicle taking care to lift all the subdeltoid tissue with the muscle flap from the underlying coracoacromial ligament. This release is brought as far lateral as is required, and can release the deltoid as far as the anterior corner of the acromion. At the completion of the procedure, the deltoid flap is secured with multiple 1-ethibond cerclage or transosseous sutures.

37.3.1.5 Subscapularis Tendon

To gain access to the anterior glenohumeral joint, the surgeon needs to appreciate the finer points of the subscapularis tendon. Preoperative assessment with history, examination and imaging is critical to understand the integrity of the subscapularis. It is important to appreciate pre-existing pathology, such as

1. A previous shoulder dislocation may have a subscapularis tear or lesser tuberosity fracture.
2. Degenerative arthritis may have a contracture of the tendon and joint capsule.
3. A failed shoulder arthroplasty, the subscapularis may be deficient as it failed to heal.

The subscapularis is exposed by reflecting the loose clavipectoral fascia and retracting the conjoint tendon medially. The management options for the subscapularis are

1. Horizontal splitting tenotomy of the muscle/tendon unit
2. Vertical tenotomy
   (a) Mid-tendon
   (b) Off the bone
3. Partial 'L shaped' tenotomy (superior tendon)
4. Lesser tuberosity osteotomy.
5. Tendon retracting (tendon sparing approach)

Mobilisation of subscapularis is advocated; however, it must be noted that the upper and lower subscapular nerves (from posterior cord) enter the anterior muscle surface, medial to the rim of the glenoid. However, the nerve supply is

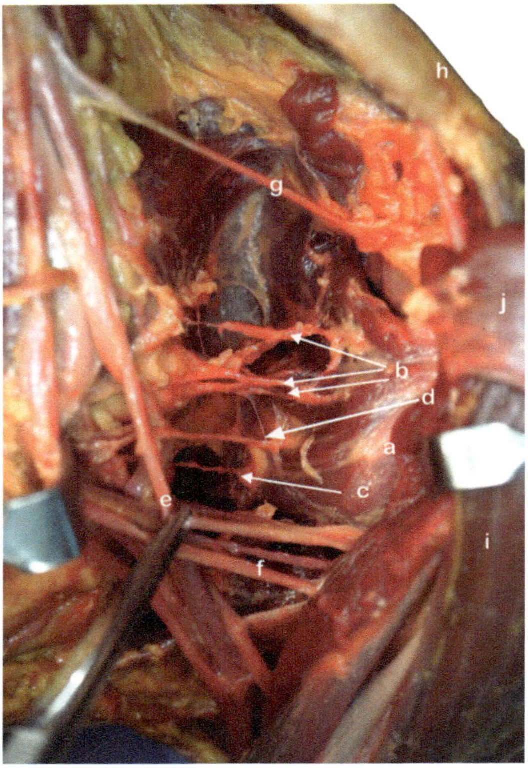

Fig. 37.11 Nerves to subscapularis: Anterior view, pectoralis major removed, pectoralis minor/conjoint tendon and deltoid reflected laterally. *a* anterior surface of subscapularis muscle belly; *b* branches of upper subscapular nerve; *c* branch from axillary nerve to subscapularis; *d* lower subscapular nerve; *e* axillary nerve (retracted); *f* musculocutaneous nerve; *g* suprascapular nerve; *h* clavicle; *i* deltoid; *j* reflected pectoralis minor

variable, including from the axillary nerve (Fig. 37.11).

Most releases are performed along the superior aspect of the subscapularis tendon and the posterior surface of the tendon where it is intimately related to the anterior joint capsule. The plane between the muscle and the anterior scapula may safely be developed well medial to the glenoid and provided valuable intraoperative orientation of glenoid version. The interval between anterior capsule and the subscapularis tendon is more easily developed medially where the two structures diverge at the level of the glenoid labrum.

Danger
The nerves to subscapularis are at risk with dissection anterior to the subscapularis muscle and medial to the conjoint tendon.

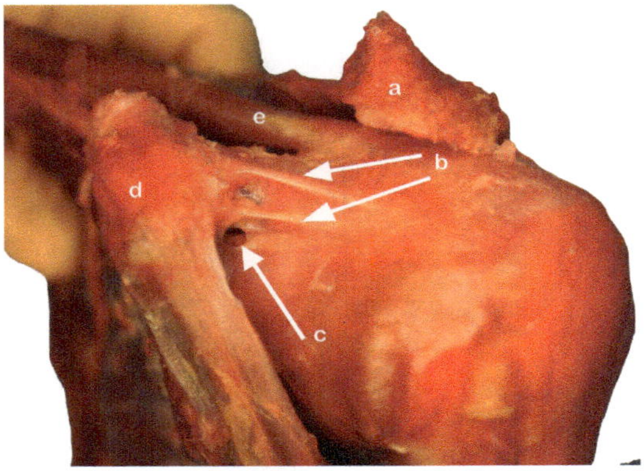

Fig. 37.12 Coracohumeral ligament: Anterior view of rotator interval, clavicle and coracoacromial ligament removed. *a* anterior acromion; *b* coracohumeral ligament; *c* superior margin of subscapularis visible through small hole in capsular reflection; *d* tip of coracoid process with attached conjoint tendon (short head of biceps and coracobrachialis; e superior margin of supraspinatus tendon

Internal rotation contracture is frequently associated with thickening of the coracohumeral ligament, a capsular reflection within the rotator interval (Fig. 37.12). Release of this structure aids in subscapularis mobilisation and glenoid exposure.

It is also important to recognise the close proximity to the axillary nerve to the glenoid and subscapularis muscle as it runs from the brachial plexus through the quadrilateral space [1]. The thickness of subscapularis separating the axillary nerve from the glenohumeral joint capsule is variable, and decreases as the nerve runs posteriorly and closer to the glenoid.

Exposure of the joint itself may be performed in several ways. For full exposure of the joint such as is required in arthroplasty a full release of the tendon is usually performed either by tenotomy, ten-

Danger
The axillary nerve may be in direct contact with the inferior capsule (Fig. 37.13)

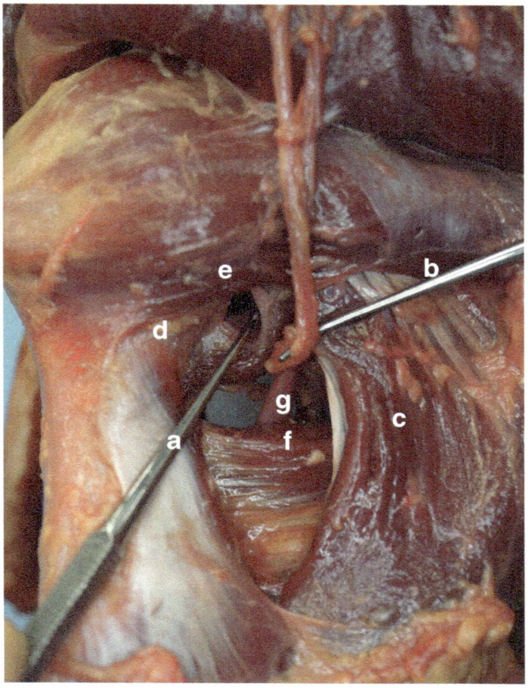

Fig. 37.13 Axillary nerve and inferior capsule: Posterior view of quadrilateral space with deltoid reflected cephalad. *a* probe in postero-inferior capsular recess; *b* probe on axillary nerve; *c* long head of triceps; *d* medial neck of humerus; *e* teres minor; *f* teres major; *g* radial nerve

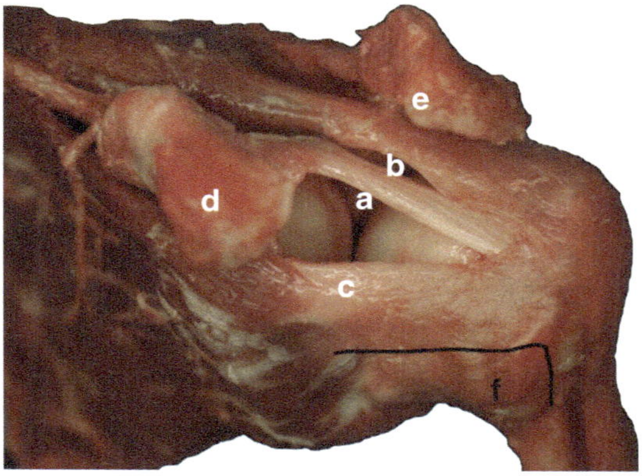

Fig. 37.14 Rotator interval and LHB: Anterosuperior view. *a* long head of biceps; *b* supraspinatus; *c* subscapularis; *d* coracoid process; *e* anterior acromion; *f* the incision in the subscapularis tendon for tendon sparing approach

don peel or lesser tuberosity osteotomy. However, there have been several descriptions in the recent literature of subscapularis sparing approaches that utilise the rotator interval [6] or tenotomy of the inferior 50 % of the suscapularis, with mobilisation of the inferior leaf of tendon by horizontally splitting the subscapularis muscle ([11], Fig. 37.14). These tendon-sparing approaches ensure preservation of subscapularis function, at the expense of limiting exposure to the joint. Because of this they can be technically demanding and difficult to recommend in patients with significant deformity or extensive osteophytes.

Osteotomy of the lesser tuberosity is the most effective way of managing the subscapularis tendon, with a stronger repair [5, 13] and improved clinical outcome [8]. It is important to define the superior and inferior margins of the subscapularis tendon. Definition of the subscapularis tendon is aided by identification of the rotator interval, which is simplest to perform by tracing the biceps tendon superiorly from the bicipital groove. The biceps tendon runs through the rotator interval to insert onto the superior aspect of the glenoid, and in tracing it superiorly into the joint the subscapularis tendon is defined medially to the tendon (Fig. 37.14).

The lesser tuberosity osteotomy starts in the floor of the bicipital groove and exits at the medial end of the subscapularis insertion. The plane of the osteotomy may be guided by the insertion of a ring spike through the rotator interval under the tendon, providing a target to aim at when using an osteotome or oscillating saw. Ideally the lesser tuberosity fragment should be between 3 and 6 mm thick, in order to preserve the cortical margins of proximal humerus for support of the implanted prosthesis. As the lesser tuberosity is mobilised, subperiosteal dissection can be extended inferiorly onto the humerus, and a continuous sheet of tissue can be raised from the medial humerus. This forms an osteoperiosteal flap which may contain, from proximal to distal: lesser tuberosity with subscapularis tendon, glenohumeral joint capsule, subscapularis muscle, teres major tendon and latissimus dorsi tendon. This stripping of the medial aspect of the proximal humerus improves external rotation as the joint is dislocated for resection of the humeral head, and eases humeral retraction when accessing the glenoid.

37.3.1.6 Coracoid Osteotomy

If increased medial access to the subscapularis tendon is required then release of the conjoined tendon by coracoid osteotomy may be performed [9]. It is important to predrill and tap the coracoid prior to osteotomy, as the coracoid tip has a tendency to fracture if drilling and tapping is attempted after the osteotomy. Though osteotomy of the coracoid improves subscapularis access, it increases risk of injury to the musculocutaneous nerve, which enters coracobrachialis on its medial aspect, and is subject to increased stretch when the protective tension in the conjoined tendon is released (Fig. 37.2).

> **Tip**
> It is wise to tag the subscapularis tendon prior to release or osteotomy with a long suture, as it may retract and be difficult to find.

37.3.1.7 Final Glenoid Exposure

After mobilisation of subscapularis, glenoid exposure may be facilitated by capsular releases. The coracohumeral ligament release should be completed. Anterior capsulectomy may extend from superior to inferior as an inverted triangle; however, it is safer to convert to a capsulotomy immediately adjacent to the anteroinferior labrum beyond 5 o'clock where the axillary nerve is in close proximity (Fig. 37.16).

> **Danger**
> If coracoid osteotomy is performed, the close relationship between the suprascapular nerve and the base of the coracoid should be considered (Fig. 37.15)

Exposure of the inferior glenoid is sometimes required. This is particularly in the case of reverse total shoulder arthroplasty or metal-backed anatomic arthroplasty where inferior screw placement in the scapula/glenoid neck is required. In

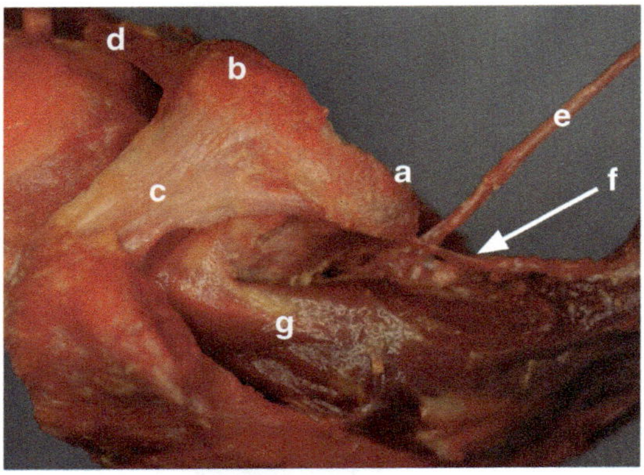

Fig. 37.15 Coracoacromial ligament and suprascapular nerve: Superior view, clavicle removed. *a* base of coracoid process; *b* tip of coracoid process; *c* coracoacromial ligament; *d* short head of biceps; *e* suprascapular nerve; *f* transverse scapular ligament; *g* supraspinatus muscle belly

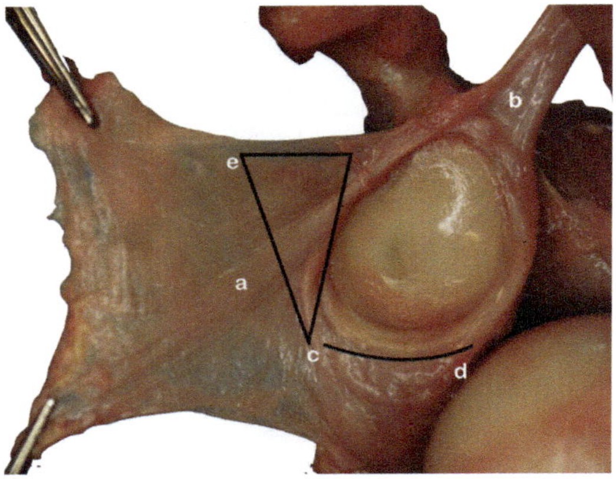

Fig. 37.16 Glenoid , labrum and attachments: Antero-inferior capsulolabral structures, capsule detached from humeral insertion and humeral head retracted posterior. Inverted triangle denotes anterior capsulectomy for glenoid exposure in arthroplasty, curved line indicates capsulotomy where axillary nerve is at risk. *a* middle glenohumeral ligament; *b* long head of biceps; *c* anterior band of inferior glenohumeral ligament; *d* posterior band of inferior glenohumeral ligament; *e* superior glenohumeral ligament

> **Tip**
> The plane between the subscapularis tendon and the anterior capsule is more readily identified medially at the level of the glenoid labrum.

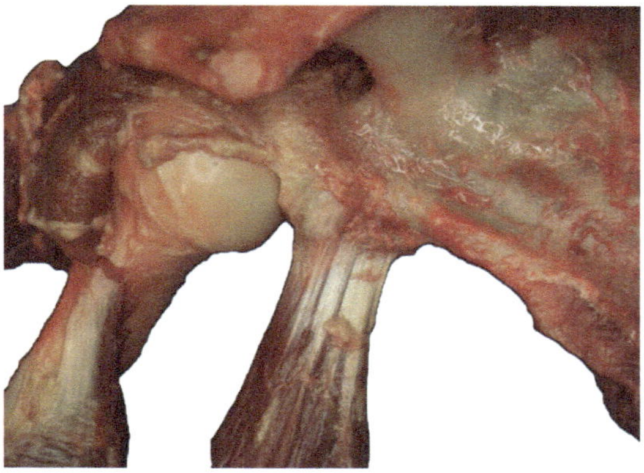

Fig. 37.17 Long head triceps: Posterior view, deltoid and infraspinatus removed. *a* glenoid neck and infraglenoid tubercle; *b* long head of triceps

addition, adequate clearance of the inferior glenoid region in reverse arthroplasty prevents soft tissue or bony impingement which may contribute to instability or notching. After inferior capsulotomy the long head of triceps may be released to facilitate inferior clearance. It has an extensive tendinous origin from the infraglenoid tubercle extending at least 2 cm medial to the labrum and may be safely released (Fig. 37.17).

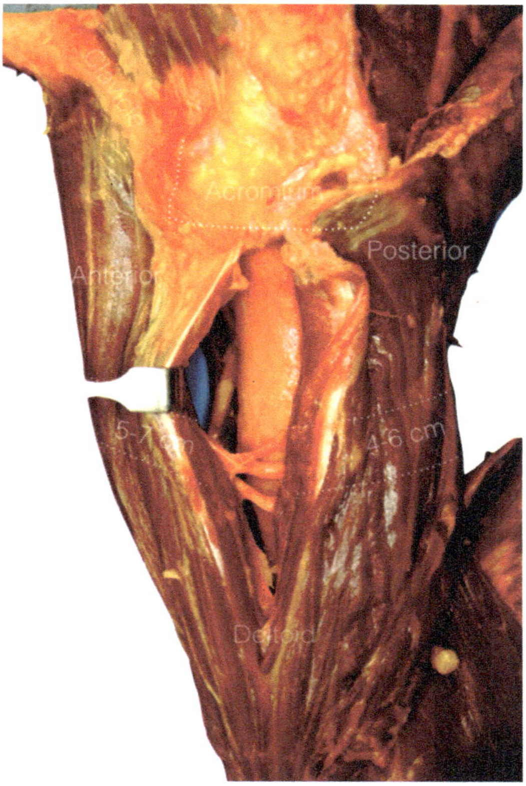

Fig. 37.18 Deltoid splitting approach: Note the axillary nerve is at risk 4–6 cm distal to the acromion. The nerve is palpable just distal to the subacromial bursa, on the undersurface of deltoid. Once identified, the distal muscle may be split to expose the humeral shaft

37.3.2 Deltoid Splitting Approach

Access to the glenohumeral joint may also be achieved by splitting the deltoid, although this can place the axillary nerve at significant risk as it runs on the undersurface of deltoid from posterior to anterior. (Fig. 37.16) The axillary nerve lying approximately 6 cm distal to the acromion where it is at risk in surgical approaches that split the deltoid (Fig. 37.18). The axillary nerve needs to be protected during deltoid splitting approaches, and can usually be felt via the proximal muscle split running on the undersurface of deltoid. Once identified the distal muscle split may be performed for access to the humeral shaft.

First described by Mackenzie in 1993 [7] and modified by Robinson in 2007 [10] the anterior superior deltoid splitting approach provides excellent access to the rotator cuff, and good access to the proximal humerus for management of simple fractures or shoulder arthroplasty. Either a longitudinal incision centred over the anterior corner of the acromion or a shoulder strap incision with elevation of a distally based cutaneous flap may be used for access. The deltoid is then split proximally to allow access of a finger into the subdeltoid bursa. The axillary nerve is palpated as a horizontal band of tissue running on the under surface of the deltoid, and then the split is continued below this point to create a superior and inferior window. A haemostat is then passed from the inferior split to the superior split and a vessel loop is passed around the intact strip of deltoid containing the axillary nerve. This is particularly important for treatment of proximal humeral fractures, where access to the shaft screws of the plate is difficult to perform though the proximal split without propagating the deltoid split inferiorly, or subjecting the axillary nerve to excessive stretch. The location of the axillary nerve and a demonstration of the split can be seen in Fig. 37.18.

Conclusion

There are many surgical alternatives for approaching the glenohumeral joint. Surgeons need to have a clear understanding of what anatomical structures need to be visualised. It is important to select the surgical approach that provides the best exposure. A healthy respect for the adjacent nerves, attention to detail regarding the surgical releases and final fixation are key to obtaining a good surgical outcome.

Acknowledgements Special thanks to Jim and Jodie Kelly, Medical Engineering and Research Facility, Queensland University of Technology, for their generous assistance with cadaveric specimens.

References

1. Galley IJ, Watts AC, Bain GI. The anatomic relationship of the axillary artery and vein to the clavicle: a cadaveric study. J Shoulder Elbow Surg. 2009;18:21–5.
2. Gill DRJ, Cofield RH, Rowland C. The anteromedial approach for shoulder arthroplasty: the importance of the anterior deltoid. J Shoulder Elbow Surg. 2004;13:532–7.
3. Henry AK. Extensile exposure. 2nd ed. Baltimore: Churchill Livingstone; 1957.
4. Hoppenfeld S, deBoer P, Buckley R. Surgical exposures in orthopaedics: the anatomic approach. Philadelphia: Wolters Kluwer Lippincott William; 2009.
5. Krishnan SG, Stewart DG, Reineck JR, Lin KC, Buzzell JE, Burkhead WZ. Subscapularis repair after shoulder arthroplasty: biomechanical and clinical validation of a novel technique. J Shoulder Elbow Surg. 2009;18:184–92.
6. Lafosse L, Schnaser E, Haag M, Gobezie R. Primary total shoulder arthroplasty performed entirely thru the rotator interval: technique and minimum two-year outcomes. J Shoulder Elbow Surg. 2009;18:864–73.
7. Mackenzie DB. The antero-superior exposure for total shoulder replacement. Orthop Traumatol. 1993;2:71–7.
8. Qureshi S, Hsiao A, Klug RA, Lee E, Braman J, Flatow EL. Subscapularis function after total shoulder replacement: results with lesser tuberosity osteotomy. J Shoulder Elbow Surg. 2008;17(1):68–72.
9. Redfern TR, Wallace WA, Beddow FH. Clavicular osteotomy in shoulder arthroplasty. Inter Orthop. 1989;13:61–3.
10. Robinson CM, Khan L, Akhtar A, Whittaker R. The extended deltoid-splitting approach to the proximal humerus. J Orthop Trauma. 2007;21:657–62.
11. Savoie FH, Charles R, Casselton J, O'Brien MJ, Hurt JA. The subscapularis-sparing approach in humeral head replacement. J Shoulder Elbow Surg. 2014;23:1–7.
12. Thompson JA. Anatomic methods of approach in operations on the long bones of the extremities. Ann Surg. 1918;68:309–29.
13. Van den Berghe GR, et al. Biomechanical evaluation of three surgical techniques for subscapularis repair. J Shoulder Elbow Surg. 2008;17:156–61.

Posterior Surgical Approaches to the Shoulder

38

Giovanni Di Giacomo, Andrea De Vita, and Alberto Costantini

38.1 Introduction

38.1.1 Anatomy of the Posterior Shoulder

The posterior region of the shoulder presents anatomical structures, which if understood in in-depth detail, allow access without risking damage to the posterior scapular region or to the glenohumeral joint.

The skin covering the bone and muscle/tendon structures makes it possible to highlight osseous landmarks rendering the posterior approach safe. The posterior deltoid muscle fascia and the posterior rotator cuff muscles fascia lie beneath the skin.

The posterior deltoid originates from the edge of the spine of the scapula and runs laterally and inferiorly toward its humeral insertion covering the posterior rotator cuff muscles.

The posterior scapular fossa is occupied by the infraspinatus muscle, which originates below the spine of the scapula and inserts in the greater tubercle of the humerus. The teres minor muscle, distal with respect to the infraspinatus muscle, arises from the dorsal surface of the axillary border of the scapula and runs laterally inserting in the lower section of the greater tubercle.

The gap between the infraspinatus muscle and the teres minor muscle is difficult to locate medially and can be more easily identified laterally.

The articular capsule and the glenohumeral joint are underneath the tendons of the infraspinatus and the teres minor muscles. The posterior capsule is reinforced from posterior band of the inferior glenohumeral ligament.

The posterior portion of the scapula presents a concavity containing the infraspinatus muscle and is separated superiorly from the supraspinatous fossa by the scapular spine, whose posterior section gives origin to the deltoid muscle. The scapular neck separates the glenoid surface (laterally) from the scapular body (medially).

G. Di Giacomo, MD
Chief of Orthopedic Shoulder,
Department of Concordia, Hospital for Special Surgery,
Via delle Sette Chiese, 90, Rome 00145, Italy
e-mail: devita.andrea@gmail.com

A. De Vita, MD (✉) • A. Costantini, MD
Orthopedic Shoulder, Department of Concordia,
Hospital for Special Surgery, Via delle Sette Chiese, 90,
Rome 00145, Italy
e-mail: devita.andrea@gmail.com

38.1.2 Internervous Plane

The internervous plane lies between the infraspinatus muscle and teres minor muscle. Innervation of the teres minor and the deltoid muscle comes from the axillary nerve while innervation of the infraspinatus muscle comes from the suprascapular nerve.

G.I. Bain et al. (eds.), *Normal and Pathological Anatomy of the Shoulder*,
DOI 10.1007/978-3-662-45719-1_38, © ISAKOS 2015

38.1.3 Posterior Approach: Indications

The posterior approach offers access to the posterior and inferior aspects of the shoulder joint [1]. It is rarely necessary, but can be used in the following instances:

1. Repairs in cases of recurrent posterior dislocation or subluxation of the shoulder [2, 3]
2. Glenoid osteotomy [4]
3. Biopsy and excision of tumours
4. Removal of loose bodies in the posterior recess of the shoulder
5. Treatment of fractures of the scapula neck, particularly those in association with fractured clavicles (floating shoulder)
6. Treatment of posterior fracture dislocations of the proximal humerus

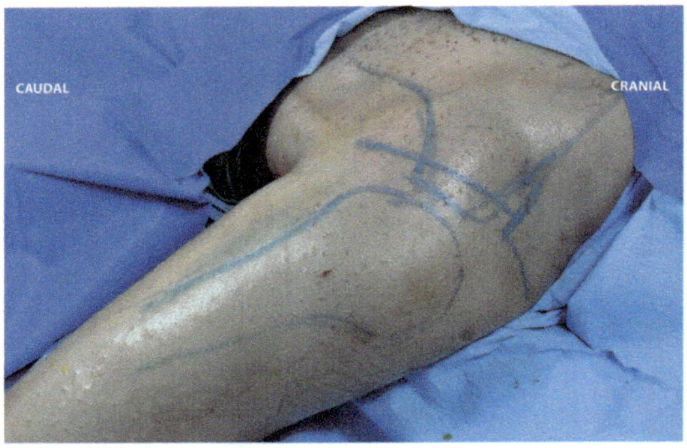

Fig. 38.1 Prone position of the patient. The arm is free and is useful to paint the landmarks on the skin before the incision

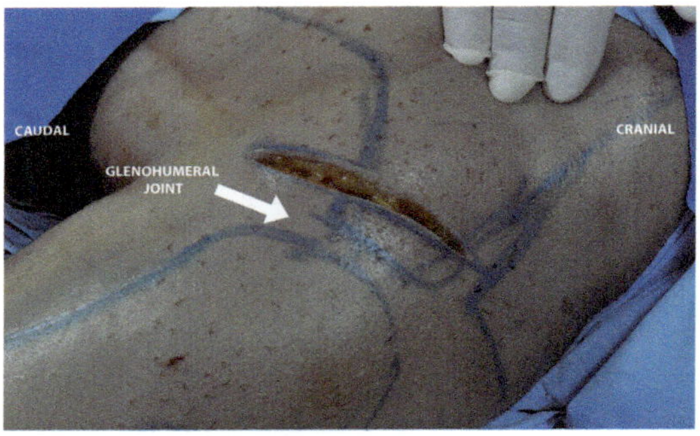

Fig. 38.2 Skin incision. A vertical incision starting from the scapula spine to axillary fold, 2 cm medial to posterior corner of acromion

38.2 Description

38.2.1 Set Up

The lateral or prone position can be used for this approach. For lateral position, place the patient on the edge of the operating table with the affected side uppermost. Drape him/her to allow independent movement of the arm. Stand behind the patient, taking care that the ear is not accidentally folded under the head. For prone position, put a pillow under the thorax for rotation of the patient's head and for freeing only the arm and moving it during surgery (Fig. 38.1).

The landmarks for the incisions are the acromion and the spine of the scapula that together form one continuous arch. The spine of the scapula extends obliquely across the upper four-fifths of the dorsum of the scapula and ends in a flat, smooth triangle at the medial border of the scapula. It is very easy to palpate.

38.2.2 Skin Incision

Either a horizontal or vertical incision can be chosen for posterior shoulder approach.

For horizontal approach, make a linear incision along the entire length of the scapular spine, extending to the posterior corner of the acromion. However, for vertical approach, make a vertical incision starting from the scapula spine to the axillary fold, 2 cm medial to the posterior corner of the acromion (Fig. 38.2). The central point of this incision is the same landmark as the posterior arthroscopic portal. Because the skin incision does not lie in the line of fibers of the deltoid muscle, the skin is elevated medially and laterally, dissecting between the deep fascia and overlying fat. A vertical incision at the lateral end of the scapular spine is more cosmetic but provides poor exposure of the joint.

Tip
To improve posterior glenohumeral joint exposure, it is useful to avoid a too lateral incision.

38.2.3 Superficial Muscle Dissection

The fascia is opened along the spine of the scapula and the muscles are exposed. The posterior part of the deltoid is now visible. The posterior deltoid muscle originates from the medial region of the scapula spine to the posterior corner of the acromion. The lateral deltoid originates from the lateral part of the acromion.

Superficial surgical dissection starts with the detachment of the posterior and small part of lateral deltoid from the scapula (Fig. 38.3). It is useful to leave some fibers of the deltoid tendon on the bone for simple reattachment at the end of the procedure. It is not simple to find the plane between the deltoid and the underlying infraspinatus muscle.

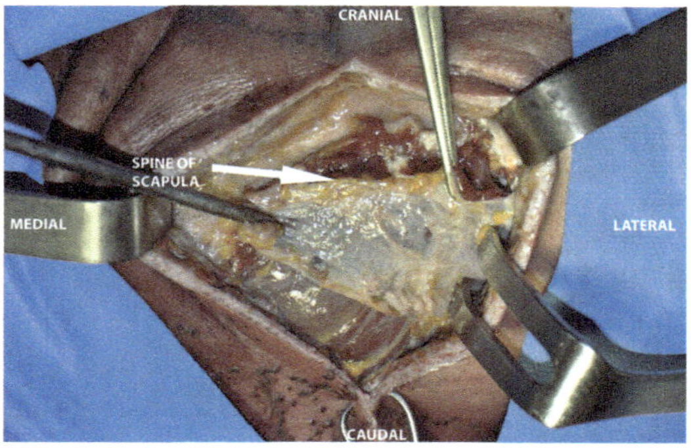

Fig. 38.3 The posterior deltoid is exposed. Detachment of the posterior deltoid from the scapula spine

Tip
To aid the dissection, it is important to identify the muscular plane that starts from the lateral end of the spine of the scapula. Once it has been found, it is not difficult to develop it if the deltoid is retracted inferiorly and the infraspinatus is exposed. Sometimes it is useful for the partial detachment of the posterior deltoid that is retracted laterally with Hohmann retractor placed between the deltoid and the humeral head (Fig. 38.4).

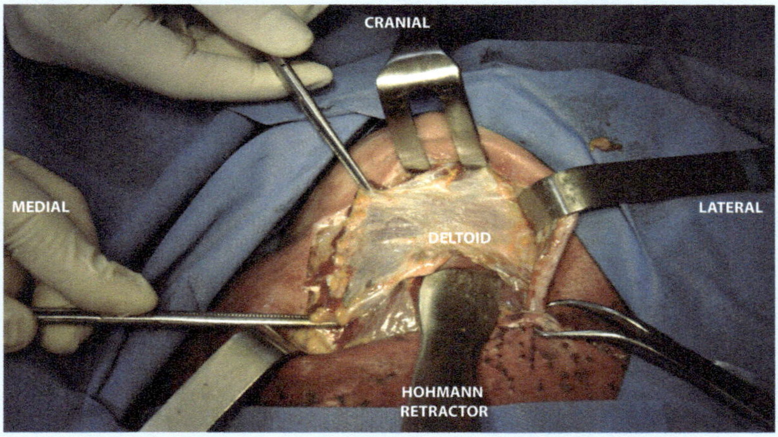

Fig. 38.4 The Hohmann retractor between the deltoid and humeral head

38.2.4 Deep Muscle Dissection

Deep dissection identifies the plane between the infraspinatus and teres minor muscles, and develops it by blunt dissection, using a finger (Fig. 38.5). This important plane can be difficult to identify. Sometimes the infraspinatus "raphe" can create confusion during dissection. The "muscular raphe" is a small fibrous band between the upper half and lower half of the infraspinatus muscle that creates a depression on the muscle. It is very dangerous to extend the dissection to the inferior border of the teres minor muscle (quadrangular space) due to the presence of the axillary nerve and the posterior circumflex humeral artery. The axillary nerve runs through the quadrangular space beneath the teres minor. Since a dissection carried out inferior to the teres minor can damage the axillary nerve (see Fig. 37.7), it is critical to identify the muscular interval between the infraspinatus and teres minor muscles and to stay within that interval. The suprascapular nerve passes around the base of the spine of the scapula as it runs from the supraspinatus fossa to the infraspinatus fossa. It is the nerve supply for both the supraspinatus and infraspinatus muscles. Gentle retraction of the infraspinatus muscle can avoid nerve apraxia after the procedure that at times may result from stretching the nerve around the unyielding lateral edge of the scapular spine.

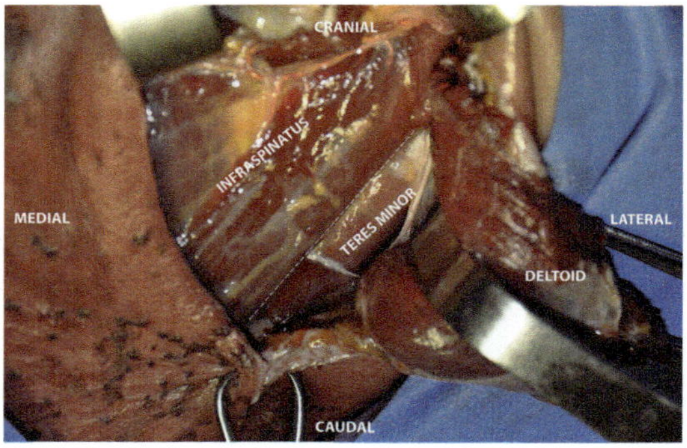

Fig. 38.5 Infraspinatus muscle and teres minor muscle are exposed. Deep dissection between infraspinatus and teres minor muscles lead to the capsular plane. *Dotted line* indicates the dissection plane

Danger
Damage to the posterior circumflex humeral artery in quadrangular space can create serious bleeding that is difficult to control.

Tip
Progressive internal rotation of the arm during dissection of the interval between the infraspinatus and teres minor muscle and gentle inferior retraction of teres minor can avoid damage to the axillary nerve.

38.2.5 Capsular Incision

After muscle retraction, the posteroinferior corner of the shoulder joint capsule is exposed (Fig. 38.6a, b). To explore the joint, incise it longitudinally, close to the edge of the scapula, or horizontally between the upper half and lower half of the capsule. We prefer horizontal incision in line with the interval dissection of muscles to expose the posterior glenohumeral joint (Fig. 38.7).

Trick
To gain better access to the posterior aspect of the shoulder joint, it can be useful to detach the infraspinatus 1 cm from its insertion onto the greater tuberosity. Retract the muscle medially, taking great care not to damage the suprascapular nerve, which enters the undersurface of the muscle just below the spine of the scapula. If the skin incision is too latral, it is possible to have more difficulty accessing the posterior glenoid neck.

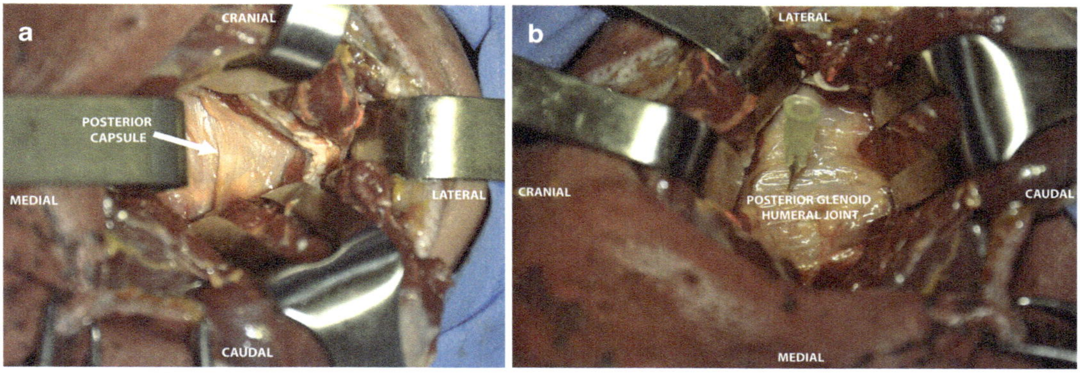

Fig. 38.6 (**a**) Posterior capsule is exposed. Progressive internal rotation of the arm is useful during dissection of the interval between the infraspintus and teres minor muscles. Gentle inferior retraction avoids nerve damage. (**b**) The needle is useful to indicate the posterior glenohumeral joint

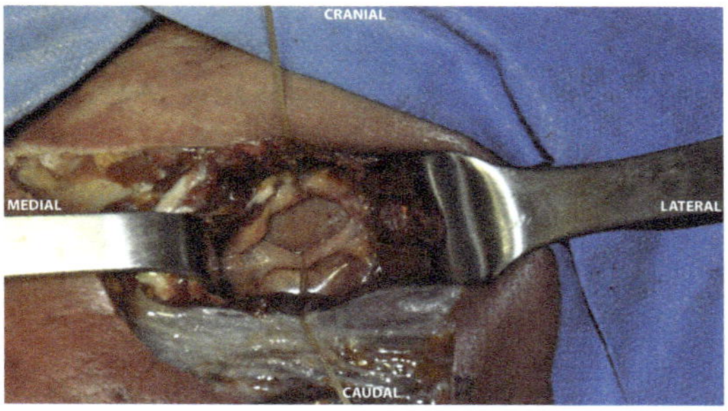

Fig. 38.7 Posterior glenohumeral joint after horizontal capsular incision

38.3 Pathoanatomy

38.3.1 Effect of Trauma

Traumatic events can create various problems in the shoulder: fractures, instability, muscle and tendon lesions. Glenohumeral instability is a relatively common condition affecting 2 % of the general population [5]. Anteroinferior glenohumeral joint instability is the most common problem after trauma. Posterior glenohumeral joint instability is far less common than anterior instability, accounting for approximately 2–10 % of all cases of shoulder instability [6]. Posterior instability of the shoulder and fracture of the glenoid sometime need arthroscopic or open treatment. Posterior approach to the shoulder is used in open procedure. Regarding posterior instability, the trauma may be macro or micro. Macrotrauma is a single event [7, 8] of an axial load to the arm while the shoulder is flexed. Microtrauma is a repetitive injury, such as straight-arm pass blocking in football or bench pressing [9, 10]. Trauma may injure the rotator cuff, labrum, and/or glenoid, contributing to posterior subluxation. Isolated sectioning of the posterior rotator cuff musculature does not increase posterior translation [11, 12].

38.3.2 Posterior Shoulder Instability and Shoulder Lesions

Posterior dislocation of the shoulder is a rare but clinically and radiologically well-defined entity. It is of interest because most are missed during initial emergency room examinations [13, 14]. There is confusion between posterior subluxation and dislocation. Posterior dislocation is an acute entity associated with trauma and with an impression defect of the humeral head (reverse Hill-Sachs lesion) [15]. Misdiagnosis can create chronic posterior dislocation. Treatment is determined by the size of the defect and the duration of the dislocation. Recurrent posterior subluxation is a distinct and separate entity, which is often not associated with trauma and one that requires completely different management, such as nonoperative treatment or posterior reconstruction of the shoulder. This condition is usually caused by epileptic fit, electric shock or trauma, such as a fall on the outstretched arm.

Many classifications of posterior shoulder instability have been described, including degree, direction, mechanism of injury, and volition [16–18]. Hawkins and McCormack [19] discussed acute posterior dislocations, chronic (fixed/locked) posterior dislocations, and recurrent posterior subluxation. Of these, recurrent posterior subluxation is the most common.

Posterior shoulder instability can be further divided dimensionally: unidirectional (posterior), bidirectional (posteroinferior) [20], and multidirectional (posterior, inferior, and anterior) [20, 22, 24]. Bidirectional and multidirectional instability are much more common than unidirectional instability.

Reverse Bankart lesion or Bony Bankart lesion, are often observed in patients with posterior instability caused by trauma.

Multidirectional instability may have an inciting traumatic event, but it is due to preexisting global capsular laxity [24]. Excessive glenoid retroversion may be a predisposing factor to posterior shoulder instability [12, 25, 26]. Last, the posterior shoulder instability can be presented on the basis of volition. A subset of patients habitually subluxate their shoulders using patterns of muscle activity [27]. These patients are different because of their abnormal psychological urge to subluxate their shoulders. It is important to identify these patients because their treatment plans differ significantly.

For a diagnosis of instability, it is important to have a complete medical history of the patient and to carry out a thorough physical examination. Patients with posterior shoulder instability primarily complain of aching pain and weakness along the posterior joint line. These symptoms intensify with the arm in 90° forward flexion, adduction, and internal rotation. The most important tests for posterior instability are the "Posterior Draw Test," the "Kim Test" for posteroinferior instability, and the "Jerk Test."

Imaging is important support for diagnosis. X-ray is useful for understanding the position of the humeral head and to quantify humeral head defect. AP view and lateral view on scapular plane describe the position of humeral head. Stryker Notch view helps to quantify the reverse Hill-Sachs lesion. West Point axillary view optimally detects osseous Bankart defects on the posterior glenoid rim. MRI is much more useful in evaluating soft tissue pathology. Computed tomography is useful in defining the size and orientation of a reverse Hill-Sachs lesion, a reverse Bankart lesion, posterior glenoid bone loss, or bony Bankart lesion, and for determining the version and morphology of glenoid.

38.4 Surgical Procedures

It is important to know that a conservative physical therapy protocol is usually indicated for the treatment of posterior instability of the shoulder [7, 8, 28–31]. The optimum duration of nonsurgical management is not based on scientific evidence, but at least 6 months of therapy is common before consideration for surgical repair [3, 21, 25].

Several exceptions exist, including evidence of bony pathology of the glenohumeral joint and traumatic instability with stable posterior dislocation of the shoulder.

Surgery is contraindicated for habitual instability because of psychological problems leading to a high failure rate [32].

38.4.1 Capsular Procedures

The capsular procedures are indicated when conservative treatment fails. Posterior capsule redundancy is the most common pathologic lesion for patients with isolated unidirectional posterior instability without a true labral tear, arthroscopic or open posterior capsular plication is recommended [33–36].

A posterior-inferior capsular shift should be performed for posterior-inferior subluxation with no anterior component and an intact rotator interval. The posterior aspect of the glenohumeral capsule is opened via a T-shaped incision creating medial caudal and cranial flaps, which are then drawn together and overlapped to reduce the volume of the glenohumeral capsule [23]. When a reverse Bankart lesion (posterior labral tear) coexists with a redundant posterior capsule, the labrum must be repaired along with the posterior capsular plication. These procedures are more successful.

38.4.2 Bone Procedures

Bone procedures are indicated for patients presenting with failed capsular plication, glenoid hypoplasia, increased glenoid retroversion (version >20°), or osteochondral fracture of the glenoid cavity [4, 8, 37], technically demanding and less successful than soft tissue procedures [29], with high complication rates [8]. The osseous procedures include the posterior bone block, posterior opening wedge osteotomy of the glenoid neck, or posterior glenoid osteochondral allografting [5, 38]. A posterior glenoid osteotomy increased glenoid antiversion. The bone block helps to reconstruct the bone loss of the posterior glenoid cavity [4, 8, 37].

However, in the case of reverse Hill-Sachs lesion, bone procedure with anterior approach is indicated. If the reverse Hill-Sachs lesion is inferior of 25 % of the humeral head volume, conservative treatment is suggested. If the Hill-Sachs lesion is between 25 and 50 %, open transfer of the lesser tuberosity with the subscapularis tendon in the defect is recommended [8]. McLaughlin [39] described the transfer of subscapularis for a defect of between 20 and 40 %; the tendon of subscapularis is secured into the defect through drill holes in the bone. Hughes and Neer [19] modified this technique by osteotomizing the lesser tuberosity with the attached subscapularis. With this method, filling the defect of the humeral head can prevent new posterior dislocation of the shoulder. If the bone defect is more than 50 %, this procedure is not recommended and arthroplasty is indicated.

Conclusion

The posterior approach to the shoulder is an uncommon surgical procedure because it is reserved only for posterior instability of the shoulder or posterior fractures of the scapula; conditions whose incidence is very low with regard to shoulder injuries. However, precise and exact knowledge of the anatomical structures of the posterior aspect of the shoulder is essential when adopting this approach, in the event that it is required, in order to avoid damage to nerves and vessels.

References

1. Rowe R, Vee LBK. A posterior approach to the shoulder joint. J Bone Joint Surg. 1944;26:580.
2. Hindenach JCR. Recurrent posterior dislocation of the shoulder. J Bone Joint Surg Am. 1947;29(3):582–6.
3. Boyd HB, Sisk TD. Recurrent posterior dislocation of the shoulder. J Bone Joint Surg Am. 1972;54:779.
4. Scott Jr DJ. Treatment of recurrent posterior dislocation of the shoulder by glenoplasty: report of three cases. J Bone Joint Surg Am. 1967;49:471.
5. Ahlgren SA, Hedlund T, Nistor L. Idiopathic posterior instability of the shoulder joint: results of operation with posterior bone graft. Acta Orthop Scand. 1978; 49(6):600–3.
6. Bradley JP, Tejwani SG. Arthroscopic management of posterior instability. Orthop Clin North Am. 2010;41: 339–56.
7. Fronek J, Warren RF, Bowen M. Posterior subluxation of the glenohumeral joint. J Bone Joint Surg Am. 1989;71(2):205–16.
8. Hawkins RJ, Koppert G, Johnston G. Recurrent posterior instability (subluxation) of the shoulder. J Bone Joint Surg Am. 1984;66(2):169–74.
9. Pollock RG, Bigliani LU. Recurrent posterior shoulder instability: diagnosis and treatment. Clin Orthop Relat Res. 1993;291:85–96.
10. Schwartz E, Warren RF, O'Brien SJ, Fronek J. Posterior shoulder instability. Orthop Clin North Am. 1987;18(3):409–19.
11. Ovesen J, Nielsen S. Anterior and posterior shoulder instability: a cadaver study. Acta Orthop Scand. 1986;57(4):324–7.
12. Sanders TG, Morrison WB, Miller MD. Imaging techniques for the evaluation of glenohumeral instability. Am J Sports Med. 2000;28(3):414–34.
13. Dubousset J. Luxation posterièures de l'epaule. Rev Chir Orthop. 1967;53:65–85.
14. Michos IB, Michaelides DP. Reduction of missed posterior dislocation of the shoulder: report of 2 cases, 1 of them bilateral. Acta Orthop Scand. 1993;64: 599–600.

15. Petersen SA. Posterior shoulder instability. Orthop Clin North Am. 2000;31:263–74.
16. Beall Jr MS, Diefenbach G, Allen A. Electromyographic biofeedback in the treatment of voluntary posterior instability of the shoulder. Am J Sports Med. 1987; 15(2):175–8.
17. Kuhn JE. A new classification system for shoulder instability. Br J Sports Med. 2010;44(5):341–6.
18. Takwale VJ, Calvert P, Rattue H. Involuntary positional instability of the shoulder in adolescents and young adults: is there any benefit from treatment? J Bone Joint Surg (Br). 2000;82(5):719–23.
19. Hawkins RJ, McCormack RG. Posterior shoulder instability. Orthopedics. 1988;11(1):101–7.
20. Antoniou J, Duckworth DT, Harryman 2nd DT. Capsulolabral augmentation for the management of posteroinferior instability of the shoulder. J Bone Joint Surg Am. 2000;82(9):1220–30.
21. Gartsman GM, Roddey TS, Hammerman SM. Arthroscopic treatment of bidirectional glenohumeral instability: two- to five-year follow-up. J Shoulder Elbow Surg. 2001;10(1):28–36.
22. Miniaci A, McBirnie J. Thermal capsular shrinkage for treatment of multidirectional instability of the shoulder. J Bone Joint Surg Am. 2003;85(12):2283–7.
23. Neer 2nd CS, Foster CR. Inferior capsular shift for involuntary inferior and multidirectional instability of the shoulder: a preliminary report. J Bone Joint Surg Am. 1980;62(6):897–908.
24. Sekiya JK, Cole BJ, Cohen SB. Arthroscopic treatment of multidirectional shoulder instability. In: Surgical Techniques of the Shoulder, Elbow, and Knee in Sports Medicine. Portland: WB Saunders Co; 2008. p. 816.
25. Edelson JG. Localized glenoid hypoplasia: an anatomic variation of possible clinical significance. Clin Orthop Relat Res. 1995;321:189–95.
26. Wirth MA, Lyons FR, Rockwood Jr CA. Hypoplasia of the glenoid: a review of sixteen patients. J Bone Joint Surg Am. 1993;75(8):1175–84.
27. Rowe CR, Pierce DS, Clark JG. Voluntary dislocation of the shoulder: a preliminary report on a clinical, electromyographic, and psychiatric study of twenty-six patients. J Bone Joint Surg Am. 1973;55(3):445–60.
28. Burkhead Jr WZ, Rockwood Jr CA. Treatment of instability of the shoulder with an exercise program. J Bone Joint Surg Am. 1992;74(6):890–6.
29. Norwood LA, Terry GC. Shoulder posterior subluxation. Am J Sports Med. 1984;12(1):25–30.
30. Pagnani MJ, Warren RF. Stabilizers of the glenohumeral joint. J Shoulder Elbow Surg. 1994;3(3):173–90.
31. Tibone JE, Bradley JP. The treatment of posterior subluxation in athletes. Clin Orthop Relat Res. 1993;291: 124–37.
32. Hurley JA, Anderson TE, Dear W, Andrish JT, Bergfeld JA, Weiker GG. Posterior shoulder instability: surgical versus conservative results with evaluation of glenoid version. Am J Sports Med. 1992;20(4):396–400.
33. Bottoni CR, Franks BR, Moore JH, DeBerardino TM, Taylor DC, Arciero RA. Operative stabilization of posterior shoulder instability. Am J Sports Med. 2005;33(7):996–1002.
34. Kim SH, Ha KI, Park JH, et al. Arthroscopic posterior labral repair and capsular shift for traumatic unidirectional recurrent posterior subluxation of the shoulder. J Bone Joint Surg Am. 2003;85(8):1479–87.
35. Williams 3rd RJ, Strickland S, Cohen M, Altchek DW, Warren RF. Arthroscopic repair for traumatic posterior shoulder instability. Am J Sports Med. 2003;31(2):203–9.
36. Wolf EM, Eakin CL. Arthroscopic capsular plication for posterior shoulder instability. Arthroscopy. 1998; 14(2):153–63.
37. Wilkinson JA, Thomas WG. Glenoid osteotomy for recurrent posterior dislocation of the shoulder: in proceedings of the British Orthopaedic Association [abstract]. J Bone Joint Surg (Br). 1985;67:496.
38. Jones V. Recurrent posterior dislocation of the shoulder: report of a case treated by posterior bone block. J Bone Joint Surg (Br). 1958;40(2):203–7.
39. Hawkins RJ, Krishnan SG, Karas SG, Noonan TJ, Horan MP. Electrothermal arthroscopic shoulder capsulorrhaphy: a minimum 2-year follow-up. Am J Sports Med. 2007;35(9):1484–8.

Part VIII

The Functional Shoulder

The Functional Shoulder

39

Gregory Ian Bain, Joideep Phadnis,
and David H. Sonnabend

39.1 Evolution

39.1.1 Quadrupeds

Most mammals are quadrupeds and use the glenohumeral joint as a weight-bearing joint, from which the forelimb moves as a flexion/extension pendulum [1]. Most of these animals do not have a clavicle as there is no need for their forelimb to be suspended away from the body. The presence of a clavicle in these mammals would likely slow them down.

39.1.2 Bipeds

Evolution of the shoulder has been driven by the development of the orthograde posture with anatomic changes required to accommodate the demands of a mobile, non-weight-bearing joint and allow the arm to become a brachiating, prehensile limb [2–4] (see Fig. 39.1). The detailed evolution of the shoulder is covered in Chap. 1. In bipeds, the lower limbs bear all the weight, leaving the upper limbs free to perform many other functions. In brachiating mammals such as humans, the clavicle is strong, the scapula wide and the coracoid enlarged. These features allow the arm to be positioned away from the body but also allow it to bear load in these positions.

There are significant biomechanical changes in the bipedal shoulder, which hangs from the axial skeleton but has the ability to elevate the humerus and the arm. The biped's shoulder girdle is required to function in compression but also in tension and shear with shoulder elevation and rotation on a daily basis.

39.1.3 Acromion and Deltoid

The acromion is much larger in humans as it serves as the main attachment for the deltoid, and by increasing its size, it effectively lateralises the deltoid. The deltoid is the muscle that has enlarged the most with evolution, indicating the importance of the deltoid to shoulder function in humans. With a larger deltoid, the tensile forces on the acromion have increased, and the coracoacromial ligament has evolved to transfer tension from the acromion to the coracoid process.

Electronic supplementary material The online version of this chapter (doi:10.1007/978-3-662-45719-1_39) contains supplementary material, which is available to authorized users.

G.I. Bain, PhD, MBBS, FRACS, FA(Ortho)A (✉)
Department of Orthopaedic Surgery and Trauma, Flinders Medical Centre, Flinders University, Adelaide, SA, Australia
e-mail: greg@gregbain.com.au

J. Phadnis, FRCS (Tr&Orth)
Department of Orthopaedic Surgery, Flinders University, Adelaide, SA, Australia
e-mail: joideep@doctors.org.uk

D.H. Sonnabend, MD, MBBS, BSc(Med), FRACS
Department of Orthopaedic Surgery, University of Sydney, Sydney, NSW, Australia

G.I. Bain et al. (eds.), *Normal and Pathological Anatomy of the Shoulder*,
DOI 10.1007/978-3-662-45719-1_39, © ISAKOS 2015

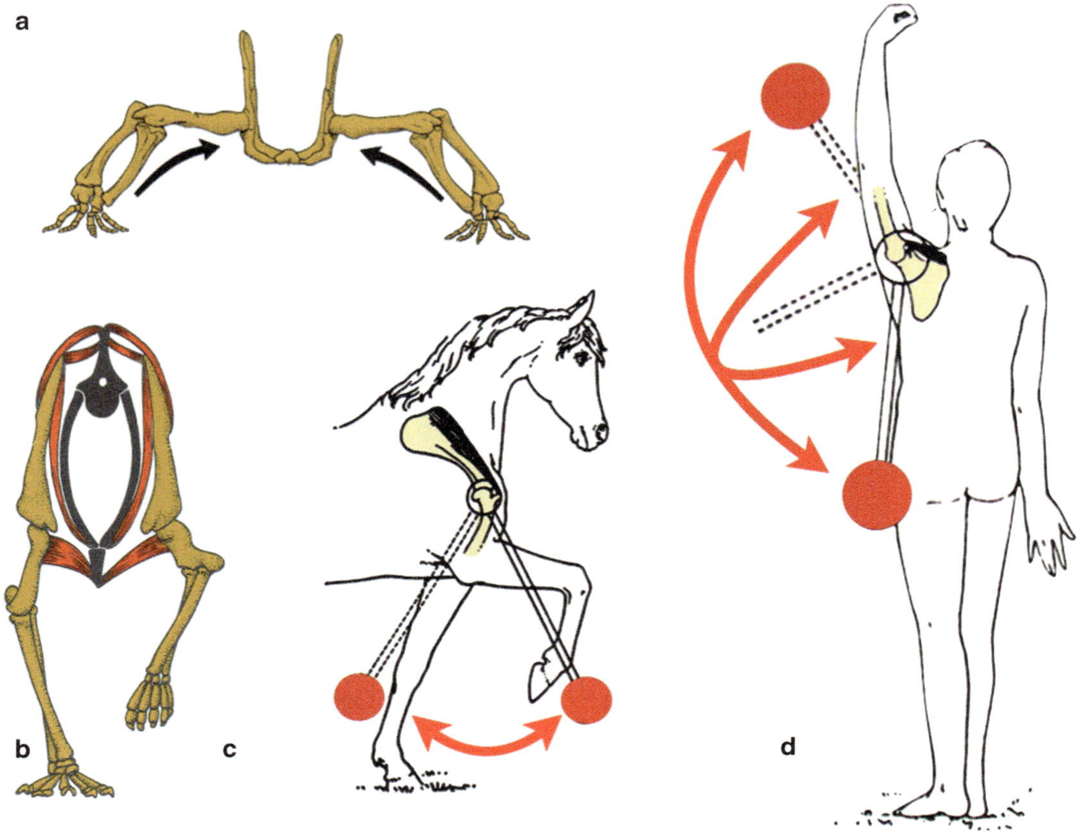

Fig. 39.1 Evolution of the shoulder girdle. (**a**) Reptiles have a sprawling posture with the forces directed medially. (**b**) In terrestrial mammals the limbs are brought under the body, and the forces are directed ventrally. Note the extensive muscle sling that stabilises the scapula and shoulder girdle. (**c**) Running quadrupeds have a forelimb that swings like a pendulum, and but do not have a clavicle which would restrict motion. (**d**) The human shoulder can swing as a pendulum and also circumduct. (**a**, **b** Modified with permission from Kardong [2]; **c**, **d** modified with permission from Professor David Sonnabend)

39.1.4 Rotator Cuff

The supraspinatus remains virtually unchanged with evolution. However, in humans, the infraspinatus is larger and has a more oblique vector making it a strong head depressor in addition to being an external rotator. In most animals, the rotator cuffs are separate muscles that attach to the humeral neck. However, in orthograde animals, there is a common rotator cuff tendon insertion [5]. This provides superior dynamic humeral head stability and humeral rotation in various positions of abduction and flexion.

39.2 The "Shoulder Crane"

The mechanical workings of the arm are analogous to a crane (Fig. 39.2). It has many working parts, which enable the arm to be elevated, so that functional above-head activities can be performed.

39.2.1 Base, Outrigger and Spinal Tower

The *base* (pelvis) and its outriggers (legs) provide stability for the entire mechanism. The

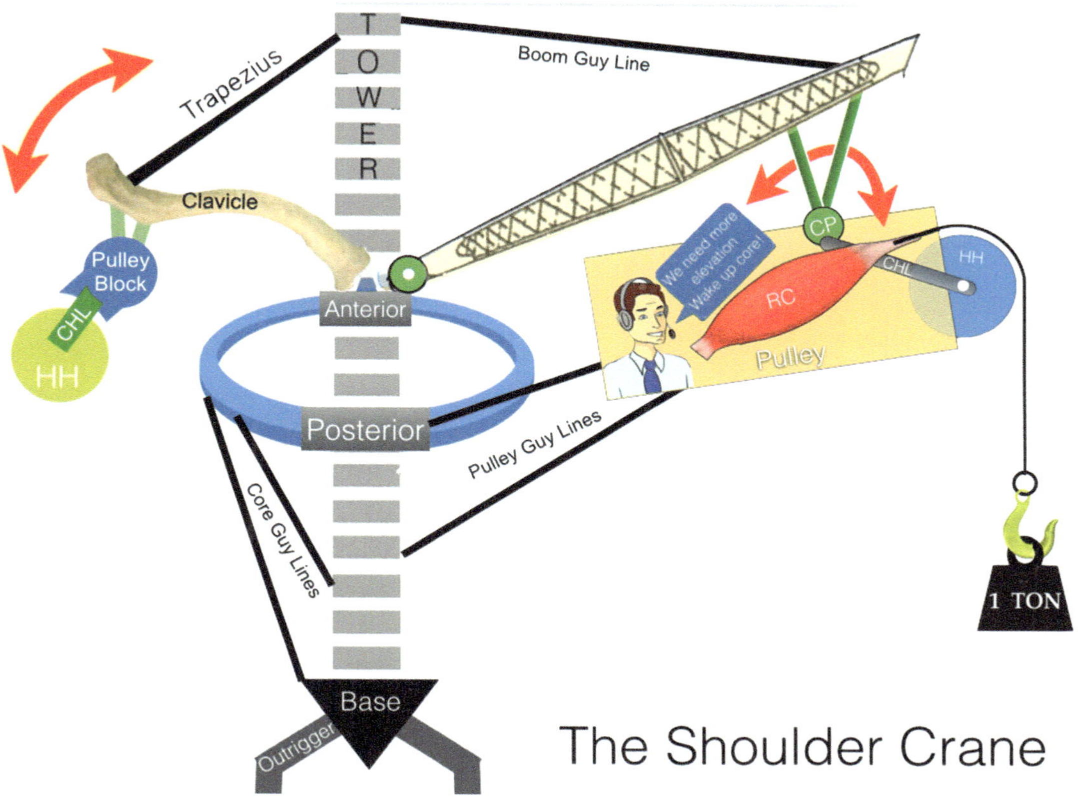

Fig. 39.2 The shoulder crane. (**a**) The crane is built on the pelvic base with legs for outriggers to provide stability and mobility. There is an articulated spinal tower with a thoracic platform on which the crane is housed. The clavicular boom articulates with the *anterior* platform, at the sternoclavicular joint and is elevated by the trapezius from the *posterior* tower. (**b**) The "suspensory cascade" extends from the outer clavicle to the humeral head and includes the clavicle/coraco-clavicular ligaments/coracoid process/coraco-humeral ligament/humeral head. (**c**) The *scapula is a pulley*, strategically positioned in the middle of the cascade of suspension, between the two ligamentous complexes. It swivels on the coraco-clavicular ligaments, to allow the scapula to accommodate to the shape of the "scapular track" and the position of the humeral head. (**d**) The *scapula* position is controlled by the multiple powerful peri-scapular muscles, that control the orientation of the glenoid and scapular body, for shoulder stability and power (Copyright Dr. Gregory Bain)

articulated spinal tower (axial skeleton) extends from the pelvic base to the skull.

39.2.2 Elevated Thoracic Platform

The thoracic cage is an *elevated platform* midway up the axial tower. On top of this platform, the entire upper limb is secured and mobilised.

The platform and articulated tower are stabilised by the "core" muscles, which control and stabilise the relative position of the platform and the axial skeleton. As outriggers, the base and tower are all articulated, making core muscular control vitally important for any upper limb activity. We all initially focus on the shoulder musculature when assessing the shoulder, but the core muscles provide the stability on which the entire thorax and upper limb must counterbalance to perform functional activities.

39.2.3 Clavicular Boom

The clavicle is the boom (or jib) of the crane, which elevates and lateralises the point of suspension away from its hinge point. The hinge is the sternoclavicular joint at the anterior aspect of the elevated platform. There is a shock absorber (articular disc) at each end of the clavicle.

39.2.4 Cascade of Suspension

The clavicular boom is elevated by the trapezius muscle ("boom guy line" muscle), which originates from the cranium and the cervical spine. Therefore, the apex of suspension is well above the thoracic platform, at the top of the spinal tower. From the apex of the tower (cranium) to the humerus, there is a cascade of osseous and intervening suspensory structures. For each articulation, there is a set of "boom guy line" muscles, which provide dynamic control of the articulation. The coracoid is suspended from the lateral aspect of the clavicular boom by the coraco-clavicular ligaments, and in turn, the coraco-humeral ligament suspends the humerus from the coracoid.

39.2.5 Suspension of the Scapular Pulley

The coraco-clavicular ligaments insert into the base of the coracoid process and thereby suspend the scapula. They form part of the superior suspensory complex of the shoulder [6]. Disruption of this suspensory ring produces AC joint instability. It is interesting to note that the conoid ligament is the *suspensory ligament* and attaches to a prominent tubercle at the posterior angle of the clavicle. With clavicle rotation, the ligament shortens and lengthens by wrapping around the clavicle like the biceps tendon wraps around the proximal radius. The trapezoid is the *lateralisation ligament* of the scapula. Its chief function is to prevent medial displacement of the scapula relative to the clavicle (Fig. 39.3a). The scapula rotates around these two ligaments, with the AC joint capsule restraining anterior and posterior translation of the scapula.

39.2.6 Scapular Pulley

A pulley is a wheel designed to support movement and change direction along its circumference. The scapula is a pulley strategically suspended in the middle of the "suspensory cascade" between the clavicle and humerus (Fig. 39.3b). The scapular pulley swivels on the coraco-clavicular ligaments, below the lateral clavicular boom, to change the direction of the rotator cuff alignment to optimise shoulder function.

The scapular body provides a wide surface area for muscle attachments. The multiple powerful peri-scapular muscles span from the spinal tower and the thoracic platform. These peri-scapular muscles control scapular rotation and translation across the thoracic cage. We refer to the area of the thoracic cage on which the scapula traverses as the "scapular track". The rotator cuff muscles control the humeral head across the "glenoid track". The scapula is essentially a sesamoid bone positioned between the peri-scapular muscles that control the scapula and the rotator cuff, which controls the humeral head. These two major groups of muscles work together to position the humerus in space.

39.2.7 Scapular Triangle and Scapular Track

The scapula is stabilised and mobilised by a functional triangle (Fig. 39.4). The sides and angles of the triangle consist of

Medial side – The fixed thoracic platform.

Anterior angle – The sterno-clavicular hinge joint.

Anterior side – The clavicular boom elevates and lateralises the scapular pulley.

Lateral angle – The coraco-clavicular ligaments suspend and swivel the scapula.

Posterior side – The scapula and its peri-scapular muscles, which power and dynamically stabilise the scapula.

Posterior angle – The peri-scapular muscles' insertions into the thoracic cage are a fixation point for the muscles.

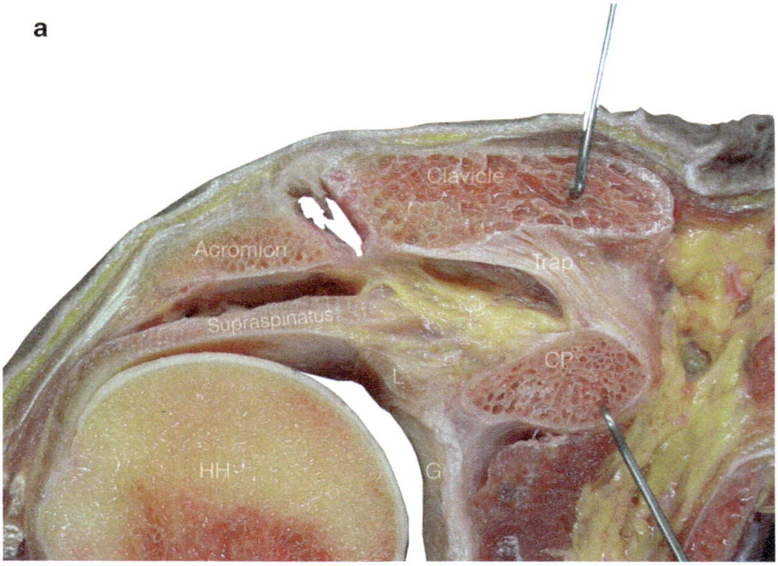

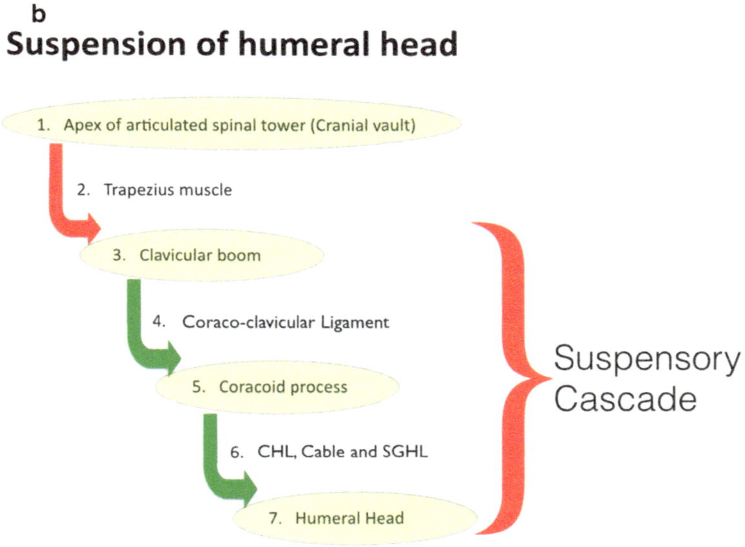

Fig. 39.3 (**a**) Coraco-clavicular ligaments. The clavicular boom suspends and lateralises the scapula. The conoid ligament suspends the scapula and the trapezoid ligament restrains medialisation of the scapula. The scapula swivels on the coraco-clavicular ligaments. (**b**) Suspension of the humeral head and arm commences. From the clavicle, the "cascade of suspension" passes down to the humeral head (Image courtesy Professor Emilio Calvo)

Note that the three sides and three corners of the triangle all have different functions. The position of the scapula is defined by the angle of elevation of the sternoclavicular joint, the length of the clavicular boom and the tension in the various peri-scapular muscles. The thoracic platform is fixed; the anterior stabilisers are hinged but static, while the posterior are dynamic. The scapular pulley mobilises on the fixed thoracic cage ("scapular track"), directed by the static anterior restraints, at the discretion of the peri-scapular muscles.

The function of this triangle is to stabilise and mobilise the scapular pulley, so that the pulley can align the rotator cuff to stabilise and mobilise the glenohumeral joint. The scapula is oriented so that the glenoid faces anterio-lateral, which determines a functional plane of the shoulder.

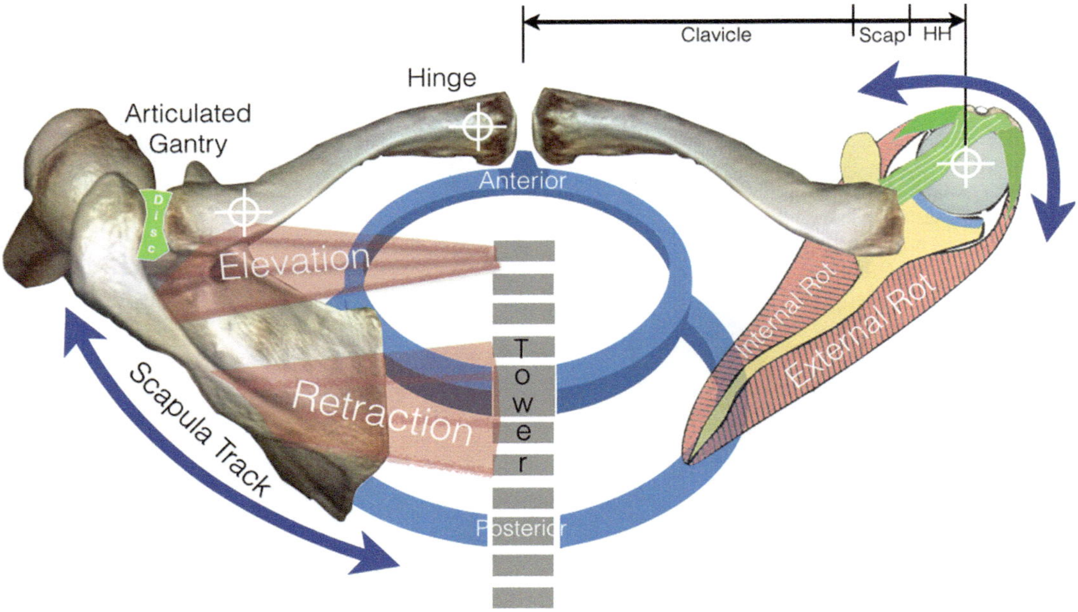

Fig. 39.4 Shoulder gantry (*Left*). The gantry is a scaffold over the top of the glenohumeral joint. It is composed of the clavicle, coracoid, CAL, acromion and scapular spine. The trapezius *muscle (posterior) elevates the gantry*, hinging on the sternoclavicular joint (*anterior-medial*). The gantry swivels at the coraco-clavicular ligaments, which allows the peri-scapular muscles to change scapula orientation. With these restraints, the scapula follows the "scapular track". The AC joint fibrocartilaginous disc buffers the compressive forces and the coraco-clavicular ligaments resist the tensile forces. Shoulder triangle (*Right*). The triangular configuration of the shoulder girdle, thoracic cage, clavicular boom and scapula. The medial side is the elevated thoracic platform. The anterior side is the clavicular boom, which elevates and maintains the lateral position of the scapula. The lateral angle of the triangle is the coraco-clavicular ligaments, from which the scapula is suspended and swivels. The posterior side is the triangle is the scapular body and the peri-scapular muscles, which mobilise the scapula. Note that the centre of rotation of the humeral head is lateralised by this triangle, but also the width of the pulley block and the radius of the humeral head. The rotator cuff insertion is lateral to the centre of rotation of the humeral head, making it an important rotator. Note how the peri-scapular muscles power the scapula, and the rotator cuff controls the humeral head (Copyright Dr. Gregory Bain)

39.2.8 Suspension of the Humeral Head: CHL

The CHL passes from the base of the coracoid process through the rotator cuff interval, reinforces the rotator cuff cable, and then inserts with the supraspinatus and subscapularis into the greater and lesser tuberosity. The coraco-humeral ligament is a significant ligamentous structure, which tethers the head to the coracoid process. The humeral head is a "ball on a string" that is suspended into the glenohumeral joint (Fig. 39.5a, b). As the rotator cuff attaches lateral to the tether, the cuff can actively control the rotation of the head in any plane, while the CHL and the gleno-humeral ligaments maintain the position of the humeral head (Fig. 39.6a, b).

39.2.9 Glenoid Bearing

The glenoid projects from the lateral aspect of the scapula and forms the major articulation of the shoulder girdle. The capsule and labral thickening of the glenoid provide primary stability. The rotator cuff muscles span from the scapular body, passing lateral to the glenoid and the coraco-humeral ligament to insert into the prominences of the humeral tuberosities. They provide dynamic stability, especially with the support of the common tendon, cable and CHL. The glenoid is perpendicular to the body of the scapula, therefore if the glenoid is correctly aligned for stability, then the body of the scapula will be correctly positioned to maximise rotator cuff power.

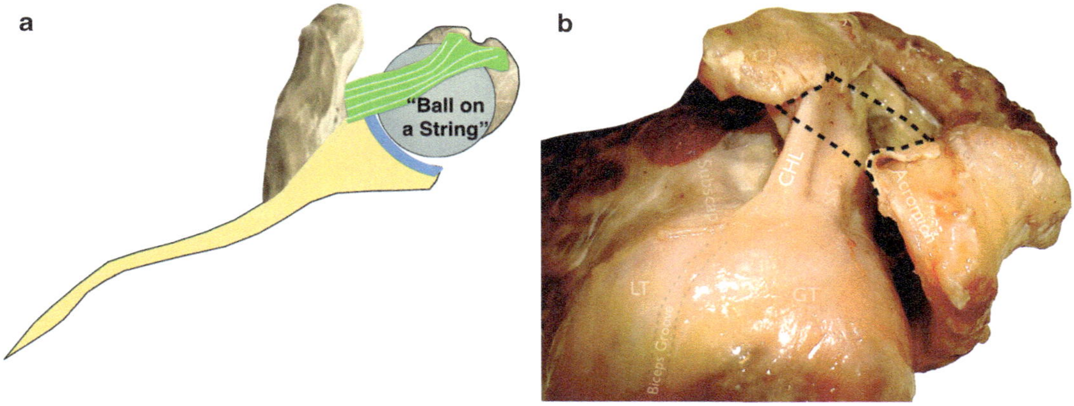

Fig. 39.5 (**a**, **b**) Coraco-humeral ligament. The coraco-humeral ligament is a significant structure with a wide attachment (usually two attachment sites) on the coracoid process. Distally it attaches to the greater and lesser tuberosity of the humeral head. It reinforces the biceps pulleys, the cable and the insertions of the subscapularis and supraspinatus. It is the second ligamentous aspect of the "cascade of suspension". (**a**) Cadaveric photograph of the CHL. The coraco-acromial ligament has been resected (*black dots*) to better appreciate the CHL and the base of the coracoid. The CHL suspends and tethers the humeral head, like a "Ball on a string." The two components of the CHL are a four bar linkage, which restricts the extremes of the motion of the humeral head throughout circumduction. (**b**) The CHL can be seen to have two attachments on the coracoid process, is closely opposed to the supraspinatus, and contributes to the rotator cuff cable (Image reproduced with permission from Di Giacomo [9]; Figure copyright Dr. Gregory Bain)

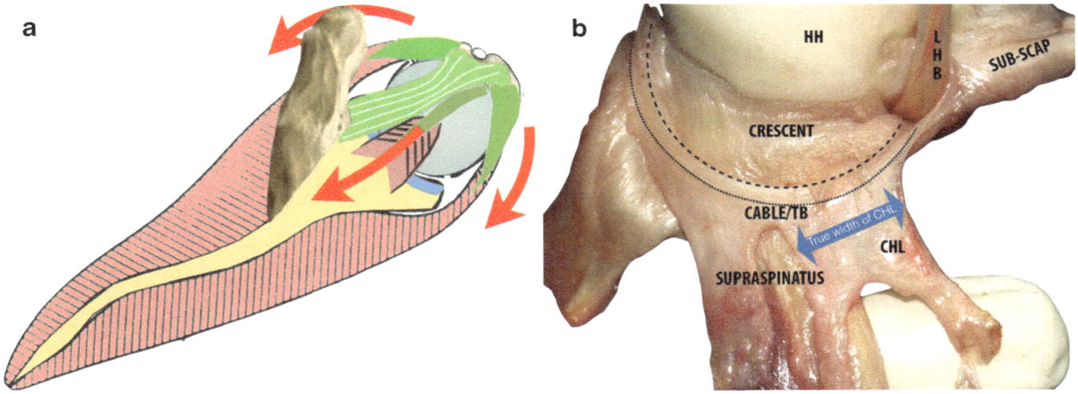

Fig. 39.6 (**a**, **b**) The CHL reinforces and inserts with the subscapularis and supraspinatus attachments. Note there are two coracoid attachments and two humeral attachments. The cable links the various components of the rotator cuff (Photographs reproduced with permission from Di Giacomo [9]; Figure copyright Dr. Gregory Bain)

39.3 Rotator Cuff

The *individual muscles of the rotator cuff* have different intra-muscular tendinous configurations (Fig. 39.7). The supraspinatus is a single tendon in a bipennate muscle, which is directly adjacent to the CHL suspension point. The subscapularis is multipenate with a thick upper tendon. The infraspinatus is multipenate with oblique and transverse heads. The upper tendons of subscapularis and the oblique head of infraspinatus are above the equator of the humeral head and are abductors. They provide rotation when the arm is at the side and contribute to abduction when the arm is elevated. The lower tendons are predominantly rotators.

The common rotator cuff insertion is important for above-head activities and is almost exclusively found in primates [5]. The rotator cuff muscles all take origin from the wide scapular body and coalesce into a common tendon cuff that inserts into the tuberosities of the proximal

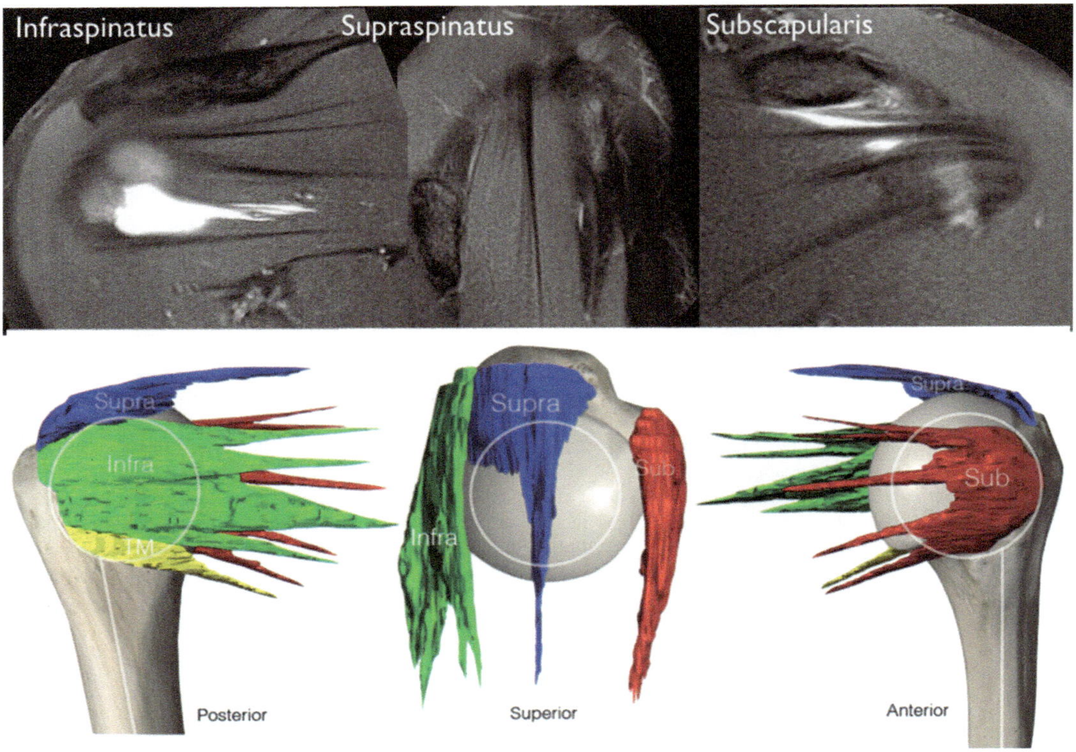

Fig. 39.7 Intra-muscular tendons within the rotator cuff. The supraspinatus tendon has a single tendon within a bipennate muscle. The subscapularis has four tendons that span the insertion. The infraspinatus has an oblique head that is an effective head depressor, and a transvers head that is an effective external rotator (Copyright Dr. Gregory Bain)

humerus (Fig. 39.8a, b). The common tendon insertion is created by a network of interlacing tendons and ligamentous structures [7] (Fig. 39.9). This includes the coraco-humeral ligament, gleno-humeral ligaments and "rotator cuff cable" (semicircular ligament of the humerus). The "rotator cuff cable" can be visualised arthroscopically, as a 1 cm wide ligamentous "suspension bridge", spanning and reinforcing the deep surface of the cuff insertion from subscapularis to teres minor [8–10] (Fig. 39.10).

The rotator cuff muscles all have a unique common *origin* from the scapula.

1. Therefore, the scapula and its peri-scapular muscles must be critical to rotator cuff function.
2. The peri-scapular muscles are required to position and stabilise the scapula. EMG studies have demonstrated that with shoulder abduction, the first muscle to be activated is the supraspinatus. However, before the arm even moves, the scapula is stabilised by the trapezius and deltoid [11]. The serratus anterior is activated as the shoulder commences motion.
3. The peri-scapular and rotator cuff muscles must synchronise to optimise function. When abducting the arm, the angle of maximal electrical activity is different for each muscle (e.g. supraspinatus 88°, mid-trapezius 95°, mid-deltoid 105°, serratus anterior 125° and lower trapezius 140° [11]).

The rotator cuff tendons have a unique common *insertion* into the rotator cable, then together into the tuberosities.

1. The individual cuff muscles create a different effect depending upon the position of the humerus (e.g. superior subscapularis can either abduct or internally rotate).
2. The cuff muscles can work together to increase power. (e.g. supraspinatus is the prime abductor but can be potentiated with upper subscapularis and infraspinatus).

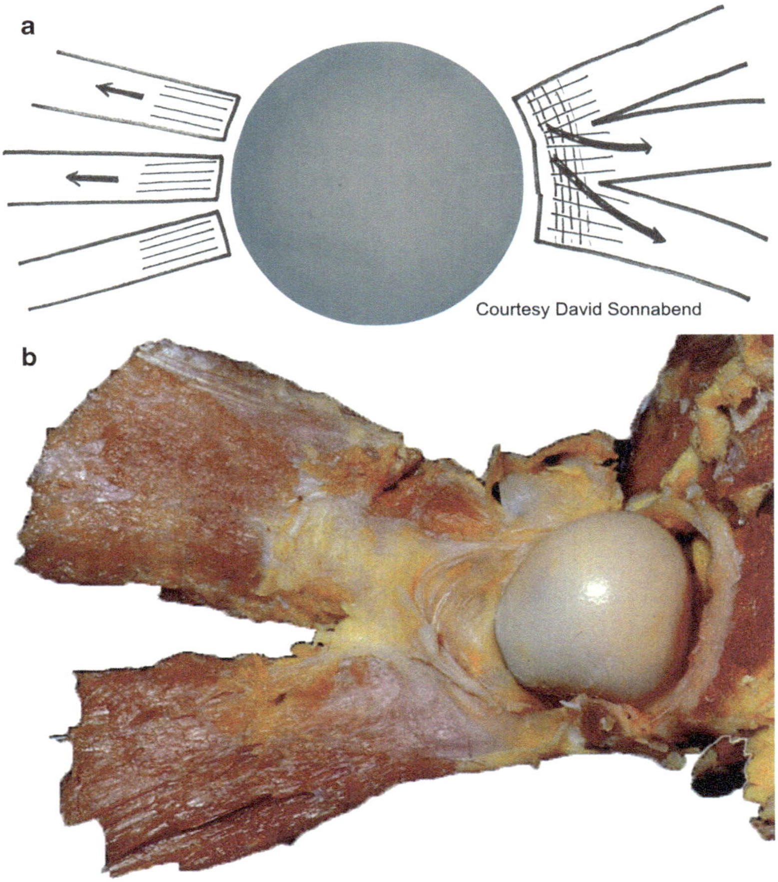

Fig. 39.8 (**a**) Common rotator cuff tendon. In most animals the rotator cuff muscle are separate to their insertion. However in orthograde animals, there is a common rotator cuff tendon that inserts into the width of the tuberosities. The cuff provides joint stability and allows the tendons to summate their forces, elevate the shoulder and provide rotation from the elevated position (Courtesy Professor David Sonnabend). (**b**) Deep surface of the rotator cuff of the tree kangaroo. The separate muscles join to from a common tendon, before inserting into the humeral tuberosity. Note the "rotator cuff cable", 1 cm from the insertion (Courtesy Professor David Sonnabend)

Fig. 39.9 The multiple interlacing layers of the rotator cuff demonstrated with polarising light (Courtesy Professor David Sonnabend)

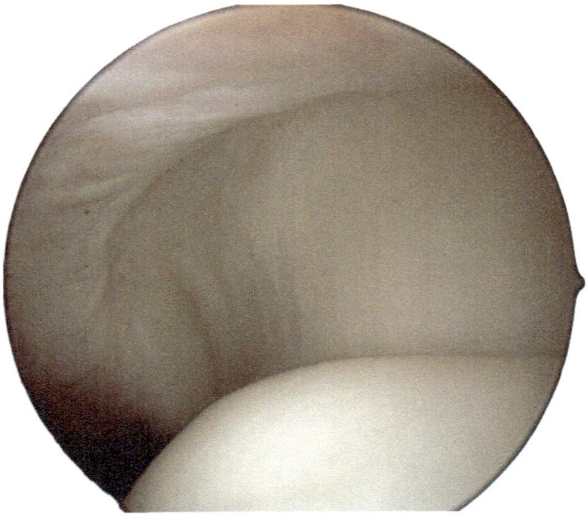

Fig. 39.10 Arthroscopic view of the cable on the deep surface of the rotator cuff

3. The cuff muscles can work together to change the direction of motion. (e.g. supraspinatus will abduct but with infraspinatus will abduct and externally rotate)

The rotator cuff and the peri-scapular muscles work together during abduction.

1. Abduction is initiated by supraspinatus, while the deltoid centres the humeral head and the trapezius and peri-scapular muscles stabilise the scapula.
2. In mid-range, the anterior and posterior deltoid and upper infraspinatus and subscapularis are recruited to abduct the arm.
3. When above 135°, the inferior trapezius rotates the scapula, which boosts the abduction, while the deltoid and supraspinatus stabilise the glenohumeral joint.

39.4 The "Gantry" (Coraco-Acromial Arch)

The coraco-acromial arch (CCA) is a gantry composed of the lateral clavicle, coraco-acromial ligament (CAL), acromion and scapular spine. The term "gantry" for the CCA was coined by Dr. Peter Hales, Perth, Australia. The word "gantry" is defined as a bridge-like framework or supportive structure (modified from www.collinsdictionary.com/). The gantry has only developed in the bipedal animals, especially in the brachiating animals, and is one of the evolutionary modifications that have allowed the arm to be elevated from the body. The gantry creates a "pseudo-articulation" between the coraco-acromial arch and the rotator cuff. The gantry has various parts, each with its own function.

The pillars of the gantry are the coracoid process and the spine of the scapula (Fig. 39.11), which provide cantilever support for the coraco-acromial arch.

The superior surface provides an extensive insertion for the trapezius muscle, to elevate the entire gantry, and provides an important contribution to abduction strength.

The lateral surface provides an attachment for the deltoid muscle, which is a strong multipenate abductor of the shoulder. It also compresses the subacromial space and therefore potentiates the rotator cuff function. The axillary nerve is adherent to the deep surface of the deltoid so that it moves with the deltoid and not the rotator cuff.

The inferior surface of the coraco-acromial arch consists of the relatively thin osseous acromion and the coraco-acromial ligament. During abduction, the deltoid contracts to narrow the subacromial space, so that the rotator cuff will abut on the arch. There is some flexibility in the arch, so that it moulds to the forces placed on it when the deltoid contracts. The CAL is strategically positioned to be a malleable "soft pivot" for the cuff and humeral head and to be able to transfer deltoid

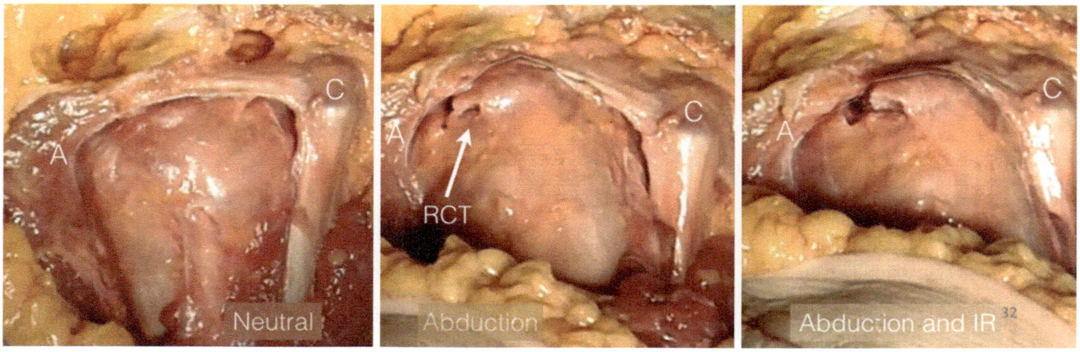

Fig. 39.11 Coraco-acromial arch. Cadaveric specimen, demonstrating impingement of the arch. (**a**) The arm is placed in neutral position, and the coraco-acromial ligament is seen to bridge over the rotator cuff. (**b**) With abduction the rotator cuff tear is catching on the coraco-acromial ligament, which is tented under load. (**c**) In abduction and internal rotation, the greater tuberosity is impinging upon the coraco-acromial ligament. *A* Acromion, *C* Coracoid process, with coraco-acromial ligament between them (Copyright Dr. Gregory Bain)

tension from the acromion to the coracoid process. The evolutionary introduction of the CAL has also allowed hominoids to stabilise the laterally projecting acromion, which in turn allowed an increased deltoid lever arm and strength.

39.5 Subacromial Pseudo-Articulation

The subacromial space is a "pseudo-articulation" between the acromion and the rotator cuff. The subacromial bursa creates a minimal friction interface between the rotator cuff and the coraco-acromial arch. With shoulder abduction from 0 to 90°, the supraspinatus mobilises the gleno-humeral joint, and the tendon traverses the sub-acromial space. From 135° to 180°, the abduction is predominantly scapulo-thoracic motion, with minimal rotator cuff translation, but the rotator cuff is up against the under-surface of the arch.

With ageing, the CAL loses its resilience and becomes stiffer, thereby placing greater force on the rotator cuff and predisposing it to impingement, degeneration and tearing. In massive rotator cuff tears, the humeral head rides high, absorbing the pseudo-articulation, to create one large acromio-glenoid humeral articulation. With rotator cuff arthropathy, an acetabulum can be created.

39.6 Scapular Track and Winging

The normal scapular motion along this "scapular track" is smooth and coordinated by the multiple peri-scapular muscles. Winging is dysfunctional scapular motion, which is identified with prominence of the scapula. The four basic groups of winging include osseous (e.g. clavicle fracture), articular (e.g. St-CI or AC joint instability), muscular (e.g. fatigue) and neurological (e.g. long thoracic nerve).

39.7 Labral Housing

The capsulo-labral complex is important for shoulder stability. The inferior labrum (5–10 o'clock) is rounded with a convex surface and designed to provide a bumper effect. It increases the glenoid depth by up to 50 % [12–14]. It has an adherent interface to the articular cartilage and has a rigid bony foundation, which prevents mobility of the labrum. The inferior labrum and the IGHL (Fig. 39.12) is a fixed organ of compression designed to provide stability [12].

In contrast, *the superior labrum* has a loose mobile interface with no bony foundation and attaches off the rim away from the glenoid articular margin. It is concave in cross section, more meniscal in nature and follows the contour of the glenoid surface [12]. The superior labrum is a mobile organ of tension. The long head of the biceps tendon is a dynamic structure that is

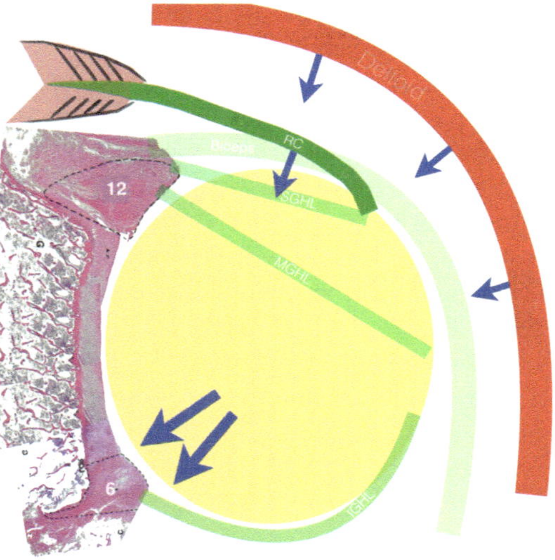

Fig. 39.12 Superior and inferior labrum. Composite histology image of the glenoid labrum at 6 and 12 o'clock. (**a**) At 6 o'clock the labrum is a fixed organ of compression. The convex bumper effect of the inferior labrum, which is mounted onto of the osseous glenoid. There is no defect between the glenoid, labrum and articular surface. (**b**) At 12 o'clock the labrum is a mobile organ of tension. The concave superior labrum attachment is "off the face" of the glenoid, and has a synovial lined cleft between the labrum and the glenoid. The superior labrum is actively controlled by the biceps tendon, and passively by the SGHL and MGHL. (**c**) The mobile superior labrum and its attachments (biceps, SGHL and MGHL) increase their tension at the extremes of motion to optimize containment and stability of the shoulder. The biceps, rotator cuff and deltoid all compress the humeral head onto the "static" inferior glenoid and labrum, to provide joint stability. H and E histology section (Modified with permission from Bain et al. [12])

anchored to the superior glenoid tubercle, but it can pull on the superior labrum and the attached glenohumeral ligaments and enhance stability throughout the range of motion [12, 15].

39.8 Function and Evolution

The crane has been described above and enables the shoulder to abduct, elevate and externally rotate. Once the arm has been placed into this position, the adductors and internal rotators can bring the hand back to the body, mouth, breast or groin. Quite an evolutionary advantage!

Now that the arms are "free" to perform other tasks, it opens many more opportunities for the remainder of the upper limb. Other evolutionary advances have subsequently included increased forearm rotation, dart thrower's wrist motion and the opposable thumb. The thumb allows us to manipulate and pinch small objects, and the fingers allow us to grasp larger objects. The wrist is a universal joint that allows us to provide controlled strength throughout wrist motion, and the forearm provides the ability to rotate the hand in space. The shoulder allows the hand to be placed in space and to bring objects back to the body. Therefore, the evolution of the shoulder girdle has enabled us to now pinch, pick, grasp, write, type, squeeze, hold, place, cup, fend, throw, hang, retrieve, nurture, suckle, groom, toilet and more recently write, type and text.

References

1. Codman EA, Akerson IB. The pathology associated with rupture of the supraspinatus tendon. Ann Surg. 1931;93(1):348–59.
2. Kardong KV. The vertebrates. In: Comparative anatomy, function and evolution. 6th ed. Boston: McGraw-Hill Companies; 2012.
3. DePalma AF. The classic. Origin and comparative anatomy of the pectoral limb. Surgery of the shoulder. Philadelphia: Lippincott Williams & Wilkins; 1950. p. 1–14. Clin Orthop Relat Res. 2008;466(3):531–42.
4. Inman VT. Observations on the function of the shoulder joint. J Bone Joint Surg Am. 1944;26:1–30.
5. Sonnabend DH, Young AA. Comparative anatomy of the rotator cuff. J Bone Joint Surg. 2009;91(12):1632–7.
6. Jeray KJ. Acute midshaft clavicular fracture. J Am Acad Orthop Surg. 2007;15(4):239–48.
7. Clark JM, Harryman 2nd DT. Tendons, ligaments, and capsule of the rotator cuff. Gross and microscopic anatomy. J Bone Joint Surg Am. 1992;74(5):713–25.
8. Kolts I, Busch LC, Tomusk H, Raudheiding A, Eller A, Merila M, et al. Macroscopical anatomy of the so-called "rotator interval". A cadaver study on 19 shoulder joints. Ann Anat Anatomischer Anzeiger: Off Organ Anatomische Gesellschaft. 2002;184(1):9–14.
9. Di Giacomo G, Pouliart N, Constantini A, De Vita A. Atlas of functional shoulder anatomy. Milan: Springer-Verlag Mailand; 2008.
10. Burkhart SS, Esch JC, Jolson RS. The rotator crescent and rotator cable: an anatomic description of the shoulder's "suspension bridge". Arthroscopy: J Arthrosc Relat Surg: Off Publ Arthrosc Assoc North Am Int Arthrosc Assoc. 1993;9(6):611–6.
11. Wickham J, Pizzari T, Stansfeld K, Burnside A, Watson L. Quantifying 'normal' shoulder muscle activity during abduction. J Electromyogr Kinesiol: Off J Int Soc Electrophysiol Kinesiol. 2010;20(2): 212–22.
12. Bain GI, Galley IJ, Singh C, Carter C, Eng K. Anatomic study of the superior glenoid labrum. Clin Anat. 2013;26(3):367–76.
13. Howell SM, Galinat BJ. The glenoid-labral socket. A constrained articular surface. Clin Orthop Relat Res. 1989;243:122–5.
14. Mileski RA, Snyder SJ. Superior labral lesions in the shoulder: pathoanatomy and surgical management. J Am Acad Orthop Surg. 1998;6(2):121–31.
15. Elser F, Braun S, Dewing CB, Giphart JE, Millett PJ. Anatomy, function, injuries, and treatment of the long head of the biceps brachii tendon. Arthroscopy: J Arthrosc Relat Surg: Off Publ Arthrosc Assoc North Am Int Arthrosc Assoc. 2011;27(4):581–92.

Index

G.I. Bain et al. (eds.), *Normal and Pathological Anatomy of the Shoulder*,
DOI 10.1007/978-3-662-45719-1,

L

M

N

O

P

GPSR Compliance

The European Union's (EU) General Product Safety Regulation (GPSR) is a set of rules that requires consumer products to be safe and our obligations to ensure this.

If you have any concerns about our products, you can contact us on ProductSafety@springernature.com

In case Publisher is established outside the EU, the EU authorized representative is:

Springer Nature Customer Service Center GmbH
Europaplatz 3
69115 Heidelberg, Germany

Batch number: 10732784

Printed by Printforce, the Netherlands